Crop Improvement, Adoption and Impact of Improved Varieties in Food Crops in Sub-Saharan Africa

FSC
www.fsc.org
MIX
Paper from responsible sources
FSC® C013604

Crop Improvement, Adoption and Impact of Improved Varieties in Food Crops in Sub-Saharan Africa

---

Edited by

Thomas S. Walker

Independent Researcher, Fletcher, North Carolina, USA

and

Jeffrey Alwang

Department of Agricultural and Applied Economics, Virginia Tech, Blacksburg, Virginia, USA

Published by CGIAR and CAB International

www.cabi.org

CGIAR

Independent Science and Partnership Council

CABI is a trading name of CAB International

CABI
Nosworthy Way
Wallingford
Oxfordshire OX10 8DE
UK

Tel: +44 (0)1491 832111
Fax: +44 (0)1491 833508
E-mail: info@cabi.org
Website: www.cabi.org

CABI
38 Chauncy Street
Suite 1002
Boston, MA 02111
USA

Tel: +1 800 552 3083 (toll free)
E-mail: cabi-nao@cabi.org

A catalogue record for this book is available from the British Library, London, UK

Library of Congress Cataloging-in-Publication Data

Crop improvement, adoption and impact of improved varieties in food crops in Sub-Saharan Africa / edited by Thomas S. Walker, Jeffrey Alwang.
pages cm
Includes bibliographical references and index.
ISBN 978-1-78064-401-1 (hbk : alk. paper) 1. Food crops--Breeding--Africa, Sub-Saharan. 2. Crop improvement--Africa, Sub-Saharan. 3. Agricultural productivity--Africa, Sub-Saharan. 4. Agriculture--Research--Africa, Sub-Saharan. I. Walker, T. S., editor. II. Alwang, Jeffrey R., editor.

SB175.C75 2015
338.10967--dc23

2015025379

Published by CGIAR Consortium of International Agricultural Research Centers and CAB International

ISBN-13: 978 1 78064 401 1 (CABI)

Commissioning editor: Joris Roulleau
Editorial assistant: Emma McCann
Production editor: James Bishop

Typeset by SPi, Pondicherry, India
Printed and bound in the UK by CPI Group (UK) Ltd, Croydon, CR0 4YY

Contents

Contributors

Tahirou Abdoulaye, IITA, PMB 5320, Ibadan, Oyo State, Nigeria. E-mail: t.abdoulaye@cgiar.org

Arega D. Alene, IITA-Malawi, P.O. Box 30258, Lilongwe 3, Malawi. E-mail: a.alene@cgiar.org

Jeffrey Alwang, Department of Agricultural and Applied Economics, Virginia Tech, 251-I Hutcheson Hall, Blacksburg, Virginia, USA. E-mail: alwangj@vt.edu

Eyram Amovin-Assagba, Africa Rice Center (AfricaRice), 01 B.P. 2031, Cotonou, Benin. E-mail: E.Amovin-Assagba@cgiar.org

Robert Andrade, Department of Applied Economics, University of Minnesota (Formerly of CIAT), 248D Ruttan Hall 1994 Buford Avenue, Saint Paul, MN 55108, USA. E-mail: andra065@umn.edu

Aden Aw-Hassan, ICARDA, Social, Economic and Policy Research Program (SEPRP), International Center for Agricultural Research in the Dry Areas (ICARDA), P.O. Box 950764, Amman 11195, Jordan. E-mail: a.aw-hassan@cgiar.org

M.C.S. Bantilan, Research Program on Markets, Institutions, and Policy, ICRISAT Center, Patancheru 502 324, Telangana, India. E-mail: c.bantilan@cgiar.org

Aliou Diagne, Universite Gaston Berger (formerly of AfricaRice), St Louis, Senegal. E-mail: aliouwork@gmail.com

Keith Fuglie, Resource and Rural Economics Division, Economic Research Service, US Department of Agriculture, 355 E. Street, Southwest, Room 6-178, Washington, DC 20024-3221, USA. E-mail: kfuglie@ers.usda.gov

Zachary Gitonga, CIMMYT, PO Box 1041-00621, Nairobi, Kenya. E-mail: z.gitonga@cgiar.org

Hugo De Groote, Socioeconomics Program CIMMYT, PO Box 1041-00621, Nairobi, Kenya. E-mail: h.degroote@cgiar.org

Moti Jaleta, CIMMYT, P. O. Box 5689, Addis Ababa, Ethopia. E-mail: m.jaleta@cgiar.org

Wellington Jogo, Bioversity International, PO Box 24384, Kampala, Uganda. E-mail: w.jogo@cgiar.org

Enid Katungi, Pan Africa Bean Research Alliance, CIAT, PO Box 6247, Kampala, Uganda. E-mail: e.katungi@cgiar.org

Florent M. Kinkingninhoun-Medagbe, Africa Rice Center (AfricaRice), 01 B.P. 2031, Cotonou, Benin. E-mail: f.medagbe@cgiar.org

Deevi Kumara Charyulu, Research Program on Markets, Institutions, and Policy, ICRISAT Center, Patancheru 502 324, Telangana, India. E-mail: d.kumaracharyulu@cgiar.org

Ricardo A. Labarta, Decision & Policy Analysis Research Area, CIAT (formerly of CIP), Km 17, Recta Cali-Palmira, Cali, Colombia. E-mail: r.labarta@cgiar.org

Catherine Larochelle, Department of Agricultural and Applied Economics, Virginia Tech, 315 Hutcheson Hall, Blacksburg, Virginia, USA. E-mail: claroche@vt.edu

A. Raja Laxmi, Research Program on Markets, Institutions, and Policy, ICRISAT Center, Patancheru 502 324, Telangana, India. E-mail: bandirajalaxmi46@gmail.com

Victor Manyong, IITA, PO Box 34441, Dar es Saalam, Tanzania. E-mail: v.manyong@cgiar.org

Jacob Marder, Dept of Agricultural and Resource Economics, University of California, Davis, One Shields Avenue, Davis, CA 95616, USA. E-mail: jmmarder@ucdavis.edu

Kai Mausch, ICRISAT-Nairobi, c/o The World Agroforestry Center, ICRAF House, United Nations Avenue, Gigiri, PO Box 39063, Nairobi, Kenya. E-mail: k.mausch@cgiar.org

Stephen Mugo, Global Maize Program, CIMMYT-Kenya, P.O. Box 1041-00621, Nairobi, Kenya. E-mail: s.mugo@cgiar.org

Rachel Muthoni Andriatsitohaina, Pan Africa Bean Research Alliance, CIAT, PO Box 6247, Kampala, Uganda. E-mail: r.muthoni@cgiar.org

Tebila Nakelse, Department of Agricultural Economics, Kansas State University, 343 Waters Hall, Manhattan, Kansas, USA. E-mail: tnakelse@ksu.edu

Jupiter Ndjeunga, ICRISAT West and Central Africa, PO Box 12404, Niamey, Niger, West Africa. E-mail: n.jupiter@cgiar.org

George W. Norton, Department of Agricultural and Applied Economics, Virginia Tech, 205-B Hutcheson Hall, Blacksburg, Virginia, USA. E-mail: gnorton@vt.edu

Sushil Pandey, Independent Researcher (Formerly of IRRI), 10 Roseville St, White Plains, Quezon City, Metro Manila 1110, Philippines. E-mail: sushilpandey056@gmail.com

Joseph Rusike, Alliance for a Green Revolution in Africa (Formerly of IITA), PO Box 66773, Westlands 00800, Nairobi, Kenya. E-mail: JRusike@agra.org

Kayode Sanni, African Agricultural Technology Foundation, PO Box 30709, Nairobi, Kenya. E-mail: S.Kayode@aatf-africa.org

Bekele Shiferaw, Partnership for Economic Policy (formerly CIMMYT-Nairobi), PO Box 2692, Nairobi 00621, Kenya. E-mail: bshiferaw@pep-net.org

D. Shyam Moses, Research Program on Markets, Institutions, and Policy, ICRISAT Center, Patancheru 502 324, Andhra Pradesh, India. E-mail: d.mosesshyam@cgiar.org

Franklin Simtowe, CIMMYT, PO Box 104-00621, Nairobi, Kenya. E-mail: f.simtowe@cgiar.org

Ali Toure, Africa Rice Center (AfricaRice), 01 B.P. 2031, Cotonou, Benin. E-mail: a.toure@cgiar.org

Maria Lourdes Velasco, IRRI, DAPO Box 7777, Metro Manila 1301, Philippines. E-mail: l.velasco @irri.org

Thomas S. Walker, 506 Windsor Forest Court, Fletcher, North Carolina, USA. E-mail: walkerts@ msu.edu

Takashi Yamano, IRRI, DAPO Box 7777, Metro Manila 1301, Philippines. E-mail: t.yamano@irri. org

Yigezu A. Yigezu, Social, Economic and Policy Research Program (SEPRP), International Center for Agricultural Research in the Dry Areas (ICARDA), PO Box 950764, Amman 11195, Jordan. E-mail: y.yigezu@cgiar.org

Chilot Yirga, Ethiopian Institute of Agricultural Research, PO Box 2003, Addis Ababa, Ethiopia. E-mail: ctizali@yahoo.com

Di Zeng, Global Food Studies, The University of Adelaide (formerly of Virginia Tech), Adelaide, SA, Australia. E-mail: di83@vt.edu

Acronyms and Abbreviations

AGRA	Alliance for a Green Revolution in Africa (BMGF)
AgSSIP	Agricultural Services Subsector Investment Project, World Bank
AICRP	All-India Coordinated Research Projects
AMU	Texas A&M University, College Station, TX, USA
ARC	Agriculture Research Corporation (Sudanese national program)
ARIMA	Autoregressive integrated moving average
ASARECA	Association of Strengthening Agricultural Research in Eastern and Central Africa
ASTI	Agricultural Sciences and Technology Indicators
ATE	Average treatment effect
ATT	Average treatment effect on the treated
ATU	Average effect on the untreated
BMGF	Bill & Melinda Gates Foundation
BSc	Bachelor of Science degree
CATIE	Centro Agronómico Tropical de Investigación y Enseñanza (Costa Rica)
CAR	Central African Republic
CBB	Cassava bacterial blight
CBO	Community-based organization
CBSD	Cassava brown streak disease
CCC	Country-by-crop combinations
CGI	Crop genetic improvement
CGIAR	Consultative Group on International Agricultural Research
CG Center	Institute within the Consultative Group on International Agricultural Research
CGM	Cassava green mite
CIAT	Centro Internacional de Agricultura Tropical (International Center for Tropical Agriculture)
CIMMYT	Centro Internacional de Mejoramiento de Maiz y Trigo (International Maize and Wheat Improvement Center)
CIP	Centro Internacional de La Papa (International Potato Center)
CIRAD	Centre de Coopération Internationale en Recherche Agronomique pour le Développement (Agricultural Research for Development) (France)
CMD	Cassava mosaic virus disease
CRP	Collaborative research project
CRS	Catholic Relief Services

CRSP	Collaborative Research Support Program of USAID (renamed Innovation Laboratory)
CSA	Ethiopian Central Statistics Authority
CSIR	Council for Scientific and Industrial Research (Ghana)
CVRC	Central Varietal Release Committee (India)
DFID	Department for International Development (UK)
DGRST	Delegation of Scientific and Technical Research (DR Congo)
DIIVA	Diffusion and Impact of Improved Varieties in Africa
DUS	Distinctness, uniformity and stability (testing)
EAP	Escuela Agricola Panamericana Zamorano (Zamorano Pan-American Agricultural School)
EARRNET	Eastern Africa Root Research Network
ECABREN	Eastern and Central Africa Bean Research Network
ECOWAS	Economic Community of West African States
EE	Expert elicitation
EIAR	Ethiopian Institute of Agricultural Research
EMBRAPA	Empresa Brasileira de Pesquisa Agropecuária (Brazil)
ESA	East and Southern Africa
FAO	Food and Agriculture Organization of the United Nations
FAOSTAT	Food and Agriculture Organization Corporate Statistical Database
FARO	Federal Agricultural Research Oryzae (Nigeria)
FE	Fixed effects
FHIA	Fundación Hondureña de Investigación Agrícola (Honduras Foundation for Agricultural Research)
FOFIFA	National Center for Rural Development (Madagascar)
FTE	Full-time equivalent
GDP	Gross domestic product
GIS	Geographic information systems
GLCI	Great Lakes Cassava Initiative
GMM	Generalized method of moments
GMO	Genetically modified organisms
GOI	Government of India
G×E	Genotype by environmental (interaction)
ha	Hectare
HDDS	Household dietary diversity score
HH	Household survey
HPRC	Hybrid Parents Research Consortium (ICRISAT)
HYV	High-yielding varieties
IARC	International Agricultural Research Center
ICAR	Indian Council of Agricultural Research
ICARDA	International Center for Agricultural Research in Dry Areas
ICRAF	International Center for Agroforestry
ICRISAT	International Crops Research Institute for the Semi-Arid Tropics
IFPRI	International Food Policy Research Institute
IITA	International Institute of Tropical Agriculture
ILRI	International Livestock Research Institute
INSAH	Institute du Sahel
INTSTORMIL	International Sorghum and Millet Innovation Laboratories of USAID
IRAT	Institut de Recherches Agronomiques Tropicales
IRHO	Institut de Recherche pour Les Huiles et Oleagineux (France)
IRRI	International Rice Research Institute
ISAR	Institut des Sciences Agronomiques du Rwanda
ISNAR	International Service for National Agricultural Research

IV	Instrumental variable (used in statistical analysis)
IVT	Institute of Horticultural Plant Breeding (Netherlands)
JICA	Japan International Cooperation Agency
KARI	Kenya Agricultural Research Institute
KSC	Kenya Seed Company
LAC	Latin America and Caribbean
LIV	Local instrumental variable
MAE	Mean absolute error
MAPE	Mean absolute percentage error
MAS	Marker-assisted selection
MDG	Millennium Development Goals
MSc	Masters of science degree
MTE	Marginal treatment effect
MV	Modern variety
NARO	National Agricultural Research Organization (Uganda)
NARS	National agricultural research system
NBRP	National BioResource Project
NCRI	National Cereals Research Institute (Badeggi, Nigeria)
NEPAD	New Partnership for Africa's Development
NERICA	New Rice for Africa (AfricaRice)
NGO	Non-governmental organization
NISR	National Institute of Statistics of Rwanda
NRA	Nominal rate of assistance
NSA	National seed authority
NSCN	National Seed Company of Malawi
NVRC	National variety release committee
NVRS	National Vegetable Research Station – Wellsbourne Project (UK)
OLS	Ordinary least squares
OPV	Open-pollinated varieties
PABRA	Pan-African Bean Research Alliance
PhD	Doctor of Philosophy degree
PPP	Purchasing Power Parity Dollars
PRAPACE	Regional potato and sweetpotato improvement network in Eastern and Central Africa (Acronym in French)
PRONAF	Project for Cowpeas in Africa
PVS	Participatory varietal selection
QPM	Quality protein maize
RAB	Research Agriculture Bureau (Rwanda)
RE	Random effects model
SABRN	The Southern Africa Bean Research Network
SACCAR	Southern Africa Centre for Cooperation in Agricultural Research and Training
SADC	Southern African Development Community
SADC–GLIP	Southern African Development Community – Grain Legume Improvement Program
SAFGRAD	Semi-Arid Food Grain Research and Development Project
SAT	Semi-Arid Tropics
SAU	State Agricultural Universities
SPIA	Standing Panel on Impact Assessment
SARRNET	Southern Africa Root Crops Research Network
SASHA	Sweetpotato Action for Security and Health in Africa
SIMLESA	Sustainable Intensification of Maize-Legume Systems for Food Security in Eastern and Southern Africa
SSA	Sub-Saharan Africa

SSCA	State Seed Certification Agency (India)
SSDC	State Seed Development Corporation (India)
SVRC	State Varietal Release Committee (India)
SYs	Scientist years
TE	Treatment effect
TFP	Total factor productivity
TIA	Trabalho de Inquerito (Mozambique national survey)
Tifton	University of Georgia, Tifton, GA, USA
TRIVSA	Tracking Improved Varieties in South Asia
UBOS	Uganda Bureau of Statistics
UNDP	United Nations Development Program
USAID	United States Agency for International Development
VCU	Value for cultivation and use
WAAPP	West Africa Agricultural Productivity Program
WCA	Western and Central Africa
WARDA	West Africa Rice Development Association
WECABREN	West and Central Africa Bean Research Network

Foreword

The simple idea that better seeds can change the lives of poor farmers has proven to be so powerful and enduring that efforts to increase the spread of improved crop varieties have now been at the core of agricultural development for more than 50 years. The early progress in breeding high-yielding semi-dwarf rice and wheat varieties provided the rationale for creating the CGIAR research system and for investing in national agricultural research systems around the world. How different would the progress in delivering improved varietal technologies over the past five decades have been had Dana Dalrymple not been collecting and analysing variety diffusion data during the 1970s and 1980s? Dalrymple's data illustrating the temporal and geographic patterns of the Green Revolution uptake of modern varieties informed strategies to increase the impact of genetic improvement. Given the valuable insights provided by Dalrymple's analysis, it remains a mystery how it is that this book represents just the second serious attempt in the past 30 years to develop a comprehensive picture of the diffusion of improved crop varieties in developing countries. Let us hope that the monitoring and analysis of diffusion becomes a routine and regular activity in future years.

Crop Improvement, Adoption, and Impact of Improved Varieties in Food Crops in Sub-Saharan Africa provides the most comprehensive, accurate and informative view of the spread of improved crop varieties in sub-Saharan Africa that has ever been produced. The coverage and quality of the data go well beyond anything available until now, and the attention given to verifying and improving data collection methods sets a new standard in establishing the credibility of diffusion estimates. The studies in the book demonstrate that access to better seeds should remain a core concern for farmers, donors and governments. The book's nuanced analysis also clearly illustrates the complexity of the story.

While there has been progress in building the capacity of national crop breeding programmes, progress has been uneven across countries and crops. The diffusion and turnover of improved varieties shows even greater variability. Two of the more striking findings are that biotechnology and the private sector are playing surprisingly limited roles in delivering technological change to African farmers. Each of these roles needs to be better understood, and the data that this book makes publicly available provide a place to start in examining those roles. It is clear that despite the many institutional and scientific changes of recent years, conventional plant breeding conducted by CGIAR and national public sector scientists will be the source of improved genetics for the overwhelming majority of sub-Saharan farmers for the foreseeable future. And, despite the progress documented in this book, there remains much to be done to improve access to better crop technologies. The book also makes a strong case for the vital importance of continuing to monitor the generation and uptake of improved varieties.

The data and analysis contained in this volume greatly exceed the expectations of the original project design as first discussed in 2008. At that time, the Bill & Melinda Gates Foundation's newly created Agricultural Development division had just begun investing heavily in crop improvement in sub-Saharan Africa; yet little reliable recent data on variety uptake were available to guide those investment decisions. The Standing Panel on Impact Assessment (SPIA), under the guidance of Derek Byerlee, Tim Kelley and Doug Gollin, should be congratulated for effectively organizing and executing this valuable study. The editors and authors of this volume have done a wonderful job of producing an important reference for agricultural development scholars, practitioners and investors.

Greg Traxler
Evans School of Public Affairs,
University of Washington

Preface

Introduction

For fifty years or so, development economists have been concerned with tracking the diffusion of improved agricultural technologies in the developing world. This focus is not based on mere curiosity. One reason for documenting diffusion is that it provides a simple measure of the success of agricultural research: when new crop varieties are taken up, or when new agronomic practices are adopted by farmers, it provides information about the effectiveness of the research and the success of research investments. Since a large fraction of agricultural research is publicly funded and since many genetic technologies diffuse freely, there may be no market signals of success. This makes diffusion data a valuable source of feedback for research planning.

A second reason for documenting diffusion is that the resulting data can be used as an input into subsequent research intended to uncover the multidimensional impacts of the research – on productivity, on farm income, even on poverty and inequality. In addition, differential patterns observed across space and time can reveal underlying constraints or problems with technology take-up. Perhaps certain technologies fail to gain a foothold in particular agroecologies, or perhaps practices beloved by researchers have failed to spread widely. This information can feed back into the research process to inform scientists and shape further research. Indeed, information on diffusion can also inform the broader development community and can shape thinking about a wide set of potential constraints to adoption – resulting, perhaps, from failures in financial markets, extension and information, or simply reflecting high transport and transaction costs.

Efforts to document the diffusion of improved crop varieties date back to the path-breaking work of Dana Dalrymple (1969, 1978, 1986a, 1986b). Dalrymple's work drew on the cooperation of national research programmes and international scientists, and it provided the data on which were based many early analyses of the Green Revolution and its impacts. But for a variety of reasons, the important task of documenting diffusion was left to languish after Dalrymple's last effort in 1986; the next major effort to document diffusion came more than a decade later. Under the leadership of Bob Evenson and drawing on the work of numerous collaborators, this study compiled data on the diffusion of improved varieties of 11 food crops, and it attempted to achieve global coverage. The project included three country case studies and several cross-cutting analyses and modelling exercises. A book (Evenson and Gollin, 2003) summarized the main findings of the project and established a 1998 baseline for crop varietal adoption and diffusion data.

The current book emerges from an effort that represents the first major follow-up of the Evenson and Gollin baseline. It grows out of the DIIVA Project (Diffusion and Impact of Improved Varieties in Africa), which was funded by the Bill & Melinda Gates Foundation (BMGF) with the goal of assessing incremental progress in sub-Saharan Africa (SSA) in the years after 1998. The DIIVA Project (and the companion TRIVSA Project, focused on South Asia) have greatly advanced our knowledge of varietal adoption and diffusion, both by expanding knowledge about areas where diffusion was previously not well documented and by improving the methodologies used for measuring diffusion.

The DIIVA Project was organized around three distinct activities: documenting key performance indicators of crop genetic improvement, collecting nationally representative survey data on varietal adoption, and assessing the impact of varietal change. The DIIVA Project covered 20 crops and 30 countries in SSA. Because some crops are locally absent or unimportant, the report does not account for every crop in every country; but coverage extends to 154 crop-by-country combinations that together account for over 70% of the gross value of agricultural production in SSA.

The study's findings represent a major advance in terms of both the scope and quality of data for SSA. In the Evenson-led study of 2003, the available data on varietal adoption and diffusion in Africa were very limited. Many of these data were based on a combination of small-scale studies of adoption and rather vague regional estimates; the specific crop-by-country estimates of varietal adoption were mostly the product of interpolation and triangulation. The current study has enormously improved the quality of the evidence. In comparisons of adoption estimates between 1998 and 2010, it is important to note that the new data is of substantially higher quality than the old data. Thus, changes in the adoption estimates may simply reflect improvements in data quality, as opposed to changes in the underlying patterns of varietal use.

We note that the entire database for the DIIVA study is publicly available, with full documentation, on the ASTI website (http://www.asti.cgiar.org/diiva). We encourage readers and researchers to visit the website and to make use of the data. In addition to the data on MV adoption data, the database includes observations on varietal releases for each crop-by-country combination and data related to the number of full-time equivalent scientists engaged in crop improvement research. This will provide a benchmark at the level of individual countries and crops so that specific crop-by-country combinations can be tracked and analysed over time. This of course assumes a comparable effort will be sustained over time at regular intervals so that progress can be assessed.

Structure and Contribution

This volume contains a wealth of information from the DIIVA study, beyond the varietal adoption estimates. For a start, it provides detailed information about the research investments in crop improvement across SSA, at the level of individual commodities. This expands on the information previously provided by ASTI and allows for an improved understanding of the differences in research intensity across commodities and countries.

On adoption, the book provides a clear and carefully articulated statement of methods. Not all of the crop-country studies used the same approach to eliciting expert opinion on adoption, but many of the studies followed broadly similar methods. These are spelled out here, making for a significant improvement over the Evenson-Gollin study, which made little effort to impose uniformity of method on the different crop-country studies.

The heart of the book is found in Part 2, which presents the commodity-based chapters. These offer a remarkable level of detail on the diffusion and adoption of different crops. Chapters 6–12 present data from the African studies of DIIVA. These are the major cereal crops (maize, sorghum, pearl millet, rice, wheat and barley); the main root crops (cassava, yam, sweetpotato, banana, and potato), and a number of different legume crops (cowpea, beans, groundnut, pigeonpea, soybean, chickpea, faba bean, field pea and lentil). Taken together, this set of chapters provides the most comprehensive examination of varietal adoption ever undertaken for Africa, and they will set the standard for future studies. These chapters shed light on wide disparities in research effort and success across crops and regions. The success of agricultural research has been uneven, and these chapters identify the challenges and specific accomplishments that have given rise to differential adoption.

Chapters 13 and 14 provide a useful summary of findings from the TRIVSA study (Tracking Improved Varieties in South Asia), a companion to the DIIVA study that focused on South Asia. The TRIVSA study was undertaken more or less in parallel with DIIVA, using similar methods and data sources. Chapter 13 summarizes findings on rice improvement and adoption in South Asia, while Chapter 14 provides comparable results for sorghum, pearl millet, chickpea, pigeonpea and groundnut.

In Part 3, the book provides three chapters that take up the challenge of moving from adoption estimates to impacts. When improved varieties diffuse, replacing previous varieties, there are potential benefits to both consumers and producers. The three studies in this part of the book show how the diffusion data can be used to estimate impacts. Chapters 15 and 16 provide methodologically similar estimates of the poverty impacts of technology improvements in maize (for the case of Ethiopia) and beans (for Uganda and Rwanda). These are important studies that are using frontier methods to take aim at very difficult questions. Estimating the poverty impacts of improved crop technologies is very difficult, as many of the benefits accrue to consumers, and the impact on producers will depend on the structure of the market (e.g. on the extent to which prices fall when production rises). These chapters combine models of individual markets with detailed and disaggregated household data, and they provide valuable insights into the different impacts of the technology on different households. Both studies find measurable and significant impacts on the well-being of the poor, with Chapter 15 showing a reduction in poverty in Ethiopia and Chapter 16 showing a reduction in food insecurity for households in Rwanda and Uganda. These careful studies are frank in their discussion of the challenges involved in estimating impacts, but they also show that even with conservative approaches, improved crop germplasm continues to have significant impacts on the well-being of poor people.

Chapter 17 reports the results of an important exercise: it attempts to show the aggregate effect of varietal improvement on agricultural productivity. Estimates from this analysis show that varietal adoption appears to have a strongly significant impact on total factor productivity in SSA, with an additional significant effect from the agricultural research effort of the Consultative Group on International Research (CGIAR). This chapter argues that improved varieties have raised average net crop yields on adopting areas by almost 50% since 1976–80. The methodological challenges here are large, but the results are generally in line with previous estimates that have shown large impacts.

Part 4 of this book provides a series of extremely useful reflections on synthetic findings across the commodity studies in Part 2 and on the methods used in the DIIVA and TRIVSA studies in Parts 2 and 3. It highlights the different substantive findings and approaches on varietal generation and research output (in Chapter 18) and on varietal adoption and outcomes, including impact (Chapter 19). By making the cross-chapter results explicit and by assessing their implications for the strengths and weaknesses of crop improvement, these chapters provide a roadmap for those wishing to invest in varietal change in Africa. Chapter 20 discusses the results of efforts to validate expert opinion estimates of varietal adoption by using household surveys. The two methods coincide well in some cases, but in other cases there are significant discrepancies. The Standing Panel on Impact Assessment (SPIA) is at present conducting further research to see how different methods of eliciting adoption data compare – and trying to validate these methods using genetic identification methods. We hope to learn more in the coming months and years to guide future research on adoption and diffusion. Finally, Chapter 21 talks about the data needs and methodological changes that face researchers trying to measure adoption and impact. This chapter offers a valuable assessment of the state of the art, and it also describes the challenges that need to be overcome.

Key Findings and Implications

Arguably the most significant finding of this report is the impressive growth achieved in terms of the share of cropped area now under modern varieties in SSA. In 1998, about 20–25% of cropped area was under modern varieties (based on a weighted average across 11 crops). By 2010, this figure had grown to 35% in 2010 (based on a similarly weighted average across 20 crops).[1] Calculated another way, the annual growth rate in the adoption of MVs was 1.45% per annum over this period.[2] This in itself is a remarkable achievement for agricultural research. Although one can still ask questions about the quality of the data, the DIIVA study provides important evidence that agricultural research is continuing to provide technologies of value to farmers. Technology adoption is, in some sense, a logically sufficient measure of impact; farmers would not use these technologies if they did not provide some advantage.

The continued growth in area under modern varieties indicates that research is continuing to provide farmers with useful technologies – and that farmers are continuing to find ways to take up these new technologies, in spite of the constraints that they face. Of course, there are crop-by-country combinations where adoption of MVs is still quite low – 14 of the crops are characterized by a mean adoption rate below 35%. It will be important to analyse the factors that have limited adoption rates for these crops. Conversely, there are crop-by-country combinations that have already achieved a relatively high (for Africa) level of MV adoption (soybean, wheat, maize, cassava, rice) or where adoption has been quite rapid – cassava, barley and maize doubled their share over this period. Here too, there may be lessons to be learned. But an important point to note is that, whether the 1998 base levels were relatively low or high, over 90% of the crop-by-country observations experienced a rise in MV adoption between the two studies. The notion that African crop farming is stagnant is not supported by the data from this study.

Over time, as the level of MVs approaches full adoption, other measures of success of crop improvement programmes, in particular the velocity of varietal change, will become more relevant. Even now, for many crops, this is an important measure of success. The DIIVA study team looked at this and found the area-weighted mean age of varieties in the field was 14 years across all crops – not much change from the earlier period. More analysis is clearly needed here to understand the causes of this. Some older 'modern' varieties are proving to be remarkably robust in the face of many new varieties being released – or alternatively, recent research has not always succeeded in producing genuinely useful technologies.

How reliable are the estimates of adoption emerging from this study? Is there any way to measure their accuracy? These questions occupied the DIIVA Project at every stage. By necessity, the DIIVA data largely draw on judgments made by expert panels. This remains the dominant method for estimating crop area under MVs at a large scale, due to the cost and complexity of collecting data on varietal diffusion through other means. Thus, the DIIVA study relied primarily on expert panel judgments (for 115 crop-by-country combinations). In a number of cases, however, these expert data were supplemented by estimates based on household surveys (for 36 crop-by-country combinations). It was possible to compare these two methods for 18 observations. Of these, ten lined up reasonably well, but household survey estimates were lower for eight observations. Unfortunately, there is no easy way of knowing which of the methods is closer to the truth. On the one hand, nationally representative household surveys might be presumed to be more reliable than expert opinion, since they are based on data collected from individual farm households. On the other hand, there may be gaps in coverage (e.g. because of the low probability of sampling from large commercial farms). Moreover, the quality of the data obtained from household respondents may not be higher: in many settings, it is not clear that farmers can accurately identify the varieties, and the vernacular names that they assign to particular varieties may make identification difficult.[3]

Taken together, we conclude only that further research is needed to reconcile the discrepancies between expert opinion data and survey data on varietal adoption. It would be valuable to know whether there are consistent patterns that would allow us to predict which approach is more accurate for a particular crop-by-country combination. This is certainly an area worthy of further analysis and research. SPIA is currently conducting research to establish cost-effective and reliable methods for measuring adoption, using DNA fingerprinting as a benchmark to assess the accuracy of alternative methods.

Given that expert panel surveys are likely to remain a major source of data in the future when conducting large scale adoption surveys, there are valuable lessons to be learned from the report's observations concerning how best to conduct expert surveys (Chapter 20).[4]

These lessons should not be lost in the vast array of data generated by this study.

The estimates of impact in Chapters 15–17 are also of considerable importance. Although impact assessment is always a challenge, the clear findings from these chapters are that varietal improvement has affected outcomes for the poor. SPIA continues to seek improved methods for estimating impacts, but for now these results stand as some of the best available estimates. They provide strong evidence that research has increased agricultural productivity in Africa and that, for the specific cases reported in Chapters 15 and 16, this has resulted in reductions in poverty and food insecurity respectively.

Issues Emerging and Future Directions for Research

The pages that follow offer a richly detailed account of varietal improvement and its impacts in Africa (and to a limited degree in South Asia). We hope that many researchers will take advantage of the underlying (and publicly available) DIIVA data to construct additional estimates of productivity and impact, and we hope that the current volume will serve as the beginning for a lively conversation over the key messages to be taken from the data.

The main results raise a number of issues that deserve further exploration. Some are easily answered. Others will require new methods or perhaps may be so challenging that they simply invite speculation. For instance:

- Is Africa finally experiencing a Green Revolution? If so, does Africa's experience look like the Green Revolutions of Asia and LAC? Arguably, we are seeing diffusion of modern varieties without seeing much intensification of accompanying inputs. In Asia, the spread of modern varieties was linked to far greater use of fertilizer and mechanization; but in Africa, the growth of these inputs has been much slower.
- Does yield growth in SSA seem to match the diffusion of modern varieties? Do we see substantial yield increases in the crops and countries where we see correspondingly large increases in adoption? This seems like an important question to ask, but perhaps a difficult one. A key challenge is that, by many accounts, crop yield data are very poor in quality. It is not clear whether many countries in Africa conduct regular yield surveys based on crop cuts. Even theoretically, it is possible that the diffusion of improved varieties need not be accompanied by an increase in yields; for instance, a new trait (e.g. drought tolerance) might allow for crop area to expand along an extensive margin where yields are lower. This could in principle result in a decline in average yield.
- A related question: In the crops and commodities where adoption levels are high, have crop yields reached levels that might be viewed as satisfactory? If adoption in some crop-by-commodity combinations is nearly complete, and if yields are still low, what should we conclude? Is this evidence that crop genetic improvement is a weak tool in the sub-Saharan context? Or should we expect that successive generations of improved varieties will increase yields where previous generations have failed? Or should we simply accept that high rates of adoption provide sufficient evidence that improved varieties are useful, even if this is not manifested in crop yields?
- What can we learn from the patterns of diffusion that might inform the research process? What characteristics seem to be associated with high levels of take-up? How can we learn from the DIIVA study to target future research more effectively?

The Need for Continued Data Collection and Analysis

The DIIVA study represents a major contribution towards measuring and understanding the diffusion of modern crop varieties. The value of the study serves as a reminder of the importance of collecting similar data on a regular basis – and of expanding the coverage across geographic areas. In the long run, varietal adoption and diffusion data should ideally become a regular component of national agricultural statistics – collected, for example, as part of national agricultural censuses. In the short run, however, this task remains in the purview of research institutions such as the CGIAR and its partners. SPIA continues to support the collection of diffusion data and to promote the institutionalization of data collection.

Among the activities that SPIA is currently engaged in, as of mid-2015:

- With numerous partners, SPIA is currently working to pioneer and validate new ways of measuring varieties in use, with the hope that these approaches can be incorporated routinely into micro studies and household surveys.
- SPIA is working to collect and report varietal adoption data from Asia.
- We are looking to expand the set of technologies for which adoption and diffusion data are collected; specifically, we seek to extend the data to include observations on improved agronomic practices (e.g. conservation agriculture); irrigation technologies; livestock technologies and practices; and a range of other changes that can potentially be linked to CGIAR research.

In this sense, we think it is important that the DIIVA project be viewed as part of an ongoing set of research activities designed to reveal the continuing diffusion of agricultural technologies, broadly defined. Much remains to be done, and SPIA welcomes partners and researchers who bring new approaches and ideas.

SPIA Chair's Acknowledgments

As will be apparent from this foreword and the document that follows, the DIIVA Project involved a major undertaking. Any project of this size necessarily involves a team effort. In this case, the team was large, including researchers at seven CGIAR centres and numerous national partner institutions. The acknowledgments section of this report lists the full cast of participants, but I would like to take this opportunity to thank, on behalf of SPIA, all of those who contributed time and effort.

The project depended in the final analysis on the efforts and expertise of many researchers based at CGIAR centres and in a range of national research institutions across Africa. We are grateful to the hundreds of scientists who contributed their time to this effort – whether through participating in panels or filling out surveys or providing their field notes, based in some cases on years of data collection. The detailed field knowledge of scientists was ultimately one of the main sources of data for the DIIVA Project. We are grateful to all these scientists for their generosity in sharing time and for their desire to provide thoughtful and objective information about patterns of adoption and diffusion.

Beyond this collective effort, however, I want to single out the outstanding contributions of several individuals who brought the DIIVA Project to fruition through their extraordinary efforts.

First and foremost, we were exceptionally fortunate to have Tom Walker leading this effort on behalf of SPIA. Tom was perhaps uniquely qualified to lead this effort, on the basis of his long and distinguished record of research on agricultural technology adoption and its impacts. Not only did Tom effectively manage this large and complex multi-partner undertaking, but he also provided expertise at every stage of the study. He provided crucial insights into methods of collecting varietal data – from experts, from farmers and from farm communities. Tom's careful probing and his efforts to check and validate the data drew on his deep and detailed knowledge of African agriculture. We are enormously grateful to Tom Walker for his leadership and expertise; without him, the project could not possibly have achieved such a high-quality outcome. Tom's contributions continued through the completion of the book, including the handling of the review process for individual chapters and the editing and cross-checking of numbers used throughout the manuscript. His thoroughness and patience have been essential to the quality of this volume.

Jeff Alwang was closely involved in the DIIVA Project from the beginning, and his involvement grew considerably as the project moved towards completion. Not only did he contribute to the valuable poverty impact studies of Chapters 15 and 16, but in addition, he was a key figure in synthesizing the study, as seen in Chapters 1, 2, 19 and 21. With a keen eye for detail, Jeff played a crucial role in editing the volume and working to shepherd it to completion. Jeff has long been an expert on agricultural technology and its impact, and SPIA is grateful to him for being willing to devote so much of his time and attention to this project.

Perhaps no one was more important to the conceptualization and completion of the DIIVA study than Greg Traxler, programme officer of the Gates Foundation. Along with Prabhu Pingali (who was based at the time at the Gates Foundation), Greg urged the CGIAR to push ahead with a new effort to collect data on varietal diffusion – and he then helped to mobilize the funding for the project. Greg's contributions went far beyond his role as a conduit for funding. Over the course of several years, Greg asked persistently about the scope and quality of data and pushed to set a high standard for the study.

Another key figure in the history of the DIIVA Project was my predecessor as SPIA Chair, Derek Byerlee, who has remained a key participant throughout the duration of the project. Like Tom Walker, Derek brings an encyclopaedic knowledge of African agriculture, based on years of fieldwork and personal experience in most of the countries covered by the DIIVA study. As a dedicated social scientist of the highest calibre, Derek played a central role in the design and implementation of the study. My own term as SPIA Chair started as the DIIVA Project came to a close, so Derek was at the helm of SPIA for almost the entire duration of the project.

Finally, two members of the SPIA secretariat staff – Tim Kelley and James Stevenson – deserve special recognition for their contributions to the project. Tim Kelley's role cannot be easily described. As the head of the SPIA Secretariat, Tim played a key administrative role in managing the study. But Tim's first-hand knowledge of the CGIAR, based on some thirty years as a researcher and research manager, was ultimately of enormous importance in the quality of the DIIVA Project and its findings. I think it is no exaggeration to say that Tim read every sentence produced by the DIIVA Project; his critical eye and high standards were matched by his constantly positive outlook. Tim played a similar role in shepherding and reviewing the earlier Evenson-led study, and this provided him with a valuable long-term perspective on the DIIVA study. In both cases, Tim's contributions proved enormously valuable.

Also at the SPIA secretariat, James Stevenson has played a key role both administratively and substantively in the DIIVA study. As a member of the project steering committee for DIIVA, James participated in every stage of the project; SPIA is fortunate to be able to draw on his skills as a researcher and his thoughtful analysis.

In closing, I would like to honor the memory of Bob Evenson, who died in February 2013. Bob's career-long efforts to document the diffusion and impact of agricultural technologies grew out of his passionate belief that science had the potential to improve the lives of the poor and of rural people. His illness prevented Bob from taking part in the planning of the DIIVA Project, but I have no doubt that he would have been delighted and impressed by the work that has been done – and eager to see it continued through the future.

Douglas Gollin
Professor of Development Economics
Department of International Development
Oxford University

Chair, CGIAR Standing Panel on Impact Assessment (SPIA)

Notes

1 If we look only at the paired comparison of 61 crop-by-country observations for the 10 continuing crops, area-weighted adoption was 27% in 1998 and 44% in 2010.

2 There are a number of qualifiers that must be kept in mind when making comparisons here, given that the number and types of crops and crop-by-country combinations varied between the two periods and that

the methods used to elicit expert opinion were not always consistent over the periods. Nevertheless, while the confidence interval may be large – perhaps more so for the earlier survey results when less scrutiny was applied to the method for eliciting expert opinion – there is no reason to believe that there is a particular upward or downward bias in these different period estimates. All one can say is, the study is using BAD – 'best available data' – and the methods used to collect those data are documented in the reports.

[3] For instance, farmers may use the same name for distinct varieties, and they may use different names for the same variety.

[4] In general, more effective elicitation was characterized by:

- close and intensive supervision of CG project-related staff,
- organization of and attendance at time-bound workshops with direct interaction with expert panel members,
- greater spatial resolution in the elicitation of estimates that were subsequently aggregated to regional and national levels,
- including more members from the informal sector and from NGOs with geographic-specific expertise in technology transfer on the panels, and feedback from CG Center breeders in the final stages of the process

References

Dalrymple, D. (1969) *Technological Change in Agriculture: Effects and Implications for Developing Nations.* USDA/Foreign Agricultural Service, Washington, DC.

Dalrymple, D. (1978) *The Development and Spread of High-Yielding Varieties of Wheat and Rice in the Less-Developed Nations.* 6th Edn. Foreign Agriculture Economic Report No. 95. USDA/USAID, Washington, DC.

Dalrymple, D. (1986a) *Development and Spread of High-Yielding Rice Varieties in Developing Countries.* USAID, Washington, DC.

Dalrymple, D. (1986b) *Development and Spread of High-Yielding Wheat Varieties in Developing Countries.* USAID, Washington, DC.

Evenson, R.E. and Gollin, D. (eds) (2003) *Crop Variety Improvement and Its Effect on Productivity: The Impact of International Agricultural Research.* CAB International, Wallingford, UK.

Acknowledgements

This book is the product of the effort of several hundred people, not all of whom are named here. The project it documents was first conceived in 2008 during a SPIA workshop organized by Flavio Avila in Brasilia, capital city of Brazil, where the need for a comprehensive assessment of the improved varieties of food crops in sub-Saharan Africa was strongly expressed. A project proposal workshop was subsequently held in May 2009 at the Bill & Melinda Gates Foundation (BMGF); a proposal was submitted by Bioversity International; and the grant was awarded later that year. The idea for this volume in the spirit of Evenson and Gollin's 2003 publication *Crop Variety Improvement and its Effects on Productivity* was proposed and discussed at the project's finalization workshop in Rome at Bioversity International in November 2012. The project ended on 30 June 2013.

Greg Traxler from the Foundation was the conceptual force behind the Diffusion and Impact of Improved Varieties in Africa (DIIVA). Project that was fully supported by the BMGF in Prabhu Pingali's Division of Agricultural Policy and Statistics. Greg served as an ex-officio member of the Project Steering Committee (PSC) that met 23 times during the life of the project. Derek Byerlee and Doug Gollin – the chairpersons of SPIA – headed the PSC, devoting considerable time and energy to its conduct and management. Tim Kelley guided all of the project's activities at SPIA, interacting with the coordinator on a daily or at least weekly basis. Tim also coordinated the reviews of project-related papers and sub-proposals. Tim was ably assisted by James Stevenson who also sat on the PSC and was involved in several project-related activities, including the storage and public distribution of the data. We are thankful for the time and inspiration of other PSC members too: Mywish Maredia, Gerry O'Donoghue and Elisabetta Gotor.

We are equally grateful to the following people who contributed to the success of the project in multiple ways – from timely consultancies, through full-fledged support at the CG Centers, to the organization of three project workshops, to reviews of papers: Giorgia Beltrame, Dana Dalrymple, Daniel Calendo, Maria Piedade Caruche, Cheryl Doss, Jessamyn Findlay, Melanie Glover, Carolina Gonzalez, Ben Groom, Guy Hareau, Hilde Koper, Roberto LaRovere, C. Magorokosho, Bernadette Majebelle, Bonny Ntare, Alastair Orr, David Raitzer, Bryon Reyes Randy Shigetani, Cecily Stokes- Prindle, Gert-Jan Stads, Graham Thiele, Rob Tripp and Stan Wood.

Others also warrant a special mention: Tim Dalton and Catherin Ragasa shared data with the DIIVA Project participants; Nienke Bintema hosted the DIIVA databases on the Agricultural Sciences and Technology Indicators Initiative (ASTI) website; James Burgess contributed substantially to preparing the databases for distribution; and Tony Murray spent many hours tailoring software to their storage and distribution. In addition: Peter Gregory drafted a proposal for funding on DNA fingerprinting; Abdoulaye Adam oversaw the design of the sampling survey; Bea Amara managed all the travel-related requests from the project; Patti Sands organized the final workshop; and Brenda Zepeda was responsible for all the financial accounts and reporting of the project. Princess Ferguson worked on several project-related publications, including this one, which was knowledgably and efficiently copy-edited by Frances East and ably published by Emma McCann, James Bishop and their colleagues at CABI Publishing.

We acknowledge the following individuals from the CG Centers and their institutional partners for their invaluable collaboration in making this book a reality:

AfricaRice: Patrice Y. Adegbola, Stella Everline Okello Adur, Daniel Andriambolanoro, Sekou Diawara, Lamin Dibba, Alioune Dieng, Sekou Doumbia, Chantal Ingabire, Dorothy Malaa Kenyi, Winfred A. Odenyo Kore, Nazir Mahamood; M. Hervé Nbedane, N. Robert Ngonde, Vivian T. Ojehomon, Mathieu Ouedraogo, Koku Domenyo Tsatsu, Alexander Nino Wiredu.

Bioversity: Edward Bbosa, Ronald Kakooza, Henry Kayondo, Ronald Mutebi, Betty Nakiyingi, Lydia Namayanja, Robert Ssebagala and Steven Ssengendo who assisted in data collection; Dr Alex Barekye, Micheal Batte, Dr Deborah Karamura and Tendo Ssali scientists from the National BioResource Project (NBRP), Bioversity, and the International Institute of Tropical Agriculture (IITA); and the district agricultural and extension officers of various districts. Our thanks also go to the farmers interviewed in this study.

CIAT: Dr Rowland Chirwa, Prof. Paul Kimani and Dr Clare Mukankusi who are three regional breeders, along with bean program coordinators and national groups of experts from Democratic Republic of the Congo, Ethiopia, Malawi, Mozambique, Rwanda, Tanzania, Uganda, Zambia and Zimbabwe. Finally, we wish to dedicate our paper in this volume to Bonny Ongom who was a Global Information Systems (GIS) research assistant at CIAT in 2006–2011. Bonny was a key collector of in-country data, which he obtained by canvassing experts in project workshops.

The authors of Chapter 16 are grateful for the support and contribution of Matthew Capelli and Stella Wambugo, research assistants at Virginia Tech and CIP, respectively. The authors also thank colleagues at Rwanda Agricultural Bureau, especially Josaphat Mugabo, Jean de Dieu Nsabimana, Jean-Claude Nshimiyimana, and Domitille Mukankubana, for contributions to the research. Helpful comments were received from participants in the DIIVA workshop in Rome, November 2012, and two anonymous reviewers. The authors are indebted to the BMGF for financial support.

CIMMYT: Roberto Larovere for study start-up with Olaf Erenstein, Dan Makumbi, and B.M. Prasanna; and for country reports: Manuvanga Kiakanua (Angola), Moti Aleta and Chilot Tizale (Ethiopia), Lutta Muhammad (Kenya), Mahagayu Clerkson (Kenya), Alexander Phiri (Malawi), Zubeda Nduruma (Tanzania), Moses Mwesigwa (Uganda), Thomson Kalinda (Zambia) and Shamiso Chikobvu (Zimbabwe).

CIP: Scientists in potato and sweetpotato programmes and officials in various NGO and extension services of Burundi, Ethiopia, Kenya, Malawi, Mozambique, Rwanda, Tanzania and Uganda, along with Maria Isabel Andrade, Merideth Bonierbale, Ted Carey, Paul Demo, Manuel Gastelo, Elmar Schulte-Geldermann, Marc Ghislain, Wolfgang Grüneberg, Stef de Haan, Diedonne Harahagazwe, Guy Hareau, Rogers Kakuhenzire, Imelda Kashaija, David Kipkoech, Juan Landeo, Jan Low, Everina Lukonge, Robert Mwanga, Obed Mwenye, Jean Ndirigwe, Jean Claude Nshimiyimana, Michel Ntimpirangeza, Jose Ricardo, Steffen Schulz, Ntizo Senkesha, Kirimi Sindi, Gorretie Ssemakula, Graham Thiele, Silver Tumwegamire and Gebremedhin Woldegiorgis.

ICARDA: The experts and their partners from the national research systems of Ethiopia, Eritrea and the Sudan whose contribution to this work deserve acknowledgment include: the late Geletu Bejiga and his staff in the ICARDA-Ethiopia country office, Mohammad Maatougui, Stefania Grando, Shiv Kumar, Muhammad Imtiaz, Fouad Maalouf, Tamer El-shater, Adefris Chere, Asnake Fikre, Birhane Lakew, Gemechu Keneni, Musa Jarso, Tamene Temesgen, Alemayehu Assefa, Abdelaziz Hashim, Gamal Elkheir. We would also like to thank the EIAR management in general and the Holeta Agricultural Research Center in particular for their logistical and administrative support and the enumerators, data entry and cleaning crew, the drivers and the development agents without whom the project would not be possible. Most importantly special thanks go to the kind and generous Ethiopian farmers who volunteered to be interviewed and spared their invaluable time.

ICRISAT: Experts and national partner institutions that participated in the interviews as well as those who provided useful support throughout the research, including: Dimphyna (Kenya), Albert Chamango and Geoffrey Kananji (Malawi), Stephen Lymo, Rose Ubwe and Mponda (Tanzania), Jasper Mwesigwa Batureine (Uganda) and Francisco Miti and Catherine Mungoma (Zambia). In West and Central Africa the main collaborators were: H. Ajeigbe, A. Amadou, O. Cisse, C.A. Echekwu, I. Faye, C.T. Hash, A. Ibro, A. Kassari, O. Kodio, A. Magagi, A. Minougou, A. Moutari, J. Naino, B.R. Ntare, M. Sanogo, A. Toure, M. Yeye and R. Zangre.

In India, the ICRISAT authors of Chapter 14 are grateful to officials of ICAR and the AICRP research projects on sorghum, pearl millet, chickpea, pigeonpea and groundnut for their close collaboration and support from the inception of the TRIVSA Project to its end. We also thank scientists of ICRISAT, DSR, IIPR, DGR, respective SAUs and HPRC members for their generous help in data sharing and for their participation in the elicitation process.

IRRI: The authors of Chapter 13 wish to acknowledge the funding support provided by the BMGF for the project TRIVSA. We are thankful to NARS collaborators from Bangladesh (Md. Adbus Salam, Md. Abu Bakr Siddique and Md. Ariful Islam), Bhutan (M. Ghimiray), Nepal (Sudeep Gautam, Hari Krishna Panta, Y.N. Ghimire and D. Gauchan), Sri Lanka (B.R. Wallisinghe, R.D.D.P. Rajapaksa, D.M.N. Dissanayake and THCS Perera), Chhattisgarh (A.K. Koshta and V.K. Chaudhary), Odisha (Debdutt Behura) and West Bengal (B. Bagchi) for their active participation in field work, data assembly and analysis.

IITA: A number of people have contributed to the successful completion of the DIIVA study on modern variety release and adoption for IITA mandate crops in sub-Saharan Africa. The authors would like to thank Jonas Mwalughali (late), Makaiko Khonje, and Robertson Khataza for their outstanding research assistance. We also wish to acknowledge the contributions of the survey team leaders who coordinated data collection efforts and expert consultations in various countries – Adebayo Bamire, Mossi Illiassou, Luke Olarinde, Asman Wesonga, Eliya Kapalasa, Toussain Kendenga, Hobayo Claude, Bisase Dennis and Tafireyi Chamboko. We are particularly grateful to the national programme and private sector leaders and researchers from over 20 countries in SSA who provided a great deal of information used in the DIIVA study through extensive consultations and discussion as well as review of institutional documents and records including: Moniz Mutunda (Instituto de Investigação Agronomica, Angola); Aly Djima, Yallou Chabi Gouro, Kodjo Siaka, Abou Ogbon, Romuald Dossou, Institut National des Recherches Agricoles du Bénin, and J. Detongnon, (Société Nationale pour la Promotion Agricole) (Benin); Issa Drabo (Institut national d'Etudes et de Recherches Agricoles) and Jacob Sanou (Institut de l'Environnement et de Recherches Agricoles) (Burkina Faso); Alexandre Congera, (Institut des Sciences Agronomiques du Burundi); Eddy Mangaptche, Ndioro Mbassa, Tandzi Liliane, Zonkeng Celicard, Meppe Paco, Nguimgo Blaise, and Njualem Khumbah (Institute of Agricultural Research for Development, Cameroon); Louise Akanvou, Philippe Gnonhouri, Amani Michel Kouakou, Francois N'gbesso, Boni N'zue and Siaka Traoré, (Centre National de Recherche Agronomique, Côte d'Ivoire); Joe Manu-Aduening, Henry Asumadu, Manfred Ewool, K. Obeng-Antwi, and Emmanuel Otoo (Crops Research Institute), Cecil Osei (Root and Tuber Improvement and Marketing Program), and Kwabena Acheremu and Adjebeng Joseph (Savanna Agricultural Research Institute) (Ghana); Diallo Amadou, El Sanoussy Bah, Cheick Conde, Camara Fadjimba, Sekouba Korouma, Diallo Pathé, and Camara Sékouna (Institut de Recherche Agronomique de Guinee) (Guinea); Joseph Kamau (Kenya Agricultural Research Institute); M.C. Dzinkambani and Vito Sandifolo (Department of Agriculture Research Services) (Malawi); Ntji Coulibaly and Mamadou Touré (Institut d'Economie Rurale) (Mali); Anabela Zacarias (Agricultural Research Institute of Mozambique) and Rogerio Cheulele (University of Eduardo Mondlane) (Mozambique); Moutari Adamou (Institut Nationale de Recherches Agronomiques du Niger); Paul Ilona (CIAT-Nigeria); Hakeem Ajeigbe (ICRISAT-Nigeria), U.S. Abdullahi (Institute of Agricultural Research); S.R. Akande and S.A. Olakojo (Institute of Agricultural Research & Training), Shokalu Olumide (National Cereals Research Institute), S.O.S. Akinyemi (National Horticultural Research Institute), Njoku Damian (National Root Crops Research Institute), M.A.B. Fakorede (Obafemi Awolowo University, Ile-Ife), and Ibikunle Olumide (Premier Seed Nigeria Ltd.) (Nigeria); Bambala Emmanuël, Lodi Lama, and Mbuya

Kankolongo (Institut National d'Etude et de Recherches Agronomiques) (DR Congo); Ndiaga Cisse and Abdou Ndiaye (Institut Sénégalais de Recherches Agricoles) (Senegal); Edward Kanju (IITA-Tanzania); Akihila Didjeira, Béré Tchabana, Etoudo N'kpenu, and Komi Somana (Institut Togolais de Recherche Agronomique) (Togo); Elvis James Mubiru (IITA-Uganda), Godfrey Asea, Antony Bua, and Nelson Wanyera (National Agricultural Research Organization) (Uganda); Martin Chiona, B.L. Kaninga, and Laston Milambo (Zambia Agricultural Research Institute); and Patrick Nyambo (Department of Research and Specialist Services) and Caleb M. Souta and Jacob Tichagwa (SeedCo Limited) (Zimbabwe).

Finally, support for the research reported in Chapter 17 was provided by the USAID-USDA Norman Borlaug Commemorative Research Initiative and the CGIAR Standing Panel on Impact Assessment. The views expressed in Chapter 17 are the authors' own and do not necessarily reflect those of the Economic Research Service or the USDA.

1 The Importance of Generating and Documenting Varietal Change in Sub-Saharan Africa

T.S. Walker[1]* and J. Alwang[2]

[1]*Independent Researcher, Fletcher, North Carolina, USA;* [2]*Department of Agricultural and Applied Economics, Virginia Tech, Blacksburg, USA*

When a farmer in sub-Saharan Africa plants a food crop, the odds are increasing that the variety sown will be an improved variety touched by science. But more likely the farmer plants a local variety that is more or less the same as that cultivated by his or her parents, grandparents and great-grandparents. For some farmers, such as groundnut growers in West Africa and sweetpotato producers in East Africa, it is likely that the variety cultivated is a product of agricultural research but that the improved variety was bred more than 40 years ago.

A lack of dynamism in varietal change in food crop production represents a wasted opportunity that is potentially high, exacting a heavy toll on poor producers and consumers alike. Crop production consumed in the household and sold in the market may represent more than 50% of the income of poor farmers. Expenditures on staple and secondary food crops may eat up more than 60–70% of the budget of poor consumers. Because crop variety improvement can increase production that in turn can lead to declining and more stable prices, it is a cost-effective intervention with a broad scope to leverage positive outcomes and impacts for hundreds of millions of poor rural and urban households in sub-Saharan Africa.

Modern varietal change is an important tool with large potential contributions to agricultural development. Unlike some other types of agricultural technology, modern varietal change is not limited by agroecology and population density, nor does it require major capital investments by potential adopters. Uptake of improved varieties can lead directly to positive consequences for food security. Modern varietal change in and of itself may not lift large numbers of people out of poverty but greater dynamism in this area can go a long way to moving poor people closer to the poverty line. Moreover, modern varietal change can set the stage for the adoption of more intensive crop production practices, such as row planting, and is a precursor to the judicious use of purchased inputs that spark multiplier effects for economic growth.

Agricultural Research: The Engine for Generating Varietal Change

Since the independence of most African nations in the 1960s and 1970s, a foundation for modern varietal change in food crops was laid down by public sector national research programmes

* E-mail: walkerts@msu.edu

(NARS) in the countries of sub-Saharan Africa. Beginning in 1968, the International Agricultural Research Centers (IARCs) have been a partner in that effort. That seems like a long time ago but it is a recent undertaking compared to the genetic improvement in export crops, such as cocoa, cotton and rubber, that occurred much earlier in the 20th century.

In spite of its youth, crop genetic improvement in food crops is not as vigorous or widespread as it should be in sub-Saharan Africa. Its effectiveness is compromised for multiple reasons. Agroecological conditions are extremely heterogeneous in many African countries, especially compared to those in South Asia where widespread diffusion of modern varieties sparked the Green Revolution, which contributed to remarkable productivity growth and poverty reduction beginning in the mid-1960s. Limited infrastructure and weak support systems in sub-Saharan Africa have constrained the uptake of improved varieties. Lack of funding for operating budgets is an important limitation that is shared by both NARS and IARCs. Largely because of declining global food prices, real resources had steadily become scarcer for crop improvement research by IARCs and NARS, especially from the early 1990s to the early 2000s (Beintema and Stads, 2006). Expansion of the mandates of the IARCs into areas such as natural resource management also contributed to the erosion of resources for genetic improvement.

Since the abrupt rise in global food prices after 2008, funding for agricultural research has improved. Donors, in general, and the Bill & Melinda Gates Foundation (BMGF), in particular, have invested heavily in food-crop genetic improvement in sub-Saharan Africa. Once again, a strong partnership between NARS and IARCS is a hallmark of that investment.

Documenting Varietal Change: The Need and Past Achievements

Without the adoption of agricultural technologies, there is no impact (Adato and Meinzen-Dick, 2007). Indeed, the area planted to a new technology is the most important determinant in the size of economic benefits (Walker and Crissman, 1996; Morris *et al.*, 2003). Cost savings per unit of output of the new technology also determine impact by influencing diffusion and creating economic benefit for each area unit of spread.

Impact analysis of varietal change has largely relied on the economic surplus approach to estimate standard rates of return to the research. These studies suggest that, although returns to research have been positive in sub-Saharan Africa, they have been lower than in other regions. In addition to monitoring for a high return on investment, however, donors want to be better informed about the impact of research on the development goals of poverty reduction, food security and environmental sustainability. In spite of increasingly numerous reviews, impact assessment of agricultural research in sub-Saharan Africa is still best described as sparse (Maredia and Raitzer, 2006).

Highly specific information on adoption and benefits from variety use provides research managers with needed ammunition for deciding on the relative resource allocation for commodities and specific lines of research. To be successful, research needs to be sensitive to users' demands. For crop genetic research, the demand for traits is of paramount importance. The opportunity costs for research funds are high, and research on adoption levels and impacts can establish which traits are in demand and where acceptable trade-offs can be made.

Globally, credible databases on the diffusion and impact of well-identified improved varieties are rare. Maize, other cereals and oilseeds are a notable example of where sales information on hybrid seed can provide solid data on varietal uptake. Vegetatively propagated crops, such as potatoes, that are legislatively required to be planted with clonal-specific certified seed represent another case. Aside from these exceptions varietal-specific information is seldom widely available for important food crops even in developed countries. For example, the United States Department of Agriculture (USDA) stopped collecting data on the adoption of improved wheat varieties in the mid-1980s. But in developed-country agriculture, improved varieties are replaced by farmers every 2–5 years; varietal change is no longer an issue that impinges on economic and social development. In contrast, not knowing about the pace and dynamics of varietal change is a luxury that developing countries in sub-Saharan Africa can ill afford because both the level of modern cultivar adoption and the velocity of improved varietal turnover are low.

Since the release of maize hybrids in Kenya in the 1960s, episodic research on adoption of modern cultivars has been conducted in sub-Saharan Africa (Gerhart, 1974). Dana Dalrymple was the first agricultural scientist to make a systematic effort to document the diffusion of improved varieties in food crops. In 1978, Dana Dalrymple completed the sixth review of the spread of the high-yielding varieties (HYVs) of wheat and rice in developing countries (Dalrymple, 1978). These semi-dwarf, short-duration varieties had entered Africa as early as the late 1960s. Dalrymple estimated that the diffusion of modern rice varieties had reached 4% by 1978. He included 15 rice-growing countries in his assessment that was based mainly on direct communication with in-country scientists working on rice genetic improvement in Africa.

By the 1970s, sub-Saharan African farmers began to benefit from recently bred varieties in several primary and secondary food crops. A firm baseline for evaluating the effectiveness of food-crop genetic improvement, however, only began to emerge in the mid- to late-1990s. A global monitoring and evaluation research agenda (referred to here as the 1998 Initiative) retrospectively assessed varietal output, adoption and production impacts in food-crop genetic improvement in developing country agriculture (Evenson and Gollin, 2003). That initiative resulted in several surprises including the realization that dynamic varietal change was not confined to the so-called Green Revolution period between the mid-1960s and the early 1980s, but that it continued deep into the 1990s. But estimated adoption levels in Africa, averaging 22%, were especially low.

The estimates reported in Evenson and Gollin (2003) were based on partial results with limited data available for a number of crops and countries. As a result, the picture of modern varietal adoption in sub-Saharan Africa was somewhat fuzzy and fragmented even at that time and, in the past decade, no comprehensive study had updated or clarified those estimates.

The DIIVA Project

Here, the baseline established by Evenson and Gollin (2003) has been updated, widened and deepened. We report on the results of a CGIAR project – Diffusion and Impact of Improved Varieties in Africa (DIIVA Project) – the first major study to focus on the diffusion and impacts of improved crop varieties in SSA. Supported by BMGF, seven CGIAR Centers (CG Centers) and their national and other partners carried out adoption research and impact assessments as part of DIIVA. The DIIVA Project, which was directed and coordinated by CGIAR's Standing Panel on Impact Assessment (SPIA) and administrated through Bioversity International, began on 1 December 2009 and ended on 30 June 2013.

A budget of slightly under US$3 million was allocated to three objectives designed to:

- Attain a wider understanding of the performance of food-crop genetic improvement in priority crop-by-country combinations in sub-Saharan Africa;
- Verify and gain a deeper understanding of the adoption and diffusion of new varieties in selected priority countries and food crops in sub-Saharan Africa;
- Acquire more comprehensive insight in to the impact of crop improvement on poverty, nutrition and food security.

The DIIVA Project is viewed as a major building block in the construction of a routine system for monitoring varietal adoption and impact in sub-Saharan Africa for the CGIAR research programmes. This work has been driven by three complementary activities that respond to three project objectives: (i) documenting the key performance indicators of crop genetic improvement; (ii) collecting nationally representative survey data on varietal adoption; and (iii) assessing the impact of varietal change.

The novelty and value of the research reported in this book stems from its wide scope in terms of crops and countries with intensive data collection via standardized protocols. This standardization permits comparisons across countries, over time and among crops in a given country. The study is also unique for its emphasis on validation and on the use of sound integrated methods for impact assessment. In particular, household- and field-level data are used to estimate productivity gains, per-unit reductions in cost of production and other household-level outcomes. These methods represent an improvement over standard surplus estimation techniques, which usually rely on data from experimental trials. Trial data do not reflect regional variability in agroecology and yield potential or idiosyncratic differences in household management of production processes.

The adoption of improved varieties of 20 food crops in 30 countries covering about 85% of food crop production in sub-Saharan Africa was assessed in the DIIVA Project. More than 200 individuals, the majority of whom were scientists from national agricultural research systems, contributed to this effort. The DIIVA database contains information on more than 3500 formally and informally released varieties and more than 1150 improved varieties that were adopted by farmers in 2010 (http://www.asti.cgiar.org/diiva).

This volume represents the full rendering of DIIVA-related research by the participant scientists who assembled the information and collected the data. Earlier publications with a narrower focus include Alene *et al.* (2011) and Walker *et al.* (2014).

Fields crops in sub-Saharan Africa are almost entirely grown in dryland agriculture. The BMGF also invested in a smaller comparative project called TRIVSA (Tracking Improved Varieties in South Asia) that supplied information on varietal generation and adoption in food crops cultivated in the rainy season in South Asia. Research from the TRIVSA Project is represented by two chapters in this book and findings from South Asia serve as a point of reference for the results from sub-Saharan Africa that are highlighted in the synthesis chapters described below.

Organization

This volume is divided into four sections. Part 1 sets the stage by first reviewing investments in food-crop improvement in sub-Saharan Africa (Chapter 2). Chapter 2 shows that, starting from a low base in the 1960s, investments in crop improvement in the region grew robustly before slowing in the 1980s. Following a long period of stagnation beginning in the 1980s, robust growth in funding returned in 2001. The chapter shows that funding increases have also been accompanied by a generalized improvement in human capacity in national systems, but that aggregate figures of investments and growth can be misleading. Growth in funding and capacity is concentrated in the larger national research systems, whereas some smaller systems have shrunk substantially. Studies of rates of return to agricultural research is sub-Saharan Africa are summarized and these show varied results but, before the mid-1990s, estimated rates of return to crop improvement were lower than those in other regions of the world.

Chapter 3 defines concepts and hypotheses that have guided the DIIVA research on inputs, outputs, outcomes and impacts. The study documents two key inputs into crop improvement by year and country: scientific capacity and research intensity. Measured outputs in the study are variety releases, and outcomes are adoption and rate of variety turnover. The impact measures employed vary by study; these include yield, productivity, household income and poverty reduction. Chapter 4 goes on to describe data, methods and crop by country coverage. The DIIVA data can be divided into three domains: assembled data on scientific capacity and varietal release/availability; elicited estimates of varietal adoption; and household survey data. The variety-specific data contain about 150 crop-by-country observations selected to cover the most important food crops in the main producing countries. Crop-by-country data were assembled to provide a broad perspective of the important food crops in the region and to allow the study to be comparable to the 1998 Initiative.

Chapter 5 provides the historical context for genetic improvement for the 11 crops in the 1998 Initiative and an exploratory analysis of the variation in inputs, outputs and outcomes across commodities and countries. Country- and crop-specific comparisons show striking differences in scientific staff capacity and research intensity, but comparisons to the rest of the developing world show that sub-Saharan African indicators of these inputs are in line with other continental regions. The 1998 estimates of variety release display high variability over time for most crops in many countries. The most salient finding is that varietal output from crop improvement programmes accelerated dramatically in the 1990s. This acceleration sets the stage for a renewed look at impacts, as a variety's uptake lags behind its release, often by many years.

impacts are likely to have become more pronounced and visible after 1998.

Varietal generation, output, adoption and turnover in food crops are addressed in nine studies in Part 2. Chapters 6–12 focus on sub-Saharan Africa. They are organized around and are synonymous with the mandated-crops of these CG Centers: International Institute for Tropical Agriculture (cassava, cowpea, maize and yams); International Center for Research in the Semi-arid Tropics (groundnut, pearl millet, pigeonpea and sorghum); International Center for Tropical Agriculture (beans); International Potato Center (potato and sweetpotato); International Center for Maize and Wheat Improvement; International Center for Agricultural Research in Dryland Areas (barley, chickpea and faba bean); and AfricaRice. This work is complemented by two comparative studies from South Asia where the commodity emphasis is on rainfed rice in multiple countries and states in India (Chapter 13) and on sorghum, pearl millet, groundnut, pigeonpea and chickpea in peninsular India (Chapter 14).

The impact of the adoption of modern varieties is assessed in case studies on maize in Ethiopia (Chapter 15) and beans in Rwanda and Uganda (Chapter 16). These studies show that impacts of adoption on productivity and cost savings are relatively large at the field level. They show that poor farmers have not been excluded from adoption; these varietal improvements seem to be accessible to all farmers. Benefits are broad-based, but vary by characteristics of adopting farmers and their agroecologies and, because areas planted are relatively small, impacts of adoption on household income and poverty are modest.

Estimates of total factor productivity with the updated DIIVA adoption data in sub-Saharan Africa are found in Chapter 17, the final chapter in Part 3. Chapter 17 shows that adoption of improved food crop varieties raised productivity of adopting areas in sub-Saharan Africa by an average of 47% and accounted for about 15% of the growth in food crop production between 1980 and 2010. By 2010, the higher productivity of improved food crop varieties had added US$6.2 billion to the annual value of agricultural production in the sub-continent.

Both substance and process are featured in Part 4, which begins with two syntheses that draw on the data and findings in Chapters 6–14. Varietal generation and output are the subjects of Chapter 18. Adoption, turnover and impact are themes for Chapter 19. What we learned about estimating varietal adoption and assessing varietal impact is discussed and summarized in Chapters 20 and 21.

References

Adato, M. and Meinzen-Dick, R. (eds) (2007) *Agricultural Research, Livelihoods and Poverty: Studies of Economic and Social impact in Six Countries*. John Hopkins University Press and IFPRI, Washington, DC.

Alene, A., Yegezu, Y., Ndjeunga, J., Labarta, R., Andrade, R. *et al.* (2011) *Measuring the Effectiveness of Agricultural R&D in Sub-Saharan Africa from the Perspectives of Varietal Output and Adoption: Initial Results from the Diffusion of Improved Varieties in Africa Project*. ASTI Conference Working Paper 7. IFPRI, Washington, DC.

Beintema, N.M. and Stads, G.J. (2006) *Agricultural R&D in Sub-Saharan Africa: An Era of Stagnation*. Background Report. IFPRI, Washington, DC.

Dalrymple, D. (1978) *The Development and Spread of the High-Yielding Varieties of Wheat and Rice among Less-Developed Nations*. 6th Edn. Foreign Agricultural Economic Report No. 95. USDA/USAID, Washington, DC.

Evenson, R.E. and Gollin, D. (eds) (2003) *Crop Variety Improvement and Its Effect on Productivity: The Impact of International Agricultural Research*. CAB International, Wallingford, UK.

Gerhart, J.D. (1974) *The Diffusion of Hybrid Maize in Western Kenya*. PhD thesis, Princeton University, Princeton, New Jersey.

Maredia, M.K. and Raitzer, D.A. (2006) *CGIAR and NARS Partner Research in Sub-Saharan Africa: Evidence of Impact To Date*. Consultative Group on International Agricultural Research. Science Council Secretariat, Rome.

Morris, M., Mekuria, M. and Gerpacioe, R. (2003) Impacts of CIMMYT maize breeding research. In: Evenson, R.E. and Gollin, D. (eds) *Crop Variety Improvement and Its Effect on Productivity: The Impact of International Agricultural Research*. CAB International, Wallingford, UK, pp. 135–158.

Walker, T. and Crissman, C. (1996) *Case Studies of the Economic Impact of CIP-Related Technologies.* International Potato Center (CIP), Lima.

Walker, T., Alene, A., Ndjuenga, J., Labarta, R., Yigezu, Y. *et al.* (2014) *Measuring the Effectiveness of Crop Improvement Research in Sub-Saharan Africa from the Perspectives of Varietal Output, Adoption and Change: 20 Crops, 30 Countries, and 1150 Cultivars in Farmers' Fields*. Report of the Standing Panel on Impact Assessment (SPIA). CGIAR Independent Science and Partnership Council (ISPC) Secretariat, Rome.

2 Investments in and Impacts of Crop Improvement Research in Africa

J. Alwang*
Department of Agricultural and Applied Economics, Virginia Tech, Blacksburg, Virginia, USA

Because agricultural productivity in sub-Saharan Africa (SSA) has historically been low and continues to lag other regions of the world, there is increasing interest in understanding how research investments in the region are associated with productivity growth. To understand this relationship, it is important to begin with an assessment of historical investments in agricultural research. Research investments in support of SSA agriculture have received wide attention in the development literature and these studies have produced a broadly consistent picture. Investments in agricultural research and development (R&D) in SSA started from a very low base immediately following independence in the late 1950s and early 1960s. Investment grew at a rapid rate in the 1960s and through the 1970s and 1980s, but slowed midway into the 1980s and declined in the 1990s. Since 2000, R&D investments in the region have increased and growth in research expenditures was robust through 2008, the last year for which comprehensive data are available.

There remain, however, inconsistencies in the analysis of R&D expenditures for SSA. For example, there has been only limited analysis of comparative investments across SSA in commodity-specific research and whether the distribution of research resources accurately reflects the distribution of commodities by area produced. It is also well known that observed growth in SSA-wide agricultural R&D from 2001 to 2008 was driven by investments in large systems[1] such as Nigeria, Ghana, Tanzania and Uganda. Little analysis has been conducted of growth patterns among medium-sized and, particularly, small systems. Evidence shows severe declines from 2001 to 2008 in several smaller systems (Agricultural Science and Technology Indicators; ASTI). These inconsistencies are partly due to information challenges: data on research expenditures come from multiple sources, many with irregular reporting practices, and many data sources have gaps. Given these challenges, it is important to summarize findings of studies on research expenditures in SSA, identify consistent patterns and explore discrepancies in reported trends.

The purpose of this chapter is to document evidence about agricultural research investments, describe patterns of change over time, and discuss the current state of knowledge and knowledge gaps. The chapter begins with a discussion of information sources and inherent challenges in assembling consistent time series from alternative sources. Important past studies are reviewed critically with an aim at synthesizing the current state of knowledge about agricultural

---

* E-mail: alwangj@vt.edu

R&D expenditures in SSA. A short discussion follows of the role of agricultural R&D in promoting output and productivity in SSA. The chapter ends by discussing areas of agreement and disagreement.

Data and Information Sources

Public investments[2] in agricultural research come through two major streams: (i) government and donor-sourced investments in national agricultural research systems (NARS) and country-based research entities; and (ii) investments made by International Agricultural Research Centers (IARCs) under the umbrella of the Consultative Group on International Agricultural Research (CGIAR). The former stream of investments has historically far exceeded the latter, but the latter, by leveraging research findings from other regions and focusing more heavily on basic research, plays an important role. In fact, Evenson and Gollin (2003) note that IARC research investments are more likely to be important to NARS in SSA compared to other regions, because more than one-half of all improved varieties in SSA came from an IARC cross (compared to 36% worldwide). Funding for the CGIAR (CG Centers) has undergone substantial change. From 1990 to around 2006, global funding for the CG remained approximately constant in real terms at around US$400 million (2005 purchasing power parity; PPP). Since 2006, funding has grown relatively steadily to the point that it approached US$1 billion by 2013. CG funding sources have also changed, with the emergence of the Bill & Melinda Gates Foundation (BMGF) as a major contributor to the CGIAR beginning in 2006. By 2013, BMGF accounted for more than 10% of total CGIAR Fund contributions, putting the Foundation among the top donors to the Fund.

A third source of public research funding is through bilateral, multilateral and donor assistance to regional research groups and directly to university and private researchers. Examples of this stream include USAID-funded Collaborative Research Support Programs (now called Feed the Future Innovation Labs). The Department for International Development (DFID)[3] provides substantial direct support to host-country researchers through UK research councils and for private funds in developing countries such as the Forum for Agricultural Research in Africa; the Canadian International Food Security Research Fund supports research partnerships between Canadian and developing country researchers; other donors, including BMGF and other major philanthropies, channel some funding for agricultural research through non-CG and non-NARS entities (Norton and Alwang, 2012). These latter funding streams are difficult to trace, fluctuate over time, and may be absorbed into CG and NARS funding reports. They also historically represent a relatively small amount compared to the first two streams.[4] As a result, most analyses of agricultural research expenditures do not explicitly include this third stream.

Whereas information on IARC investments has been available for many years through individual centre annual reports and various compendia, information on national research investments has historically been difficult to come by. An initiative for collecting and compiling indicators on agricultural R&D began in 1981 as a joint venture of the International Service for National Agricultural Research (ISNAR)[5] and the International Food Policy Research Institute (IFPRI). This initiative uses data from primary (extensive surveys in developing countries) and other sources to compile a benchmarked and comparable data series for many developing countries. The initiative, now known as ASTI (Agricultural Science and Technology Indicators Initiative) has comprehensive data series from 32 SSA countries, which together contribute more than 90% of the region's agricultural gross domestic product (GDP). Although this information is comprehensive, it is not complete, and further efforts are needed to fill data gaps.[6]

Measures

An adequate measure of inputs into R&D requires understanding of the institutional structure of agricultural research in SSA. Historically, most public agricultural research in the region was conducted in government agencies but research in institutions of higher education has grown.[7] Inputs into the agricultural R&D process thus include both expenditures (put on a common PPP basis) and scientist full-time equivalents (FTEs). Comparisons across countries of different sizes and resource endowments require normalization, such as research intensity (expenditures divided by agricultural GDP) or scientists per unit of agricultural GDP. Research intensity has intuitive appeal as an indicator because in 2003 the African Union's New Partnership for Africa's Development set a target research intensity of 1% for its member nations (Beintema and Stads, 2011).

Brief Overview of Agricultural R&D in Africa

Real agricultural GDP growth in SSA averaged about 2% per year since 1961 and accelerated to 3.4% in 2001–2008 (Fuglie and Rada, 2013). Prior to 2000, agricultural productivity growth lagged behind population growth, implying lower food availability per person for the four decades between independence and the start of the new millennium. Causes of lagging productivity growth include slow production of new crop varieties suitable for conditions in SSA, poor performance of input and output markets, lack of agricultural support services, discriminatory agricultural policies, and slow rates of adoption of improved varieties. Lagging productivity is partly explained by unique challenges in SSA: near-complete reliance on rainfed agriculture and extreme spatial variability in biotic and abiotic constraints. Also, prior to 2000, the impact of agricultural R&D on SSA agricultural productivity growth was estimated to be lower than elsewhere in the world. In fact estimates show that before 2002, agricultural output growth in the region was driven by increased resource use, not by enhanced agricultural productivity (Fuglie and Rada, 2013).

Since the mid-1980s, however, annual total factor productivity growth in SSA agriculture has accelerated somewhat (to about 1%). This expansion coincides with increased availability and adoption of new agricultural technologies, including new crop varieties. In fact, by the late 1990s nearly 20% of the area planted to food crops in SSA was sown to improved varieties (Evenson and Gollin, 2003). SSA-wide adoption is uneven; among major foods approximately 18% of area is planted to improved varieties,[8] but prior to the more comprehensive estimates presented in Part 2 of this volume, it was widely perceived that adoption of modern varieties of some important food crops, such as pearl millet and groundnut, was low to negligible (Fuglie and Rada, 2013). Thus, while adoption and spread of modern varieties is growing, there is substantial potential for future spread.

The ultimate determinant of the supply of modern varieties is investment in agricultural R&D, particularly investments in crop improvement research. Several studies have examined expenditures and scientist FTEs involved in agricultural R&D in SSA. The main references for this research, summarized in Table 2.1, are Pardey *et al.* (1997), Evenson and Gollin (2003), Maredia and Raitzer (2006), Beintema and Stads (2011), and Fuglie and Rada (2013). Primary data sources include the ASTI indicators for NARS research expenditures and various incarnations of these data, CGIAR Center-based records, which account for CG investments, and ad-hoc surveys of IARC and NARS scientists and research administrators.

Agricultural research expenditures in SSA

As of the mid-1990s, studies of agricultural R&D in Africa showed significant increases from a very low base in public investments through the 1960s and 1970s. This expenditure growth slowed in the mid-1980s and into the 1990s (Pardey *et al.*, 1997; see Table 2.1). Data also show substantial inter-year variability, but little analysis has been conducted on this variability, except to note that year-to-year funding variability in SSA exceeds that in the rest of the developing world. During the latter part of the 20th century, instability in funding for SSA R&D was much higher than other regions of the world. SSA accounted for 5% of global public R&D in 2000, down from 7% in 1981. The slowdown in spending during 1980–2000 is likely to have lasting impacts region-wide (Beintema and Stads, 2011).

The observed slowdown in spending on agricultural R&D in the mid-1980s was partially caused by reductions in bilateral and multilateral grants and loans as donors turned to other investments, but national systems saw

Table 2.1. Summary of recent literature on agricultural R&D in sub-Saharan Africa.

Source	Years	Focus	Data source	Main findings
Pardey *et al.* (1997)	1961–1991	Research expenditures by country, composition of expenditures; explicit focus on SSA	IFPRI/ISNAR	Government research constitutes lion's share of expenditures (91% in 1961 to 85.6% in 1991) University-based research expanded by 7.1% per year 1961–1991 Increase in researchers from 2000 to 9000 (1961–1991) Shift from expatriate researchers (90% to 11% of total research staff 1961–1991); nearly 65 % of NARS researchers have post-graduate degrees A few very large systems dominate Real research expenditures grew rapidly during 1960s, moderately in 1970s and ceased growth in 1980s through to early 1990s NARS are becoming increasingly dependent on external (to country) spending Strong heterogeneity – some systems continued to grow, while others shrank
Evenson and Gollin (2003)		Adoption and impacts of improved varieties; crop-specific estimates	Various, depending on crop; most crop-specific data are from CG Center cross-sectional surveys	Sorghum: low investments in African R&D for sorghum despite its economic importance (data from late 1980s and early 1990s) Maize: lower research intensity in eastern and southern Africa compared to rest of the world; limited private sector involvement in maize breeding compared to rest of the world Millet and groundnuts: few details on research resources
Maredia and Raitzer (2006)	1970s – 2004, with a focus on 2000–2004	Research expenditures of CGIAR, estimates of research impacts	CGIAR Center reports; case studies of research impact	40% of annual CG research budget devoted to SSA since inception Virtually all CG Centers are heavily invested in SSA IITA, ICRAF, ILRI and CIMMYT have highest research expenditures in SSA System-wide cumulative investment of US$4.3 billion ($ 2004) by 2004 in SSA Change in composition of CG investments away from crop-productivity enhancements; dramatic decline in the real value of crop productivity, enhancing research since the mid-1980s

Beintema and Stads (2011)	Focus on 2001–2008	Research expenditures and staffing	ASTI	Strong growth in real resources devoted to agricultural R&D since 2001 in SSA Agricultural GDP growth in SSA lags behind overall growth even through 2008 Uneven patterns of spending: large systems drive overall trends and some very small NARS are vulnerable because of low spending and staffing levels Institutional arrangements for agricultural R&D vary from country to country but single agency dominates in most; in smaller countries, bulk of research is being conducted by disperse government agencies and universities Government role is shrinking over time Researchers in higher education are growing and more than doubled from 1991–2008; the share of higher education in public R&D research staff grew from 15 to 24% during the same period
Fuglie and Rada (2013)	1981–2005	Research expenditures, spread of modern varieties, impacts of research	ASTI, supplemented by CG Center reports	R&D investments have had a strong impact on total factor productivity region-wide Prior to mid-1980s, growth in agricultural output in SSA was due to increased use of inputs (land, labour and capital), not growth in productivity Since mid-1980s, total factor productivity growth in SSA averaged about 1% per year Policy environment affects linkages between research investments and productivity growth

CIMMYT, Centro Internacional de Mejoramiento de Maiz y Trigo (International Center for the Improvement of Maize and Wheat); ICRAF, International Center for Agroforestry; IITA, International Institute of Tropical Agriculture; ILRI, International Livestock Research Institute.

dwindling support from their own governments. Lagging support is attributable to a number of factors, including moderating commodity and energy prices, increased attention in public sector spending for social sectors, low perceived returns to agricultural investments in the region due to policy-related factors such as tax and marketing policies, and lack of broad public support for research.

African NARS capacity started from a very low base following independence when staffing was thin and most senior scientific staff comprised expatriates (Pardey *et al.*, 1997). Employment in SSA NARS grew at a robust average annual rate of 5% from 1961 to 1991, and by 1991 the region employed around 9000 full-time agricultural scientists (Table 2.2). Overall NARS growth was accompanied by a gradual shrinking of the share of expatriate scientists (to about 11% of the total) as national investments increased training of local scientists. SSA NARS funding relies disproportionately on donor funding and the dependence increased through the 1990s; donor contributions accounted for about 35% of total investments in 1996 (Pardey *et al.*, 2007). Donor support represents a larger share of total R&D expenditures for the poorest countries, particularly for smaller poor countries.

Region-wide agricultural research expenditures began to grow again in 2001 and growth between 2001 and 2008 averaged more than 2% annually (Beintema and Stads, 2011). By 2008 the overall level of spending for the region reached US$1.7 billion (2005 PPP; see Table 2.2). Investments in agricultural research are manifest in different measures. For example, scientist quality has improved over time. As of 2008, 73% of SSA agricultural scientists in SSA research systems had an advanced degree. This should be compared to about 65% in 1991 and 45% in the early 1980s (Beintema and Stads, 2011; Pardey *et al.*, 1997).

Intra-Regional Differences

The aggregate spending picture presented above obscures important differences within SSA. These differences include stark heterogeneity in national system size and quality, and different patterns of investment over time and within systems. A salient characteristic of the agricultural research complex in SSA is uneven size with a few very large systems predominating[9] (Pardey *et al.*, 1997). This unevenness makes it difficult to make generalizations from aggregate trends; the aggregates obscure major differences across countries and for individual years. For example, investments in the Nigerian system grew during the 1960s and 1970s as oil revenues boomed, but shrunk dramatically during the 1980s to the point where they were (in 1991) less than one-half what they were in the 1970s. Some systems had relatively even growth, such as Kenya, Burkina Faso and Ethiopia, while others, including Nigeria, Ghana and Madagascar, had rapid growth followed by a decade-long decline in real research expenditures. Size disparities across the region were reduced somewhat between 1961 and 1991 as the number of mid-sized SSA research systems (those with between 100 and 400 researchers) grew – from 3 in 1961 to 18 in 1991. However, national system size and quality remain uneven and generalizations about regional growth patterns are difficult to make.

As of 2008, eight countries – Nigeria, South Africa, Kenya, Ghana, Uganda, Tanzania, Ethiopia and Sudan – with large research systems account for about 70% of SSA's agricultural R&D spending

Table 2.2. Long-term trends in research expenditures and FTE capacity (31 ASTI countries).

Year	Expenditures (millions 2005 PPP $)	Rate of growth	Researchers	Rate of growth
1971	963		3,060	
1981	1,218	1.7	5,819	5.4
1991	1,335	0.6	9,065	3.8
2001	1,432	1.0	9,824	1.3
2008	1,741	2.4	12,120	2.8

Source: Beintema and Stads (2011).

and 64% of its FTE researchers. In addition, over one-third of the region-wide expenditure growth during 2001–2008 was driven by increases in Nigeria, Ghana, Sudan, Tanzania and Uganda, and the funding (FTE) concentration in the eight large systems has grown from 53% since 1991. This imbalance distorts the overall picture and creates doubt about the sustainability of R&D investments in the remaining countries. Currently, some countries have stagnant systems with too few resources to guarantee long-term survival.[10] In the smaller and less well-funded systems, national commitment to continued funding is questionable and the systems are vulnerable to cut-backs from external sources. More regional cooperation may be necessary to strengthen these relatively small systems (Beintema and Stads, 2011).

National systems also exhibit heterogeneity in the composition of spending. Ghana and Nigeria expend large proportions of their budgets on scientist salaries, while other countries such as Uganda and Tanzania spend far higher proportions on operations (Uganda) and capital investments (Tanzania). Although agricultural research staff in SSA grew rapidly from 1961 through to 1991, R&D expenditures grew at a slower rate and, in some systems, real research expenditure growth was negative during the 1990s. Differences in spending patterns are reflected in uneven scientist quality: in 1991, 63% of the SSA national scientists with PhDs worked in three systems: Nigeria, South Africa and the Sudan.

The robust increase in scientist numbers in SSA from 1961 to 2008 was not accompanied by growth in other areas of research funding; therefore, region-wide resources allocated per scientist have declined over time.[11] Several factors explain this decline, including changing proportions of expatriate researchers, changes in educational attainment of researchers and changes in funding for support staff. The decline became most pronounced in the 1980s but all of the aggregate fall occurred prior to 1990 (Fuglie and Rada). Since 1990, aggregate research resources per scientist have grown slightly in SSA, possibly reflecting other indications of recommitment by several governments to agricultural research. The shallowing of research resources prior to 1990 combined with irregular overall funding levels had clearly negative effects on the efficiency and effectiveness of agricultural R&D in the region (Pardey *et al.*, 1997) but these effects may be somewhat mitigated by changes since 2001 (Fig. 2.1).

Despite their relatively optimistic assessment of recent funding trends, Beintema and

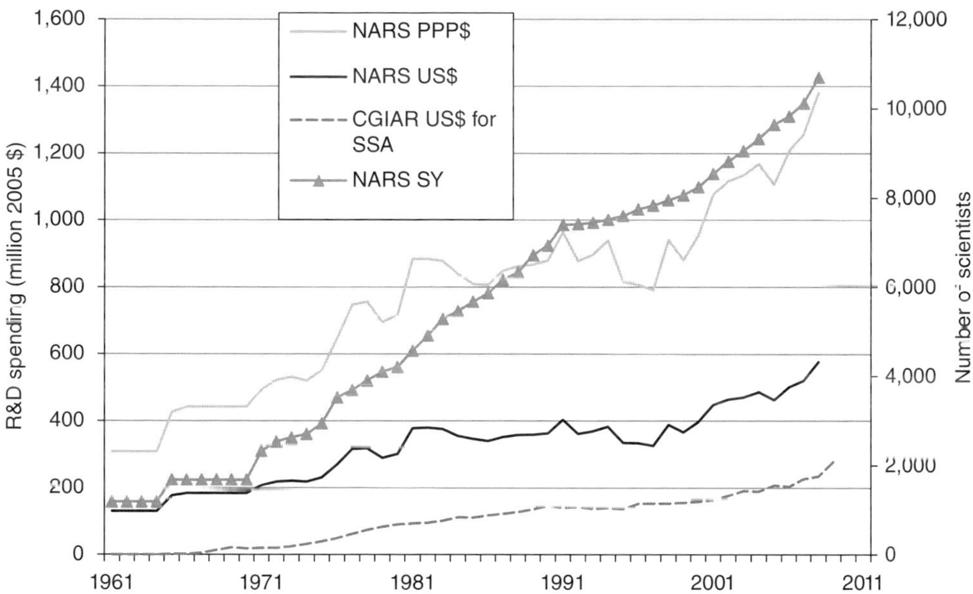

Fig. 2.1. Total agricultural R&D spending by source and number of scientists in SSA NARS (excluding South Africa), 1961–2008. SY, scientist years. (Source: Fuglie and Rada, 2013.)

Stads (2011) find that only 8 of 31 countries in their analysis had funding intensity ratios that exceeded the region-wide target of 1%. This disparity is further evidence of within-SSA heterogeneity research investments. Funding reductions experienced during the 1980s and 1990s are having a lasting effect on research capacity; current intensity ratios are still below levels in the late 1980s.

Beintema and Stads argue that total research expenditures and staffing levels are a poor indicator of funding sufficiency; only three of the eight 'large systems' (with funding of more than US$50 million per year) met the targeted funding intensity ratio. Sudan and Ethiopia, two of the larger African systems, have intensity ratios among the bottom for the 31 countries included in the analysis; the size of the systems distorts the region-wide funding analysis presented above. Ratios of staff per million dollars expended varied dramatically by country suggesting major differences in composition of systems or in the relationship between staffing levels and overall system costs. The Guinea system had over 57 FTEs per million dollars spent, whereas Côte d'Ivoire and South Africa had fewer than 3 FTEs per million in R&D expenditures. Smaller systems tend to have more FTEs per dollar of expenditure, which indicates – consistent with the overall findings – that some of the smaller systems face viability issues.

An analysis of funding sources reveals a mixed bag: some SSA countries are heavily dependent on outside sources of funding; others, particularly the larger systems, rely on resources from their own government. Many NARS have diversified their funding sources to include a mix of government funding, dedicated commodity taxes, sales of goods (e.g. seeds) and services (e.g. extension) and donor funding. Those systems more dependent on commodity tax revenues (e.g. Mauritius) find commodity price variability to be a problem for sustainability. The overall conclusion of the report with respect to funding is one of diversity. Across SSA, multiple funding models can be found, and these models evolve over time as country and external conditions change.

IARC contributions

International agricultural research centres have played an important historical role in strengthening agricultural research and contributing to variety releases in SSA. Evidence shows that IARC investments complement NARS investments; NARS investments were estimated to be about 15% higher than they would have been in the absence of IARC funds (Evenson and Gollin, 2003, Chapter 21). This finding might indicate that the slowdown in aggregate funding for agricultural research in SSA documented by Pardey *et al.* (1997) reflects changes in donor emphasis and the reaction of individual NARS to this changed emphasis. Growth in research spending since 2000 (documented further below) reflects donor interests along with a growing SSA-wide consensus in support of agricultural R&D.

The CGIAR presence in SSA is broad based because virtually all the centres have had a major presence in the region since the early 1970s or since the date when a particular centre joined the CGIAR system (Maredia and Raitzer, 2006)[12]. Since its inception, the CGIAR as a whole has invested more than 40% of its global research budget to SSA (Fuglie and Rada, 2013). The CGIAR investment share devoted to SSA has remained high over time but the composition of the research budget has changed. The proportion of CGIAR research expenditures on productivity-enhancing technologies (mostly crop improvement research) has shrunk dramatically over time. During 1972–1976, more than 80% of the CGIAR SSA research budget was devoted to crop productivity-enhancing research; by 2002, this share fell to less than 33% (Maredia and Raitzer, 2006). The change in composition is partially driven by the emergence of non-commodity focused centres and reflects a system-wide change in emphasis. Over the same period, the CGIAR share devoted to environmental improvement rose from zero to almost 15%, policy research experienced a similar growth in prominence, whereas biodiversity research rose from zero to almost 7%. In terms of overall resources devoted to SSA, IITA, followed by ICRAF and ILRI, each spent more than US$15 million annually during 2000–2004, and CIMMYT, ICRISAT and IFPRI spent between US$10 and US$15 million annually during the same period (see Maredia and Raitzer, 2006, Figure 4, p.13).

As a result of these factors, nominal values of expenditures on productivity-enhancing research allocated to SSA in 2002 were identical to levels in the mid-1980s (about US$60 million) so the real value of expenditures on this research has fallen dramatically. Because the CG Center is often a leader in setting broad research priorities for the region, shifts away from crop-improvement research are likely to have occurred for national systems. These shifts have clear consequences for variety releases and productivity changes.

Crop-specific research investment patterns

In SSA, six food crops – sorghum, maize, millet, cassava, cowpea and groundnuts – account for about 73% of cropped food area, and, if rice, banana, beans and yams areas are included, they account for nearly 90% (Fuglie and Rada, 2013). Data show that research investments in R&D in SSA for these key crops lag behind those of other regions of the world. Lower investments by national governments in SSA even extend to the crops that represent the largest shares of land area. Evidence also shows fewer releases of MVs (modern varieties) in SSA compared to the rest of the developing world since the early 1960s. However, the rate of MV release for SSA has picked up since the late 1970s (Evenson and Gollin, 2003, multiple chapters).

Although sorghum is widely planted in Africa, resources for crop improvement research for sorghum in the region are limited.[13] Outside of Ethiopia, Sudan and Kenya, SSA NARS generally devote only between one and five scientists to sorghum; and region-wide there are only about 170 scientists engaged in sorghum research.[14] Interestingly, Nigeria, which together with Sudan is the largest producer of sorghum (in terms of planted area and production levels), had only six (1991–1992 data from ICRISAT) scientists involved in sorghum research (Sudan had 21).

NARS size unevenness in SSA is further reflected in differences in resources devoted to specific crops. Despite its obvious importance to SSA smallholders, African NARS have invested far fewer resources (quantity and scientist quality) in sorghum compared to countries in Asia (Deb and Bantilan, 2003). The case of millet, another important consumer crop in West Africa is similar to that of sorghum.[15] Pearl millet represents more than 95% of the millet planted in SSA and yields in the region are low. Despite its importance as a consumer crop, pearl millet commands few R&D resources; an estimated 250 scientists across SSA were involved in pearl millet research in the late 1990s, with many of these scientists sharing time on other crops (Bantilan and Deb, 2003). Evidence for these important crops suggests an imbalance: together they account for about 33% of cropped area in SSA, yet less than 5% of the region's FTE researchers were engaged in sorghum and millet research as of 1991.[16] No evidence since then has shown this imbalance to have changed.

Patterns of crop improvement for maize differ by subregion within SSA. Maize breeding programmes in East and Southern Africa (ESA) have lower research intensities, fewer scientist numbers and are more centralized than similar programmes in Asia and Latin America.[17] Whereas some national programmes have decentralized their maize breeding[18] to reflect agroecological heterogeneity, decentralized breeding probably suffers from acute resource shallowing. The SSA region also is characterized by less involvement of private sector breeding in ESA (an estimated 45 FTEs compared to 109 in the public sector) and West and Central Africa (WCA; 51 senior and intermediate-level researchers compared to 112 in the public sector) compared to other regions of the world. This outcome is probably due to the relative lack of commercialized maize sectors in Africa.

Sub-Saharan Africa accounts for over one-half of the world's cassava production and an estimated 95% of the crop in the region is dedicated to human consumption. Prior to establishment of the IARCs, cassava research, unlike most other major food crops, commanded virtually no NARS resources throughout the developing world. Within a few years of establishment of CIAT and IITA, several NARS established cassava R&D programmes. In cassava R&D, the IARCs have had a major impact because many of the scientists working in national research systems were trained by CIAT and IITA.[19] Estimates show about 49 cassava breeders working in NARS, universities and the private sector in 1998. This figure compares favourably with Asia (23 total) and Latin America and the Caribbean (16 total), and breeding intensity in SSA in 1998 was comparable with other parts of the world (approximately 0.6 FTEs per million tonnes of production). Although research intensities for cassava in SSA do not differ substantially from those in the rest of the world, these estimates are very low by any standard because few, if any, other crops are characterized by a research intensity at or below 1.0 scientist per million tonnes of production.

Of the less important (on an SSA-wide scale) foods, rice research in West Africa has been limited by relatively small numbers of scientists in the NARS (fewer than 46 FTEs by 1998). In spite of this limited capacity, more than 319 improved varieties had been released by 2003 with more than 40% of them having some contribution from CG germplasm or parents. Numbers of bean breeders have grown in SSA NARS from two in 1980 to more than 40 by 1998.[20] Wheat, although accounting for a relatively small percentage of SSA cropland, had more than 104 FTE scientists in 1997, up from 62 in 1992. NARS expenditures on wheat research for SSA reached about 3.7 million ($1990) in 1990.[21]

Research Impacts

Studies examining the impacts of agricultural research include those focusing on intermediate measures such as variety release or land area covered by improved varieties, impacts or rates of return from specific research programmes and aggregate rate of return studies. Studies of impacts of agricultural research on agricultural productivity in SSA have employed various methods including direct econometric estimation (e.g. Fuglie and Rada, 2013), summaries of findings from existing studies (e.g. Maredia and Raitzer, 2006) and meta-analyses of econometric studies (e.g. Alston *et al.*, 2000).

Trends reflecting the spread of modern varieties in Africa are now relatively well known (Evenson and Gollin, 2003; Fuglie and Rada, 2013; and several chapters from this volume).[22] The spread of these varieties in SSA has lagged behind that in other areas, but a recent increase in this spread is noted in several chapters of this volume. By 1998, the share of SSA area planted to modern varieties was 23%, far lower than Asia (83%), the Middle East and North Africa (56%), and Latin America (51%) (Renkow and Byerlee, 2010). CGIAR contributions to modern variety development in SSA were, however, thought to be larger than its contribution in other regions, suggesting strongly that the role of the CGIAR in genetic improvement in SSA is high (Renkow and Byerlee, 2010). Of the major cereal crops in SSA (sorghum, maize and millet, which account for approximately 86% of cereal cropped land), about 20% was planted to improved varieties in 2005 (see Table 2.1). Overall, although data are incomplete, approximately 18% of food-cropped area in SSA is now planted to improved varieties and the vast majority of these are from CGIAR sources (see Table 2.1). Much of the increased adoption has occurred since the mid-1980s and evidence shows that for some crops in some areas adoption rates are increasing. For example, Alene *et al.* (2009) document that about 60% of maize area in WCA is now under improved varieties. Like the discussion of research investments, the aggregate picture for adoption rates in SSA masks important successes.

Numerous studies have been conducted on impacts of agricultural research in SSA. Block (1995) used a precursor to the ASTI data set and found that R&D expenditures explain about one-third of the productivity growth in SSA between 1983 and 1988. Masters *et al.* (1998) examined 32 case studies of the relationship between research expenditures and agricultural output in SSA. They found that 24 of these studies reported annual returns over 20% and many were far higher, with most gains arising in the late 1980s and 1990s.[23] For a comprehensive account of evidence accumulated prior to 2000, see Alston *et al.* (2000). Their meta-analysis included 47 studies of assessments of research impacts from SSA conducted between 1958 and 1997 that generally found quite high rates of return to individual research programmes (Alston *et al.*, 2000). Notably, the vast majority of these studies focused on research conducted by NARS (40) and only four focused exclusively on research conducted by a CG Center (Maredia and Raitzer, 2006). Maredia and Raitzer found that the CGIAR's impact in terms of benefits and costs in SSA was generally lower than system-wide estimates of impact and the major impacts in SSA have emerged from research on biological control (almost all due to control of the cassava mealy bug). Renkow and Byerlee (2010) noted that the relatively low spread of modern varieties in SSA meant that the CGIAR contribution to yield growth from 1965 to 1998 was much lower than in other regions. This picture has changed, however, with several recently emerging SSA success stories. Recently documented crop genetic improvement success stories for Africa include maize (Alene *et al.*, 2009), cowpea (Kristjanson *et al.*, 2002), common bean (Kalyebara *et al.*, 2008) and rice (Diagne, 2006).

Fuglie and Rada (2013) focused on mechanisms by which national agricultural research investments affect productivity and identified two pathways of impact: research investments can help diffuse a CGIAR technology, which subsequently raises farm productivity, and they can affect total factor productivity (TFP) through other, unspecified means, such as by furthering diffusion of non-CGIAR technologies, influencing policy changes, or by encouraging farmers to improve their resource management. The Fuglie and Rada (2013) econometric analysis found both pathways to be important; national expenditures on agricultural research as well as other policy reforms – such as enhanced education and investments in infrastructure – helped diffuse CGIAR-sourced technology and other non-specific crop technologies. Both factors helped raise TFP in agriculture. CGIAR-generated technologies were associated with a 45–82% increase in TFP over the period. The study examined limited measures of research expenditure complementarity and found that increased investments in NARS led to significantly more diffusion of CGIAR-sourced technologies, but it did not examine whether increased CGIAR investments enhanced the productivity of NARS research.

Evidence of impacts of agricultural research in SSA on non-efficiency objectives is more limited. These alternative objectives include poverty reduction, improved environmental sustainability, gender empowerment and others. Renkow and Byerlee (2010) summarized studies of non-efficiency outcomes of CGIAR research in SSA, and showed limited evidence of impacts on poverty and on the environment. Chapters 15 and 16 on distributional impacts in this volume find that impacts of improved maize (Ethiopia) and beans (Rwanda and Uganda) on poverty are rather modest. In these studies, resource-scarce farmers are able to adopt the new varieties, and variety adoption is accompanied by increased net income from farming. However, small farm size limits the magnitude of income gains; therefore, the direct effect on the adopting household is relatively small. Market-mediated effects, however, can be larger, depending on the conditions in the respective markets (Chapter 15 and Chapter 16, this volume).

Discussion

The overall trends in agricultural R&D expenditures in SSA are clear. Starting from a low base in the early 1960s, aggregate funding grew throughout the 1960s, slowed in the 1970s, and underwent an even more dramatic slowdown in the 1980s and through the 1990s. Since 2000, steady increases in funding have come from external donors and national governments. Growth in research expenditures has, however, been uneven, with individual countries showing patterns that differ from mean trends. Even during the post-2001 period of overall growth, 13 of 30 ASTI countries had negative compound growth rates. Uneven growth in research funding creates a region characterized by several very large NARS whose resource allocations and other decisions dominate the overall picture. This dominance has fallen over time with the growth of a number of medium-sized systems, but system-wide trends still mask substantial intra-regional variation.

An important question that none of the research addressed is whether there has been 'convergence' in spending. Convergence[24] occurs when systems with high research intensity at a starting point grow their intensity at a slower rate than systems with low intensities. Under convergence, we would expect research intensities to approach a steady state where the intensities become relatively equal for all countries. To examine the tendency toward convergence, we graph for each country research intensity at a suitable start point (1965) and examine the percentage change in intensity from this point through the end of the data series (2008). Figure 2.2 shows evidence of convergence over the entire period. All but one country with moderate-high intensities above 0.0075 in 1965 experienced negative intensity growth from 1965 to 2008, whereas those with starting intensities below 0.004 showed higher rates of growth.

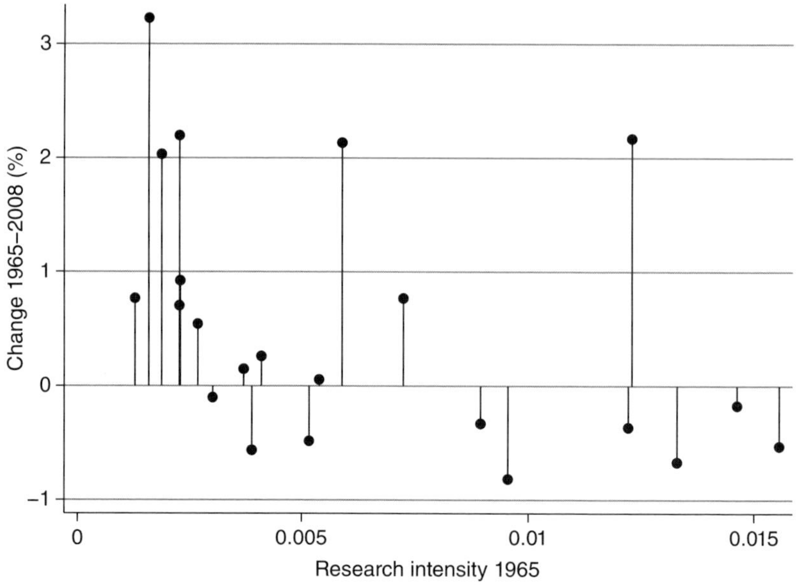

Fig. 2.2. Convergence in SSA research intensity, 1965–2008. (Source: Own analysis using data from Fuglie and Rada (2013) supplemented with World Bank data.)

This pattern is consistent with a pattern of increased investments over time in areas where research is lacking and, possibly, of diminishing returns in well-funded systems.

Further analysis shows that the 1980s were characterized by strong convergence in research intensities across agricultural R&D systems in SSA (Fig. 2.3a), while patterns of convergence disappeared after 1990 (Fig. 2.3b). The explanation for this changing pattern is not obvious but the finding implies that, until around 1990, countries in the region as a whole were investing in research in a pattern that reflects simple economic considerations (e.g. higher investments in relatively under-funded systems). This pattern was broken following the slowdown in research investments experienced in the early 1990s and growth in intensity was actually negative in the decade prior to 2000. Investment patterns since 2000 show renewed tendency toward convergence in Fig. 2.3c.

In fact, the region-wide within-year variability in research intensity has grown substantially since the early 1960s (Fig. 2.4) indicating a growing spread of high- and low-intensity countries in SSA. Variability in research intensity across countries increased modestly until the mid-1990s and then grew dramatically since the late 1990s. This trend confirms the Beintema and Stads (2011) conclusion that the recent growth in agricultural research expenditures in SSA is not broad-based. A few countries are increasing investments substantially, whereas others are not.

Support for agricultural R&D has experienced fits of increase and decrease for the entire SSA region and, in an even more pronounced fashion, for individual countries. This variability may contribute to lower than expected research productivity. Uncertainty about longer-term funding prospects has clear potential to damage multi-year research efforts and may bias researchers toward engaging in projects with shorter-term payoffs. Lags between research expenditures and impacts on agricultural productivity are quite long and, whereas the impacts of variable funding on productivity are less well-known, evidence shows funding slowdowns experienced beginning in the 1980s may persist in lowering agricultural productivity even today.

Since the early 1960s, a dramatic shift in scientific capability has occurred in the SSA region, with African scientists now representing a large majority of agricultural researchers. Region-wide, more than 70% of researchers now have advanced degrees (30% have PhDs).

Evidence shows, however, that this scientific capacity is being spread thinly; scientist numbers as well as proportions of budgets spent on scientific salaries have grown, leading to a shallowing of the resource pool for operating expenditures. Some smaller systems have lost researchers and pressure continues to be high to increase training pools and salaries for scientists. Although high salaries are needed to retain the most productive scientists, more information is needed

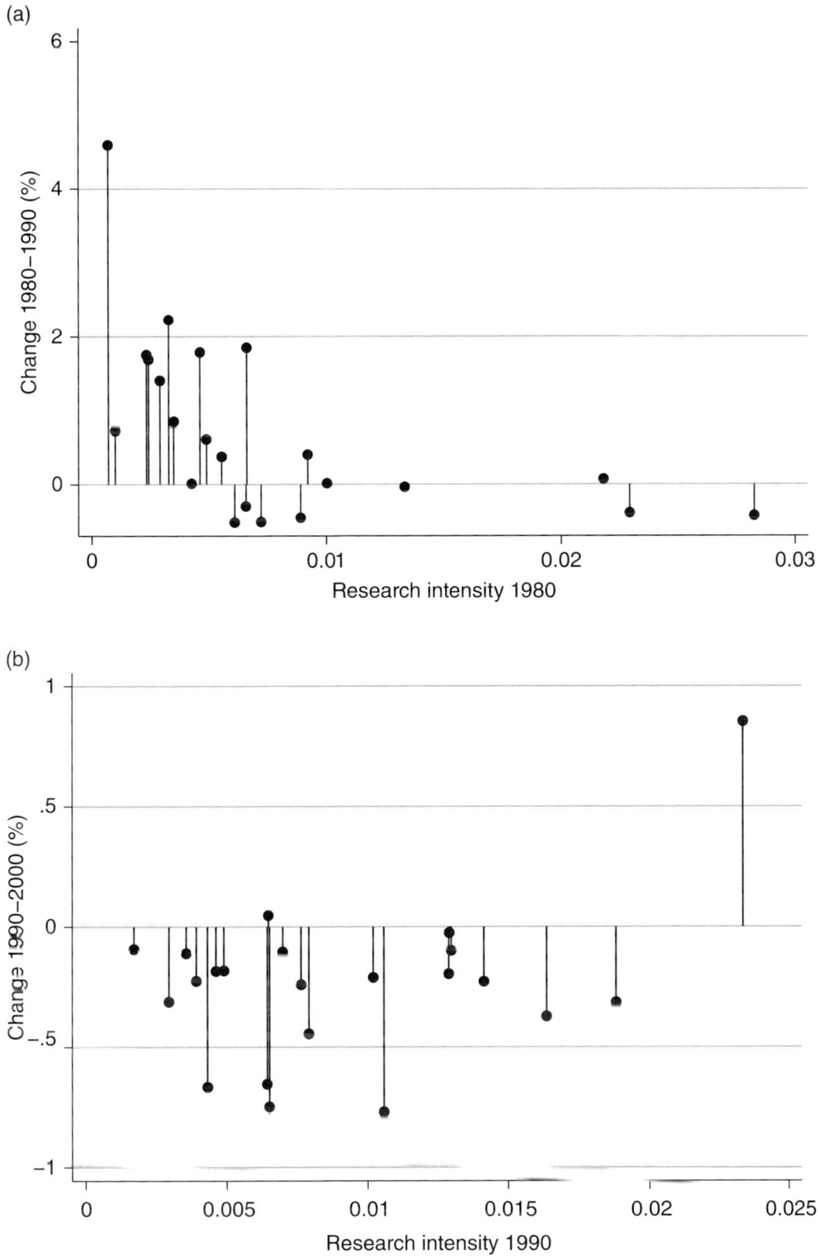

Fig. 2.3. Convergence patterns for research intensity in SSA countries, sub-periods. (Source: Own analysis using data from Fuglie and Rada (2013) supplemented with World Bank data.)

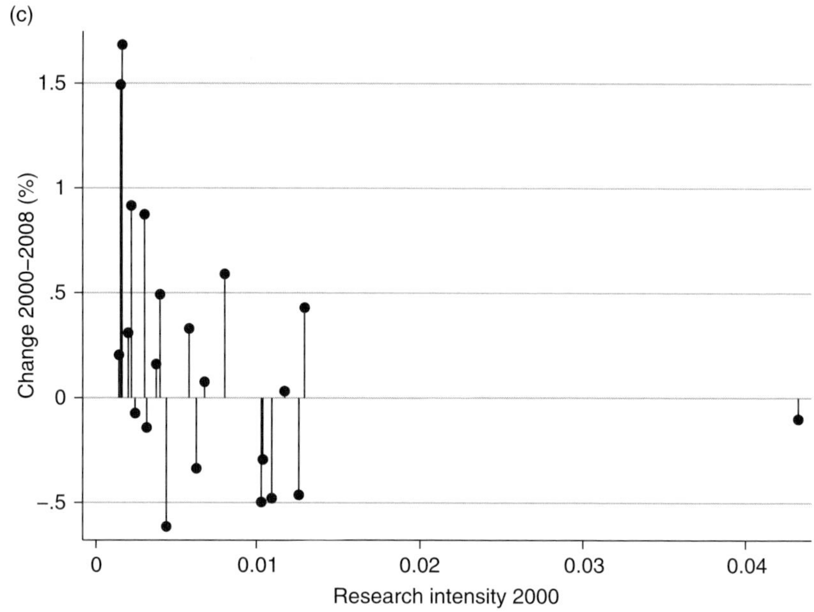

Fig. 2.3. Continued.

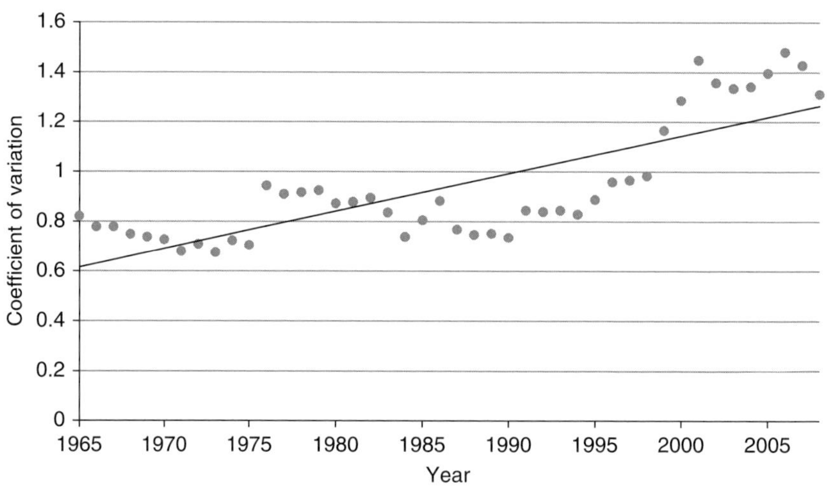

Fig. 2.4. Annual SSA-wide coefficient of variation in research intensity, 1965–2005. (Source: own analysis using data from Fuglie and Rada (2013) supplemented with World Bank data.)

on the consequences of dedicating high proportions of research budgets to staff salaries.

The CGIAR has had a major influence on the development and direction of NARS in SSA by training NARS scientists, providing germplasm and collaborating closely with NARS-led research. NARS and CGIAR research expenditures have been complementary; increased CG spending is associated with additional resources from national governments suggesting a potential crowding-in effect. Whereas a few of the smaller systems are still dependent on donor funding, the region as a whole has undergone a transition toward alternative funding streams.

Since 2000, agricultural R&D in SSA has become increasingly interlinked across the region. This integration has been promoted by regional groups including the Forum for Agricultural Research in Africa (FARA); the Association for Strengthening Agricultural Research in Eastern and Central Africa (ASARECA); and others. These groups help coordinate research across the region through scientist networks. Commodity-specific networks such as the Pan-African Bean Research Alliance (PABRA), comprising 24 country members, have also strengthened region-wide coordination, collaboration and research information sharing. Perhaps their primary strength is to allow specialization of individual national agricultural research systems in certain fields. As a result, it becomes more possible to obtain economies of scale in research and cross-network sharing is particularly beneficial for small countries that might lack a critical mass.

A clear implication of the analysis summarized in this chapter is that it is dangerous to make summary statements about region-wide trends. As with most economic phenomena, a region-wide analysis masks important differences. It is clear that to understand impacts of investments in SSA agricultural R&D focused studies are required and only through the aggregation of focused results can the whole picture be understood.

Notes

[1] Nigeria, South Africa and Kenya account for about half of total agricultural R&D spending in the region; Ghana, Uganda and Tanzania are also relatively large systems.

[2] Data on private research investments in SSA are limited and, as a result, an assessment of private-sector research is not included here. Public research historically dominated agricultural research in SSA and, while government agencies accounted for about 73% of full time equivalent (FTE) researchers in 2008, this number had fallen from 82% in 1991 (Beintema and Stads, 2011). Beintema and Stads note that most privately funded research in Africa is conducted in government agencies and universities (and thus, these expenditures would appear among other indicators of public-sector expenditures), and privately conducted research represents only 2% of total research funding for SSA. Private sector research is probably most evident in the hybrid maize sector; by 1998 farmers in Kenya, Zambia and Zimbabwe were heavy users of private-sector hybrids. In contrast, private sector maize hybrids have had hardly any traction in West and Central Africa.

[3] Currently about 50% of DFID funding for agricultural research in developing countries goes directly for core support for the CGIAR; the other half flows through other mechanisms. Although a large part of this remainder will show up in CG and NARS accounts, a substantial portion will not.

[4] BMGF funding now represents a relatively large proportion of total agricultural research funding in SSA. The BMGF helped form the Alliance for a Green Revolution in Africa (AGRA), which provides large-scale support in the areas of plant breeding and soil health.

[5] ISNAR was subsequently absorbed into IFPRI.

[6] As an example, Fuglie and Rada (2013), discussed in detail below, merged ASTI data with information on area cropped under CGIAR sourced varieties.

[7] Beintema and Stads note that from 1991 to 2008 the proportion of agricultural researchers at African universities grew from 14 to 24%. Despite this growth, university researchers, on average, spend less than 25% of their time on research.

[8] Estimates come from Fuglie and Rada (2013) and include information from 32 countries.

[9] South Africa (1339), Nigeria (1013) and Kenya (819) were the largest systems, and Tanzania (546) was the only other SSA NARS with more than 500 employees in 1991.

[10] Nigeria, Ethiopia, Sudan and Kenya each have more than 1000 FTE scientists, and 16 other NARS have FTEs in the 100–500 range.

[11] Fuglie and Rada show that region-wide research expenditures per scientist per year fell in real terms by more than 50% from 1961 to 2008.

[12] IRRI is the exception as the West African Rice Development Association (WARDA), now AfricaRice, conducts rice research for much of West Africa. IRRI has, however, invested considerably in rice research in Madagascar, Tanzania, Mozambique and other rice-growing countries in southern Africa. Historically, rice has had a more diversified pattern of institutional investment in SSA than any other food crop.

[13] Evenson and Gollin (2003), Chapter 9, U.K. Deb and M.C.S. Bantilan, Impacts of Genetic Improvement in *Sorghum*.
[14] The authors note that it is difficult to accurately account for resources devoted to pearl millet and sorghum because most scientists share their time on the two crops.
[15] Evenson and Gollin (2003), Chapter 10, M.C.S. Bantilan and U.K. Deb, Impacts of Genetic Enhancement in Pearl Millet.
[16] Of course millet and sorghum production environments show less variability than those of other crops so that fewer researchers might be optimal but this cannot explain the huge discrepancy documented here.
[17] Evenson and Gollin (2003), Chapter 77, M. Morris, M. Mekuria and R. Gerpacio, Impacts of CIMMYT Maize Breeding Research. This chapter addresses maize breeding in Eastern and Southern Africa (ESA) and compares it to the rest of the world.
[18] Kenya, for example, breeds maize for six distinct agroecological regions.
[19] 23 PhD- and 28 MSc-level scientists in SSA were trained in cassava research by IITA in 1970–1998; 13 additional cassava specialists were trained at CIAT in 1972–1994 (Evenson and Gollin, 2003).
[20] Evenson and Gollin (2003), Chapter 12, N.L. Johnson, D. Pachico and C.S. Wortmann, The Impact of CIAT's Genetic Improvement Research on Beans.
[21] Evenson and Gollin (2003), Chapter 4, P.W. Heisey, M.A. Lantican and H.J. Dubin, 'Wheat'. Data on research expenditures are from Byerlee and Traxler (1995); estimates of scientists in NARS are from Bohn *et al.* (1999).
[22] Fuglie and Rada (2013) document crop-by-crop and country-by-country annual spread of modern varieties for SSA; various chapters in this volume extend and deepen this analysis.
[23] Rates of return estimates depend clearly on the assumptions underlying them, methods used, etc. As a result, it is difficult to draw general conclusions from summaries of rates of return studies.
[24] In neoclassical economic growth models, the driving force behind convergence is diminishing returns at the margin to increased capital in an aggregate production function. In cases where returns are increasing convergence is not expected. The patterns observed here are consistent with a diminishing marginal return to research expenditures, but other factors may explain the observed convergence.

References

Agricultural Science and Technology Indicators (ASTI) Internationally comparable data on agricultural R&D investments and capacity for developing countries. International Food Policy Research Institute, Washington, DC. Available at: http://www.asti.cgiar.org (Accessed 22 April 2015).

Alene, A.D., Menkir, A., Ajala, S.O., Badu-Apraku, B., Olanrewaju, A.S. *et al.* (2009) The economic and poverty impacts of maize research in West and Central Africa. *Agricultural Economics* 40, 535–550.

Alston, J., Chan-Kang, C., Marra, M., Pardey, P. and Wyatt, T. (2000) *A Meta-analysis of Rates of Return to Agricultural R&D: Ex Pede Herculem?* IFPRI Research Report 113. International Food Policy Research Institute, Washington, DC.

Bantilan, M.C.S. and Deb, U.K. (2003) Impacts of genetic enhancement in pearl millet. In: Evenson, R.E. and Gollin, D. (eds) *Crop Variety Improvement and Its Effect on Productivity: The Impact of International Agricultural Research*. CAB International, Wallingford, UK, pp. 215–240.

Beintema, N. and Stads, G.-J. (2011) *African Agricultural R&D in the New Millennium Progress for Some, Challenges for Many*. International Food Policy Research Institute, Washington, DC.

Block, S. (1995) The recovery of agricultural productivity in sub-Saharan Africa. *Food Policy* 20(5), 385–405.

Bohn, A., Byerlee, D. and Maredia, M.K. (1999) Investments in wheat improvement research in developing countries. In: Maredia, M.K. and Byerlee, D. (eds) *The Global Wheat Improvement System: Prospects for Enhancing Efficiency in the Presence of Spillovers*. CIMMYT Research Report No. 5. CIMMYT, Mexico City.

Byerlee, D. and Traxler, G. (1995) National and international wheat improvement research in the post-Green Revolution period: evolution and impacts. *American Journal of Agricultural Economics* 77, 268–278.

Deb, U.K. and Bantilan, M.C.S. (2003) Impacts of genetic improvement in *Sorghum*. In: Evenson, R.E. and Gollin, D. (eds) *Crop Variety Improvement and Its Effect on Productivity: The Impact of International Agricultural Research*. CAB International, Wallingford, UK, pp. 183–214.

Diagne, A. (2006) Diffusion and adoption of NERICA rice varieties in Côte d'Ivoire. *The Developing Economies* 44(2), 208–231.

Evenson, R.E. and Gollin, D. (eds) (2003) *Crop Variety Improvement and Its Effect on Productivity: The Impact of International Agricultural Research*. CAB International, Wallingford, UK.

Fuglie, K.O. and Rada, N.E. (2013) *Resources, Policies, and Agricultural Productivity in Sub-Saharan Africa.* ERR-145, US Department of Agriculture, Economic Research Service, February 2013. USDA, Washington, DC.

Kalyebara, R., Andima, D., Kirkby, R. and Buruchara, R. (2008) *Improved Bean Varieties and Cultivation Practices in Eastern-Central Africa: Economic and Social Benefits*. CIAT, Cali, Colombia.

Kristjanson, P., Tarawali, S., Okike, I., Singh, B.B., Thornton, P.K. *et al.* (2002) *Genetically Improved Dual-purpose Cowpea: Assessment of Adoption and Impact in the Dry Savannah of West Africa.* Impact Assessment Series No. 9, International Livestock Research Institute, Nairobi.

Maredia, M.K. and Raitzer, D. (2006) *CGIAR and NARS Partner Research in Sub-Saharan Africa: Evidence of Impact to Date.* Commissioned Paper, CGIAR Science Council Secretariat, Rome.

Masters, W., Bedingar, T. and Oehmke, J. (1998) The impact of agricultural research in Africa: aggregate and case study evidence. *Agricultural Economics* 19, 80–81.

Norton, G.W. and Alwang, J. (2012) *Synthesis Paper: Donor Methods to Prioritize Investments in Agricultural Research and Development*. Secretariat of the Global Donor Platform for Rural Development, Bonn, Germany.

Pardey, P., Roseboom, J. and Beintema, N. (1997) Investments in African Agricultural Research. *World Development* 25, 409–423.

Pardey, P., James, J., Alston, J., Wood, S., Koo, B. *et al.* (2007) *Science, Technology and Skills*. INSTEPP, University of Minnesota, Minneapolis, Minnesota.

Renkow, M. and Byerlee, D. (2010) The impacts of CGIAR research: a review of recent evidence *Agricultural Economics* 35, 391–402.

3 Relevant Concepts and Hypotheses in Assessing the Performance of Food Crop Improvement in Sub-Saharan Africa

T.S. Walker*
Independent Researcher, Fletcher, North Carolina, USA

This book addresses the performance of national and international food crop improvement in sub-Saharan Africa (SSA) from 1970–2010. In this chapter, the main elements of performance are discussed from the perspective of evaluation that centres on inputs, outputs, outcomes and impacts. Each of these four areas of assessment contains a brief description of relevant concepts and definitions followed by a discussion of related hypotheses. Many of the hypotheses featured in the proposal for the Diffusion and Impact of Improved Varieties in Africa (DIIVA) Project (Bioversity International, 2009). Their treatment here is not exhaustive but rather the purpose is to provide the reader with a substantive touchstone for identifying research content that is common to the chapters of this book. Inputs, outputs and outcomes are the subjects of Chapters 5–14. Impacts are addressed in Chapters 15–17.

Inputs: Scientific Capacity and Research Intensities

Performance in crop improvement depends on scientific capacity, operating budget, research infrastructure, research-extension linkages, agroecological diversity, access to relevant germplasm and the understanding of farmer demand for technologies that suit their circumstances. Relative to other investments in economic development, genetic crop improvement programmes are not costly activities, but they require a recurring expenditure on an adequate number of skilled scientists combined with sufficient operating budgets to get the job done year after year. 'Adequate' and 'sufficient' are not rigorously defined in the literature; however, comparative evidence across countries and crops and over time establishes some orders of magnitude that are discussed in Chapter 5–14 and in Chapters 18 and 19 include the synthetic concluding chapters of this volume.

From the multiple inputs that go into crop improvement, we focus only on one, the number of research scientists by discipline. This restricted emphasis is conditioned by several considerations. The main intent of the DIIVA Project was to estimate adoption of modern crop varieties in sub-Saharan Africa. Outputs and outcomes were the primary concerns. Moreover, as described in Chapter 2, investments in crop improvement by multiple agencies at the national level are periodically and extensively monitored by the Agricultural Science and Technology Indicators (ASTI) Initiative, funded by the Bill & Melinda Gates Foundation (BMGF) and other donors and housed at the International Food Policy Research

* E-mail: walkerts@msu.edu

Institute (IFPRI; http://www.asti.cgiar.org/). Research scientists by discipline are a key input into crop improvement and they are more visible than research infrastructure or even operating budget. Until recently, ASTI did not collect disciplinary scientific strength by education level at the disaggregated level of the crop or commodity.[1] Lastly and most importantly, estimates on scientific staff strength were gathered by crop at the national level in 1998 (Evenson and Gollin, 2003), affording the opportunity to build on this benchmark.

Operationally, testing hypotheses about scientific strength requires definition of the boundaries of crop improvement and crop-related research content of scientists. Additionally, numbers of scientists need to be standardized across programmes of varying size to draw meaningful comparisons.

The boundaries of crop improvement

Crop improvement, as used here in the DIIVA study, embraces plant breeding's closely allied disciplines, such as genetic resources, molecular biology and tissue culture. It also covers pathology, entomology, agronomy and any other discipline – such as social science and postharvest technology – that helps to identify priorities in the development of genetically improved materials. Natural resource management is excluded as is soil science, unless the research focuses on genotype by environmental interactions. Therefore, the definition of crop improvement used here focuses on genetic research – broadly defined and potentially fully supported.

Full-time equivalent scientists

Scientific staff strength is equated to full-time equivalent (FTE) scientists (see Chapter 2, this volume). 'Scientists' are defined as public sector, private sector and university staff who work in crop improvement research and who have an educational level equivalent to a Bachelor of Science (BSc) degree or above. Research technicians and staff working in seed production and related transfer and extension activities are excluded but scientists active in producing breeders' seed are included.

Full-time equivalency means that the scientist works 100% of their research time in genetic improvement of the crop of interest. Because of its highly variable nature over time and across space, time in administration is not considered in this calculation.

Two examples illustrate the meaning of full-time equivalency. A breeder in a public sector programme works entirely in a coarse cereal improvement programme. Her time is equally divided between the improvement of pearl millet and sorghum. As a sorghum or pearl millet scientist, her FTE estimate is 0.5. A university breeder spends 30% of his time on maize improvement and 70% on teaching. His FTE estimate as a maize scientist is 0.3.

Scientists, especially those in smaller countries, often work on more than one crop in more aggregated cereal, pulse, oilseed, and root and tuber improvement programmes. Therefore, the number of scientists working on a crop is substantially larger than the number of FTE scientists.

Research intensities

Research intensities are a means to standardize estimates of FTE scientists across countries of varying sizes. Research intensity is typically expressed as the number of FTE scientists per million tonnes of production. Standardization by value of production and hectares of growing area is also common. As we shall soon see in the commodity chapters in Part 2 of this volume, estimates of research intensity are almost always very high for the smallest producing countries and very low for the largest producing countries. Hence, estimated research intensities are not that informative about whether very small countries are investing too much in research and the largest producers are investing too little in research within the same crop. They become more informative in cross-sectional (within the same year) across crop comparisons and in comparisons over time within the same crop as both the numerator (number of FTE scientists) and denominator (number of tonnes of production) change.

Hypotheses about scientific strength in crop improvement

Stated positively, the following input-related hypotheses are addressed in this book:

1. The number of FTE scientists in national food crop improvement programmes in SSA is increasing.
2. Research intensities in national food crop improvement programmes are also increasing.
3. Disparities in research intensities across crops and regions are not substantial.
4. Private sector participation in research is increasing in the genetic improvement of cereal hybrids.
5. University participation in crop improvement research is becoming increasingly visible from a small base.
6. The disciplinary distribution of FTE scientists in crop improvement reflects an increasing capacity in biotechnology.

From the most recently published ASTI research on investment in research and development (R&D) in SSA (Beintema and Stads, 2006) and, as discussed in Chapter 2, we realized in 2009 when the DIIVA Project was formulated that several findings related to inputs would not be favourable for food crop improvement.[2] For example, the number of FTE scientists had declined in some programmes, although the overall trend in numbers was positive. Diminishing operating budget per scientist was increasingly an issue in crop improvement. Ageing scientific capacity was also problematic, especially in several programmes in West Africa. As discussed in Chapter 2, we knew considerably more at the start of the DIIVA Project about inputs into crop improvement in SSA than about outputs, outcomes and impacts from crop improvement.

Outputs: Modern Varieties Available for Use

'Output' refers to the expansion that can be attributed to genetic improvement in the potential availability of valuable genotypes for cultivation. Ideally, attribution is measured in a with-and-without comparison, i.e. the difference between what is potentially available with genetic improvement and what is available without an investment in plant breeding.

Released varieties

By its nature, crossing and selection is a winnowing process characterized by a search for a smallish number of genotypes perceived to be valuable. Elements of perceived value are encoded in government registry and release practices that place an imprimatur on breeders' elite selections. Official release is tantamount to saying that 'liberated' varieties are potentially valuable for cultivation in the sense that they have satisfied rigorous criteria, such as threshold yield advantages, compared to check varieties in multi-locational testing on research stations over time. In well-functioning systems of varietal release and registry, information on the quantity and location of breeders' seed is published. In this book, varietal release – the most immediate and observable indicator of progress in crop improvement – establishes an initial base for estimating varietal output.

Varietal release is not a perfect indicator and, in specific cases, may not even be a good measure of varietal output in agriculture within developing countries. Both private sector and public sector improved varieties may be available for adoption but may not appear in release registries. Escapes from breeding programmes may be widely adopted but not well identified.

Almost all countries have well-described procedures for varietal release but just a few – such as Ethiopia and Kenya – have compiled comprehensive release registries for downloading on the Internet. An exhaustive review of varietal registration in 24 rice-growing countries shows that nine do not have an established release and registry system in place (Sanni *et al.*, 2011). In some countries with established systems, release committees do not meet periodically and are financially constrained because of pressures on government operating budgets.

Moreover, changes in the release practices over time may give the illusion of increased varietal output when in fact its true trajectory has not changed. Comparing release lists over two points in time also suggests that older improved varieties can reappear at a later date in the registry, giving the impression of recent output when in fact the cultivar was generated much earlier.

One can also cite cases such as Guinea (with a rare institutional setup of multiple institutions releasing varieties of the same crop), where more than 100 rice varietal releases in the 1980s and 1990s has resulted in limited discernible adoption. However, rice in Guinea is an outlier in the joint varietal release and adoption database. The estimated simple correlation between total historical releases and the percentage of adoption for improved varieties in 2010 is a statistically significant but modest 0.17. The 'weighted by area' association is markedly higher at 0.47 for the crop-by-country observations in the DIIVA Project.

The relationship between varietal release, adoption and subsequent impact is probably not symmetric. Large numbers of releases can result in substantial or no adoption, but zero or negligible releases rarely result in appreciable adoption of improved varieties. Absence of release activity is synonymous with negligible output from plant breeding. Performance in crop improvement needs to be measured and varietal release, for all of its imperfections, is still an important benchmark for assessing progress in varietal output. Research that focuses only on official government releases can, however, seriously understate the potential availability of improved varieties for adoption by farmers. For this reason, release is interpreted broadly in this volume to include escapes from breeding materials, other informally available non-released improved varieties, private-sector hybrids that may not be officially released, and the results of participatory plant breeding that may be in the stage of early adoption with farmers.

Hypotheses about varietal output from crop improvement

Relevant output-related hypotheses that are informative about and favourable to the performance of food crop improvement include the following:

1. The stock of released and non-released improved varieties that is potentially available to farmers for use is increasing both absolutely and relatively over time.
2. Output stability is increasing over time as peaks and troughs in varietal generation are less evident in the more recent past.
3. Varietal output reflects the evolution of plant breeding over time, a lower CG Center presence, and more private sector and university participation.

The first hypothesis about the incidence of varietal output over time comes from 1998 where positive trends in the number of released cultivars were documented for most crops in most countries (Evenson and Gollin, 2003). Those data also exhibited high variability over time. Varietal production was episodic: years of positive output in the midst of longer periods of no releases.

The third hypothesis, which is multi-faceted and derived from the examination of the pedigree of improved varieties and hybrids, is also based on findings from the 1998 global initiative. Evidence for the maturing of plant breeding over time was one of its most important findings that applied to numerous programmes in Asia and Latin America. Initially, crop improvement in most countries began with the importation of finished varieties from other countries for testing and release together with an evaluation of local landraces for prospective release. With the growth in crop improvement at the International Agricultural Research Centers (IARCs) in the 1970s and 1980s, breeders in national programmes had the option to select progenies from crosses made by the CG Centers (Institutes within the Consultative Group on International Agricultural Research). Further programme advancement is synonymous with breeders making their own crosses from well-identified parental material. A final stage in development could be reached in the 2000s with the use of biotechnological techniques, such as marker-assisted selection (MAS), in the generation of improved varieties targeted for the enhancement of specific traits (Collard *et al.*, 2005). Over time progression in the use of tools involved in how varietal output was generated speaks to a programme's capability to engage in genetic improvement.

As national breeding programmes advance in their capabilities, CG Center-related materials would become less visible in profiles of varietal output, especially in larger and stronger programmes.

Greater private-sector participation in cross-pollinated crops, such as maize, would also contribute heavily to national programme strengthening and the redefining of the CG Center role in genetic improvement. Likewise, more university input would complement public-sector breeding in food crops in general and in self-pollinated crops in particular.

Outcomes: Adoption and Turnover of Improved Varieties

Arriving at reliable estimates of adoption of improved varieties in SSA was a major aim of the DIIVA Project and all the papers presented in Parts 2, 3 and 4 of this volume. The use of different definitions of adoption can lead to misleading results across crops and countries and over time.

Improved varieties

The majority of improved varieties could also be called high-yielding varieties (HYVs). But when the emphasis is on shorter duration or disease resistance, productivity of HYVs may not be significantly greater than local landraces unless shorter duration translates into drought escape or disease infestation occurs. Furthermore, the seasonal production potential in most regions of sub-Saharan Africa is constrained by infertile soils, inaccessible fertilizers and infrequent rainfall (Giller, 2012); therefore, we prefer to use 'improved varieties' to 'high-yielding varieties' to describe the products of genetic improvement. Productivity considerations in the form of a good agronomic background still loom large in crop research in rainfed agriculture. They are addressed in the last section on impacts in this chapter.

It is perhaps important to note that none of the improved cultivars in farmers' fields in food crops in 2010 were the products of transgenic varietal change. In other words, none of the improved varieties were genetically modified organisms (GMOs).[3] They were all the results of conventional crop improvement within their species of origin. Since 2000, transgenic varietal change is limited to Bt cotton in a few countries such as South Africa and Burkina Faso.

Having broadly introduced the subject, we can now address the central concept of interest: what constitutes an improved variety? A robust definition begs several sub-questions:

- Would the variety be available to farmers without research in crop improvement?
- Are breeding and selection embodied in readily identified materials that farmers are growing but that have not received formal release?
- Is the seed of improved open-pollinated varieties (OPVs) renewed periodically?
- Is the seed of hybrids renewed annually?
- Should very old released varieties be included?

Responding to these questions often requires accompanying information on the seed sector and careful inspection of the varietal release database in the studied country. Ideally, landrace materials in their country of origin should not qualify as improved varieties even though their seed is purified and they are formally released. Farmers would most likely be producing these materials with or without a crop improvement programme, although their identification and performance does require some effort in selection. Productivity gains are more likely to be related to the effect of cleaner seed than of genetic change. Additionally, variety-specific comparisons show that productivity gains from released in-country landraces are substantially lighter than heavier yield differences estimated for more modern materials characterized by greater breeding content (Dalton and Guei, 2003).

For most crops, some released landraces are still in the basket of improved varieties in our adoption estimates, but these have been excluded for bean and sweetpotato where they are readily identified. Released landraces from other countries are included because of the assumption that such materials would not be available to farmers without the intervention of adaptation trials by the national crop improvement programme.

As discussed in the previous section on outputs, our definition of an improved variety is inclusive of escapes, products of participatory varietal selection from improved materials, and breeding outputs in countries that do not have a functioning formal release and registry system. Focusing only on released varieties would understate the performance of investments in crop improvement.

The issue of the frequency of seed renewal for improved OPVs and hybrids in open-pollinated crops is one that needs to be addressed in the future. This aspect acquires heightened importance when comparing estimated adoption levels in maize between regions in sub-Saharan Africa. Where survey data were available, i.e. maize in Ethiopia, an OPV was considered a modern variety if the age of the seed was three years or less (Jaleta *et al.*, 2013).

Unfortunately, survey data on seed vintage in OPVs were only available in this one case. Older OPVs released in the 1970s and 1980s in countries where seed renewal from government or private sector sources is limited were also deleted from the list of improved varieties. Matuba, which was released in 1984 in Mozambique where civil war prevailed until 1992 and where seed programmes are still not institutionally well developed, is an example of an OPV that was now considered to be a local variety. In general, the problem of outcrossing in defining an improved variety is most pronounced in maize, pearl millet and pigeonpea among the 20 crops in this study.

Likewise, we have not confronted the issue of varietal age with much analytical rigour in defining improved varieties. Arbitrarily, we have used 1970, the year the high-yielding rice variety IR-8 was introduced into SSA (Dalrymple, 1986), as the cut-off point to define improved varieties. The use of 1970 as a cut-off point means that improved materials bred in the colonial era are not included. Using an earlier cut-off date, such as 1950 or 1960, would have resulted in markedly greater estimated adoption levels in groundnut and in rice in several countries in West Africa. In groundnut, the estimates based on a large survey in northern Nigeria and expert opinion in Mali suggests that two varieties bred in the colonial period are still extensively grown. The variety 55-437 is estimated to cover about 40% of groundnut-growing area in Nigeria; 47-10 is believed to be cultivated on about the same percentage of groundnut area in Mali (Ndjeunga *et al.*, 2012). In rice, using a cut-off date of 1960 or 1965 would bring several popular introduced purified landraces into play. Inclusion of these materials in the set of modern cultivars would result in a sharp rise in the adoption level in some large rice-growing agroecologies in West Africa (Dalton and Guei, 2003). They were listed but not included as improved varieties in the 1998 Initiative; therefore, we opted for the same course of action to promote consistency in comparing estimates over time. The adoption results for other crops and countries are not that sensitive to the use of an earlier cut-off date.

Adoption estimates

The source of the adoption information potentially affects estimates in terms of their variance and bias. With 20 crops in multiple countries, a uniform application of a protocol to elicit information on adoption is desirable (Walker, 2010). One also wants to use the same protocol in the future to generate valid time-series estimates. The protocol that was used in the 1998 Initiative was adhered to as strictly as possible and featured the elicitation of adoption estimates based on expert opinion. That protocol was administered by seven different institutional partners resulting in some variation but, in general, usable estimates were obtained. In the small minority of cases where such information was incomplete, survey estimates were relied on if they were nationally representative. The expert opinion protocol is outlined in Chapter 4 and its validation is described in Chapter 20.

National adoption estimates refer to area harvested of all improved varieties in the numerator divided by total area harvested of the crop in the denominator. Harvested area is a desirable measure because it has relatively easy interpretations in terms of production impacts. In contrast to what is desirable, expert panels and focus-group respondents in community surveys are more comfortable in giving adoption estimates in per cent of farmers rather than in per cent of area. The per cent of farmers using improved varieties is easily measured in surveys of farmers and does not require estimates of area planted and harvested. It says something about the access of different individuals to the new technology. However, per cent farmers almost always results in higher adoption estimates than those based on per cent area. Furthermore, the area measure imposes the added discipline that area shares between traditional and improved varieties have to add to 100, and area shares among specific improved varieties have to sum to their aggregate total.

Varietal turnover

The level of adoption of improved cultivars only tells part of the story about the performance of investment in crop improvement. The velocity of varietal change is an important outcome especially for countries where levels of adoption are already high. The rate of change, or replacement, of older varieties by newer cultivars is informative about the performance of genetic improvement programmes. Past research suggests that if newer materials are not replacing their earlier generation counterparts, returns to genetic improvement stagnate (Brennan and Byerlee, 1991). The permanency of first-generation improved varieties in farmers' fields points to a problem of declining productivity in the search for and release of new varieties in crop improvement.

Varietal turnover is measured by the age of varieties weighted by their area in production. The date of release is usually assumed to initiate the age calculation of the variety when it becomes available to the public for adoption. Therefore, age is measured from the current year to the year of release unless farmers have access to the variety prior to the date of release. Only improved varieties enter into the calculation irrespective of their adoption level. In calculating a weighted (improved) variety age, the age of each improved variety is weighted by its relative share in the total area of improved varieties. Varietal age will fall irrespective of whether younger varieties replace older improved varieties or traditional varieties because their share will increase in the basket of improved varieties.

Area-weighted age estimates under 10 years indicate rapid varietal change and robust progress in plant breeding from an economic perspective. However, the adoption level also needs to be factored into the evaluation. Having rapid varietal turnover with less than 10% adoption does not imply significant economic progress in plant breeding. Estimates of varietal turnover that exceed 20 years indicate that more recent materials are having a hard time competing with earlier materials. Rising varietal age is associated with declining marginal returns to plant breeding.

Past studies have documented large disparities in varietal turnover rates in different agricultural settings. Irrigated wheat farmers in the Yaqui Valley of Mexico replace their varieties every 3–4 years on average. The breaking down of disease resistance and the steady increase in yield gains are positive incentives for rapid varietal change. In the corn belt of the USA, farmers switch to newer hybrids every 2–3 years. In contrast, potato growers in specialized compact regions of outstanding production potential in the USA have limited incentives to replace Russet Burbank with newer varieties. Russet Burbank is difficult to grow but it is highly productive and has strong market demand. For potato growers in Canada and the USA, estimated varietal age has fluctuated between 40 and 50 years since the 1990s, indicating a low rate of return to most state and national programmes in North America (Walker, 1994; Walker *et al.*, 2011).

Spill-over varieties

Spill-over varieties are improved cultivars that are adopted by farmers in two or more countries. The positive incidence of spill-over varieties is associated with wider adaptability of genetic materials indicative of more homogeneous demand preferences and less marked environment-by-genotype interactions. In principle, spill-overs in varietal change demonstrate remunerative returns in investing in plant breeding internationally, regionally and nationally in larger countries because smaller countries can benefit from the work of others.

Hypotheses about varietal adoption and turnover

Attributing positive outcomes to crop improvement in SSA would be confirmed by finding the following statements true for most crop and country observations:

1. The level of adoption of improved varieties and hybrids is steadily increasing over time and is substantially higher in 2010 than in 1998.
2. Spill-over varieties are found in all food crops and they lay claim to a sizeable share of adopted area.
3. The share of materials related to CG Centers is higher in varietal adoption than in varietal output.
4. Disadoption of improved varieties on aggregate is rare and is not caused by economic restructuring and liberalization.

5. Adoption of improved varieties is positively influenced by market demand, the potential of the production environment and the crop's multiplication ratio.
6. Varietal turnover is relatively high and is increasing over time.

Most of these six conditioning statements require order of magnitude thresholds for testing, although such thresholds are arbitrary. About one hectare in four was planted to an improved variety in the ten food crops in the 1998 Initiative. By 2010, increased uptake of improved varieties in these benchmark crops at an annual rate of gain of 1–2% per annum would seem like a reasonable expectation. Average varietal age in the range of 10–15 years would also be consistent with relatively high turnover of improved cultivars in 2010. Citing several major spill-over varieties by crop should lend qualitative support for the third hypothesis that speaks to the scope for wide adaptability.

Greater shares in adoption than in release reinforce the claim that IARC-related materials are relevant in meeting the demands of farmers in developing countries. IARC-related materials may also be better promoted than other materials in public-sector and non-governmental organization (NGO) extension programmes. In any case, finding more influence on outcomes than on outputs suggests synergies in genetic improvement between national programmes and international centres.

The fifth hypothesis addresses the concern that market liberalization has resulted in increased price ratios between fertilizer and food crop output thereby discouraging the use of improved varieties that are potentially heavier and more responsive users of fertilizer than traditional cultivars. This concern applies mainly to maize, which is the most extensive user of fertilizer among food crops in SSA.

Although the determinants of adoption are not explicitly treated in Parts 2 and 3 of this volume, cropwise variation in economic orientation, production potential and multiplication ratio give rise to the last hypothesis. Such inter-crop differences can be encapsulated in the expectation that the uptake of maize hybrids produced in favourable highland conditions with a longer growing season on volcanic soils and partially sold in the market will be distinctly higher than the adoption of improved clones of cassava produced largely for household consumption on sandy soils in the lowland rainfed tropics.

Impacts: Yield, Net Revenues and Poverty

Chapters 5–14 of this volume address the same content of inputs, outputs and outcomes in a uniform manner relying heavily on descriptive analysis presented in tables and figures. Chapters 15–17 (Part 3) are more eclectic in their treatment of impacts of food crop improvement in SSA in terms of both content and approach. Nevertheless, the same impact-related themes weave their way through the three chapters in Part 3. A focus on farm-level productivity impacts in the form of increased yield and net revenue per hectare is common to the two impact crop-by-country case studies in Chapters 15 and 16 and to the aggregate analysis of impact for SSA as a whole in Chapter 17. When combined with economic modelling, these quantified direct effects provide the foundation for estimating the national poverty impact of improved maize varieties and hybrids in Ethiopia in Chapter 15 and improved bean varieties in Rwanda and Uganda in Chapter 16.

Context also plays a role in the focus of impact assessment. For example, beans, sown in small plots in two seasons, are a very important food crop in Rwanda and Uganda. For that reason, the consequences of varietal change for dampening food insecurity feature prominently in Chapter 16.

A favourable assessment of the consequences of crop improvement would be supported by increased yields, augmented net revenues, reduced poverty and enhanced food security attributed to the adoption of new varieties. Similar to estimated adoption outcomes, the estimates of the size of impacts is as or even more important than the direction of the signs. In predominantly rainfed environments with mostly poor quality soils, the size of the effects may be too small to generate widespread and deep improvements in welfare. Indeed, the main impact-related hypothesis that is relevant to this work would state that varietal change generates marked changes in rural household welfare that can be quantified with survey data.

Yield and productivity

The increase in productivity per hectare is often the most frequent manifestation of the adoption of improved varieties in regions of higher and more assured production potential where farmers are using improved inputs such as chemical fertilizers and pesticides. Production of food crops by small farm households in SSA does not fit the above textbook expectation of finding large productivity effects attributed to varietal change. In SSA, production is rainfed and unassured. Drought is a frequent visitor to farmer fields. The demand for shorter duration varieties that escape drought is high. All things being equal, shorter duration translates into lower yields in good rainfall years. The use of chemical inputs is low. Without accompanying changes in input use, farmers cannot leverage varietal change into abrupt gains in yields in favourable weather. Therefore, varietal change in and of itself is unlikely to result in substantial productivity change (Sanders *et al.*, 1996; Bulte *et al.*, 2014).

The above pessimistic scenario about the prospects for productivity enhancement from varietal change does not apply to all economic contexts and ecological environments in SSA. Higher production potential environments where soil fertility is not as constraining and where agriculture is supported by favourable input policies engender brighter prospects for productivity growth from varietal change. The tropical highlands in several countries in East Africa fit these conditions for more transparent and higher productivity consequences from varietal change.

Productivity prospects also depend on the technology. Switching from traditional or even improved varieties of bush beans to more intensive climbing beans, requiring greater investment, should be accompanied by easily detectable differences in productivity.

What are reasonable expectations on the size of the yield gain? For first-generation improved varieties – those varieties that replace traditional landraces – yield differences in farmers' irrigated fields in Asia were of the order of 50–100% (Barker and Herdt, 1985). Adoption of improved varieties stimulated the adoption of improved inputs that led to an improvement in production potential. In much of SSA, it seems reasonable to expect that relative yields would increase by 10–30% in most rainfed environments where fertilizer is not readily available. In higher production potential environments with fertilizer availability, first-generation improved varieties should be able to leverage productivity gains in the region of 30–50%. In unassured production zones where farmers attach a high value to yield stability, the increase in expected productivity could be as small as 10%.

The productivity of most oilseed and pulse traditional landraces only approaches or slightly exceeds one tonne per hectare in most countries in SSA. Hence, a 30% yield change implies a productivity change of about 300 kg per hectare. For cereals, yields of traditional varieties can range from 1.0 to 2.0 tonnes; thus, a 50% increase in productivity in the best of circumstances could be as much as 1000 kg per hectare. Yields of roots and tubers have less dry matter and therefore are considerably higher than other crops in SSA ranging from 5 to 10 tonnes per hectare. However, except for potatoes, improved inputs are seldom applied to roots and tubers. Improved clones could be expected to generate increases of 500–3000 kg per hectare.

Net revenue

Adopting improved varieties almost always implies an investment by the farmer. The change in net revenue or net benefit per hectare is the monetary difference between net revenue of the improved variety and the variety that the farmer is replacing. The difference in net revenue is calculated in a partial budget setting where the only items that matter are changes in input and output prices, grain and straw production, and input use levels between the varieties in question as one replaces the other. For a farmer to switch from a traditional to an improved variety, the marginal rate of return on investment should exceed a threshold of 40–100% (Anderson *et al.*, 1976; CIMMYT, 1988). In today's prices, attributing a net revenue difference to an improved variety that exceeds US$150 per hectare is sizeable. Net revenue differences of US$50–100 are more typical of stand-alone varietal change in more marginal rainfed production environments in SSA.

Poverty

Estimation of yield and net revenue differences allow the authors of Chapters 15 and 16 to assess the poverty impact of improved varietal change on household income. Reductions in the incidence, severity and depth of poverty are calculated nationally (Foster *et al.*, 1984). Finding that adoption of improved varieties resulted in a reduction in the headcount index of poverty that approached or exceeded 1% nationally would be equivalent to a very large poverty impact.

Food security

Increased yields from improved varieties may enhance food security directly by stretching household consumption over more months in the same cropping year. Some of those months may occur in the hotter and drier hunger season. Improved varieties may also have the potential to capitalize on good rainfall years conducive to heavier production that opens up opportunities for interyear storage of staple food crops. Indirectly, and just as importantly, higher yields will result in lower prices that reduce food insecurity. Given that many semi-subsistence, small-producing households are net consumers in that they buy more than they sell, the indirect price effect of increased production on food security should figure prominently in traded staple commodities, such as maize.

Hypotheses about impacts of improved varieties

Because the emphasis in the DIIVA Project was on the uptake of improved varieties, hypotheses were not that well articulated at the start of the work on impact assessment in 2010/2011. Recognizing that hindsight bias may play a role in their formation (Pinker, 2014), the impact-related results in Chapters 15–17 are brought to bear on the following hypotheses:

1. The yield advantage of improved varieties is characterized by wide variation across crops and countries in SSA.
2. Where the adoption of improved varieties does not result in the use of complementary inputs, such as chemical fertilizer, the yield advantage of improved varieties over local landraces will be smaller than 25%.
3. Across sub-Saharan Africa as a whole from 1980 to 2010, the estimated yield advantage from the adoption of improved varieties is superior to 30% and figures as an important contributor to technological change documented from the analysis of time-series data on varietal diffusion.
4. Adoption of improved varieties will be strong enough to reduce poverty by 1% nationally in at least one crop-by-country case study.
5. Adoption of improved varieties will result in reducing food insecurity by at least 10% among producing households in at least one crop-by-country case study.

Notes

[1] For the subset of the same observations, the DIIVA and ASTI estimates on number of research scientists are compared in Walker *et al.* (2014).

[2] Since 2009, the earlier ASTI evaluation by Beintema and Stads (2006) has been updated and expanded in Beintema and Stads (2011).

[3] The absence of transgenic varietal change in food crops in SSA is not a good thing (see Paarlberg, 2008).

References

Anderson, J.M., Moscardi, E.R., Perrin, R.K. and Winkelmann, D.L. (1976) *From Agronomic Data to Farmer Recommendations: An Economics Training Manual*. CIMMYT, Mexico (DF).

Barker, R. and Herdt, R.W. with Beth Rose (1985) *The Rice Economy of Asia.* Resources for the Future. Washington, DC.

Beintema, N.M. and Stads, G.J. (2006) *Agricultural R&D in Sub-Saharan Africa: An Era of Stagnation*. International Food Policy Research Institute, Washington, DC.

Beintema, N.M. and Stads, G.J. (2011) African agricultural R&D in the new millennium: progress for some, challenges for many. *IFPRI Food Policy Report*. International Food Policy Research Institute, Washington, DC.

Bioversity International (2009) *Tracking Varietal Change and Assessing the Impact of Crop Genetic Improvement Research in Sub-Saharan Africa: A Proposal Submitted to the Bill and Melinda Gates Foundation.* Bioversity International, Rome.

Brennan, J.P. and Byerlee, D. (1991) The rate of crop varietal replacement on farms: measures and empirical results for wheat. *Plant Varieties and Seeds* 4, 99–106.

Bulte, E., Beekman, G. Di Falco, S., Hella, J. and Lei, P. (2014) Behavioral responses and the impact of new agricultural technologies: evidence from a double-blind field experiment in Tanzania. *American Journal Agricultural Economics* 96(3):813–830. doi:10.1093/ajae/aau015.

CIMMYT (1988) *From Agronomic Data to Farmer Recommendations: An Economics Training Manual.* CIMMYT, Mexico (DF).

Collard, B.C.Y., Jahuer, M.Z.Z., Brouwer, J.B. and Pang, E.C.K. (2005) An introduction to markers, quantitative trait loci (QTL) mapping and marker-assisted selection for crop improvement: the basic concepts. *Euphytica* 142(1–2), 169–196.

Dalrymple, D. (1986) *Development and Spread of High-Yielding Rice Varieties in Developing Countries.* USAID, Washington, DC.

Dalton, T.J. and Guei, R.G. (2003) Productivity gains from rice genetic enhancements in West Africa: countries and ecologies. *World Development* 31(2), 359–374.

Evenson, R.E. and Gollin, D. (eds) (2003) *Crop Variety Improvement and Its Effect on Productivity: The Impact of International Agricultural Research*. CAB International, Wallingford, UK.

Foster, J., Greer, J. and Thorbecke, E. (1984) A class of decomposable poverty measures. *Econometrica* 52, 761–766.

Giller, K. (2012) Conservation: no silver bullets for African soil problems. *Nature* 485(7396), 41. doi:10.1038/485041c.

Jaleta, M., Yirga, C., Kassie, M., de Groot, H. and Shiferaw, B. (2013) *Knowledge, Adoption and Use Intensity of Improved Maize Technologies in Ethiopia. CIMMYT's Report on Objective 2. Part I.* CIMMYT, Nairobi.

Ndjeunga, J., Hash, C.T., Faye, I., Sanogo, M., Echekwu, C.A. *et al.* (2012) *Assessing the Effectiveness of Agricultural R&D in West Africa: Cases of Pearl Millet, Sorghum and Groundnut Crop Improvement Programs*. Diffusion and impact of improved varieties in Africa (DIIVA) project, Objective 1 Report. ICRISAT, Niamey, Niger.

Paarlberg, R.L. (2008) *Starved for Science: How Biotechnology is Being Kept Out of Africa*. Harvard University Press, Cambridge, Massachusetts.

Pinker, S. (2014) *The Sense of Style.* Viking, New York.

Sanders, J.H., Shapiro, B.I. and S. Ramaswamy. (1996) *The Economics of Agricultural Technology in Semiarid Sub-Saharan Africa.* Johns Hopkins University Press, Baltimore, Maryland.

Sanni, K., Toure A. and Diagne, A. (2011) *Variety Release and Registration Practices for Rice in Africa Countries.* Study Report. AfricaRice, Cotonou.

Walker, T.S. (1994) *Patterns and Implications of Varietal Change in Potatoes*. Social science department working papers series no. 1994–3. International Potato Center (CIP), Lima.

Walker, T.S. (2010) *Guidelines for Data Collection for Objective 1 of the DIIVA Project*. Fletcher, North Carolina.

Walker, T., Thiele, G., Suarez, V. and Crissman, C. (2011) *Hindsight and Foresight about Potato Production and Consumption*. Social Sciences. Working Paper 2011-5. International Potato Center (CIP), Lima.

Walker, T., Alene, A., Ndjeunga, J., Labarta, R., Yigezu, Y. *et al.* (2014) *Measuring the Effectiveness of Crop Improvement Research in Sub-Saharan Africa from the Perspectives of Varietal Output, Adoption, and Change: 20 Crops, 30 Countries, and 1150 Cultivars in Farmers' Fields*. Independent Science and Partnership Council Secretariat, Rome.

4 Coverage, Data and Methods in Assessing the Performance of Food Crop Improvement in Sub-Saharan Africa

T.S. Walker*
Independent Researcher, Fletcher, North Carolina, USA

Introduction

This chapter was written to provide a reference for readers who want to understand the context underlying the substantive results reported in this volume. It describes the data and how they were collected.

A major objective of the Diffusion and Impact of Improved Varieties in Africa (DIIVA) Project was to provide comprehensive information on the geographical spread of improved crop varieties in sub-Saharan Africa (SSA). Information on inputs, outputs and outcomes associated with diffusion of modern varieties was also sought. Data collection began in 2010 and continued into 2012. Here, we use 2010 as the point of the reference to describe the DIIVA data set. Comparable data assembled in the 1998 Initiative and reported in Evenson and Gollin (2003) are described in the next chapter. In general, the data collected in the late 1990s were more variable from CG Center to CG Center (Institutes within the Consultative Group on International Agricultural Research) and from crop to crop. In contrast, the DIIVA Project in 2010 represented a concerted effort to canvass and assemble uniform data on inputs, outputs, outcomes, inputs and impacts.

The DIIVA data can be divided into three domains: assembled data on scientific capacity and varietal release/availability; elicited estimates of varietal adoption; and household survey data. Data were assembled from existing sources on scientific capacity in 2010 and on improved varietal output from 1970 to 2010. Other than the need for intensive in-country interaction and supervision, these data on inputs and outputs did not entail any notable methodological difficulties because participants followed consensus guidelines described later in this chapter (Walker, 2010).

Arriving at a method to generate reliable estimates of improved variety adoption was the main challenge facing the DIIVA Project. The project settled on expert opinion panels to obtain these estimates. Reasons for this choice and the process of how those estimates were generated receive considerable attention in this chapter.

Household surveys were carried out for several crops and in a few countries to provide the raw material for impact assessment. These nationally representative household surveys are a rich source of information on varietal adoption and are also used as a basis for testing the reliability of the expert opinion estimates reported in Chapter 20. One consideration in the design of these household surveys was their ability to validate the adoption estimates from expert opinion.

Before describing the assembly of data on scientific inputs and varietal output, the methods

* E-mail: walkerts@msu.edu

used eliciting expert opinion on varietal adoption, and the content of the household surveys, we briefly present crop and country coverage in the next section. Comparative information is given in Chapter 5 on crop and country coverage in the 1998 Initiative.

Crop and Country Coverage

The DIIVA databases contain about 150 crop-by-country observations selected to cover the most important food crops in the main producing countries.[1] The planned design of coverage was balanced in the DIIVA Project Proposal (Bioversity International, 2009); but, for multiple reasons,[2] the number of observations varies somewhat by type of data.

In Table 4.1, coverage is described for the national-level adoption data. Twenty crops and two large maize-producing regions result in 21 crop categories. About half of these were included in the '1998 Initiative' and are described in Table 4.1 as 'continuing'. The other half is 'new' indicating where a baseline on varietal diffusion has been constructed for the first time.

The area harvested within the 20 study crops in SSA totals about 140 million hectares. These 20 primary and secondary food staples make up about three-quarters of the total crop area in SSA including annuals and perennials.[3]

The number of country observations varies from one each for lentil, wheat, banana and field pea to 17 for cassava and 20 for maize in East and Southern Africa (ESA) and West and Central Africa (WCA) combined. Maize is split regionally because of its relevance as a food crop, and because the ESA and WCA are so distinct in their uptake of hybrids in relation to improved open-pollinated varieties (OPVs). The private sector is dynamic and now dominant as the source of modern varieties (MVs) in several important maize-producing countries in ESA, but is only recently emerging in the production of hybrids in a few West African countries.

Overall, the countries included in Table 4.1 covered 83% of the harvested area of the 20 crops in SSA in 2010.[4] Only three crops were sparsely represented at a level below 60% of area coverage. Beans in Kenya, sorghum in Ethiopia and sweetpotatoes in Nigeria were arguably the most important omissions among the country-by-crop combinations covered in the DIIVA Project. Sesame and cocoyam are two of the other most extensively grown food crops that were not included in the DIIVA Project (Fuglie and Marder, Chapter 17, this volume).

Breadth of coverage by database is addressed in Table 4.2. The proposal envisaged coverage of 104 crop-by-country observations. Field pea, banana and yam were brought in during the course of the project. Moreover, AfricaRice, the International Center for Agricultural Research in Dry Areas (ICARDA) and the International Institute of Tropical Agriculture (IITA) covered many more countries than was initially planned. INTSTORMIL (the International Sorghum and Millet Innovation Laboratories of

Table 4.1. Description of crop coverage in the DIIVA database.[a]

Crop	Description	Number of countries	Share (%) of total SSA area under production for the DIIVA countries in 2010[b]
Faba bean	New[c]	2	100
Cowpea	New	18	98
Maize-ESA[d]	Continuing	9	97
Yam	New	8	95
Lentil	New	1	95
Barley	Continuing	2	91
Cassava	Continuing	17	90
Soybean	New	14	86
Maize-WCA[d]	Continuing	11	85
Wheat	Continuing	1	84
Chickpea	New	3	80
Pearl millet	Continuing	5	80
Pigeonpea	New	3	79
Rice	Continuing	19	79
Sorghum	Continuing	8	78
Banana	New	1	71
Potato	Continuing	5	65
Groundnut	Continuing	10	63
Bean	Continuing	9	59
Sweetpotato	New	5	54
Field pea	New	1	46
Total/Weighted mean		152	83

[a]Refers to the national aggregate adoption data; [b]This is based on FAOSTAT for 2010; [c]Refers to crops that were not covered in the 1998 Initiative; [d]For maize: ESA = East and Southern Africa; WCA = West and Central Africa.

Table 4.2. The number of crop-by-country observations in the DIIVA Project by type of database.

Database	Description	Number
Proposed (intended)		104
Scientist years (SYs)	Full-time equivalent scientist input database	151
Varietal release	Output database listing released varieties	149
National adoption	Aggregate adoption database in 2010	152

USAID) partnered with the DIIVA Project to improve coverage in sorghum, which now includes the Sudan (Zereyesus and Dalton, 2012). Hence, the database contains about 50% more crop-country observations than was proposed (Table 4.2).

Expanded coverage by AfricaRice added a few very small producers, such as the Central African Republic (CAR) and Guinea Bissau, resulting in a total national coverage of 30 countries. The median country in the national adoption database contributes five crop observations. Four countries, CAR, Eritrea, Madagascar and Sierra Leone, have only one crop observation. At the other end of the range, Uganda supplies 11 of a possible 20 crop observations.

For 62 observations, data are available for comparative analysis between 1998 and 2010 on scientific strength, varietal output and MV adoption (Chapter 5, this volume).

Assembling Data on Scientific Capacity and Varietal Output

Scientific capacity

All participating CG Centers collected cross-sectional data for 2009 or 2010 on the number of full-time equivalent (FTE) scientists working in national programmes defined broadly as encompassing the public, private and university sectors. For most crops and countries, almost all researchers were employees of the public sector. Data were also assembled on CG Center investments in FTE scientists for selected years.

Data on education, age and disciplinary orientation comprised the minimum data set on scientific capacity. Some CG Centers invested in an expanded database on scientific capacity. For example, CIP (the International Potato Center) compiled information on gender, age and experience of scientists as well as on research infrastructure (Labarta, Chapter 9, this volume).

The benchmark on scientific capacity also varied across CG Centers by the level of aggregation in data collection. Richer and more detailed data in this aspect were gathered at the level of the individual scientist (Ndjeunga *et al.*, 2012). A coarser benchmark was established for crops such as maize in ESA where large NARS and private sector companies made data collection at the individual scientist level a more onerous task (De Groote *et al.*, Chapter 11, this volume).

As discussed in Chapter 2, the DIIVA Project was not the first to gather information on levels of, and trends in, scientific capacity in SSA. Since the late 1980s, economists at the International Service for National Agricultural Research (ISNAR) and now at the International Food Policy Research Institute (IFPRI), working under the Agricultural Sciences and Technology Indicators (ASTI) Initiative, have collected comprehensive information on agricultural research in SSA.

Although DIIVA focused on specific crop improvement programmes and ASTI addresses country-level sectoral agricultural research as a whole, the substantive findings in Chapters 6–12 resonate well with those from a recent analysis of the latest round of ASTI inquiries (Beintema and Stads, 2011). In general, ASTI researchers collect data on all institutional agencies engaged in agricultural research and aggregate the information to the national level, whereas relevant budgetary information is documented annually. Data collection for the DIIVA Project was at a lower, more disaggregate level – its sources of information were the scientists in, and leaders of, commodity improvement programmes. Many of these contacts were long-standing partners of the participating CG Centers.

Varietal output

As described in Chapter 3, national varietal release registries were the starting points for quantifying varietal output. The registries were complemented by an assessment of improved varieties that were available to farmers but not officially released. The existence of these available but non-released improved varieties was most evident during the elicitation of adoption estimates from expert opinion panels that are described in the next section.

The minimum data set for varietal output consists of the five descriptors: (i) official name of the improved variety; (ii) year of release or of first availability to farmers; (iii) institutional source of the material; (iv) genetic background (usually pedigree or related ancestry information); and (v) release classification from the perspective of type of material, NARS input, IARC input and institutional source.

The first four descriptors are easy to understand. They identify the improved cultivar. The release classification is more complicated because it entails institutional information on the role of NARS and CG Centers in plant breeding. Getting the release classification right required considerable judgment by the participating CG Centers on the source of plant breeding materials and their use in crop improvement. Hence, information on the release classification speaks to the institutional development of crop improvement in the country for the crop of interest. In 1998, the number of categories in the release classification across the eight participating CG Centers, including the International Rice Research Institute (IRRI), ranged from 4 to 12. Arguably, the most novel classification was that of the International Center for the Improvement of Maize and Wheat (CIMMYT) and IITA, which recorded the percentage contribution from each institute in released material when the variety of interest was related to an IARC. The following categorization by ICARDA for barley was representative of many Centers:

- ICARDA-cross, ICARDA-selection;
- ICARDA-cross, NARS-selection;
- NARS-cross, ICARDA-parent;
- ICARDA germplasm accession;
- NARS-cross, NARS-parent;
- NARS-landrace;
- Other international sources.

Each participating CG Center was encouraged to come up with its own varietal release classification depending on the attributes of the crop and the breeding context. In general, the more disaggregate the descriptive classification the better. Major dimensions of the classification include the role of NARS, IARC and private-sector participation in the breeding process. Potentially, biotechnology offers the possibility of expanding the number of categories in the classification if any varietal releases are related to marker-assisted selection or in the medium-term future to transgenic varietal change.

Because most national release lists are more informative than the five pieces of information in the minimum database, all Centers compiled quite extensive but somewhat heterogeneous information on varietal output. Salient characteristics of the improved variety were the most typical data that exceeded the requirements of the minimum database.

It is important to recognize that the varietal output database covers four decades from 1970 to 2010. Most CG Center participants, such as IITA (Alene *et al.*, Chapter 6, this volume), did not simply update the existing 1970–1998 database for 'continuing' crops; they redid varietal output in the earlier years to be compliant with the latest registries. Promising varieties that were scheduled for release in the late 1990s but not subsequently released or made available to farmers were deleted from the earlier database.

Meeting the Methodological Challenge of Generating Consistent Estimates on Improved Varietal Adoption

Reliable estimates of adoption of improved varieties in food crops in sub-Saharan Africa were of paramount importance for the DIIVA Project. Adoption estimates could have been generated in several ways including large-scale, nationally representative household surveys, rapid rural appraisals featuring systematic field visits during the cropping season, expert opinion and information on seed sales. A handful of considerations loomed large in the choice of an appropriate method:

1. For meaningful cross-sectional comparative analysis, adoption estimates in all crops and countries should as much as possible be generated by the same method.

2. For meaningful time-series analysis, the method used to estimate adoption in 2009–2010 should not change substantially from the method deployed in the 1998 Initiative. Likewise, the method used in the future should build on methods-related experience in 1998 and 2009–2010.
3. Although resources for estimating adoption of specific improved varieties were ample, they were not unlimited.
4. The time of participants was another potentially binding constraint in the choice of methods for estimating adoption. Irrespective of the method chosen, it was apparent that their successful application required close and sustained supervision. However, methods still varied substantially in their time intensity because participants were involved in several other projects.
5. The choice of methods had to be adapted to crop context. For example, the small variation in phenotypic differences in improved cassava clones required field visits and skilled interpretation of resulting cultivar photos to distinguish one variety from another. Information on seed sales as a source of adoption estimates was ideally suited to maize in East and Southern Africa where private-sector hybrids were dominant.

If the 1998 Initiative had carried out household surveys, if resources were not limiting, if participants had agreed to collaborate on and had time to supervise 50–75 multiple-crop surveys in study countries, if surveys provided the most cost-effective mechanism to generate adoption estimates, and if so-called experts did not know that much about the level of aggregate and cultivar-specific adoption, then household surveys would have replaced expert opinion as the method of choice for estimating adoption of improved varieties in the DIIVA Project. None of the above conditions was true at the start of the DIIVA Project; therefore, expert opinion was the preferred option for adoption estimation for the majority of the country-by-crop observations.

Not all the adoption estimates were derived from expert opinion. One hundred and ten crop-by-country combinations in the DIIVA adoption data set of 152 observations are based on expert opinion (Table 4.3). Highly focused, nationally representative surveys account for 36 observations – 16 of these were financed and canvassed by the DIIVA Project, and 20 drew on complementary research by other CG Centers and donors, especially AfricaRice's Japan Project; several others (for maize in ESA, adoption studies were more readily available) were inferred from recent literature; and one observation, maize in Tanzania, relied on investing in the collection of variety-specific seed production information.

The protocol for eliciting expert opinion and its validation are presented in Chapter 20 that pulls together the experiences reported in Chapters 6–14. Prospects for the emerging method of DNA fingerprinting are also commented on in Chapter 20 together with an evaluation of the strengths and weaknesses of household surveys that are described in the next section.

Structuring Household Surveys to Provide a Foundation for Validating the Adoption Estimates and Assessing Impacts

Nine large-scale adoption surveys were funded and undertaken by the DIIVA Project. Their coverage and sampling features are described in Table 4.4. Although multi-purpose in nature, their primary intent was to validate the adoption estimates generated by the national expert panels. Eight of the nine surveys were nationally representative; cassava's inquiry was regional for south-west Nigeria.

The 15 crop observations in the surveys described in Table 4.4 were complemented by a more limited survey that canvassed four regions in Uganda to assess adoption of recently released clonal material in banana (Kagezi *et al.*, 2012). We also used output from a recent IFPRI–CSIR (Council for Scientific and Industrial Research, Ghana) survey on adoption of maize and rice

Table 4.3. Source of the national adoption estimates by number of observations.

Source	Number
Expert opinion	110
DIIVA adoption survey	16
Non-DIIVA adoption survey	20
Inferred from the literature	5
Seed production and trade	1
Total	152

Table 4.4. Description of the sampling features of the diffusion MV validation impact surveys conducted by participants in the DIIVA Project.

Crop	Country	Geographic basis for sampling	Sample size: Primary sampling unit (PSU)	Sample size: Households per PSU	Number of households[a]	Community survey
Barley	Ethiopia	The three major regions where barley is grown	123 kebeles	12	1469 (1280)	Yes
Bean	Rwanda	Ten major agroecological regions	80 communities	18	1440	Yes
Bean	Uganda	Four major geographic regions	19 districts, 108 communities	18	1908	Yes
Cassava	Nigeria	All five States in Southwest Nigeria	80 enumeration areas	10–12	841	Yes
Groundnut	Nigeria	Ten major groundnut-producing States	243 villages	10	2739	Yes
Groundnut	Tanzania	Seven main-producing regions	77 wards, 104 villages	15–16	1622 (1046)	Yes
Maize	Ethiopia	Production potential from 118 maize-growing districts	156 kebeles	15–16	2455	No
Pigeonpea	Tanzania	Seven main-producing regions	77 wards, 104 villages	15–16	1622 (816)	Yes
Potato	Ethiopia	The three major regions where potato is grown	123 kebeles	12	1469	Yes
Potato	Rwanda	Ten major agroecological regions	80 communities	18	1440	Yes
Rice	Nigeria	All 36 States in Nigeria	589 enumeration areas	10	5445	Yes
Sorghum	Tanzania	Seven main-producing regions	77 wards, 104 villages	15–16	1622 (902)	Yes
Sweet-potato	Rwanda	Ten major agroecological regions	80 communities	18	1440	Yes
Sweet-potato	Uganda	Four major geographic regions in Uganda	19 districts, 108 communities	18	1908	Yes
Wheat	Ethiopia	Eight wheat-growing agroecologies	125 kebeles	15–18	2096 (1839)	No

[a]The first number denotes total sample size; numbers in parentheses are households growing the crop.

Sources: Yigezu *et al.* (2012) for barley; Alene and Mwalughali (2012) for cassava; Diagne *et al.* (2013) for rice; Jaleta *et al.* (2013) for maize; Katungi and Larochelle (2012) for bean; Ndjeunga *et al.* (2013) for groundnut in Nigeria; Mausch and Simtowe (2012) for groundnut, sorghum and pigeonpea in Tanzania; Labarta *et al.* (2012) for potato and sweetpotato; and Yirba *et al.* (2012) for wheat.

MVs in Ghana (Ragasa *et al.*, 2013a,b) to validate adoption estimates.

Previous adoption surveys, if they existed, were largely restricted to small project areas in the other crop and country settings. Both NARS and IARC participants requested a national survey to complement their project-specific inquiries that often addressed only the initial uptake and very early adoption of well-defined introduced materials.

The average cost of the nine surveys was about US$100,000. During the Project Implementation Workshop, project participants were encouraged to pool their resources and canvass joint surveys. They were reluctant to do so initially. But the reality of a fixed budget for survey work, combined with the desire for greater country coverage in their crops of interest, subsequently spawned a more collaborative approach. ICARDA and CIP worked together with EIAR, the Ethiopian national programme, to carry out a survey on MVs of barley, faba bean and potatoes in Ethiopia in mostly shared agroecologies across the three crops. CIP and Centro Internacional de Agricultura Tropical (CIAT) jointly undertook surveys with their NARS partners in Rwanda on beans, potatoes and sweetpotatoes, and in Uganda on beans and sweetpotatoes. The International Crops Research Institute for the Semi-Arid Tropics (ICRISAT) also carried out a multi-crop survey on groundnut, pigeonpea and sorghum in Tanzania and on groundnut in Nigeria.

The guidelines for the survey recommended a stratified cluster sampling (Walker and Adam, 2011). Most of the participants followed this recommended framework. Sample size varied from 841 households in the cassava survey in five states of south-western Nigeria, to 5445 households in the rice survey also in Nigeria where all 36 states were covered. Households interviewed per village ranged from 10 to 18. Because varietal adoption can be highly sensitive to genotype by environmental (G×E) interactions, sampling more villages and fewer households within a village was emphasized. Community interviews based on focus groups preceded the household interviews in most of the surveys.

Oral responses on seed usage and on area planted to specific varieties provided the raw material for the subsequent calculation of adoption estimates. The cassava survey team complemented their household interviews with field measurements that featured varietal photographs using mobile phones (Alene and Mwalughali, 2012). These were analysed by research scientists who were able to assess varietal identity from the pictures displaying morphological plant characteristics. Without high resolution photographs from mobile phones, the identification of specific varieties would have been impossible.

Most household surveys also featured a field module useful in gathering plot-specific data on improved varieties by cultivar. Most of the surveys elicited information on the following priority adoption thematic areas:

- Demographic data by family member with information on gender, literacy and education level;
- Agriculture as primary, secondary or tertiary occupation and a qualitative assessment of the main sources of household income;
- A schedule of fields cultivated in the most recent cropping season and planted to the crop of survey interest with field characteristics such as village soil descriptors and location in the toposequence;
- Improved cultivar-specific area by field and season and local varieties as a group;
- Historical profile of improved cultivar-specific adoption (year of first use, source of seed of first use, identification of the replaced variety and trend in use over time);
- Pair-wise trait comparison with the replaced variety (superior, inferior, no difference, don't know) by characteristic;
- Plotwise data on cash input use; and
- An inventory of improved varieties used prior to the most recent cropping year but not sown in that year; reasons for their disuse and abandonment.

Prior to the initiation of the household adoption surveys, four of the prospective inquiries were selected as case studies for impact assessment on varietal change (Ndjeunga *et al.*, 2011; Groom *et al.*, 2013; Larochelle *et al.*, 2013; Zeng *et al.*, 2013). The emphasis on impact assessment led to expanded and new modules on assets and wealth, consumption and food security in survey design. The multipurpose surveys on varietal change in beans and sweetpotatoes were also collected in two rounds in Rwanda and Uganda to reduce interviewee fatigue and to sharpen the appreciation of the seasonality of production and consumption (Larochelle *et al.*, 2013). The surveys on rice in Nigeria, sorghum and pearl millet in Nigeria,[5] and maize in Ethiopia were carried out in a single interview format. The impact assessment case studies on maize in Ethiopia and bean in Rwanda and Uganda are presented in Chapters 15 and 16.

Notes

[1] The data are available online at: http://www.asti.cgiar.org/diiva.
[2] Incomplete data collected on improved wheat varieties adopted on large irrigated farms in Kenya, Zambia and Zimbabwe and the lack of Food and Agriculture Organization (FAO) and/or national production data in very small-producing countries for cowpea and soybean are two prominent considerations that led to an unbalanced coverage across the three databases. However, this effort is substantially more balanced than the '1998 Initiative'.
[3] Fuglie and Marder (Chapter 17, this volume) provide a more comprehensive view of crop coverage in the DIIVA Project from the inclusive perspective of all food and cash crops sown in sub-Saharan Africa.
[4] For banana, area coverage in 2010 refers to East Africa. For the purposes of the project and this chapter, production in South Africa is not included in SSA. South Africa was included in the '1998 Initiative' for maize and wheat.
[5] The impact assessment case study on coarse cereal varieties in Nigeria draws on a survey funded by the Bill & Melinda Gates Foundation (BMGF) that was conducted in 2009 prior to the initiation of the DIIVA Project (Ndjeunga *et al.*, 2011).

References

Alene, A.D. and Mwalughali, J. (2012) *Adoption of Improved Cassava Varieties in Southwestern Nigeria, Objective 2 Technical Report*. International Institute of Tropical Agriculture (IITA), Lilongwe, Malawi.

Beintema, N.M. and Stads, G.J. (2011) *African Agricultural R&D in the New Millennium: Progress for Some, Challenges for Many*. IFPRI Food Policy Report. International Food Policy Research Institute, Washington, DC.

Bioversity International (2009) *Tracking Varietal Change and Assessing the Impact of Crop Genetic Improvement Research in Sub-Saharan Africa: A Proposal Submitted to the Bill and Melinda Gates Foundation*. Bioversity International, Rome.

Diagne, A., Kinkingninhoun-Medagbe, F.M., Ojehomon, V.T., S.B. Abedayo O.O., Amovin-Assagba E. and Nakelse T. (2013) *Assessing the Diffusion and Adoption of Improved Rice Varieties in Nigeria*. Diffusion and Impact of Improved Varieties in Africa (DIIVA) – Objective 2 Report. AfricaRice Center, Cotonou, Benin.

Evenson, R.E. and Gollin, D. (eds) (2003) *Crop Variety Improvement and Its Effect on Productivity: The Impact of International Agricultural Research*. CAB International, Wallingford, UK.

Groom, B., Diagne, A. and Coruche, M.P. (2013) *Measuring and Assessing the Impacts of Crop Genetic Improvement in Africa: The Case of Rice*. Objective 3 Report for the Diffusion and Impact of Improved Varieties in Africa (DIIVA) Project. AfricaRice in partnership with IRRI, and SOAS, Cotonou, Benin.

Jaleta, M., Yirga, C., Kassie, M., de Groot, H. and Shiferaw, B. (2013) *Knowledge, Adoption and Use Intensity of Improved Maize Technologies in Ethiopia.* CIMMYT's Report on Objective 2. Part I. CIMMYT, Nairobi.

Kagezi, G., Mpiira S., Jogo, W., Karamura, E., Gotor, E., Garming, H. and Johnson, V. (2012) *Adoption of New Banana Varieties in Uganda Report for the DIIVA Project.* Diffusion and Impact of Improved Varieties in Africa (DIIVA) – Objective 1 Report. Bioversity International, Rome.

Katungi, E. and Larochelle, C. (2012) *Validating Adoption of Modern Bean Varieties in Rwanda and Uganda using Household and Community Surveys*. Diffusion and Impact of Improved Varieties in Africa (DIIVA) – Objective 2 Report. CIAT, Kampala, Uganda.

Labarta, R., Wambugu, S., Yirga, C., Mugabo, J., Dieu Nsabimana, J. *et al.* (2012) *The Adoption of Improved Potato and Sweetpotato Varieties in Ethiopia, Rwanda and Uganda: Comparing Expert Opinion Estimates and the Results of Nationally Representative Surveys.* Diffusion and Impact of Improved Varieties in Africa (DIIVA) – Objective 2 Report. International Potato Center (CIP), Nairobi.

Larochelle, C., Alwang, J., Norton, G.W., Katungi, E. and Labarta, R.A. (2013) *Ex-Post Impact of Adopting Improved Bean Varieties on Poverty and Food Security in Uganda and Rwanda*. Diffusion and Impact of Improved Varieties in Africa (DIIVA) – Objective 3 Report. Virginia Tech, Blacksburg, Virginia and CIAT, Cali, Colombia.

Mausch, K. and Simtowe, F. (2012) *Adoption Rates and Their Reliability Evidence Based on Expert Panel Discussions, Community Focus Groups and Household Survey in Tanzania – Groundnuts, Pigeonpea and Sorghum*. Diffusion and Impact of Improved Varieties in Africa (DIIVA) – Objective 2 Report. ICRISAT, Nairobi, Kenya.

Ndjeunga, J., Umar, J., Ahmed, B., Aba A., Ibro, A. *et al.* (2011) *Adoption and Impacts of Modern Sorghum and Pearl Millet Varieties in Northern Nigeria*. ICRISAT, Niamey, Niger.

Ndjeunga, J., Hash, C.T., Faye, I., Sanogo, M., Echekwu, C.A. *et al.* (2012) *Assessing the Effectiveness of Agricultural R&D in West Africa: Cases of Pearl Millet, Sorghum and Groundnut Crop Improvement Programs*. Diffusion and Impact of Improved Varieties in Africa (DIIVA) Project – Objective 1 Report. ICRISAT, Niamey, Niger.

Ndjeunga, J., Ntare, B.R., Ajeigbe, H., Echekwu, C.A., Ibro, A. and Amadou, A. (2013) *Adoption and Impacts of Modern Groundnut Varieties in Nigeria.* Diffusion and Impact of Improved Varieties in Africa (DIIVA) Project – Objective 2 Report. ICRISAT, Niamey, Niger.

Ragasa, C., Dankyi, A., Acheampong, P., Nimo Wiredu, A., Tripp, R. *et al.* (2013a) *Patterns of Adoption of Improved Maize Technologies in Ghana*. Draft report. IFPRI, Washington, DC.

Ragasa, C., Dankyi, A., Acheampong, P., Nimo Wiredu, A., Tripp, R. *et al.* (2013b) *Patterns of Adoption of Improved Rice Technologies in Ghana*. Draft report. IFPRI, Washington, DC.

Walker, T.S. (2010) *Guidelines for Data Collection for Objective 1 of the DIIVA Project*. Fletcher, NC.

Walker, T.S. and Adam, A. (2011) *Guidelines for Data Collection for Objective 2 of the DIIVA Project.* Fletcher, NC.

Yigezu, Y.A., Yirga C., Aw-Hassan, A. and Labarta, R. (2012) *Barley in Ethiopia: Verifying the Adoption Estimates from a Panel of Experts with Secondary Data and Household and Community Focus Group.* Diffusion and Impact of Improved Varieties in Africa (DIIVA) – Objective 2 Report. ICARDA, Aleppo, Syria.

Yirba, C., Jaleta, M., De Groote, H. and Gitonga, Z. (2012) *Diffusion and Impact of Improved Maize and Wheat Varieties in East and Southern Africa*. CIMMYT's Report on Objective 2. Part II – Wheat in Ethiopia. CIMMYT, Nairobi, Kenya.

Zeng, D., Alwang, J., Norton, G.W., Shiferaw, B. and Jaleta, M. (2013) *Ex-post Impacts of Improved Maize Varieties on Poverty in Rural Ethiopia.* Diffusion and Impact of Improved Varieties in Africa (DIIVA) – Objective 3 Report. Virginia Tech, Blacksburg, VA and CIMMYT, Nairobi.

Zereyesus, Y.A. and Dalton, T.J. (2012) *The Impact of Sorghum Varietal Improvement Research in Sudan.* Diffusion and Impact of Improved Varieties in Africa (DIIVA) – Objective 3 Report. Kansas State University, Manhattan, KS.

5 Genetic Improvement of the Crops in the 1998 Initiative: Historical Context and Exploratory Analysis

T.S. Walker*

Independent Researcher, Fletcher, North Carolina, USA

Introduction

The 1998 Initiative provided a point of reference for the Diffusion and Impact of Improved Varieties in Africa (DIIVA) Project, but it was a messy baseline. Roughly, the same types of data were gathered by participating CG Centers (Institutes within the Consultative Group on International Agricultural Research); however, uniform methods and protocols were not used. This variation across crops is described in Appendix 5.1.

Could a pooled analysis of these somewhat disparate data sets lead to a viable benchmark for comparing results over time? Economists at CIMMYT (Centro Internacional de Mejoramiento de Maiz y Trigo; International Center for the Improvement of Maize and Wheat) and the West Africa Rice Development Association (WARDA; now AfricaRice) did undertake an analysis of their data sets for sub-Saharan Africa (SSA). Hassan *et al.* (2001) and Heisey and Lantican (2000) analysed the maize and wheat data sets for SSA in considerable detail in an effort to tease out lessons for the improvement of maize and wheat breeding programmes in the region. Dalton and Guei (2003a) also published a richer version of their chapter in Evenson and Gollin on the results of rice improvement in West Africa (Dalton and Guei, 2003b). But a pooled analysis across the crops in the 1998 Initiative was not carried out.

A reading of the results in the 13 commodity chapters allowed Evenson and Gollin (2003b) to synthesize several salient empirical facts, but their assessment was not based on a pooled data analysis that is necessary for a reliable comparative evaluation across crops of differing characteristics and production contexts. Moreover, the late 1990s multi-institutional effort was global in scope. At that time, there was not much demand for a pooled data analysis that focused on only one region, sub-Saharan Africa. The data were perhaps viewed as being too CG-Center specific and nuanced to have potential for sharing at a higher level.

The objectives of the DIIVA Project are congruent with those of the 1998 Initiative, but they are not a perfect match. In particular, the DIIVA Project is not as International Agricultural Research Center (IARC)-centric; the emphasis is on crop improvement as a whole at the country level irrespective of the source of genetic materials.[1]

In spite of differences in data and objectives, establishing the relevancy of the 1998 data for the forthcoming DIIVA-related results was viewed as desirable. Findings on the strength of the national agricultural research system (NARS), modern varietal output and improved varietal

* E-mail: walkerts@msu.edu

adoption are reported from an exploratory analysis of the 1998 baseline. In terms of content, this chapter is the prototype for Chapters 6–14 that follow in Part 2. A concluding section revisits the more important empirical findings of the pooled data analysis from the three earlier sections and draws implications for measuring varietal change in SSA.

A majority of the 11 crops in the 1998 Initiative were characterized by sufficiently complete data to establish a reliable benchmark in one or more of the key aspects of crop improvement that are assessed in this volume. A few crops did not pass a test of sufficiency largely because their geographic coverage in the 1998 Initiative was not representative for production in SSA in the late 1990s. Sufficient and insufficient crops are identified in the next section that compares crop and country coverage between the 1998 baseline and the 2010 estimates described in this volume.

Commodity and Country Coverage Compared

The observational unit in the 1998 Initiative is the same as in the DIIVA Project, crop-by-country combinations. For a given commodity, a priority country is one where the commodity is economically important in contributing to food security at the national level or a country that accounts for a sizeable share of food production in SSA. Sizeable was not defined rigidly in the DIIVA Project but most, if not all, crop-by-country combinations were characterized by a production share that exceeded 1%.

We would expect that coverage should be more extensive in the DIIVA Project than in the 1998 Initiative because the former focuses on SSA, whereas the latter had a global orientation. Furthermore, the 1998 Initiative had a narrower conceptual emphasis on IARC-related genetic change, whereas the DIIVA Project had a broader orientation of improved varietal change irrespective of the source. For example, smaller countries with weaker NARS could have been selected in the 1998 Initiative if adoption was substantial and if varietal change could have been attributed to a CG Center.

Cherry picking countries with high CG-Center-attributed rates of varietal change irrespective of the size of their production was congruent with one of the major objectives of the 1998 Initiative that sought to document the profitability of IARC-related inputs in crop improvement. Being comprehensive makes good sense in a rate of return analysis where all benefits are juxtaposed to all costs. But the emphasis in the DIIVA Project was on the representativeness of the estimates across important producing countries of each food crop studied in SSA.

Congruence between the 1998 and 2010 data sets

In the 1998 data set, crop-by-country coverage was not uniform across the strength of NARS, varietal release and cultivar-specific adoption for barley, lentil, beans, pearl millet, groundnut and sorghum. For example, aggregate measures on strength of NARS are available for 123 crop-by-country observations in the 1998 Initiative; meaningful data on varietal release are restricted to only about 80 observations.

Arguably, the most important database pertains to cultivar-specific adoption and those heterogeneous data – some are very fragmentary – are given for 105 crop-by-country combinations in 1998 (Table 5.1). The data in the first two columns of Table 5.1 a number of these combinations suggest that differences in coverage between the two periods are not an issue. Indeed, the total number of crop-by-country combinations was greater in 1998 than in 2010 and 57 overlap; this points to the potential for country-specific time series analysis. The numbers in the rest of Table 5.1, however, point to severe problems in using the 1998 data as a benchmark for pearl millet, groundnut and sorghum that were characterized by cherry picking in the 1998 Initiative. Most of the countries included in 1998 but not covered in 2010 (column 5 in Table 5.1) each contributed to less than 1% of groundnut, pearl millet and sorghum production in SSA in 1998.

The authors of the commodity chapters in Evenson and Gollin (2003a) on coarse cereals and groundnuts relied heavily on existing studies conducted in the 1990s in arriving at estimates in SSA. No new data were formally collected. Fortunately, the International Crops Research

Table 5.1. Comparing country coverage in 1998 and 2010 in sub-Saharan Africa for the continuing crops from the perspective of aggregate adoption.

Commodity	2010	1998	Overlapping in 1998 and 2010	New in 2010	In 1998 but not covered in 2010
Maize	18	24	10	8	3
Cassava	11	19	11	0	8
Rice	10	7	7	3	0
Beans	10	7	7	3	0
Groundnut	10	9	4	6	5
Sorghum	7	14	5	3	10
Wheat	5	5	5	0	0
Potato	5	8	4	1	4
Pearl millet	5	9	1	4	8
Barley	1	1	1	0	0
Lentils	2	2	2	0	0
Total	84	105	57	28	38

Institute for the Semi-Arid Tropics (ICRISAT) had invested in impact assessment studies and therefore had considerable literature to review. Unfortunately, for the purposes of the DIIVA Project, the absence of new data resulted in a very shallow benchmark because coverage in groundnut and pearl millet accounted for only about 10% of production in SSA. For all intents and purposes, the major groundnut- and millet-producing region in West and Central Africa was not covered in the varietal release and adoption databases. The value of the 1998 data as a benchmark for sorghum is somewhat higher as five of the country observations overlap.[2]

Summing up, there is sufficient coverage to carry out a meaningful comparative analysis between the two periods for nine of the 11 original commodities in the 1998 Initiative. For pearl millet and groundnut, any results and implications from the analysis that follows should be taken with several large grains of salt. Groundnut observations in 1998 only accounted for 6% of SSA production; pearl millet was somewhat better with a 10% share of production. Each of the other crops summed to over a 50% share. Cassava, maize and wheat exceeded 85%.

Analysis of the Strength of NARS Data

In this section, we focus on two aspects of the strength of NARS data set that was common to all CG Centers in 1998: the number of full-time equivalent (FTE) scientists by crop-by-country observations and their associated research intensities in scientists per million tonnes of production.

Scientific staff strength and estimated researcher intensities

The 123 country-by-country summed to a total of 872 FTE scientists in the late 1990s. The largest contingents were from the two staple food crops: maize with 290 scientists and cassava with 178 scientists. These numbers resulted in an average country size of 7 scientists ranging from 17–18 for maize and wheat in East and Southern Africa (ESA) to less than 3.5 for sorghum (Table 5.2).

Research intensity is conventionally expressed as numbers of FTE scientists divided by million metric tonnes of production. The pooled data across crops show a typical negative exponential relationship in Fig. 5.1 for countries that are not severe outliers in research intensity. Smaller-producing countries invest proportionally more than larger-producing countries per million tonnes of production.

Mean estimates of researcher intensity are given in Table 5.2. The first set of estimates belie the problem facing small countries with less than 80,000 tonnes of production when they invest in human capacity in agricultural research (Brennan, 1991). This dilemma is especially

Table 5.2. Estimated researcher intensities by food crop in sub-Saharan Africa in the late 1990s.

Crop	Country observations	Production (000 tonnes)	FTE Scientists	Scientists per country	Researcher intensity: scientist per million tonnes production			Consistency with expectations
					Mean of country averages	Weighted average (sum production)	Weighted average (without Nigeria)	
Beans	11	1,955	41	4	48	21	21	Higher
Cassava	19	56,805	178	9	16	3	7	Lower
Maize ESA	12	21,804	220	18	14	10	10	Lower
Maize WCA	10	10,177	107	11	44	10	18	Higher
Pearl Millet	28	11,161	107	4	171	10	14	Markedly higher
Potato	9	2,070	45	5	52	22	22	Expected
Rice	7	6,230	36	5	72	6	10	Lower
Sorghum	21	15,903	72	3	62	5	7	Lower
Wheat	6	5,378	101	17	72	19	19	Higher

ESA, Eastern and Southern Africa; FTE, full-time equivalent; WCA, Western and Central Africa.

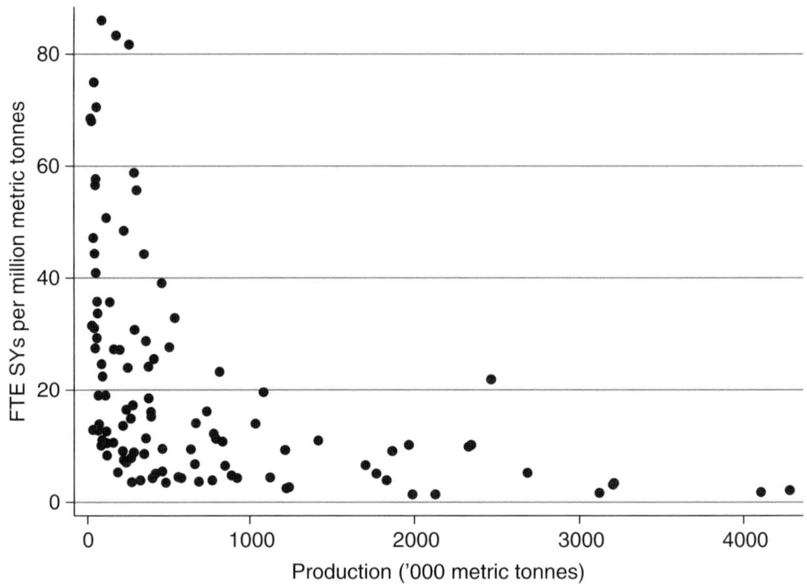

Fig. 5.1. The relationship between researcher intensity and production for those country-by-crop observations with less than 100 researchers per million tonnes production and with less than 5 million tonnes of production. FTE, full-time equivalent; SY, scientist years.

pronounced in pearl millet with a research intensity exceeding 100 and is also visible in wheat and sorghum, which have several small producers in the country-by-crop data set.

The next estimates in Table 5.2 are weighted by production to control for the problem of unusually high research intensities among small-producing countries. Researcher intensity is lowest for cassava and is also substantially less than 10 scientists per million tonnes of production in rice and sorghum.

Nigeria stood out as a country with consistently low researcher intensity. Mean readings of the ratio of FTE scientists to million tonnes of production were 0.1 for cassava, 0.5 for sorghum, 1.7 for rice, 1.8 for pearl millet and 2.6 for maize, which benefited from some private-sector participation. Nigeria ranked the lowest in each of the five commodity groups in which it was a major contributor.

Nigeria's outlier status may partially be explained by overestimated production in the FAO data, an incomplete accounting of FTE scientists or by a lack of commitment to agricultural research relative to the contribution of food crops to value of agricultural production. In any case, it is useful to recalculate the researcher intensities without Nigeria. The weighted averages increase substantially for each of the five crops for which Nigeria figures prominently as a major producer (Table 5.2). The estimated ratio for cassava more than doubled; the comparable estimate for rice increased by two-thirds. The weighted average estimate without Nigeria probably best reflects mean differences across the food crops.

At the other end of the spectrum, Ethiopia, Kenya and Sudan were characterized by estimated researcher intensities higher than other countries. In the late 1990s, Kenya and Ethiopia were recognized as countries that invested heavily in agricultural research proportional to the value of their agricultural production (Beintema and Stads, 2006).

In general, estimated researcher intensities did not vary widely across crops in the same countries. Across-country variation seemed to be a more important source of total variation than within-crop variation in the same country. Nigeria was one of the main drivers of the importance of across-country variation. The inclusion of several very small countries in the database also added to cross-country variation in estimated researcher intensity.

Overall, the estimated research intensities in Table 5.2 both confirmed expectations and generated surprises. The set of lower than expected commodity groupings included rice, sorghum, cassava and maize in ESA. Both rice and maize in ESA could qualify as surprises because rice imports loom large in West Africa, and maize in ESA benefits from considerable private-sector participation in agricultural research, particularly in southern Africa.

In contrast, pearl millet, beans and wheat had higher estimated research intensities than expected. An over-investment in wheat research in small countries is well described in the literature (Maredia and Eicher, 1995) so its high position relative to other crops was expected. The research intensity for pearl millet is overestimated in the data set because joint coarse cereal programmes most likely favour sorghum; our 50/50 allocation for the number of scientists reported in Bantilan and Deb (2003) is probably tilted towards pearl millet at the expense of sorghum in coarse cereal programmes. The true estimate for pearl millet should not differ that much from those of cassava and sorghum in the next-to-last column of Table 5.2.

The value for beans, estimated at 21 scientists per million tonnes of production, may not seem that high but the interpretation of bean scientists as breeders for crop improvement was very narrow. Given that breeders probably constitute at most half of the scientists, an estimate of 40 would be the highest in Table 5.2.

An estimate for potato of 22 may seem high, but potato has the most diversified disciplinary programme across the food crops. Only about 30% of the FTE scientists in potato crop improvement programmes were breeders across the nine potato-growing countries. Twenty per cent of scientists were engaged in seed production and tissue culture.

Comparing researcher intensities in SSA with those in other regions of the world

Realizing that many food-crop-producing countries in SSA are relatively small compared to other parts of the world, the estimates in Table 5.2 beg the question of how researcher intensities compare across regions. In comparing estimates across regions for the same commodity, no systematic differences were detected in the crop chapters of Evenson and Gollin (2003a).

The results in Table 5.3 respond to this issue for potatoes where a detailed global database was available from the 1998 Initiative. SSA did not seem to be substantially above or below average in this inter-regional comparison. The vast areas of the two largest Asian producers, India and China, however, generate low average research intensities.

Producing countries in SSA are not staffed at inferior levels relative to other regions based on two definitions of what constitutes a crop improvement programme (Table 5.3). Several of the 'Other Asian' producers are as small as or smaller than the potato producers in SSA. 'Other Asia' is characterized by higher staffing intensities than any other region in Table 5.3. But statistical differences in researcher intensity do not emerge across the two regions when one controls for the size of production. Estimated researcher intensities in SSA seem to be in line with those in other regions of the world for potato improvement.

Analysis of the Varietal Release Database

In this section, we re-examine conclusions from the earlier Evenson and Gollin overview and

Table 5.3. Comparing estimated researcher intensity in potatoes in SSA to other regions of the world.

Region	Researcher intensity by programme definition	
	Broad[a]	Narrow[b]
China and India	11.3	4.0
Other Asia	33.0	16.2
Latin America	15.2	9.3
Sub-Saharan Africa	21.5	11.4
Mean per observation	13.5	5.7

[a]Broad includes seed production, social science, post-harvest and other disciplines that are included in crop improvement. [b]Narrow is restricted to breeding, molecular biology, tissue culture, pathology and entomology.

summary in Chapter 3 of their edited volume which in turn relied heavily on the centre-by-commodity reports in Chapters 4–16. In Chapter 3, Evenson and Gollin (2003b) focused on two aspects of varietal release: the incidence of total releases over time and institutional source and utilization of materials in the development of improved varieties over time. For total releases for most crops, time referred to three decades: the 1970s, 1980s and 1990s. Institutional sources were aggregated into and described for three types of modern varieties: IARC crosses, NARS crosses with IARC ancestors and NARS crosses with NARS ancestors.

Varietal release over time: contents of the database

CG Center data sets on varietal release were available for assembly for nine crop/region categories: barley, beans, cassava, lentils, maize in ESA, maize in Western and Central Africa (WCA), potatoes, rice and wheat. A total of 82 crop-by-country observations were assembled for analysis. These were equivalent to 1393 cultivar-specific records. The vast majority of these had complete information on the minimum data set; a small minority had missing components.

The minimum data set for analysis consisted of crop-by-country combinations with historical information on varietal release. Four pieces of information were common to all CG Center databases and were assembled for each combination by cultivar: name of release, date of release, a qualitative description of in-country breeding level and effort, and a 0–1 indicator of the contribution of an IARC.

The period for analysis spanned the early 1960s to the late 1990s for most crops with a few country exceptions. The historical period was divided into three eras: the 1960s and early-to-mid 1970s; the late 1970s and the 1980s; and the 1990s. The first period prior to 1975 corresponds to output in the late colonial and early independence period. The middle period from 1975 to 1989 shows initial and developing progress from investing in food crop genetic improvement in NARS. The last period corresponds to results for NARS programmes at maturity.

Incidence of release over time

The summary data in Table 5.4 show that about as many varieties (611) were released in the 1990s ending in 1997 or 1998 as were in the late 1970s and 1980s (577). Hence, the annual release rate was roughly twice as high in the 1990s as in the previous 15-year period. With the exception of barley, the annual release rate increased for all crops, but the difference in the rate of release was not significant for most crops. Substantially higher release rates occurred in three crops: cassava, beans and maize in ESA.

Table 5.4. Number of varietal releases by time period and crop.

		Releases			
		Time period			
Crop	Number of country observations	The 1960s and early 1970s	Late 1970s and 1980s	The 1990s	Total
Barley	1	1	5	3	9
Beans	15	3	51	94	148
Cassava	18	9	84	114	207
Lentils	3	0	3	13	16
Maize ESA	12	33	86	167	286
Maize WCA	10	24	92	58	174
Potato	9	13	40	27	80
Rice	7	28	92	64	184
Wheat	8	95	129	74	298
Total	82	205	577	611	1393

For maize in ESA, releases are most likely understated in the earlier two periods prior to the 1990s (Hassan *et al.*, 2001). Given the importance of private-sector hybrids in Southern Africa in the late 1990s, release was equated to cultivars available for commercial production at that time. Improved varieties and hybrids that were released in the 1960s, 1970s and 1980s without seed for sale in the late 1990s could have been excluded from the release database. Moreover, the privatization of public-sector parastatals in the late 1980s and 1990s also could have contributed to losses in documentation of release lists (Hassan *et al.*, 2001).

With the possible exception of lentils in Lesotho, the data in Table 5.4 correspond to actual releases and do not include projections, which can result in a substantially expanded list. For example, projected rice releases in West Africa for 2000–2004 were projected at 122, equivalent to about two-thirds of varietal output in Table 5.4 (Dalton and Guei, 2003b). Many of these varieties were subsequently not released (Chapter 10, this volume).

About 15% of releases in Table 5.4 took place before 1975. Releases were particularly sparse for barley, beans, cassava and lentils, suggesting that research on these crops was not that salient prior to and immediately following independence or, in the case of cassava, potential research products were delayed in getting to farmers (Nweke, 2009). In contrast, a good number of releases for wheat, rice, maize and potatoes suggest that research efforts in genetic improvement were on-going in these commodities, albeit in a few countries, at the time of the beginning of the international agricultural research system in 1968. For example, Ethiopia with 16 released varieties, Kenya (22), South Africa (22), and Zimbabwe (23) were major contributors to wheat varietal output in this period. Many of these semi-dwarf varieties from the Rockefeller Foundation in Mexico were targeted for spring bread wheat.

Release profiles by crop

The crop-specific trajectories of annual releases are profiled in the dropline graphs presented in Figs 5.2–5.5. They illustrate both the absence of transparent trends and the presence of anomalies in many of the crop-release data sets. The graphs index the period from 1975 to 1997/98.

Wheat and rice

Aggregate data across all crop releases display an increasing trend in annual releases from the 1970s to the 1980s to the 1990s (Table 5.4). But neither wheat nor rice played a major role in contributing to this positive generalized trend (Fig. 5.2). Wheat is characterized by many releases relative to its production, area and economic importance in SSA. Even inclusion of 1991, a year of 0 releases, does not diminish the finding that wheat manifests the most stable behaviour in the incidence of release of any crop. A good number of varieties were released almost every year, indicating programmatic stability responding to technological change in spring bread wheat and demand for durable leaf-rust resistance.

From a detailed analysis of the 1997 wheat release database, Heisey and Lantican (2000) show that the number of releases of spring bread wheat is larger than their share of that cultivated in Ethiopia and South Africa where other types of wheat are produced. The number of spring-durum wheat varieties is smaller than their share in cultivated area in Ethiopia, which is the only country in SSA where durum wheat is produced. Releases of winter facultative wheat in South Africa is about what one would expect from a congruence benchmark. They also note the possible influence of private-sector participation in raising the incidence of varietal release in South Africa and Zimbawbe in the 1990s. Heisey and Lantican (2000) hypothesize that a fall in the release rate in Ethiopia in the late 1980s – no varieties were released in the five-year period between 1987 and 1993 – could be partially explained by the emergence of new rust races to which Ethiopian germplasm was susceptible. Government instability from the tyrannical Mengistu Regime is an alternative explanation for the drought in varietal releases in the late 1980s and early 1990s in Ethiopia, which is the largest producer of wheat in SSA aside from South Africa (Chapter 11, this volume).

In contrast, the release data for rice in the other panel of Fig. 5.2 display as much or more

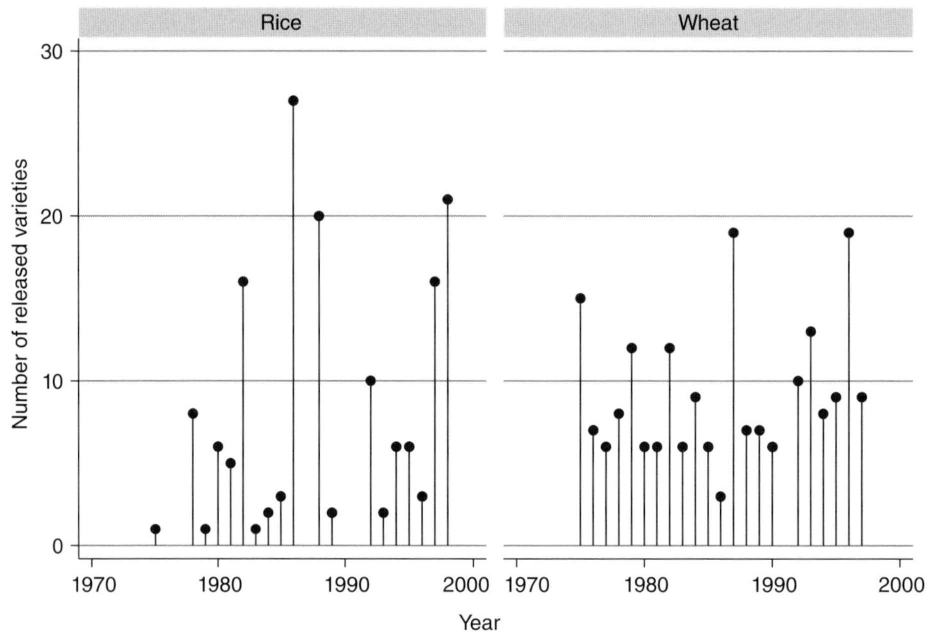

Fig. 5.2. Annual release profiles in rice and wheat from 1975 to 1997/98 in SSA.

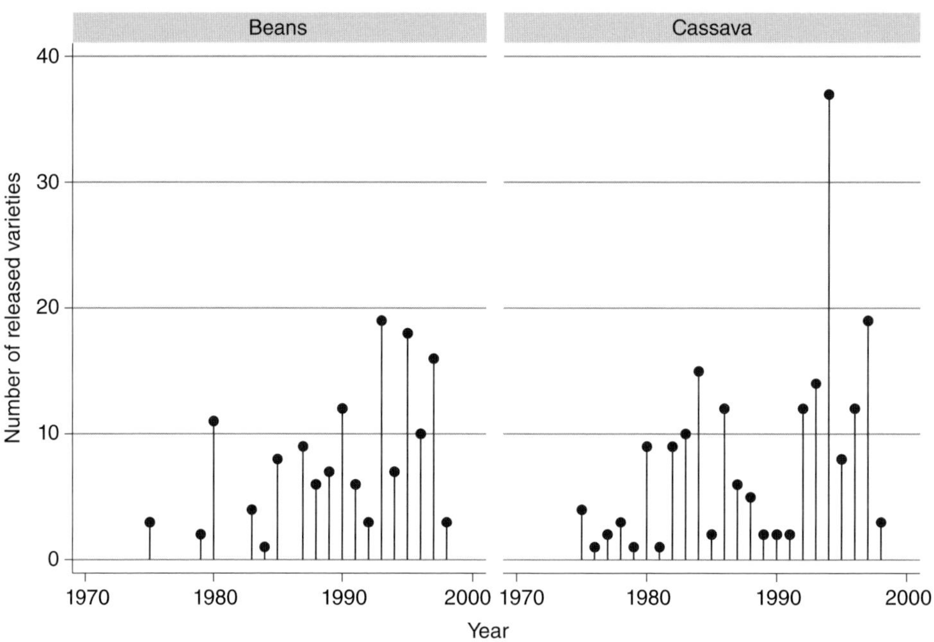

Fig. 5.3. Annual release profiles in beans and cassava from 1975 to 1997/98 in SSA.

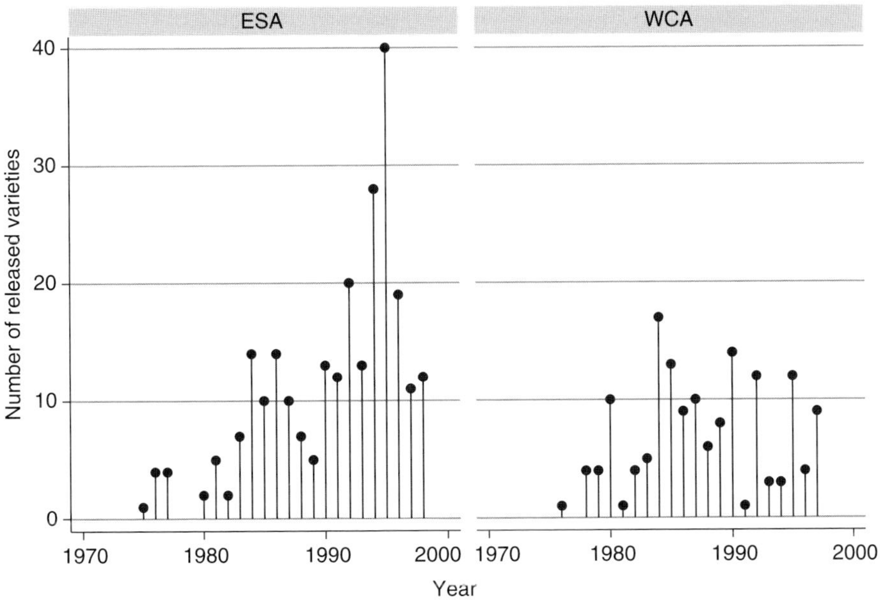

Fig. 5.4. Annual release profiles in maize in Eastern and Southern Africa and in West and Central Africa from 1975 to 1997/98.

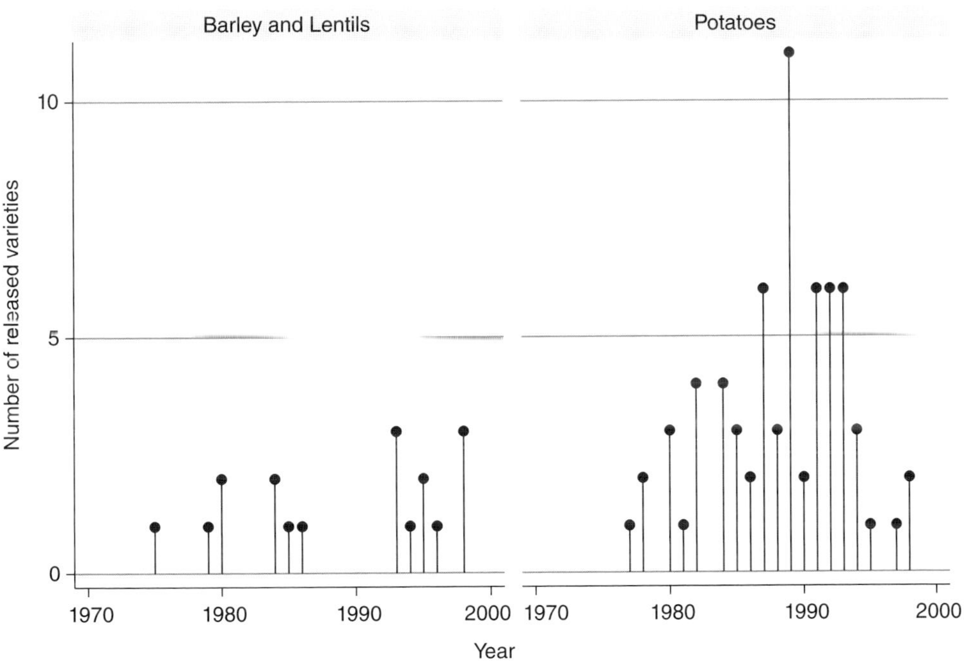

Fig. 5.5. Annual release profiles in barley and lentils and in potato from 1975 to 1997/98 in SSA.

variation than any crop in the data set. However, both crops had one release characteristic in common: a moderate-to-high incidence of release prior to the mid-1970s. Rice farmers in West Africa, in particular, benefitted from the work of several international, regional and national programmes before the CG System started to have a pronounced effect. Institut de Recherches Agronomiques Tropicales (IRAT) and later Centre de Coopération Internationale en Recherche Agronomique pour le Développement (CIRAD) have been very active in genetic improvement in West Africa since the 1960s and 1970s (Dalton and Guei, 2003b). The programmes of the National Cereals Research Institute (Nigeri; NCRI) involved in work on rice date from the early 1950s; the first official variety was released in 1954. In collaboration with the British, Sierra Leonean scientists had been working since 1934 in increasing regional rice production in the mangrove agroecology. The locus of their activities, which were curtailed in the 1990s because of civil war, was the Rokupr Rice Research Station. Several of the released rice varieties in the ROK series are widely adopted in Sierra Leone, Guinea and Guinea Bissau. They have been the subject of adoption studies and impact assessments (Adesina and Zinnah, 1993; Edwin and Masters, 1998).

Rice farmers in West Africa have benefited from the work of four CG Centers: the International Institute of Tropical Agriculture (IITA), the International Rice Research Institute (IRRI), WARDA (now AfricaRice), and, to a significantly lesser extent, Centro Internacional de Agricultura Tropical (CIAT). Large-scale bilateral assistance has also leveraged improved outcomes in varietal release. Most notably, the North Korean government has contributed to the development of the Kilissi agricultural research station in Guinea to serve the regional rainfed lowland production environment (Dalton and Guei, 2003b). The Guinean government released 12 CK-numbered varieties from this collaboration in the late 1980s and 1990s.

Returning to Fig. 5.2, several anomalies are apparent in the historical records for rice, which has a profile marked by annual releases exceeding 15 improved cultivars on four occasions. The spike in releases in 1982 could not be attributed to the release decisions taken by one country, but positive outliers in 1986, 1988 and 1998 were almost wholly the results of a single country's behaviour. Nigeria released 17 FARO-numbered varieties in 1986; Sierra Leone listed 18 varieties ranging from ROK 16 to ROK 33 in 1988; and Cote d'Ivoire placed 14 varieties on the release list targeted at the upland agroecology in 1998. Cote d'Ivoire released a total of 20 varieties in 1998. For both Sierra Leone and Cote d'Ivoire, these years of hectic release activity were sandwiched between periods of inactivity or negligible activity in listing varieties for release. Fruition of long-term UK collaboration in Sierra Leone largely explains the burst in varietal output in that country in 1988.

The existence of spikes in release activity in relatively small-producing countries is evident in all the crop release profiles. We return to the theme of feast or famine in release events later in this section.

Beans and cassava

Beans with 15 and cassava with 18 have the largest country contingents in the varietal release database. Their release profiles are markedly different (Fig. 5.3). Beans did not seem to have as many institutional building blocks for successful genetic improvement as other food crops in SSA in 1975. For example, bean research in Tanzania, SSA's second largest producer, only began in 1959 and focused mainly on introduced types and lines that were selected for export purposes in the 1960s (Hillocks *et al.*, 2006).

The rate of varietal release in *Phaseolus* oscillated around a low-level equilibrium in the 1970s and early 1980s and then took off in the late 1980s and kept rising in the 1990s (Fig. 5.3). The initial impact of improved varieties was most felt in climbing beans in Rwanda in the late 1980s and early 1990s (Pachico, 2014). The diffusion of improved varieties of bush beans that dominate production in SSA occurred later. If release is equated to output, the data suggest that beans were arguably as productive as any of the study crops in the 1990s. This impressive quantitative performance was accomplished in the time of the Rwanda Genocide, which resulted in a hiatus in varietal releases for 5 years. The founding of the Pan-African Bean Research Alliance (PABRA) in 1996 most likely had a lot to do with the positive developments and the maintenance of an upward trend in varietal release for beans in 1997 and 1998.

Like beans, cassava releases started to increase in the mid-1980s from a very small base in the late 1970s (second panel, Fig. 5.3). Unlike beans, cassava breeders had over 35 years of genetic research to draw on in their quest to find disease resistance in good agronomic and market backgrounds (Nweke, 2009).

Important sources of mosaic-disease resistance were combined successfully at the regional Amani Research Station for East Africa in Tanzania.[3] Further work in Nigeria at the Moor Plantation research station in the late 1950s added to the foundation that culminated in the release of IITA's Tropical Manihot Selection (TMS) high-yielding, mosaic-resistant clones in the mid-1970s. The earliest TMS named clone in the data set was released by Nigeria in 1976. The Biafran War in Nigeria in 1964 and the slow response of national governments to invest in cassava research following independence were cited by Nweke (2009) as major contributors to meagre varietal output in the 1960s and early 1970s.

The release profile for cassava belies the message of an upward trend over time but it also conveys a glaring anomaly in the data set. In 1994, Chad released 27 improved cassava varieties. About half of these releases were selected from IITA materials. Aside from 1994, Chad only released two other varieties during the period of study from 1975 to 1997/98.[4]

Maize in East and Southern Africa and in West and Central Africa

The maize variety SR52 was the third oldest release in maize database in both regions.[5] It was a landmark cultivar with widespread practical and institutional impact in Southern Africa (Eicher, 1995). Maize genetic research began in 1932 in Zimbabwe. Double-cross hybrids were released in the late 1940s. SR52, an innovative triple-cross hybrid, was clearly worth the wait. It demonstrated the advantage of a widely adaptable hybrid even in the drier production conditions of low soil fertility. SR52 was a stable-yielding, long-duration hybrid. It was preceded by the replacement of low-yielding but preferred flint types by higher-yielding, softer dent types in the 1920s and was superseded by shorter-season hybrids adapted to small-farmer conditions in the 1980s and 1990s (Howard *et al.*, 2000).

The maize release profile for ESA in the first panel of Fig. 5.4 contains two anomalies, one implicit and the other explicit. Because release was defined as varietal availability in the late 1990s, varietal output in the 1970s and early 1980s is likely to be underestimated. For example, South Africa only debuted in the database in 1981 and had 11 entries in the 1980s compared to more than 50 in the 1990s. In terms of the size of the underestimate, South Africa appears to be the extreme case.

The more visible outlier pertains to the 40 releases in the region in 1995. Several countries contributed to this large varietal output. Malawi released 12 of its 20 entries in the database in 1995. Zambia officially put out 10 varieties in 1995, about equally shared between the public and private sectors.

In contrast to ESA, the profile of maize research in WCA was shallow in the first half of the 20th century. The majority of the 24 releases prior to 1975 originated in Nigeria and the D.R. Congo; however, 22 of the 24 occurred in early 1970s. Therefore, a colonial research legacy may not have directly translated into materials for release in the early post-independence period. The Belgians established a network of research sites in the Congo targeted not only at export crops but also at important staples, such as maize, in the 1930s (Roseboom *et al.*, 1998). In Nigeria, screening and breeding for rust resistance began in the 1950s and later that work was expanded to maize streak and downy mildew which were becoming increasingly important over time (Iken and Amusa, 2004). The initial focus on rust resistance may have been a blind alley and might partially explain why releases were so few and far between in the 1960s.

A second contrast between the two regions in the late 1990s concerns the extent of hybridization of released materials. In WCA, only about 35 of 185 releases in the database were hybrids; in ESA, only about 65 of 300 releases were reported to be open-pollinated varieties.

Unlike the release profiles for the other crops, maize in WCA was not characterized by highly visible peculiarities. Most countries showed a fairly continuous release history throughout the study period, although many were characterized by release gaps extending 10–12 years. In this regard, Nigeria was the main outlier in the database. No releases were recorded from 1986 to 1997.

Potato, barley and lentils

Varietal change in potatoes in the last quarter of the 20th century in SSA, while not blessed with as rich an institutional milieu as rice, was enhanced by multiple types of collaboration that have affected release outcomes. In the early 1970s, three late-blight resistant varieties, at the time recently released from Mexico, were imported into Uganda and Kenya. Although these varieties never laid claim to much area in Mexico, they quickly became popular in the East African highlands. Before the 1994 Genocide in Rwanda, Sangema was the dominant variety in Rwanda and was arguably the most economically important variety in the region in the 1970s and early 1980s. Even today, Rosita, a synonym for Sangema, is the prevailing potato variety in Malawi and Mozambique.

The late blight resistance in these varieties is attributed to their development in the Toluca Valley, the confirmed centre of origin of this disease that is prevalent throughout the world in rainy-season potato production. They were developed under the leadership of John Niederhauser, the winner of the 1990 World Food Prize, with his Mexican colleagues.

Special project collaboration resulted in two varieties bred and officially released in Kenya in the early 1970s. Those varieties were selected for high partial resistance to late blight and are one of the few examples of NARS-bred varieties in potato improvement in the region.[6]

For potato, the tailing off of releases in the second panel of Fig. 5.5 in the mid- and late 1990s warrants comment. The 1994 Genocide in Rwanda largely explains this downturn in the incidence of release. Rwanda was an important hub for CIP's (the International Potato Center's) collaboration with NARS in genetic improvement. That collaboration was strengthened from the mid-1970s and early 1980s with a United States Agency for International Development (USAID) special project that resulted in several releases in Rwanda with spillovers to the D.R. Congo and Burundi. Several scientists at the Rwandan Institute for Agricultural Sciences' (ISAR's) potato-research station in Ruhengeri perished in the Genocide. One regionally important scientist died later because of it. The national programme was a shadow of its former self in the mid- and late 1990s.

Barley and lentils account for four crop-by-country observations in the data set. Ethiopia is by far the largest producer of both crops in SSA. The more recent activity in the first panel of Fig. 5.5 is attributed mostly to releases of improved lentil varieties. Barley, an ancient crop in Ethiopia, has a more continuous and older release record that started in 1973. Most releases are landrace selections from thousands of contenders in the Ethiopian Agricultural Research Organization's (EARO's) diverse germplasm bank. In 1984, an improved hulled variety, HB 42, was selected with the crossing of Ethiopian and introduced materials (Bekele *et al.*, 2005). It is one of the earliest bred varieties in the database.

Release intensities

Several measures of release intensities are presented in Table 5.5. On average across the main seven crop descriptions, the mean number of varieties per country programme ranged from slightly less than 10 in potatoes and beans to more than 35 in wheat during the entire period of analysis. Similar to estimates of researcher intensity per million metric tonnes, smaller-producing countries are usually characterized by more releases relative to their size than larger-producing countries. Therefore, the simple mean of varieties released per country programme is not a very informative measure of farmers' potential access to improved varieties.

Weighting by production in the second column of Table 5.5 provides a more representative measure of change potential from improved varieties. Although smaller countries may release more varieties per unit area, they do not usually release more improved genotypes than larger-producing countries. All the weighted means are larger than the simple means in Table 5.5, suggesting that size of production was positively associated with the incidence of release. The simple correlation between production in 1998 and the sum of historical releases was positive for six of the seven crop entries in Table 5.5 and was pronounced for wheat. Cassava was the only case of a weakly inverse relationship between production and release incidence in the database. That larger cassava-producing countries did not produce significantly more varieties than smaller cassava-producing countries is a surprising finding.

Table 5.5. Varietal rates and intensities by crop from the 1960s to the late 1990s.

Crop	Mean varieties released per country programme	Weighted mean varieties by country production	Mean annual release rate[a]	Release intensity (million tonnes)		Release intensity (million dollars)[b]	Releases per scientist (Scientist year)[c]	
				Simple	Weighted		Simple	Weighted
Beans	9.9	13.1	0.40	140	70.52	0.12	4.29	3.48
Cassava	11.5	12.9	0.44	21	3.28	0.04	1.17	0.94
Maize ESA	23.8	37.9	0.88	42	13.12	0.05	1.25	0.83
Maize WCA	17.4	23.3	0.65	76	17.10	0.07	0.63	0.34
Potato	8.9	9.7	0.34	121	38.64	0.19	0.73	0.67
Rice	26.3	34.9	0.92	55	29.53	0.12	2.69	2.66
Wheat	37.3	65.2	1.11	142	54.42	0.22	0.97	1.10
Average	17.4	28.14	0.63	80	12.38	0.12	1.66	1.01

[a]From 1974 to 1997/97. [b]Assuming prices of US$250 for maize, wheat and rice; US$600 for beans; US$200 for potato; and US$80 for cassava per metric tonne. [c]Refers to the 1990s.

Release rates per year in Table 5.5 mirror the total release data in Table 5.4. National programmes in maize ESA, rice and wheat approached or exceeded a rate of one variety released per annum. Smaller crops such as beans and potatoes were considerably below half the rate of those cereals. Cassava also fell considerably short of 50% of that rate.

Cassava as an exception is a theme that weaves its way throughout the columns of Table 5.5. Unweighted (simple) mean intensity estimates in column 5 are inflated by very high values from small countries. The weighted mean estimates in the next column convey a better sense of the potential richness of improved varieties per unit of production. Cassava with only an average of about three varieties per million tonnes of production is several orders of magnitude less than the other crops. Beans, wheat, rice and potatoes score well on this criterion.

From a normative perspective, equalization of release intensity ratios in terms of value of production would seem to be a suitable initial target for which to strive in national programmes in crop genetic improvement. The rough calculation in the 7th column of Table 5.5 shows that the lower output price of cassava does not lead to a change in its position; it still ranks last in release intensity per million dollars of value of production. Nevertheless, differences in release intensity narrow when a normative standard of economic congruence is used relative to a measure from the perspective of production.

The estimated release intensity per million dollars in value of production is also low for maize relative to wheat, potatoes, beans and rice. This gap in release intensity suggests that the release experience in the widespread staple food crops of cassava and maize in SSA has not been commensurate to their economic importance relative to the other four crops in the study. Apparently, the rapidly increasing participation of the private sector in ESA was not sufficient to offset this general tendency.

Estimates on releases per scientist in the last two columns of Table 5.5 suggest that the average varietal output of cassava scientists is roughly the same as that of scientists for the other crops. Therefore, an under-investment in cassava relative to the other crops seems to be the main factor conditioning the low release intensities in columns 5, 6 and 7 of Table 5.5.

The high estimate for rice is not attributed to definitional differences in the data. The aforementioned rich institutional mix impacted positively on high varietal release rates in specific countries. Many of the rice programmes were also 'bare bones'. For example, the Nigerian programme, the largest-producing country in the region, only had slightly more than five FTE scientists in the late 1990s (Dalton and Guei, 2003b).

Instability in varietal release

Stability in breeding effort is one of the most important criteria for success for crop genetic improvement at both the national and international levels. It is not expected that improved varieties will be released annually; it is expected that improved varieties will be released routinely over time in a productive crop improvement programme.

Instability is measured by three indices in Table 5.6 for the period 1975 to 1997/98, which spans 23–24 years for the majority of the crops. Components of instability are estimated with number of years of zero releases in that time span, the standard deviation of annual releases, and the coefficient of variation of annual releases.

Both the mean estimates of years with zero releases and the mean coefficients of variation tended to be larger than expected. Cassava and beans scored poorly on these two instability

Table 5.6. Instability in varietal release by index and crop from 1975 to 1997/98.

	Indices of instability		
Crop	Number of years with zero releases	Standard deviation of annual releases	Coefficent of variation of annual releases (%)
Beans	20.5	1.1	311.0
Cassava	21.7	1.4	343.0
Maize ESA	15.4	1.4	227.1
Maize WCA	15.9	1.3	201.4
Potato	18.6	0.9	277.4
Rice	17.9	2.4	256.7
Wheat	12.8	1.3	155.3
Average	17.5	1.4	253.1

indicators mainly because many small-producing countries are marginally involved in crop genetic improvement. On average, bean programmes only released varieties in about 5 years in 25 for those countries with data until 1999.

In contrast, for wheat, a legacy of releases prior to 1975 set the stage for a more stable release rate in several countries. On average, release during a year was a more typical event than non-release. Relative to their mean release levels, releases from cassava programmes fluctuated more than any other commodity grouping in Table 5.6.

The relatively high reading for the estimated standard deviation for rice is attributed to the four-country very large release events that were described earlier in the rice release profile. The low reading for potato indicates that no country released more than 3–5 varieties in any given year (with the exception of Sudan, a very small producer, in 1989).

Breeding effort and IARC contribution to varietal release

The other elements of the minimal varietal release data set pertain to breeding effort and IARC contribution. Mean estimates of those characteristics are presented in Table 5.7. The scale for breeding level varies from 1 (selection of landraces/elite introduced material), to 2 (selection from progenies and populations from IARC crosses) and to 3 (NARS-bred varieties based on NARS crosses). IARC contribution is a 0–1 variable with a positive response synonymous with IARC networked finished cultivars, selections from IARC populations, and selections from crosses made with IARC parents and ancestors.

Table 5.7. Mean estimates of characteristics of releases by aspect and crop.

Crop	Aspect	
	Breeding level	IARC content
Beans	1.33	0.58
Cassava	1.13	0.83
Maize ESA	2.70	0.23
Maize WCA	1.35	0.50
Potato	1.44	0.68
Rice	2.16	0.47
Wheat	2.42	0.60
Average	1.79	0.56

As expected, breeding effort divides the commodities into two groups. Varieties in rice, wheat and maize in ESA are mainly characterized by mean values that fall in the interval between NARS selection and NARs crossing.

For potatoes, cassava and beans, NARS-bred varieties were rare outcomes in the database occurring in only a few cases. The estimates for maize in WCA seem surprisingly low. Although hybrids were beginning to be produced in WCA, many varieties were released from relatively finished introduced and landrace materials during the study period.

On average, the CG Centers contributed to over half of the approximately 1400 varieties in the database. Proportionally, that contribution was highest in cassava, suggesting a dearth of alternative suppliers, and was lowest in maize ESA where greater private sector participation potentially increases the role of alternative suppliers and also makes it harder to determine institutional attribution. Excluding temperate South Africa from the maize ESA database leads to an increase in IARC contribution from 0.23 to 0.37, a substantial rise that is not sufficient to change the commodity rankings in this category in Table 5.7.

Over time, the mean rate of IARC participation rose from a low of about a third of total releases prior to 1975 to a high of about three-fifths of varietal releases in the late 1990s (Table 5.8). Most crops experienced a positive upward movement in the IARC content of varietal releases except for cassava and maize in WCA that attained high rates between 1975 and 1989. For both crop groupings, those rates plateaued in the 1990s.

Unlike their counterparts in Table 5.8, the estimates on breeding level in Table 5.9 do not show any trend over time. This counterintuitive finding seems to contradict the results presented in Evenson and Gollin (2003a). Beans are the only commodity for which one can make a plausible case that mean estimated breeding level is increasing over time. This rejection of seemingly confirmed wisdom from Evenson and Gollin (2003a) is discussed later in this chapter.

Table 5.8. Mean values of IARC content over time by crop with the number of released varieties in each time period.

Crop	Before 1975	Between 1975 and 1989	From 1990 to 1997/98	From 1960 to 1997/98
Barley	1	0.6	0	0.44
	1	5	3	9
Beans	0.33	0.37	0.71	0.59
	3	51	94	148
Cassava	0.22	0.80	0.69	0.71
	9	75	132	216
Lentils	Not defined	0	1	0.73
Maize ESA[a]	0	0.20	0.38	0.26
	30	60	86	176
Maize WCA	0.42	0.63	0.65	0.61
	24	92	72	188
Potatoes	0	0.6	1	0.6
	13	40	27	80
Rice	0.27	0.37	0.59	0.43
	33	92	71	196
Wheat	0.44	0.69	0.75	0.63
	96	127	73	296
Weighted mean	0.31	0.55	0.60	0.65
Observations (#)	209	545	566	1320

[a]Excluding South Africa.
The number of observations appears below the mean estimates for each period.

Table 5.9. Mean estimates of breeding level over time by crop with the number of released varieties in each time period.[a]

Crop	Before 1975	1975 to 1989	The 1990s	Total
Barley	1	2.2	1	1.67
	1	5	3	9
Beans	1	1.2	1.6	1.4
	3	44	73	130
Cassava	1.6	1.5	1.4	1.5
	9	82	149	207
Lentils	na	na	1	1.00
	0	0	8	8
Maize WCA	1.3	1.7	1.5	1.6
	10	58	40	106
Potatoes	1.2	1.3	1.8	1.4
	13	40	25	80
Rice	2.1	2.3	1.9	2.2
	33	92	71	189
Wheat	1.9	2.6	2.7	2.4
	96	128	67	296
Total	1.8	1.9	1.7	1.9
	165	449	436	1025

[a]The scale for breeding level is 1 = limited selection and direct use of finished material; 2 = intermediate effort equivalent to progeny selection from introduced population; and 3 = considerable effort equivalent to crossing and selection in a mature breeding programme.
The number of observations appears below the mean estimates for each period. na, not applicable.

Exploratory regression analysis

The 1998 release database is not easily analysed with a framework of multiple regression. The dependent variable is number of releases, which comes from a time series. The independent variables are cross-sectional, estimated in the late 1990s or are taken from secondary data such as national production. Unfortunately, there is no truly exogenous variable in the data set because almost everything depends on everything else. For starters, the quantity and quality of releases potentially contributes to national production.

The selective results that follow should be thought of as associations in the ceteris paribus context of multiple regression analysis. The core model tests for additive effects in crop groupings over time in the context of a country database of about 85 potential observations. Country-specific effects are estimated from a variety database of 1400 observations. Most of the results confirm differences noted in Tables 5.5–5.7.

Carry-over effects of varietal release from the first period

The data in Table 5.10 raise the issue of whether a 'head-start' for some countries and crops in the first period also translated into increased releases in the second and the third periods. The number of releases in both later periods separately was regressed on the number of releases in the first period and on crop-specific dummy variables in Table 5.10. Potato is arbitrarily assumed to be the reference point from which the effects of other crops are measured. The estimates are weighted by the size of country production in 1998.

The estimated coefficients for the second period (1975–1989) show that there was a significant carry-over effect from period 1 to period 2 (Table 5.10). The estimated effect was not that large a proportional 1% increase in varieties released in the first period was associated with a third of 1% increase in the second period. By the 1990s, i.e. the third period, the carry-over effect of the first period had vanished. Indeed, the estimated coefficient on the early period releases is negative.

Although the adjusted coefficient of determination dropped from 0.50 to 0.34 with the same specification for the two periods, estimated commodity differences were sharper in the second period. Shifting from the base level of potato with about five varieties released per programme to wheat was accompanied by an increase of 29 varieties per programme. Releases in maize in ESA also showed a healthy increase over potato's performance in the 1990s that was adversely affected by the Genocide in Rwanda in 1994.

Overall, a finding of carry-over effects into the second period shows the importance of starting early in as time-intensive a process as is crop genetic improvement. More importantly, the insignificant effect in the third period is interesting because it suggests that initial advantages were not maintained into the 1990s. Improved materials were more freely accessible internationally because the CGIAR probably played an important role in equalizing release potential or at least enhancing the capability of smaller, weaker partners in the playing field.

Table 5.10. Carry over effects of early period releases to later periods.

	Periods	
Independent variables[a]	From 1975 to 1989	The 1990s
Releases before 1974	0.946	−0.802
	(3.75)**	(1.49)
Beans	2.806	2.854
	(0.44)	(0.21)
Cassava	4.924	0.537
	(1.07)	(0.05)
Maize ESA	4.509	23.501
	(0.96)	(2.34)*
Maize WCA	7.463	2.481
	(1.49)	(0.23)
Rice	9.843	11.006
	(1.86)	(0.97)
Wheat	9.218	29.100
	(1.42)	(2.10)*
Constant	1.824	5.163
	(0.40)	(0.53)
Observations	78	78
R-squared	0.54	0.40

Absolute value of t statistics in parentheses.*significant at 5%; **significant at 1%. [a]The omitted crop is potatoes.

Multiple associations with the incidence of release

Several associations with levels of varietal release are expected in the data set. First, the crop-specific differences discussed in Tables 5.5 and 5.6 should favour enhanced releases from wheat, rice and perhaps maize in ESA. Secondly, size of country production should be positively correlated with the incidence of release. Thirdly, breeding level is also expected to be positively correlated with release because programmes with greater capacity should be in a position to release more varieties. Fourthly, IARC content could be negatively associated if IARC materials are disproportionately used by smaller programmes and if larger programmes have greater capacity to exploit different institutional sources of materials. Fifthly, the number of scientists should be positively associated with releases, although a single year's observation at the end of the period may be a poor proxy for scientific capacity during the whole period. Lastly, private-sector participation should be accompanied by more releases; however, private-sector participation is mainly concentrated in maize in Southern Africa in the current data set and is very covariate with the binary variable maize ESA.

Most of these expectations are confirmed in the regression equation that is specified in Table 5.11. Larger-producing countries are likely to release more varieties than smaller-producing countries. Differences among crops are statistically significant. Wheat is characterized by almost 28 more releases than beans, the base crop in Table 5.11. Private sector investment in breeding seems to be a positive force for varietal release. Increases in breeding level are associated with substantially more releases.

The estimated coefficient on IARC content is signed negatively and is statistically significant. In other words, higher IARC content is a marker for countries that are less likely to release improved varieties, everything else equal. For example, South Africa is the country with the highest number of varietal releases and with one of the lowest estimates of IARC content largely because of the prevalence of temperate maize and winter facultative wheat, which are atypical of maize and wheat production in the rest of SSA.

The South African observations substantially influence the estimated outcomes in Table 5.11. The results of re-estimating the equation without the South African observations is given in the second column of estimated coefficients in Table 5.11. The coefficient of determination drops from 0.83 to 0.69 and the size of several of the coefficients also declines substantially. However, all the estimated coefficients retain their statistical significance; hence, the negative and significant coefficient for IARC content is not driven solely by the presence of South Africa in the data set. This negative estimate does not mean that IARC activity results in fewer released varieties but rather that countries with proportionally more IARC content in their varietal releases have less capacity than others to release varieties. This nuanced interpretation points to the equalizing role of IARC activity in levelling the playing field from the perspective of varietal output.

Table 5.11. Multiple correlates of total releases.

Independent variables[a]	Observations: All	Observations: Without South Africa
Production share	0.337	0.210
	(6.20)**	(4.79)**
Cassava	1.275	0.288
	(0.17)	(0.05)
Maize ESA	−8.704	−4.182
	(1.03)	(0.67)
Maize WCA	−5.634	−0.930
	(0.76)	(0.15)
Potato	−4.714	−3.959
	(0.45)	(0.52)
Rice	−0.262	9.998
	(0.03)	(1.49)
Wheat	28.042	19.724
	(3.05)**	(2.70)**
IARC-related	−17.568	−11.604
	(3.68)**	(3.22)**
Private breeding	13.644	10.434
	(3.12)**	(3.17)**
Breeding level	16.359	7.644
	(6.51)**	(3.25)**
Constant	−2.385	7.298
	(0.27)	(1.11)
Observations	78	75
R-squared	0.83	0.69

Absolute value of t statistics in parentheses.*Significant at 5%; **significant at 1%. [a]The omitted crop is beans.

Multiple correlates of instability

The description of multiple correlates of instability in Table 5.12 also agrees closely with the crop-wise tabular estimates presented in Table 5.6. The dependent variable in this multiple regression analysis of correlates is the number of years of zero releases. Positive estimated coefficients imply increasing instability; variables with negatively signed coefficients imply increasing stability. Wheat was characterized by an annual release pattern that was more stable than that of any other crop. This behaviour is confirmed in Table 5.12 because switching from the base crop potato to wheat is associated with a gain in 9.5 years of positive varietal releases. None of the other crops' estimated coefficients were significantly different from potato's record on the number of zero-release years, which averaged 19.5.

FTE scientist years were also included as a regressor in the estimated specification reported in Table 5.12. Scientific strength in number of scientists was negatively and significantly associated with instability in the release pattern over time. But the size of the estimated coefficient is relatively small. Adding an additional scientist reduced the level of instability by only 0.14 years; however, this result, hinting at an insulating influence of scientific strength on instability in varietal output, is reassuring.

Table 5.12. Multiple correlates of instability: number of years of zero releases.

Independent variables (production/crop/country)	Estimated coefficients (t values)
Beans	0.819 (0.27)
Cassava	2.688 (1.27)
Maize WCA	−0.698 (0.30)
Maize ESA	−1.450 (0.63)
Rice	−1.021 (0.76)
Wheat	−9.471 (3.83)**
Scientist years	−0.144 (6.09)**
Constant	19.517 (9.35)**
Observations	72
R-squared	0.79

Absolute value of t statistics in parentheses. The omitted crop is potato.
*Significant at 5%; **significant at 1%.

Analysis of Modern Variety Adoption

In spite of the aforementioned variation in the CG Center databases on the adoption of improved varieties, some findings can be teased out from an analysis of the available data that were cultivar-specific for wheat, potato, rice and maize in ESA. This descriptive analysis is restricted to tabular aggregates for the other eight potential commodity groupings. Four thematic areas are covered in this section: (i) the level of modern variety adoption; (ii) determinants and correlates of modern variety adoption; (iii) turnover in the use of modern varieties; and (iv) spillovers of modern varieties from one country to another.

The level of adoption for modern varieties

Three estimates for the adoption of modern varieties are given in Table 5.13 for the 105 commodity-by-country observations in the 1998 data set. Modern cultivars were pervasive in the production of wheat and, to a lesser extent, in potato and rice. Their use was also quite common in maize in both ESA and WCA, accounting for more than one-third of area in each sub region. At the other extreme, the use of improved cultivars was negligible in lentils and rare in barley.

The adoption level for cassava at 22% seems unfavourable, but it is an impressive showing for a commodity that is vegetatively propagated and characterized by low multiplication ratios and release intensities. In contrast, the penetration of modern varieties into sorghum, pearl millet, bean and groundnut fields was less than expected.

IARC-related materials were heavily felt in the composition of modern varieties adopted. Use of IARC-related materials was pronounced

Table 5.13. Inferences on improved varietal adoption by crop in the late 1990s.

Commodity	Improved cultivars (%)	IARC-related materials (%)	Coverage (%)	Improved cultivars: conservative assumptions (%)
Wheat	66	57	85	56
Potato	56	41	68	44
Rice	45	21	57	26
Maize WCA	37	19	94	35
Maize ESA	36	13	90	34
Cassava	22	18	83	18
Sorghum	23	11	54	13
Beans[a]	?	15	67	10
Barley	11	0	90	10
Groundnut	30	4	6	2
Pearl millet	19	19	10	2
Lentils	0	0	80	0

[a]IARC only.

in wheat and cassava where more than 8 of every 10 hectares planted to modern materials were related to genetic inputs from the CG Centers. IARC material intensity in adopted area was also probably high in beans, but data on non-IARC improved cultivar adoption were not presented in Evenson and Gollin (2003a) and a cultivar-specific database was not available.

The adoption estimates in Table 5.13 are reasonably accurate for SSA as a whole if the country observations comprise a large share of commodity area and production. High levels of coverage were achieved in wheat, maize, cassava, barley and lentils. Coverage in groundnut and pearl millet was unsatisfactory. The other commodities ranked somewhere between these two extremes. Under the conservative but plausible assumption that countries not included in the data set had very low or negligible levels of improved varieties, a threshold minimal level of adoption is estimated in the last column of Table 5.13. This lower-bound estimate conveys the notion that modern variety adoption was at least this level in the late 1990s.

In comparing the two adoption estimates in columns 2 and 5, we arrive at an adoption interval that should contain the actual level of adoption. With the exception of wheat, it is unlikely that omitted producing countries were characterized (on average) by higher levels of adoption than those found in column 2. Therefore, the estimates in column 2 are interpreted as upper bounds and those in column 5 reflect lower bounds for the adoption interval.

The span of the adoption interval is less than 10% in absolute terms for wheat, maize and cassava. This tight adoption interval imparts confidence to the estimates of modern cultivar adoption in the 1998 data set for those three crops. At the other extreme, the adoption estimate for modern groundnut cultivars could have been as low as 2% or as high as 30%. Rice also is characterized by a wide range in Table 5.13.

Correlates of modern variety adoption

The multiple correlates of modern variety adoption are examined in Table 5.14 with the country-level data set of 85 potential observations. Per cent adoption of modern varieties as a group for each crop-by-country observation is regressed on crop binary variables, varietal output in the form of releases in the late 1990s, the country of Ethiopia, substantial private-sector participation in the provision of seed, and the level of IARC content of released varieties.

From the discussion of Table 5.13, we expect that wheat and potato would have significantly higher adoption rates than the other commodities. That expectation is confirmed in Table 5.14

Table 5.14. Multiple correlates of adoption of modern varieties.

Independent variables[a]	Estimated coefficients (t values)
Post-1989 releases	1.198
	(7.01)**
Ethiopia	−23.941
	(2.60)*
Private distribution	19.069
	(2.23)*
Beans	−60.634
	(3.20)**
Cassava	−56.526
	(5.74)**
Maize ESA	−62.546
	(5.22)**
Maize WCA	−48.707
	(3.75)**
Potato	4.723
	(0.30)
Rice	−29.731
	(2.49)*
Constant	70.644
	(7.11)**
Observations	68
R-squared	0.78

Absolute value of t statistics in parentheses; *significant at 5%; **significant at 1%.
[a]The omitted crop is wheat.

where the base crop is wheat with a 70% adoption rate. The average uptake of improved varieties in beans, cassava, maize and rice is significantly lower than this estimated diffusion rate. Shifting from wheat to potato is accompanied by about an insignificant 5% increase in the rate of adoption of modern varieties.

Perhaps the most interesting outcome in Table 5.14 is the positive and significant association between varietal output and the late 1990s adoption level of improved varieties. Ceteris paribus, an additional released variety in the 1990s was associated with a 1.2% rise in the level of improved cultivar adoption. The size and statistical significance of this effect is stronger than anticipated. It is reassuring to find that varietal output is positively correlated with adoption, although country and crop observations with higher varietal output are likely to have invested more in public-sector extension, which can play an important role in varietal adoption. In any case, this positive finding helps to justify the DIIVA Project's emphasis on monitoring varietal output.

Ethiopia was included as a regressor because of its agroecological and genetic uniqueness. Ethiopia contains agroecologies, such as Highland Vertisols, that are not found elsewhere in SSA. It is also the primary or secondary centre of diversity of several food crop species. Making genetic progress in the centre of diversity can be especially challenging. For example, most durum wheat is grown in Ethiopia and it is also home to the widest collection of local potato varieties cultivated in SSA. For these reasons, we expect that the estimated coefficient on Ethiopia is signed negatively in Table 5.14 and that is indeed the case. At mean levels of the observations, shifting the location of wheat production to Ethiopia is associated with a 25% fall in modern variety adoption level to 45%.

In the late 1990s database, CIMMYT and IITA reported significant private-sector participation in the distribution of hybrids for about ten maize-producing countries. Identifying those countries with private-sector (often multinational) participation accounts for some of the variation of adoption level. Switching to maize in one of the countries endowed with private-sector distribution is associated with a 19% increase in the adoption level of modern varieties.

It was difficult to make the case for including other variables from the data set in the regression equation because most of these appeared in earlier regressions of varietal releases. As an exception to this generalization, the IARC content of improved varieties was included to test the hypothesis that adoption intensity was positively correlated with IARC content. The estimated coefficient on IARC content, which is not presented in Table 5.14, was positive but small and insignificant. This hypothesis should be accepted or rejected at the level of the crop, which should provide a richer, more contextual interpretation than that furnished by an aggregated pooled analysis.

Similar to the regression equation for varietal release, the changes brought about by dropping the maize, wheat and bean observations from South Africa were also evaluated. Explained variation declined but the level of statistical significance of the estimated coefficients was unchanged.

Turnover of modern varieties

Although SR52 was cultivated in over 95% of maize area in Zimbabwe in the late 1960s, researchers were assessing impacts from later varietal change in hybrids in the 1980s and 1990s (Bourdillon *et al.*, 2003). No sales of SR52 were registered in the 1990s database. High rates of varietal turnover are desirable and are associated with productivity in crop improvement. So-called first-generation modern varieties can be difficult to replace if more recent releases are not superior in satisfying farmers' changing demand for characteristics.

Cultivar-specific data in wheat, rice and potato opened up the opportunity to estimate measures of varietal turnover. Sufficient data were also available for maize in ESA to estimate turnover rates for all countries, except South Africa.[7]

The turnover measure that is commonly used is weighted average age of modern cultivars in farmers' fields from their date of release. Weights are derived from cultivar-specific data on area adopted. Turnover measures are more informative as the adoption level of modern varieties rises. For that reason, we also estimate a second weighted average that encompasses all cropped area. Area planted in local varieties is arbitrarily assigned an age of 50 years in that measure.

Heisey and Lantican (2000) estimated weighted average varietal age for improved wheat varieties from the 1998 data set for seven countries in SSA. Estimated age varied from a low of 2.5 in Zimbabwe to 15.9 in Ethiopia. Six of the seven had estimates that fell in range of 10–16 years. Those estimates were roughly the same for comparable data in 1990, indicating stagnant progress in varietal turnover. In Zimbabwe, varietal turnover was very rapid because a small number of large, homogeneous wheat producers were effective in communicating and realizing their demands for varietal change. Elsewhere, Heisey and Lantican felt that moderately high varietal age reflected the slow diffusion of varieties in farmers' fields.

Ten years seems like a realistic target to aim for in a cereal like wheat with a high multiplication ratio; however, few country-by-crop observations on average satisfy this target in Table 5.15. Surprisingly, potato, the crop with the lowest multiplication ratio, was the commodity with the highest varietal turnover.

Weighted average age of only 10 years in improved potato varieties is a truly impressive and perhaps idiosyncratic performance. Once the assumed age of local varieties is factored into the calculation, the difference between wheat and the other three crops in varietal adoption becomes apparent in the second weighted age comparison in Table 5.15. In particular, maize and potato are characterized by a bimodal distribution of adoption. For some countries, adoption of improved materials is low to negligible; for others, diffusion approaches full adoption.

Several older improved varieties added more than 5 years to their country's weighted average age profiles. In maize, improved varieties that were becoming traditional included the composite A511 in Ethiopia, Kenya's dominant hybrid H614D, Cargill's Zimbabwean hybrid CG4141 that is discussed in the next sub-section, and two very old Angolan open-pollinated varieties (OPVs), Branco Redondo and Catete Branco.

In wheat, the problem of first-generation dominant improved varieties was not that much of an issue in most countries. Only Debeira in Sudan, Lowerie II in Zambia and Samwhit 5 in Nigeria contributed more than 5 years to their country's weighted age profile.

Dominant, old improved varieties were endemic across the rice-growing agroecologies of West Africa in the 1990s data set (Dalton and Guei, 2003b). In 21 country by agroecological observations, 14 varieties increased their country's age profile by more than 5 years. Farmers had been cultivating most of these old timers since the 1960s and early 1970s. Improved

Table 5.15. Weighted average varietal age by crop in 1997/98.

		Weighted average varietal age (years from release)	
Crop	Number of country observations	Area of modern varieties only	Total area assuming 50 years age for local varieties
Maize ESA	10	15.4	31.2
Potato	9	10.4	27.9
Rice	7	18.6	29.6
Wheat	8	12.6	19.9

varieties that appear ripe for replacement include FARO 1, 8 and 9 in rainfed lowland rice production in Nigeria, ROK 3 in the uplands and ROK 5 in the mangrove swamps of Sierra Leone, JAYA in the irrigated lowlands of Senegal, and KHAO GAEW in the deep-water production environment in Mali. As Dalton and Guei (2003b) hint at, several of these are purified landrace, limited selection materials that because of their age stretch the limit of the definition of improved varieties.

Spill-overs

Spill-over varieties, first released in one developing country and subsequently released in another, exerted a significant effect on release and adoption outcomes in potato. Their effect in the other crops was less pronounced, but examples could be cited for most crops in the study. Spill-over cultivars, either outside, most prominently from Mexico, or inside the region, mainly from Rwanda, made up a quarter of the modern varietal releases in the potato database, and they accounted for about half of the area planted to improved clones. Hence, they were roughly twice as important in adoption as they were in release.

The Cargill Hybrid CG4141 was another notable example of a spill-over cultivar. This hybrid formally entered the market in 1980 in Zimbabwe and was subsequently sold in six other countries in ESA in the mid-1980s. By the late 1990s, farmers were still purchasing CG4141 in the six countries. CG4141 was the top-selling Cargill hybrid in Zimbabwe in 1996–1998 when it was the second-ranking improved cultivar in area in the ESA region.

Summing Up

The analysis of the pooled data for SSA from the 1990s Initiative in documenting varietal output, diffusion and impact of the CG Centers mandated-commodity genetic improvement programmes was fraught with problems of incomplete country coverage, disparate and fragmented information for the three key databases, and, in a small minority of cases, fuzzy database and conceptual definitions. In spite of these difficulties, the pooled analysis did generate some findings and also showed that the late 1990s data set could be used as a point of reference with varying levels of informational value for the ten food-crop categories that were studied and that are currently the backbone of the DIIVA Project.

The 1990s data set as a benchmark has the highest value for wheat, potatoes and rice and the lowest for groundnut, sorghum and pearl millet. Without a benchmark, data collected in the DIIVA Project for the latter three crops needs to be scrutinized very carefully. The quality of the 1990s data for maize and cassava is also high but is compromised by the lack of detailed cultivar-specific information on adoption for some countries. Beans, lentils and barley did not attain this level of quality but, at least, they were characterized by reasonably good country coverage and by complete historical data on varietal release that was not restricted to CG-Center-related materials.

The findings are not sufficiently solid to be interpreted as empirical facts but they provide a framework for discussion in the context of confirming expectations or generating surprises. We begin with varietal releases. The varietal output data are consistent with an increasing rate of annual release from the 1960s to the late 1990s. This positive trend in the rate of release over time is one of the shared findings across the commodity chapters in Evenson and Gollin (2003a). Finding a generalized upward trend in varietal output was expected. However, beans, cassava and maize in ESA were the only commodity groupings that truly fit the positive-trend stereotype. Varietal output for the other crops peaked in the 1980s and was maintained at roughly the same level in the 1990s.

Political instability adversely affected varietal output in some crops in key countries in the 1990s. The 1994 Genocide in Rwanda took a severe toll on potato varietal output in countries of the Zaire-Nile Divide for more than a decade. The West African Rice Research Station that was established in 1935 at Rokupr released two dozen improved ROK-labelled varieties in the late 1970s and 1980s; civil war in Sierra Leone choked off the supply of varieties and destroyed research infrastructure in the 1990s. With the exception of civil unrest in a diminishing number of countries, most prominently Zimbabwe, the 2000s are a period of enhanced economic stability setting the stage for improved prospects for varietal output.

Some crops were characterized by higher than expected numbers of releases prior to 1975. A few countries could draw on stable lines of research that existed prior to and continued immediately following independence to generate early varietal output. These early positive performers also released substantially more varieties in the period from the mid-1970s to the late 1980s; however, the advantage of an early start vanished in the 1990s. We speculate that the IARC crop improvement programmes contributed to offsetting differences in initial advantage in research endowments. We also find evidence that the IARC content of the 1415 varieties in the data set is inversely related to total releases at the national level during the entire period. This (seemingly perverse) finding suggests that IARC activity tends to be proportionally concentrated in countries that would otherwise have less capacity to release varieties. Hence, these findings point to an equalizing influence of IARC activity on national varietal output both temporally and spatially.

The higher and more stable release rate in wheat was anticipated. Yield-enhancing technological change induced by the semi-dwarf varieties has been more evident in spring bread wheat than in any other crop type with the exception of irrigated rice, which is not a major production agroecology in most rice-growing countries in SSA. Wheat research has been well supported relative to its value of production because investing in research is perceived as a relatively cheap policy instrument to contribute to import substitution. More than for any other crop, the chronic problem of leaf rust and the dominant strategy of vertical resistance also increase the demand for varietal turnover and varietal output in wheat.

In contrast, the very low release intensity for cassava was unanticipated. Cassava ranked last in average varietal output by a wide margin on any criterion of release intensity. Cassava did have a colonial legacy of genetic research to draw on in the 1960s but governments were slower to invest in this important staple than in grain crops where technological change was perceived to be more of a reality. Other crops, especially rice, have had a substantially richer institutional milieu in the form of national, regional and international organizations that have been actively involved in promoting crop improvement over the past 50 years in SSA.

The relative neglect of cassava is manifested by arguably the oddest finding in this chapter. For cassava, the size of country production was not positively correlated with the number of releases.

For the other crops, production size was highly correlated with varietal output. Two other expected correlates of varietal output were also confirmed. Private-sector breeding, almost entirely in maize in Southern Africa, was positively and significantly associated with varietal output, which was defined as what was available for farmers to plant in the late 1990s when private-sector activity seemed to make a net addition to national varietal output and did not replace or displace public-sector research. NARS with greater breeding capacity as evidenced by greater selection pressure per release were also characterized by significantly more releases than NARS that relied mainly on the direct use of introduced elite and other finished materials.

Unexpectedly, we found no evidence for increasing NARS capacity in breeding level in terms of how released varieties were selected over time. Finding positive evidence for a progression from direct use of finished and landrace materials, to selection from introduced progeny, to crossing and selection was one of the empirical facts that was highlighted in the global assessment of Evenson and Gollin (2003a). In contrast, based on cultivar-specific information, breeding capacity has not increased in a practical sense of process-related attributes of how varieties were selected in SSA since the 1980s. Beans was the only commodity to register a substantial increase in breeding sophistication in the 1990s, probably because of activity of the PABRA network that emerged during that period and because genetic improvement in beans was characterized by a very low level of breeding capacity in the 1970s and 1980s.

This finding should be interpreted with caution because breeding level depends on crop context. Moreover, transparent codes were not used throughout the data set, which in the case of maize in ESA was limited to IARC-related materials. The image that breeding in the sense of crossing is still very limited in SSA is a hypothesis that is revisited in Chapter 18. If true, it sets SSA apart from other regions of the world.

As expected, IARC content did increase in varietal output over time for most crops. For cassava and maize in WCA, IARC content was high in the

1980s and it needed to approach 90% to register an increase in the 1990s. Maize in ESA was characterized by proportionally fewer releases related to IARC activity than the other crop groupings.

Crop-release behaviour was often erratic over time in small countries. Plausible reasons for instability in releases were put forward to explain bursts of activity sandwiched between long periods of inactivity.

The exploratory regression analysis of the variation in the number of total releases confirmed several correlates of varietal output. Production share, private-sector breeding and breeding level were positively and significantly associated with varietal output. The inclusion of South Africa in the analysis substantially influenced these results but it did not alter their fundamental nature.

Several expectations were also confirmed in the evaluation of the strength of NARS database. Estimates of researcher intensity decline exponentially as the size of production increases from less than 50,000 to more than 5 million tonnes. Research intensity is lower in cassava than in other crops even when the relatively inferior output value of cassava is factored into the calculation. The variation in scientific research strength is more due to countries than to crops, i.e. the evidence suggests that staff strength exhibited more variation across countries within a crop than across crops within a country. Large countries, such as Ethiopia, Kenya and South Africa, have invested proportionally more than many others nations in SSA in agricultural research. This behaviour is reflected in the data set in positive and statistically significant estimated country coefficients for those countries.

The evaluation of staff strength also generated one partial surprise: the degree to which the very low researcher intensity in Nigeria affected the mean outcomes in the food crops in which it was a very large producer. In particular, Nigeria's researcher intensity in cassava was, arguably, the lowest ever documented anywhere in the world. The implication for DIIVA participants is transparent: getting right Nigeria's allocation of its scientists to food crop genetic improvement is one of the most important and challenging aspects of the project. Only with reliable data can the validity of competing explanations for Nigeria's outlier behaviour in researcher intensity be tested.

The diffusion analysis did not provide much in the way of surprises. Relative to their mean levels, maize, wheat, cassava, potatoes, lentils and barley were characterized by a proportionally small range between the upper and lower bounds of modern variety adoption. Because of very incomplete country coverage, the adoption estimates for modern varieties in the 1998 data set for groundnut and pearl millet are fuzzy and of limited value as points of reference for the DIIVA Project. The analysis of the variation in country-level adoption drew attention to the outstanding diffusion performance in spring bread wheat, the moderately high-level adoption of recent 20th century cultivars in potatoes (that are more widely diffused in several countries of SSA than they are in North America and several European countries), the uniqueness of Ethiopia, and the expected strong positive interaction between private sector participation and adoption of modern varieties.

The analysis of varietal turnover and spillovers did reveal several unanticipated findings. Estimates of varietal age of 18–20 years did not seem that different from the rice-growing countries of South and South-east Asia where one old dominant variety often prevails. Older improved varieties and hybrids, such as SR52 in maize, are being replaced but not at a rate that one would expect from highly productive crop improvement programmes. Few crop-by-country observations were able to comply with a target of a maximum of 10 years in weighted average varietal age. Potato, the crop with the lowest multiplication ratio and the commodity with the bleakest prospects for an institutionally efficient seed programme, was characterized by the highest estimate rate of varietal turnover as indicated by the lowest weighted average varietal age.

The lack of turnover in old varieties is pronounced in rice in selected countries in all of its five production agroecologies in West Africa. Improved landrace materials purified and released in the 1960s have had remarkable staying power and appear ripe for replacement.

Spill-over varieties were visible in the 1990s data set in almost all crops. Their consequences were most marked in potato where they accounted for 25% of the releases and about 50% of improved cultivar area. CG4141, a short duration maize hybrid, released in Zimbabwe in 1980 also warrants mention. That Cargill cultivar was subsequently marketed in six other countries in the mid-1980s and was still planted on more than 200,000 hectares across the region in the late 1990s.

In closing, this analysis has also provided some comfort to DIIVA participants that some of the underlying assumptions of the project are reasonable in focusing effort on measuring and assessing interactions in the three key databases. For example, the observation that adoption in 1998 was significantly and positively correlated with earlier releases may seem too obvious to get excited about. However, the technology generation, varietal release and varietal diffusion processes are laden with context. That releases subsequently translate into adoption is not a foregone conclusion. Similarly, the finding that the strength of scientific staff is negatively associated with instability in release behaviour does not appear to be that important, relevant or even interesting. But it is findings such as these that confirm the quality of the late 1990s data set as a benchmark for the DIIVA Project. They also reinforce the rationale for the Project.

Notes

[1] In impact assessment, the focus of the late 1990s initiative was on economic consequences. In contrast, the DIIVA Project addressed impacts on poverty and food security along with an evaluation of some economic outcomes.

[2] This issue is revisited in Appendix 5.1.

[3] The Amani station was also known for its excellent work on tolerance to cassava brown streak that was endemic in the first half of the 20th century. Virus resistance in conferring adaptability on sweetpotato cultivars was another practical success story. Since the 1960s, the most widely grown sweetpotato variety in East Africa was selected at the Amani station. That cultivar is known as Tanzania in several sweetpotato-growing countries in East Africa.

[4] Chad's flurry of release activity in 1994 may have been the outcome of a bilateral funding project. Whatever the case, it requires explanation.

[5] Angola released two open-pollinated varieties in the late 1950s.

[6] That breeding work was led by Raoul Robinson, a plant pathologist, who is an ardent disciple of John Niederhauser. Wikepedia: http://en.wikipedia.org/wiki/Raoul_A_Robinson (Accessed 28 April 2015).

[7] The seed sales data on hybrid maize in Zimbabwe seemed incomplete as only about a quarter of the declared hybrid area had cultivar-specific data from which area could be attributed to hybrids. The bulk of the area probably originated from replanting of hybrids.

References

Adesina, A. and Zinnah, M.M. (1993) Technology characteristics, farmer's perceptions, and adoption decisions: a Tobit model application in Sierra Leone. *Agricultural Economics* 9, 297–311.

Bantilan, M.C.S. and Deb, U.K. (2003) Impacts of genetic enhancement in pearl millet. In: Evenson, R.E. and Gollin, D. (eds) *Crop Variety Improvement and Its Effect on Productivity: The Impact of International Agricultural Research*. CAB International, Wallingford, UK, pp. 215–240.

Beintema, N.M. and Stads, G.J. (2006) *Agricultural R&D in Sub-Saharan Africa: An Era of Stagnation. Background Report*. International Food Policy Research Institute, Washington, DC.

Bekele, B., Alemayehu, F. and Lakew, B. (2005) Food barley in Ethiopia. In: Grando, S. and Macpherson, H.G. (eds) *Food Barley: Importance, Uses, and Local Knowledge*. Proceedings of the International Workshop on Food Barley Improvement, 14–17 January 2002. Hammamet, Tunisia. ICARDA, Aleppo, Syria.

Bourdillon, M., Hebinck, P., Hoddinott, J., Kinsey, B., Marondo, J. *et al.* (2003) *Assessing the Impact of High-Yielding Varieties of Maize in Resettlement Areas of Zimbabwe*. FCND Discussion Paper No. 161. International Food Policy Research Institute, Washington, DC.

Brennan, J.P. (1991) *Economic Criteria for Establishing Plant Breeding Programs*. CIMMYT Economics Working Paper 92-01. CIMMYT, Mexico (DF).

Dalton, T.J. and Guei, R.G. (2003a) Ecological diversity and rice varietal improvement in West Africa. In: Evenson, R.E. and Gollin, D. (eds) *Crop Variety Improvement and Its Effect on Productivity: The Impact of International Agricultural Research*. CAB International, Wallingford, UK, pp. 109–133.

Dalton, T.J. and Guei, R.G. (2003b) Productivity gains from rice genetic enhancements in West Africa: countries and ecologies. *World Development* 31(2), 359–374.
Edwin, J. and Masters, W.A. (1998) Returns to rice technology development in Sierra Leone. *Sahelian Studies and Research* I, 67–74.
Eicher, C. (1995) Zimbabwe's maize-based Green Revolution: preconditions for replication. *World Development* 23(5), 805–818.
Evenson, R.E. and Gollin, D. (eds) (2003a) *Crop Variety Improvement and Its Effect on Productivity: The Impact of International Agricultural Research*. CAB International, Wallingford, UK.
Evenson, R.E. and Gollin, D. (2003b) Crop genetic improvement in developing countries: overview and summary. In: Evenson, R.E. and Gollin, D. (eds) *Crop Variety Improvement and Its Effect on Productivity: The Impact of International Agricultural Research*. CAB International, Wallingford, UK, pp. 7–38.
Hassan, R.M., Mekuria, M. and Mwangi, W. (2001) *Maize Breeding Research in Eastern and Southern Africa: Current Status and Impacts of Past Investments Made by the Public and Private Sectors 1966–97*. CIMMYT, Mexico (DF).
Heisey, P. and Lantican, M.A. (2000) *International Wheat Breeding Research in Eastern and Southern Africa 1966–97*. CIMMYT Economics Working Paper 00-01. CIMMYT, Mexico, (DF).
Hillocks, R.J., Madata, C.S., Chirwa, R., Minja, E.M. and Msolla, S. (2006) Phaseolus bean improvement in Tanzania, 1959–2005. *Euphytica* 150, 215–231.
Howard, J., Rubey, L. and Crawford, E. (2000) *Getting Technology and the Technology Environment Right: Lessons from Maize Development in Southern Africa*. Staff Paper No. 00-57. Department of Agricultural Economics, Michigan State University, East Lansing, Michigan.
Iken, J.E. and Amusa, N.A. (2004) Maize research and production in Nigeria. *African Journal of Biotechnology* 3(6), 302–307.
Maredia, M.K. and Eicher, C.K. (1995) The economics of wheat research in developing countries: the hundred million dollar puzzle. *World Development* 23(3), 401–412.
Nweke, F. (2009) *Controlling Cassava Mosaic Virus and Cassava Mealybug in Sub-Saharan Africa*. IFPRI Discussion Paper 912. International Food Policy Research Institute, Washington, DC.
Pachico, D. (2014) *Towards Appraising the Impact of Legume Research: A Synthesis of Evidence*. Report prepared for the Standing Panel on Impact Assessment, CGIAR Independent Science and Partnership Council. Science Council Secretariat, Rome.
Roseboom, J., Pardey, P.G. and Beintema, N.M. (1998) *The Changing Organizational Basis of African Agricultural Research*. ETPD working paper no. 37. International Food Policy Research Institute, Washington, DC.

Appendix 5.1. Uniformity and Quality of the Three Core Databases across Commodities in the 1998 Initiative

Having confirmed in Table 5.1 that the 1998 data set is sufficiently broad in terms of coverage to be an adequate benchmark for 8 of 11 food crops, we address the related issues of the uniformity and quality of data collected by the CG Centers in this Appendix. Although all Centers in the 1998 Initiative addressed the same issues and were coordinated by the Impact Assessment and Evaluation Group (IAEG), resources were not sufficient to coordinate a standardized data collection effort. At one extreme, ICRISAT relied almost entirely, as noted earlier, on a literature review to quantify the objectives of the 1998 Initiative in SSA. At the other end of the spectrum, CIMMYT and IITA closely coordinated their data collection efforts in maize and substantially exceeded minimal data set requirements in assembling and gathering information on varietal release. IITA made use of the same pro-forma that they used in maize for cassava, imparting a high degree of uniformity across both crops.

Much of the minimal data set collected in DIIVA Project described in Chapter 4 was also collected in the 1998 Initiative. But there are important gaps in each of the three main databases. More importantly, differences in aggregation compel the analyst all too often to use the tabular estimates in Evenson and Gollin (2003) instead of applying the same procedures to the disaggregated raw data. Several hypotheses cannot be tested without the disaggregated data; therefore, one is constrained by the minimal common data collected across the ten food crops in a pooled analysis.

Centers in the 1998 Initiative in a few instances collected data on more aspects than what is contained in the DIIVA Project. Information on training outputs, on spill-over varieties originally released in one country and later released in another, and IARC input to national programme crossing blocks was presented by two or more Centers in Evenson and Gollin (2003a). Eliciting information in these and other areas was considered at the DIIVA Project Proposal workshop, but was accorded a lower priority than the so-called minimal data sets on varietal release, improved cultivar adoption and strength of NARS.

Varietal release

For most Centers, varietal output is the richest database in the 1998 Initiative. With the exception of ICRISAT, all the CG Centers submitted databases on varietal output that included all improved varieties released irrespective of source (Table 5.A1). Information on cassava, maize and wheat substantially exceeds the demands of the minimal varietal release data set for the DIIVA Project. In particular, those databases are rich in their presentation of information on characteristics and pedigree.

Strength of crop improvement programmes

The databases on the strength of crop improvement programmes were a mixed blessing in 1998. All CG Centers presented reliable information at roughly the same level of detail on investments and costs of IARC crop improvement programmes over time. Human resources in NARS crop improvement programmes were extensively discussed in Evenson and Gollin (2003a) but, in terms of information retrieval, a full-fledged database at the level of a named scientist was only available for potato (Table 5.A2). Groundnut was the worst offender: no data were presented on the strength of NARS crop improvement. In contrast, information on pearl millet and sorghum was characterized by several years of coverage in response to medium-term special projects financed regionally in SSA.

Strictly speaking, data on disciplinary coverage was only available for rice, potato, lentil and barley. Human capital data for cassava and maize in West Africa was characterized by numerical aggregates tallied for broad utilization groups invested in research, production, distribution and administration.

Varietal diffusion

In contrast to varietal output, information on improved cultivar-specific adoption is scanty and heterogeneous across the CG Centers (Table 5.A3). Reliable and comprehensive cultivar-specific adoption data were available only for rice, beans, potatoes and maize. Data for the other six crops were characterized by one or, more often, multiple aspects of incompleteness. The disjointedness

Table 5.A1. Describing the texture and uniformity of data on varietal release by crop.

Continuing commodity	Summary or database	Comprehensive or only IARC	Information: rich or sparse
Cassava	Database	Comprehensive	Rich
Maize	Database	Comprehensive	Rich
Groundnut	Summary	IARC	Sparse
Rice	Database	Comprehensive	Sparse
Pearl millet	Summary	IARC	Sparse
Sorghum	Summary	IARC	Sparse
Potato	Database	Comprehensive	Sparse
Beans	Database	Comprehensive	Rich
Wheat	Database	Comprehensive	Rich
Barley	Database	Comprehensive	Sparse
Lentils	Database	Comprehensive	Sparse

Table 5.A2. Describing the texture and uniformity of data on scientific capacity by crop.

Continuing commodity	Summary or database	Breadth of disciplinary coverage	Years: single or multiple	Degree coverage
Cassava	Database	Scientists	Single	Senior, International, Junior
Maize	Summary	Scientists	Single	No
Groundnut	No data	–	–	–
Rice	Summary	Disciplinary	Single	No
Pearl millet	Summary	Disciplinary	Multiple	Yes
Sorghum	Summary	Disciplinary	Multiple	Yes
Potato	Database	Disciplinary	Single	Yes
Beans	Summary	Breeders	Multiple	No
Wheat	Summary	Scientists	Single	No
Barley	Summary	Disciplinary	Single	Yes
Lentils	Summary	Disciplinary	Single	Yes

Table 5.A3. Describing the texture and uniformity of the adoption data by crop.

Continuing commodity	Summary or database	Cultivar specific	Non-IARC releases
Cassava	Database	Yes	Included
Maize	Database	Yes	Included
Groundnut	Summary	No	Aggregated
Rice	Summary	No	Aggregated
Pearl millet	Summary	No	Aggregated
Sorghum	Summary	No	Aggregated
Potato	Database	Yes	Included
Beans	Summary	No	IARC only: time series
Wheat	Database	Yes	Included
Barley	No data	–	–
Lentils	No data	–	–

of the varietal adoption database reflects the lack of standardized data collection instruments and, to a lesser extent, the retrieval of data that adequately addressed varietal diffusion. For example, researchers made a valiant effort to try to collect cultivar-specific information for cassava, but reliable returns were not forthcoming for several important countries. Most CG Centers relied heavily on expert opinion of NARS scientists in eliciting estimates of aggregate and cultivar-specific adoption. The exception was maize in East and Southern Africa where information on seed sales was used to construct profile of adoption of hybrids and improved OPVs by country.

The diffusion data were richest in rice. Complete estimates were elicited by country in each of the five main producing agroecologies in West Africa: rained upland, rainfed lowland, irrigated lowland, mangrove and deep-water floating.

Summary assessment

Overall, the above discussion suggests that the cassava, maize, rice, potato and wheat databases support an analysis of several of the hypotheses in Chapter 3. We are on much shakier ground for groundnut and pearl millet. The tabular data on sorghum in Evenson and Gollin (2003a) are worth re-visiting. Barley, beans and lentils occupy an intermediate position in terms of data quality and uniformity.

6 The Effectiveness of Crop Improvement Programmes from the Perspectives of Varietal Output and Adoption: Cassava, Cowpea, Soybean and Yam in Sub-Saharan Africa and Maize in West and Central Africa

A.D. Alene,[1]* T. Abdoulaye,[1] J. Rusike,[1] V. Manyong[1] and T.S. Walker[2]
[1]*International Institute of Tropical Agriculture (IITA, various locations);*
[2]*Independent Researcher, Fletcher, North Carolina, USA*

Introduction[1]

In this chapter, varietal output, adoption and change are assessed for five of the 20 food crops covered in the Diffusion and Impact of Improved Varieties in Africa (DIIVA) Project. The chapter evaluates the performance of genetic improvement programmes in cassava, cowpea, soybean and yam for sub-Saharan Africa and maize improvement in West and Central Africa.

It is hard to overstate the importance of these crops for the livelihoods of the rural and urban poor in sub-Saharan Africa (SSA). Cassava is the most widely grown root crop and the second most important food staple after maize that provides more than half of the dietary calories for more than 200 million people (Nweke *et al.*, 2002). Cowpea, referred to as 'the poor man's meat' in the Sahel, is the most popular pulse crop in West and Central Africa. Taken together, the two species of African yam are the most economically relevant tubers with a value of production that ranks them in the top 2–3 food staples in SSA. Globally, soybean is the most rapidly expanding agricultural commodity in the 20th and early 21st centuries. Although area expansion in Africa is not as pronounced as in Latin America and Asia, soybean production in the region crossed the threshold of 1 million tonnes in 2008. Maize is the staple food crop of choice for millions of producers and consumers in West and Central Africa.

In spite of their economic importance and potential, these crops did not receive much research attention during the colonial period in West and Central Africa in the first half of the 20th century. Few of the limited breeding materials and elite varieties survived the transition from colonial rule to national independence. Exceptional survivors came from a handful of cassava and maize improvement initiatives supported by international organizations such as the Ford and Rockefeller Foundations. By the time the international agricultural research system

* E-mail: a.alene@cgiar.org

known as the CGIAR was created in 1971 following the founding of four International Agricultural Research Centers (IARCs) in the 1960s, the research shelf was bare for cowpea, soybean and yam, and was very poorly stocked for cassava and maize.

Since its establishment in 1967 with a mandate for the improvement of cassava, cowpea, maize, soybean and yam, the International Institute of Tropical Agriculture (IITA) has played a leading role in international efforts aimed at developing and disseminating well-adapted elite materials to farmers growing these crops. As two of the four founding centres, CIAT (the International Center for Tropical Agriculture) and CIMMYT (the International Maize and Wheat Improvement Center) have also played important roles in international efforts aimed at developing and disseminating cassava and maize varieties. Maize research at CIMMYT has contributed significantly to maize varietal output and change in West and Central Africa through parent germplasm sharing and capacity strengthening. Similarly, CIAT has played a key supportive role in providing germplasm for cassava research at IITA.

Other international programmes have also contributed materials and engaged in capacity strengthening in the generation of promising varieties. A prominent example is USAID's Bean and Cowpea Collaborative Research Support Program (CRSP). None the less, few credible, alternative international suppliers for the genetic improvement of these five crops exist. Moreover, with the exception of maize in Nigeria, private-sector participation in crop improvement is negligible or is in its infancy.

For these reasons, progress in crop improvement depends largely on the performance of plant breeding in public-sector national agricultural research systems (NARS) that are central to this chapter and to DIIVA-related research on varietal output and adoption. The DIIVA Project was timely for IITA because it provided a means to generate systematic and up-to-date information on the development, dissemination and adoption of improved varieties. The absence of such a baseline had limited IITA's efforts to assess the economic and poverty reduction impacts of national and international agricultural research in SSA. Not only are variety diffusion data lacking, but they are also the most expensive data to collect on a regular basis. For example, identification of improved cassava clones cannot be inferred from farmer-survey responses; therefore, field visits are required with subsequent expert input on varietal determination on the basis of morphological characteristics in photographs (Alene *et al.*, 2012). The DIIVA Project complemented emerging efforts and provided a cost-effective means to measure progress in crop improvement.

Results are reported by crop for each of the three databases described in Walker (Chapter 4, this volume): (i) recent cross-sectional data on the strength of human resources in NARS by discipline; (ii) historical data on varietal release; and (iii) recent cross-sectional data on varieties-specific levels of adoption elicited from expert panels. A closing section summarizes common findings and recurring themes across the five crops. Before discussing the crop-wise findings, survey design and data collection are described below.

Survey Design and Data Collection

In 2010, IITA carried out a survey of national crop improvement programmes in over 20 target countries in SSA. IITA initially identified 26 crop-by-country combinations (CCCs) involving 11 countries for cassava, 9 countries for maize, 5 countries for cowpea and one country for soybean. Exploiting the synergy between DIIVA and the larger effort at IITA, the project has actually covered a total of 68 crop-by-country combinations: 17 countries for cassava, 11 countries for maize, 18 countries for cowpea, 14 countries for soybean and 8 countries for yam (see Appendix Table 6.A1). With this expanded coverage, the IITA crop-by-country combinations account for about 45% of the total observations in the DIIVA Project in SSA.

Crop coverage was based on the significance of adoption and potential impacts in the mandate region. Given that one priority in the DIIVA Project was to update the 1998 database, cassava and maize as continuing crops were the obvious choices for the study. The other IITA mandate crops with significant expected adoption of new varieties to qualify for variety release and adoption were cowpea and soybean. Yams were initially not included because of the perceived paucity of released varieties and adoption experience, but were later added as data became available from the larger IITA-led survey of national crop improvement programmes.

Country coverage for each of the five crops was based on: (i) expected or documented varietal releases as a proxy for technological change; (ii) share of production such that the selected countries accounted as a group for over 75% of total area in the mandate region in 2009; (iii) importance of the crop in food consumption; and (iv) availability of previous baseline survey data on variety adoption. In 2009, the production share of the respective countries surveyed exceeded 90% across the five crops: 95% for cassava in SSA, 93% for maize in West and Central Africa, 98% for cowpea in SSA, 100% for soybean in SSA and more than 95% for yam in SSA.

The survey was conducted from April to December 2010 using a structured questionnaire. Separate research teams were set up for the surveys in Anglophone and Francophone countries. Each country was visited by one of the research teams that compiled secondary data on variety release and human resource investments from variety registers, annual reports and other sources. The survey questionnaire focused on gathering detailed information on the following aspects of the performance of national crop genetic improvement programmes: (i) human resource investments in genetic improvement; (ii) name, origin, germplasm content and agronomic characteristics of improved varieties released in the country; and (iii) estimated area planted to different varieties in the country.

Data collection was guided by the illustrative protocols given in Walker (2010). In the first part of the questionnaire, data were assembled on the full-time equivalent (FTE) scientist (BSc and above) years invested in 2009 by all public and private sector (if any) programmes in each surveyed country.

Substantial time was spent compiling data on varietal release in the second part of the questionnaire. Information on the names, origins, germplasm content and agronomic characteristics of improved varieties released in each country was obtained from variety registers (e.g. for Nigeria), annual reports, variety release reports, journal articles on variety registration, expert consultations and a range of other sources. The list included both official and unofficial releases. In many countries, unofficial releases represent a significant share of the total number of varieties released. This is due to a lack of variety-release procedures for some crops (e.g. cassava) and countries, as well as poorly functioning variety-release bodies.

Information on the IITA content of improved varieties was also collected as part of a major effort to assess IITA's role in terms of transforming the scientific capacity of the NARS from mere acceptance of nearly finished technologies to the ability to screen and adapt technologies, then to a final stage in which they have full scientific capabilities to undertake genetic improvement involving breeding and selection. Each improved variety released in a country was classified into one of three possible categories based on whether and how IITA germplasm was used: (i) Non-IITA, i.e. no IITA germplasm was used (e.g. local landraces); (ii) IITA-parent, i.e. parent germplasm from IITA, with crossing, selection and testing done by NARS; and (iii) IITA-bred, i.e. crossing and/or selection done by IITA (including evaluation and genebank conservation of landraces), with local adaptation and testing done by NARS.

The main sources of information on varietal output were official variety registers and other national commodity varietal release databases. These data were supplemented by expert knowledge from different research institutions. The major constraint to accurate reporting on the complete research outputs and genetic attribution is the expert exposure and experience with crop-specific breeding taking place in the country from the distant past to the present. Largely, as a result of staff turnover, some of the experts involved in the survey were relatively new in the national programmes and their commodity teams. Lack of experience could have led to historical omissions of important varieties or to over-optimism about the prospects for materials under trial.

Methodological differences as well as differing data sources between the 1998 and the 2010 surveys (e.g. Nigeria's crop variety register was published in 2009, therefore a variety register was unavailable in 1998) limited the scope for a comparative analysis of trends in variety release. Because varietal release data were elicited through a mailed questionnaire in 1998, there were no fora for face-to-face interaction, consultations and triangulation of the data. Many varieties reported in the 1998 survey as released were actually in the pipeline then and were formally released only after 1998. In the 1998 study, a total of 267 maize varieties were reported as released in West and Central Africa between 1965 and 1998 (Manyong *et al.*, 2003). The 2010 survey found that only 183 were released in that same time period. This big discrepancy can be explained by the aforementioned differences in methodology, as well as the inclusion of varieties that were only in the pipeline. These differences also implied that the historical record of varietal release had to be compiled not only for the new crops of cowpea, soybean and yams, but also had to be reconstructed for the continuing crops of cassava and maize. The analysis of varietal output in this chapter and this volume for the IITA-mandated crops is based on the reconstructed varietal release database.

Sequential estimation of varietal adoption was the focus of the last part of the questionnaire. Expert-opinion estimates on variety adoption were elicited through extensive discussions and consultations with programme leaders, breeders, agronomists, economists, and extensionists of public- and private-sector research institutes and seed-production agencies.

Elicitation was accomplished through a series of steps that were followed to minimize the inherent subjectivity associated with expert opinions. First, the experts were asked to identify and list the major technology recommendation domains (e.g. agroecological zones, geographic regions, etc.) for the crop in the country. Second, they were asked to estimate the share of each recommendation domain in the total area planted during the recent agricultural year. Third, they were asked to estimate, for each domain, the relative importance of improved and traditional varieties as a group, in terms of their percentage share in total area planted to the crop. Fourth, the experts were asked to list and rank, for each recommendation domain, the most important improved varieties in terms of the size of the area planted. Finally, the experts were asked to estimate, again for each recommendation domain, the percentage share of each listed and ranked variety in total area under all improved varieties so that the shares summed to 100% for each domain. National level variety-specific adoption estimates were then derived via the weighted aggregation of the domain level estimates across all recommendation domains with the domain-specific area shares used as weights.

Cassava

IITA initiated cassava research in the early 1970s with a focus on developing high-yielding varieties with resistance to major pests and diseases such as cassava mosaic virus disease (CMD), cassava bacterial blight (CBB) and cassava green mite (CGM). In addition to breeding for high yield and resistance to major pests and diseases, the cassava research programme at IITA involved developing biological control and integrated pest management options to reduce losses due to insect pests. Cassava breeding was initiated using breeding materials from the Moor plantation near Ibadan and a limited number of East African landraces with resistance to CMD as well as CBB developed through interspecific hybridization in the 1930s. Germplasm was also collected from Latin America and Asia along with local varieties from within Nigeria. This work resulted in several elite genotypes that had resistance to CMD and CBB as well as high and stable yields and good consumer acceptability. This was the first major breakthrough in the genetic improvement of cassava. The development of these resistant varieties, and their delivery to national programmes for testing under specific local conditions during the late 1970s and 1980s, has led to the widespread and successful deployment of CMD- and CBB-resistant cassava in SSA (Nweke *et al.*, 2002). Another major breakthrough in the breeding programme was the pyramiding of new sources of resistance to CMD, identified from West African landraces, with the resistance genes to the earlier Tropical Manihot Selection (TMS) varieties, providing greater and more durable resistance. The 'new generation' of cassava germplasm combines enhanced CMD resistance with improved postharvest qualities, multiple pest/disease resistance, wide agroecological adaption and greatly improved yield potential where yield increases of 50–100% without the use of fertilizer were demonstrated in many African countries.

With increasing severity and incidence of cassava brown streak disease (CBSD) and its spread into high altitude environments, recent cassava research efforts have aimed at screening the mid-altitude germplasm with CMD resistance for combined resistance to CBSD. Genotypes with combined resistance to both diseases have been identified and these are either being multiplied or tested in on-farm trials in collaboration with the Catholic Relief Services (CRS) under the Great Lakes Cassava Initiative (GLCI). Additional seeds from CBSD-resistant parent materials were produced and evaluated under high disease pressures in Uganda and Tanzania and potential resistant/tolerant genotypes with good agronomic characteristics and quality preferences by farmers have been identified. The characteristics of most new genotypes developed by the cassava breeding programme in recent years reflect the vision of an expanded future role of cassava in food, feed and industrial applications. This improved germplasm is regularly shared with the NARS as specific genotypes (certified as virus-tested) or improved seed populations for evaluation and selection under local environmental conditions. Cassava improvement programmes in Africa receiving these breeding materials from IITA have developed/selected varieties that outperform the local varieties and officially released several of them to farmers, whereas others are at various stages of utilization. The survey results showed that during the period 1970–2010, ITTA and NARS released a total of 367 improved cassava varieties in SSA.

Scientific strength of cassava improvement programmes

In 2009, 14 FTE researchers were working on cassava improvement in IITA (Table 6.1). The institute's investment in cassava was greater than for the other four crops discussed in this paper. It represented about 10% of the 139 FTE researchers engaged in cassava improvement in the public sector in SSA. With the possible exception of Ghana, it is unlikely that cassava would be the leading crop in terms of FTE scientists of any of the 17 public-sector programmes listed in Table 6.1.

Cassava improvement programmes in SSA feature a diversified portfolio of disciplines. On average, countries in SSA allocated 1.6 FTE researchers to breeding; 1.1 to agronomy; 0.9 to entomology/nematology/virology; and over 0.5 FTEs to genebank conservation, tissue culture and postharvest. Allocations to pathology, seed production, molecular biology, social science and food science fell in the range of 0.1 to 0.7 FTEs. Given the importance of pests and diseases in cassava production, 0.9 and 0.7 FTEs allocated to entomology/nematology/virology and pathology, respectively, was lower than expected.

Investments in cassava improvement varied widely across countries, ranging from 1 FTE researcher in Burundi to slightly more than 22 FTE researchers in Ghana. Of the 17 countries, 7 countries had fewer than 5 FTE researchers working on cassava improvement. A low level of investment is commensurate with light production where the crop is not of primary importance in most of these countries such as Burundi and Zimbabwe. This level of investment is, however, too low for countries such as Tanzania where cassava is an economically important commodity with an estimated annual value of production of over half a billion US dollars. A hiring freeze that was in effect between 1992 and 2002 as well as the early retirement of senior staff probably contributed to the apparent underinvestment in cassava improvement in Tanzania.

At the high end of the investment spectrum, Uganda and Ghana employed slightly more than 20 FTE researchers. As part of its major biotechnology research effort aimed at addressing key biotic and abiotic production constraints affecting major food crops, 4 FTE molecular biologists are deployed on cassava in Uganda.

The overall picture that emerges in Table 6.1 is one of fragmented cassava research capacity attributed to substantial differences in sources of funding, the importance of the crop to the economy and the size of the national agricultural research system. For instance, the Kenya Agricultural Research Institute (KARI) is a relatively well-funded institute, receiving constant support from the Kenyan Government, attracting large sums of donor funding and generating its own revenues to finance their breeding activities (Beintema and Stads, 2011; Beintema and Rahija, 2011). Funding access more than compensates

Table 6.1. Full-time equivalent staff by major specialization working on cassava improvement in sub-Saharan Africa in 2009.

CGIAR/ NARS Programme	Major specialization											
	Germplasm conservation	Breeding	Pathology	Molecular biology	Entomology/ Nematology	Agronomy	Seed production	Tissue culture	Postharvest	Social science	Food science	Others	Total
CGIAR (IITA)	0.2	6.0	0.0	2.0	1.2	4.2	0.0	0.0	0.0	0.0	0.3	0.0	14
NARS	10.3	27.5	12.0	7.6	15.0	18.8	9.5	13.3	13.9	6.9	2.0	1.8	139
Angola	0.0	2.6	0.5	0.4	0.4	2.0	2.5	0.4	0.3	0.0	0.5	0.0	9.4
Benin	0.3	0.0	0.0	0.0	0.0	0.0	0.5	1.7	0.0	0.0	0.0	0.0	2.4
Burundi	0.0	0.0	0.0	0.0	0.0	0.9	0.0	0.0	0.0	0.0	0.0	0.1	1.0
Cameroon	1.5	1.4	1.6	0.8	1.8	0.9	0.7	1.6	0.5	0.7	0.0	0.0	11.5
Côte d'Ivoire	0.2	0.9	0.3	0.3	0.1	0.9	0.4	0.9	0.4	0.4	0.5	0.0	5.1
DR Congo	0.0	3.0	2.0	0.0	3.2	1.0	0.0	1.0	2.0	2.0	0.0	0.0	14.2
Ghana	0.8	5.9	1.3	0.7	1.2	2.1	2.7	2.3	4.3	1.2	0.0	0.0	22.5
Guinea	0.0	1.0	0.0	0.0	0.6	1.2	0.0	0.0	0.2	0.0	0.0	0.0	3.0
Kenya	0.0	2.0	2.0	1.4	1.6	0.6	2.0	0.0	0.8	0.4	0.0	0.0	10.8
Malawi	0.0	1.2	0.0	0.0	0.2	1.0	0.2	0.0	0.0	0.5	0.0	0.5	3.6
Mozambique	0.3	2.5	1.0	0.0	0.0	2.0	0.0	1.8	1.0	0.3	0.0	0.0	8.9
Nigeria	2.0	2.0	2.0	0.0	0.0	2.0	0.0	0.5	2.0	0.5	0.0	0.0	11.0
Tanzania	1.0	0.0	0.2	0.0	0.0	0.5	0.5	0.0	0.0	0.0	0.0	0.0	2.2
Togo	1.1	0.4	0.1	0.0	0.0	0.5	0.1	0.8	0.8	0.2	0.0	0.0	4.0
Uganda	3.0	3.0	0.0	4.0	4.0	2.0	0.0	2.0	0.0	0.3	1.0	1.2	20.5
Zambia	0.2	1.6	1.0	0.0	2.0	0.0	0.0	0.4	1.6	0.4	0.0	0.0	7.2
Zimbabwe	0.4	0.0	0.0	0.0	0.0	1.2	0.0	0.0	0.0	0.0	0.0	0.0	1.5
Average	0.6	1.6	0.7	0.4	0.9	1.1	0.6	0.8	0.8	0.4	0.1	0.1	8.2

for low levels of production and allows Kenya to join the ranks of six countries that have invested more than 10 FTE scientists in cassava improvement.

Turning to educational qualifications, Ghana stands out with 13.9 FTE PhD scientists. The efforts of the World Bank and the Dutch Government figured prominently in this high level of educational attainment by Ghanaian cassava scientists. Between 2000 and 2008, the World Bank funded the Agricultural Services Subsector Investment Project (AgSSIP) that included a significant training component. Seventeen scientists from the Council for Scientific and Industrial Research (Ghana) (CSIR) received PhDs and 15 received MSc degrees in various fields, most of them agriculturally related. Over the same period, nine researchers from CSIR received various degrees from universities in the Netherlands under the sponsorship of the Dutch Government.

Like Ghana, Uganda is the other positive outlier in terms of educational qualifications. Uganda has quadrupled its research spending during the last decade. This has led to an enhanced research capacity, which, in turn, is dominated by the PhD and MSc staff (Beintema and Rahija, 2011).

Cassava is the most important crop under research in Nigeria, accounting for 10% of the total crop FTE researchers (Beintema and Rahija, 2011). In spite of this emphasis, Nigeria employed only six PhD FTE researchers – half that of Ghana – in cassava genetic improvement. Nigeria's 13% share of PhD scientists is substantially lower than its one-third share in cassava area amongst the 17 countries. Much of the lack of congruence between educational qualification and area shares is attributed to an increasing but still relatively low number of FTE scientists. Since 2001, shifting of agricultural research staff composition in Nigeria towards junior staff trained at the BSc level probably played a role in dampening the demand for PhD crop scientists in public-sector research (Flaherty *et al.*, 2010).

The research systems in Zambia, Togo, Burundi, Zimbabwe and Tanzania employed less than 0.6 PhD FTE researchers in cassava breeding. The more MSc and BSc intensive composition of their NARS, together with low numbers of well-qualified staff, create significant constraints on the ability of these countries to conduct high-quality research and to attract external funding. Demand and supply forces shape the research profile observed at any NARS at a particular point in time (Pardey *et al.*, 1991). High opportunity cost for qualified researchers with highly demanded specialty skills gives them a good deal of international mobility and allows them to leave the public sector and join more lucrative jobs in the private sector. Hence, small research systems always suffer from high attrition rates.

Cassava has one of the lowest research intensities relative to the volume of production of any food crop in SSA. The weighted average ratio of FTE scientists per million tonnes of production is only 1.3 across the 17 surveyed countries. The largest cassava producers, Nigeria, Angola, Benin, Democratic Republic of Congo (DR Congo), Tanzania and Malawi, each employ less than 1 FTE researcher for every one million tonnes of cassava production.

More typically, the estimates show falling research intensity ratios with increasing cassava production across countries. This inverse relationship is observed empirically in many studies (Bohn and Byerlee, 1993; Bohn *et al.*, 1999). But atypically, estimated research intensities are not that high even for the smaller producers. With a research intensity ratio of 13, Kenya was the only country investing in more than 10 FTE scientists per million tonnes of production in 2009.

Varietal output of cassava improvement

During the past four decades between 1970 and 2010, a total of 367 varieties of cassava were released in SSA, equivalent to a mean release rate of 9–10 varieties per year for all of the 17 survey countries (Table 6.2). Most (68%) were released in West and Central Africa, which is a traditional cassava-growing region and is also where cassava genetic improvement started in the 1930s. Later, breeding initiated in East and Southern Africa accounted for about one-third of the total varietal releases in SSA.

With 65 varieties, Nigeria is the leading country in the number of releases in SSA. Nigeria is also characterized by the steadiest release performance over time with multiple releases in each of the past four decades.

Table 6.2. The IITA content of improved cassava varieties in sub-Saharan Africa, 1970–2010.

	Number of varieties released				Percentage of total release		
Country	IITA-bred	IITA-parent	Non-IITA	Total	IITA-bred	IITA-parent	Non-IITA
Angola	5	2	7	14	36	14	50
Benin	3	2	0	5	60	40	0
Cameroon	0	27	1	28	0	96	4
Côte d'Ivoire	10	0	7	17	59	0	41
DR Congo	15	13	1	29	52	45	3
Ghana	14	2	5	21	67	10	24
Guinea	29	3	3	35	83	9	9
Nigeria	53	12	0	65	82	18	0
Togo	29	0	5	34	85	0	15
West and Central Africa	158	61	29	248	64	25	12
Burundi	5	0	0	5	100	0	0
Kenya	14	10	1	25	56	40	4
Malawi	5	4	4	13	38	31	31
Mozambique	3	6	14	23	13	26	61
Tanzania	2	8	16	26	8	31	62
Uganda	7	6	0	13	54	46	0
Zambia	0	3	5	8	0	38	63
Zimbabwe	0	0	6	6	0	0	100
East and Southern Africa	36	37	46	119	30	31	39
Sub-Saharan Africa	194	98	75	367	44	26	20

About three-quarters of the 367 varieties in Table 6.2 are dated with a year of release; the other quarter corresponds to informal releases. When exposed to new varieties during participatory varietal selection or farmer-to-farmer diffusion, farmers usually retain and cultivate the varieties even before they are officially endorsed for use by the government. The majority of varieties recorded in Côte d'Ivoire, Kenya and Cameroon were unofficially released or their release dates were not known.

Undated releases are the result of slow and bureaucratic varietal-release procedures where many varieties have not passed the stage of release and are already being cultivated by the farmers. For instance, Lagos, an informally released variety, is the leading improved cultivar in Togo, covering 18% of cassava area. Similarly, 8034 (IRAD8034) in Cameroon, TMS 4(2)1425(IM93) in Côte d'Ivoire and TMS 30572 (Magyera) in Kenya are other informally released varieties that are the most popular cultivars. Even in Nigeria, 30 of the 65 varieties in the output database are classified as informal releases.

In general, the production of improved varieties has increased over time. The number of recent releases between 1997 and 2010 was 145, compared to 120 between 1970 and 1997. Since the late 1990s, several countries, including Angola, DR Congo, Ghana, Malawi, Mozambique, Tanzania and Uganda, registered considerable progress by more than doubling the number of released varieties from the earlier period. However, no clear trend has emerged on the number of releases in Nigeria, Côte d'Ivoire and Zimbabwe between these two time periods. A negative trend was observed in Benin, Guinea and Togo.

IITA-related germplasm continues to be used extensively by public-sector cassava breeding programmes throughout SSA (Table 6.2). Of the 367 cassava varieties released in sub-Saharan Africa during the last four decades, 292 (or 80%) were IITA related and 18 (or 5%) were NARS-bred using IITA parent materials. Between 1970/1997 and 1998/2010, the number of releases by NARS with no IITA-content increased from 17 to 24 (up 41%), whereas IITA-based varieties trended up from 82 to 117 (up 42%).

IITA-related varieties have increased by more than 100% from 130 in 1970–1997 to 162 releases in 1998–2010. Almost all the countries appearing in both the 1998 and the 2010 surveys have increased the number of releases with IITA ancestry, attesting to the success of collaborative breeding by NARS and IITA between 1998 and 2010.

The total number of releases has increased slowly but steadily from a very low base since the 1970s (Fig. 6.1). IITA-bred varieties increased from just a single variety released in 1970–1979 to 17 in 1980–1989. Releases of IITA-bred varieties then picked up in the 1990s and subsequently dropped slightly in 2000s. Twelve varieties with IITA parents were released in the 1980s. About the same number of varieties was selected in the 1990s from IITA-bred progenies or by NARS crosses of IITA parental materials. Varietal releases from NARS selection of progenies and their bred materials featuring IITA parents increased substantially in the 2000s. Growth in non-IITA varieties, however, was gradual and generally low in the 1980s and 1990s. Between the 1990s and 2000s, the number of releases of non-IITA varieties also trended upwards.

In general, the estimates in Fig. 6.1 are consistent with the gradual strengthening of cassava improvement programmes in SSA as a whole. Before 1998, the bulk of releases came from NARS selection of landraces or of finished bred materials from IITA. Since 1998, progeny selection of IITA materials and selection from NARS crosses rival landraces and elite bred materials as sources of released varieties.

The former sources require greater effort and skill than the latter sources to generate a positive release outcome. Nonetheless, not all 17 national crop improvement programmes have participated equally in this development process. Only Nigeria, Cameroon and Kenya have released three or more varieties where NARS crosses with at least one IITA parent was the source of the variety. Although half of the country programmes have demonstrated the capacity for release of one or more varieties from IITA-bred progenies, almost all countries, including Nigeria, still rely heavily on IITA elite materials for selection and release.

Summing up, considerable variation exists among countries within the region in breeding capacity. It was hypothesized that, upon receiving IITA or CIAT germplasm, countries with strong national breeding programmes are much more likely to further select and/or cross, whereas countries with weaker programmes tend to release varieties containing IITA germplasm with little additional improvement. Contrary to this expectation, almost all the varieties released in Nigeria and Uganda, the countries with highest human capital investment in cassava genetic improvement in the region, were IITA-related.

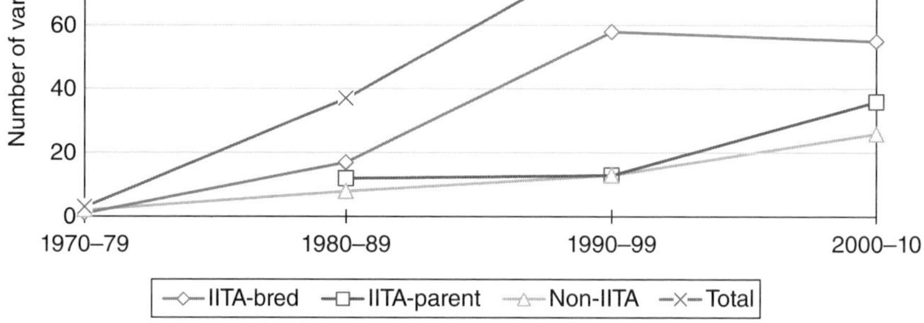

Fig. 6.1. Trends in cassava variety releases by IITA content, 1970–2010.

NARS with the strongest research capacity in Ghana produced only four varieties without IITA content. Overall, IITA remains an important player in the release of new cassava varieties in SSA.

During the past four decades, Nigeria had the highest mean varietal release rates of about 1.63 per year followed by Guinea with 0.88 and Togo with 0.85. The lowest release rates were observed in Zimbabwe, Zambia, Uganda, Benin, Burundi, Malawi, Angola and Côte d'Ivoire with less than 0.5 per annum. Over the same period, years with zero releases were common, ranging from 27 in Cameroon to 39 in Zimbabwe, Guinea and Burundi. An absence of released varieties over several years signifies inactivity or lack of progress in a plant-breeding programme.

The coefficient of variation for varietal release fell significantly across almost all countries in SSA from 1970/1997 to 1998/2010, signifying more steady progress in cassava research in SSA. Higher and more stable varietal output could be traced to more funding from the New Partnership for Africa's Development (NEPAD's) agricultural research programme and other donor agencies and participating in the international research networks, such as Eastern Africa Root Research Network (EARRNET), Southern Africa Root Crops Research Network (SARRNET), the Association of Strengthening Agricultural Research in Eastern and Central Africa (ASARECA), the Southern Africa Centre for Cooperation in Agricultural Research and Training (SACCAR) and the Institute du Sahel (INSAH). These networks and organizations provide NARS with material, technical assistance and financial assistance, as well as capacity building.

Adoption of improved cassava varieties

If cassava improvement is to deliver tangible benefits to farmers in SSA, improved varieties that have been developed and released must also be taken up by farmers and planted in their fields. Table 6.3 summarizes the area planted to improved cassava varieties in SSA by country, variety type and IITA contribution in 2009. In 2009, improved varieties accounted for 40% of the total cassava area in the countries surveyed,

Table 6.3. Adoption of improved varieties of cassava in sub-Saharan Africa, 2009.

		Proportion of total cassava area planted to:		Materials containing IITA germplasm or directly related to IITA activities	
Country	2007–2009 cassava area ('000 ha)	Local varieties (%)	Improved varieties (%)	('000 ha)	MV area (%)
Angola	839	69	31	70	8
Benin	242	34	66	160	66
Burundi	61	71	29	18	29
Cameroon	205	64	36	70	34
Côte d'Ivoire	349	80	20	51	15
DR Congo	1,850	51	49	898	49
Ghana	842	64	36	262	31
Guinea	133	80	20	24	18
Kenya	60	56	44	24	40
Malawi	182	39	61	43	23
Mozambique	941	81	19	31	3
Nigeria	3,593	55	46	1,635	46
Tanzania	800	70	30	126	14
Togo	121	61	39	23	19
Uganda	398	65	35	139	35
Zambia	193	56	45	29	15
Zimbabwe	49	48	52	0	0
All	10,957	61	40	3,628	33
Sub-Saharan Africa	12,136	64	36	3,628	30

MV, modern variety.

equivalent to 36% of the cassava area in SSA. Relatively higher adoption rates were observed in Benin (66%), Malawi (61%), Zimbabwe (52%) and DR Congo (49%).

In contrast, Guinea has 35 improved varieties recommended for cultivation by farmers but the rate of adoption was among the lowest, with only 20% of its cassava area under improved varieties. Local varieties also dominate production in Mozambique, Côte d'Ivoire, Burundi and Tanzania.

Expert estimates of adoption show that IITA-related varieties (bred by IITA or developed with IITA material used as a parent) occupied about 3.6 million hectares, equivalent to 30% of the total cassava area in SSA in 2009. Materials containing IITA germplasm covered virtually all of the area planted to improved cassava in Benin, Burundi, DR Congo and Nigeria. About 45% of cassava area in Nigeria is under IITA-related varieties. Benin appears to have the highest incidence of IITA-related varieties with 66% of the cassava area planted to varieties containing IITA genetic materials.

Of the nearly 1 million hectares under cassava in Mozambique, only 30,000 hectares (or 3%) were planted to cassava possessing at least some IITA germplasm or directly related to IITA activities. Only as recently as 2009 were three IITA-related varieties, MZMG04/433, MZMG04/1855 and MZMG04/763, recommended for cultivation.

Changes in the adoption of improved cassava varieties are evident when the results are compared to results of the 1998 Initiative (Johnson *et al.*, 2003). Between 1998 and 2009, the area under cassava production in SSA increased from 9 to 12 million hectares, mainly as a result of persistent drought and loss in soil fertility that led governments and farmers to re-value the crop for its adaptability to adverse growing conditions. During the same period, total area planted to improved varieties rose from 1.6 (18% of 1998 cassava area) to 4.4 million hectares (36% of 2009 cassava area).

Almost all the 17 countries in Table 6.4 showed a positive trend in the area planted to improved varieties. Between 1998 and 2009, the area planted to improved varieties more than doubled in 9 of the 17 countries. These include Angola, Benin, DR Congo, Kenya, Malawi, Nigeria, Togo, Zambia and Zimbabwe. The big jump in the area planted to improved varieties in Malawi, Angola, Kenya and Zambia coincides with the period when SARRNET and EARRNET were working in collaboration with national research programmes in the Southern Africa Development Community (SADC) and ASARECA member countries. The networks were backstopping activities in breeding, agronomy and value addition.

The presence of IITA headquarters in Nigeria builds synergy between IITA and Nigerian NARS in cassava research and dissemination of improved varieties. Tanzania is the only country that has experienced negligible change in area planted to improved varieties. Since the mid-1980s, the cassava sector in Tanzania has been threatened by pests and diseases – particularly white flies, mealy bug and CMD (Nweke, 2009). New CMD-resistant varieties, such as Aipin Valencca, Msitu and TMS 4(2)1425, were developed and distributed to farmers but the adoption of these varieties was very low (Kavia *et al.*, 2007).

Spill-overs of improved varieties are evident in the cultivar-specific adoption estimates in Table 6.5. Two disease resistant and high-yielding IITA varieties, TMS 30572 first released

Table 6.4. Comparing adoption of improved cassava varieties between 1998 and 2009.

	Area planted to improved varieties (%)	
Country	1998	2009
Angola	14	31
Benin	8	66
Burundi	na	29
Cameroon	31	36
Côte d'Ivoire	16	20
DR Congo	24	49
Ghana	25	36
Guinea	17	20
Kenya	16	44
Malawi	8	61
Mozambique	na	19
Nigeria	19	46
Tanzania	31	30
Togo	12	39
Uganda	30	35
Zambia	0	45
Zimbabwe	8	52
All	22	40
Sub-Saharan Africa	18	36

na = data not available.

Table 6.5. Economically important improved cassava varieties in sub-Saharan Africa, 2009.

Country	Variety	Release year	Adoption (% area)
Angola	Precoce de Angola	1994	9
	Maria Cudianeca	1997	4
	Nghanarico	1994	4
	Mundele Paco	1999	2
	TMS 42025 (Linda)	1999	2
	MZ 96/00910	2004	2
	TMS 00236 (Formosa)	2006	2
	Vermute	2008	2
	TMS 60142 (Perdiz)	2004	1
	MZ 96/001323	2005	1
	Manuela	2000	1
	TMS 40142 (Quizaquinha)	2004	1
	Regional 1	2000	1
	All MVs (National)		31
Benin	RB 89509	1989	32
	Ben 86052	1986	23
	TMS 30572 (30572/5)	1984	7
	TMS 30555	1984	4
	All MVs (National)		66
Burundi	MM96/5280 (Rugero)	2001	16.4
	MM96/0287 (Ngarukiye)	2001	6.4
	MM96/7204	2001	4.9
	Abbey-Ife (TMS 30404)	1999	1.4
	All MVs (National)		29
Cameroon	8034 (IRAD8034)	Informal	8
	8017 (IRAD8017)	Informal	8
	8061 (IRAD8061)	Informal	6
	TMS 96/1414	Informal	4
	TMS 92/0326	Informal	4
	Excel	Informal	2
	Champion	Informal	2
	658	Informal	1
	244	Informal	1
	All MVs (National)		36
Côte d'Ivoire	TMS 4(2)1425 (IM 93)	Informal	6
	Yavo (TME 7)	1999	3
	TMS 30572	Informal	2
	IM89	Informal	2
	IM84	Informal	2
	Oliékanga	1999	2
	Bocou 2 (I88/00158)	Informal	1
	Bocou 1 (CM52)	Informal	1
	88/263	Informal	1
	TMS 30395	Informal	
	TMS 30555	Informal	
	All MVs (National)		20
DR Congo	Sadisa (91/203)	1999	13.9
	Mvuama (83/138)	1997	6.9
	RAV (85/297)	1997	5.9
	Nsansi (I95/0160)	2004	5.5
	Butamu (MV99/0395)	2004	3.0
	Lueki (92/377)	2000	2.8
	Disanka (TMS I95/0211)	2004	2.6

Continued

Table 6.5. Continued.

Country	Variety	Release year	Adoption (% area)
	Zizila (MV99/038)	2004	2.0
	TME 419		1.1
	Antiota (TME 2)		1.0
	All MVs (National)		48.5
Ghana	Afisiafi (TMS 30572)	1993	14.9
	Gblemoduade (TMS 50395)	1993	3.1
	Tek Bankye (Mutant)	1997	3.0
	Bankyehemaa (TMS 97/4414)	2005	2.5
	Dokuduade (TMS 97/4489)	2005	2.1
	Esambankye (TMS 97/3982)	2005	2.0
	IFAD (DMA 002)	2004	1.3
	Abasafitaa (TMS 4(2)1425	1991	1.3
	Agbelifia (TMS 97/4962)	2005	1.2
	UCC Cape Vars Bankye (UCC 505)	2005	1.2
	All MVs (National)		35.8
Guinea	TMS 30572	1993	3.2
	TMS 92B/0033	1993	2.0
	TMS 91/02312	1993	1.7
	Tokoumbo	1993	1.6
	TMS 91/0730	1993	1.5
	Faranah	1993	1.3
	Caricass	Informal	1.3
	TMS 91/02324 (Nimaga)	2006	1.2
	All MVs (National)		20.3
Kenya	TMS 30572 (Migyera)	Informal	24
	SS4	Informal	8
	Serere	Informal	4
	Nase 4	Informal	2
	MM96/5280	Informal	1.2
	All MVs (National)		44.2
Malawi	Manyokola	1980	37.8
	Sauti (CH92/077)	2002	16.8
	Mkondezi (MK91/478)	1999	3.7
	Maunjili (TMS 91934)	1999	2.3
	All MVs (National)		61.3
Mozambique	Nikwaha	2000	4.5
	Chigoma mafia	2000	3.8
	Chihembwe	2007	1.9
	All MVs (National)		19
Nigeria	TMS 30572 (Nicass 1)	1984	17.8
	TMS 4(2)1425 (Nicass 2)	1986	8.7
	NR 8082 (Nicass 14)	1986	7.2
	TMS 92/0326 (Nicass 27)	2006	2.8
	TME 419 (Nicass 20)	2005	2.8
	TMS 30555 (Nicass 10)	1976	2.4
	TMS 98/0581 (Nicass 24)	2005	2.0
	TMS 98/0505 (Nicass 22)	2005	1.8
	All MVs (National)		46
Tanzania	Mkombozi (MM 96/4684)	2009	9
	Kiroba	1998	4
	Munba	2003	4
	Meremeta (MM96/4619)	2009	3
	Naliendele	2003	1

Continued

Table 6.5. Continued.

Country	Variety	Release year	Adoption (% area)
	Belinde (MM 96/3075B)	2009	1
	Rangimbili (MM 96/8233)	2009	1
	Kasala (95NA/00063)	2009	1
	Kyaka (MM 96/8450)	2009	1
	Suma (I 91/0067)	2009	1
	Nyakafulo (MM 96/5725)	2009	1
	All MVs (National)		29
Togo	Lagos	Informal	18
	312/524 (Sorad)	1970	11
	Gbazekoute	Informal	9
	TMS 92/0326; TMS 95/0166; TMS 96/1642	1996	1
	All MVs (National)		39
Uganda	NASE 3 (Migyera, TMS 30572)	1993	17.5
	Nase 4 (SS4)	1999	6.8
	Nase 1 (TMS 60142)	1993	4.9
	Nase 2 (TMS 30337)	1993	4.2
	Nase 9 (TMS 30555 (TMS 30555-17)	1999	1.6
	All MVs (National)		35
Zambia	Bangweru	1993	21
	Mweru (L9-303/151)	2001	12
	Nalumino	1993	5
	Chila (L9-304/151)	2001	3
	Manyopola	2000	3
	All MVs (National)		46
Zimbabwe	M7	1994	34.5
	XM6	1993	17.5
	All MVs (National)		52

in Nigeria in 1984 and TMS 4(2)1425 in 1986, are enjoying wider adoption in SSA. TMS 30572 occupies 17% of the total cassava area in Nigeria, 17% in Uganda, 7 % in Benin and 3% in Guinea. Though not officially released, the same variety, TMS 30572, is also cultivated extensively in Kenya where it covers 24% of the cassava area. TMS 30572 is also grown in Côte d'Ivoire. TMS 4(2)1425 is grown in three countries, occupying 8% in Nigeria, 1% in Ghana and 6% in Côte d'Ivoire. Another common IITA-bred variety, TMS 92/0326, which was first released in Togo in 1998, has spilled over to Cameroon and Nigeria, commanding 4% and 3% of the cassava area, respectively.

In Benin, RB 89509 (32%) and Ben 86052 (23%) are the most cultivated varieties. Both were released in 1980s from IITA parents and were reported to be high yielding. MM96/5280, also known as Rugero, is an IITA-bred variety and was released in 2001 in Burundi and covers 16% of the area under cassava. With 46% yield advantage over the local check variety, 8034 (IRAD8034) and 8017 (IRAD8017) each occupies 8% of cassava area in Cameroon. In DR Congo, the variety Sadisa (TMS 91/203) covers 14% of the cassava area. It is known for its CBB resistance, green mite tolerance, high-yield, high dry matter and high quality flour (cream). In Malawi, a sweet non-IITA variety called Manyokola (or Mbundumali) was formally released in 1980 and accounts for 38% of the total cassava area. The variety Sauti (CH2/007) that was released in 2002 is the second economically important variety in Malawi and was planted to 17% of cassava areas in 2009. In Tanzania, a sweet variety known as Mkombozi (released officially in 2009 but which has been cultivated since early 2000s) accounts for 9% of the cassava area. It is known for its resistance to CMD, tolerance to cassava brown streak disease (CBSD) and high yield. M7 (35%) and XM6 (18%) are the two dominant high-yielding varieties in Zimbabwe. Across SSA, high yield and disease resistance are the preferred attributes of cassava varieties.

Cowpea

Cowpea is an important food and fodder legume and an essential component of cropping systems in the tropical and subtropical areas of Africa, Asia and Latin America. In the dry savannahs of West and Central Africa, farmers traditionally cultivate two main types of cowpea: early-maturing varieties grown for grain and late-maturing varieties that are grown for fodder production. With 25% protein in its grains, cowpea is an important source of low-cost nutrition to the urban and rural poor who cannot afford meat and milk products. Cowpea haulms contain over 15% protein and constitute a valuable source of fodder. In view of its early maturity and ability to fit into a niche crop in multiple cropping systems involving maize, sorghum and millet, cowpea has quickly become one of the most important food, as well as cash, crops in Africa.

Cowpea is extensively covered in the DIIVA Project. The 18 survey countries account for more than 98% of area and production in SSA where cowpea is produced on more than 10 million hectares. Cowpea production is heavily concentrated in Nigeria and Niger, with the area ranging from 3.5 million hectares in Nigeria to 5.5 million hectares in Niger. With an estimated 45% share of the global cowpea production and over 55% of the production in Africa, Nigeria is the world's largest producer and consumer of cowpea. Nigeria and Niger together account for more than 50% of world cowpea production.

Burkina Faso is also an important producing country with over 0.6 million hectares. At the other extreme, 6 of the 18 survey countries are very small producers with harvested area of less than 100,000 hectares.

Given its global mandate for cowpea improvement, IITA has developed and distributed improved cowpea varieties to a large number of national programmes in Africa, Asia and Latin America. To meet the regional preferences for specific seed types and adaptability to different environments, IITA's general strategy is centred on developing a range of breeding lines and varieties with diverse maturities and plant and seed types characterized by broad adaptability in backgrounds featuring high yield and resistance to major diseases, insect pests, and the parasitic weeds *Striga* and *Alectra*.

In addition to IITA and NARS efforts, donor-supported collaborative networks have played an important role in developing and promoting the use of improved cowpea varieties in the region. In particular, USAID's Bean/Cowpea CRSP has catalysed and supported research on cowpea improvement in Cameroon and Senegal.

Scientific staffing of cowpea improvement programmes

The successful implementation of research programmes depends on the availability of well-trained professionals. In 2009, IITA employed 4.5 FTE researchers in cowpea improvement (Table 6.6). Proportionally, these international scientists represented only about 5% of the 76 FTE researchers working in NARS in the 18 cowpea-growing countries covered in the DIIVA Project (Appendix Table 6.A1).

The level of scientific investment in cowpea genetic improvement is low not only internationally but also nationally. Only Nigeria, Senegal and Burkina Faso had at least 10 FTE researchers in their cowpea improvement programmes in 2009. For Nigeria and Burkina Faso, exceeding a critical threshold of 10 researchers is warranted by the size of their national production. With an estimated 45% share of the global cowpea production and more than 55% of the production in Africa, Nigeria is the world's largest producer (and consumer) of cowpea, followed by Niger (15%) and Burkina Faso (5%). Relative to national production, Niger and Cameroon show the lowest levels of investment of the cowpea-growing countries in Table 6.6.

As expected, germplasm conservation and breeding at about 40% contribute a sizeable share to the disciplinary composition in cowpea improvement programmes. Agronomy, seed production, entomology and pathology are also well represented in Table 6.6. But only four programmes have a specialized capacity in pathology at 0.5 FTE scientists or more. Investment in entomology is somewhat higher but, given the importance of insect pests in cowpea production, this area has received less emphasis than anticipated. The three largest national programmes

Table 6.6. Full time equivalent staff by major specialization working on cowpea improvement in sub-Saharan Africa, 2009.

CGIAR/ NARS Programme	Major specialization												
	Germplasm conservation	Breeding	Pathology	Molecular biology	Entomology/ Nematology	Agronomy	Seed production	Tissue culture	Postharvest	Social science	Food science	Others	Total
CGIAR (IITA)	0.2	3.5	0.0	0.0	0.0	0.8	0.0	0.0	0.0	0.0	0.0	0.0	4.5
NARS	7.0	24.3	6.1	3.6	8.4	12.6	6.1	0.5	2.6	3.9	0.7	0.4	76
Benin	1.1	2.1	0.3	0.0	1.2	1.2	1.3	0.0	0.8	0.2	0.1	0.4	8.5
Burkina Faso	0.3	3.9	1.3	1.6	3.0	0.0	0.3	0.0	0.0	0.0	0.0	0.0	10.4
Cameroon	0.2	0.1	0.0	0.1	0.1	0.2	0.2	0.0	0.0	0.2	0.0	0.0	1.1
Côte d'Ivoire	2.3	0.3	0.1	0.0	0.1	0.4	0.4	0.0	0.4	0.0	0.0	0.0	4.0
DR Congo	0.2	0.2	0.1	0.0	0.0	0.2	0.3	0.0	0.0	0.0	0.0	0.0	0.8
Ghana	0.0	1.8	0.1	0.2	0.8	0.8	0.5	0.0	0.1	0.3	0.0	0.0	4.5
Guinea	0.0	0.2	0.3	0.5	0.3	0.2	0.1	0.3	0.3	0.1	0.0	0.0	2.0
Malawi	0.0	0.8	0.2	0.0	0.2	0.2	0.0	0.0	0.0	0.2	0.0	0.0	1.6
Mali	0.0	2.3	0.1	0.0	0.0	1.0	0.0	0.0	0.0	0.0	0.0	0.0	3.4
Mozambique	0.0	0.4	0.0	0.0	0.0	0.0	0.0	0.0	0.0	0.0	0.0	0.0	0.4
Niger	0.2	1.8	0.3	0.0	0.6	0.8	0.7	0.0	0.3	1.3	0.6	0.0	5.4
Nigeria	1.0	4.7	1.5	0.2	1.0	4.3	1.0	0.2	0.7	1.4	0.0	0.0	16.0
Senegal	0.0	1.5	1.0	1.0	0.5	0.5	0.3	0.0	0.0	0.0	0.0	0.0	4.8
Tanzania	0.2	1.5	0.0	0.0	0.0	0.3	0.0	0.0	0.0	0.4	0.0	0.0	2.4
Togo	1.4	0.7	0.7	0.0	0.7	0.7	1.0	0.0	0.0	0.0	0.0	0.0	5.2
Uganda	0.0	1.0	0.1	0.0	0.1	0.6	0.0	0.0	0.0	0.0	0.0	0.0	1.8
Zambia	0.0	0.3	0.1	0.0	0.0	1.3	0.3	0.0	0.0	0.0	0.0	0.0	1.9
Zimbabwe	0.1	0.9	0.0	0.0	0.0	0.0	0.0	0.0	0.0	0.0	0.0	0.0	1.0
Average	0.4	1.3	0.3	0.2	0.5	0.7	0.3	0.0	0.1	0.2	0.0	0.0	4.2

also allocated a small proportion of their disciplinary research portfolio to molecular biology.

The majority of the 76 researchers had earned at least a master's degree in various areas of specialization. Countries with a high level of doctoral training are Nigeria (with 6.5 FTE PhD scientists), Burkina Faso (6.3) and Ghana (3.2). Only four of the very small programmes lacked a PhD presence in their national cowpea improvement programmes.

Varietal output of cowpea improvement

Since 1970, 202 improved cowpea varieties have been released nationally in SSA (Table 6.7). There are more than ten released varieties in Nigeria, Benin, Ghana, Mali, Mozambique, Guinea, Niger and Cameroon.

Half of the reported 202 varieties are IITA-bred materials or genebank accessions released directly following adaptation tests by NARS. The bulk of the other half (36%) comes from non-IITA materials, mainly national landrace selections. The remaining 14% of the total releases were developed by NARS using IITA germplasm as a parent.

The majority of the released varieties were developed under collaborative research projects like CRSP, the Semi-Arid Food Grain Research and Development (SAFGRAD) Project, Project for Cowpeas in Africa (PRONAF), the Grain Legume Improvement Program (SADC-GLIP) and the Latin America Regional Legume Promotion Program under EMBRAPA (Empresa Brasileira de Pesquisa Agropecuária) in Brazil. The contribution of national programmes and other research centres is more pronounced in the 1980s and 1990s. Non-IITA varieties constitute recommended landraces and other improved varieties whose pedigree is devoid of IITA materials. These improved cultivars are maintained and propagated by alternative suppliers in the cowpea improvement process, e.g. national and international public universities and private commercial seed companies.

Table 6.7. The IITA content of improved cowpea varieties in Sub-Saharan Africa, 1970–2010.

	Number of varieties released				Percentage of total release		
Country	IITA-bred	IITA-parent	Non-IITA	Total	IITA-bred	IITA-parent	Non-IITA
Benin	14	3	2	19	74	16	11
Burkina Faso	3	5	1	9	33	56	11
Cameroon	6	2	2	10	60	20	20
Côte d'Ivoire	7	0	1	8	88	0	13
DR Congo	4	0	4	8	50	0	50
Ghana	10	2	3	15	67	13	20
Guinea	8	2	2	12	67	17	17
Mali	6	3	5	14	43	21	36
Niger	7	0	8	15	47	0	53
Nigeria	21	2	17	40	53	5	43
Senegal	0	2	6	8	0	25	75
Togo	3	0	0	3	100	0	0
West and Central Africa	89	21	51	161	55	13	32
Malawi	2	0	0	2	100	0	0
Mozambique	4	2	19	25	16	8	76
Tanzania	4	2	0	6	67	33	0
Uganda	1	1	0	2	50	50	0
Zambia	0	3	2	5	0	60	40
Zimbabwe	1	0	0	1	100	0	0
East and Southern Africa	12	8	21	41	29	20	51
Sub-Saharan Africa	101	29	72	202	50	14	36

West and Central Africa accounts for 161 varieties representing over 80% of the total varietal releases between 1970 and 2010. Mozambique is the only country in Eastern and Southern Africa that has released more than 20 varieties within the analysis period. Coincidentally, in 1992, Mozambique took over from Botswana the project coordination activities for the implementation of the SADC-GLIP, which helped spearhead the release of IT82E-18 and five other varieties in the *Timbawene* family. Botswana, Lesotho, Swaziland, Tanzania and Zambia are the other member countries where the SADC project facilitated the development and selection of improved cowpea germplasm. In 1993, a number of cowpea varieties were recommended for formal release in these countries.

There are 28 informal releases, so named because the date of release is not recorded or these varieties have not been formally cleared by release committees. These unofficial releases are concentrated in Benin, Mozambique and Nigeria.

Enhanced NARS capacity and improved efficiency in research peaked during the 1990s (Fig. 6.2). At that time, the transfer of material, design and capacity blossomed after two decades of investment. Augmented capacity is reflected in the varieties developed from IITA-parent germplasm that were crossed by the national programmes themselves. In this period, non-IITA varieties (which mainly include local landraces) declined from their maximum in the 1980s (Fig. 6.2). IITA-bred lines or varieties crossed with IITA-parental materials trended upwards in the 1990s. The decline in aggregate releases since 2000 should be a cause for concern.

Like almost all IARC crop improvement programmes, specific IITA-related releases reflect an evolving breeding strategy since the early 1970s. First-generation cowpea breeding produced varieties that were high yielding, early maturing, erect, disease-resistant and photoperiod insensitive compared to the parent germplasm or accessions from which they were derived (IITA, 1992). In developing these varieties, other producer and consumer preferences such as seed features of size, colour or texture were not thoroughly incorporated. As a result the developed materials were less popular to West African consumers and semi-subsistence producers. Nevertheless, they provided an excellent base for further improvement in second-generation breeding. Cowpea consumers in West Africa and other regions as well have high preference for large brown, white or cream cowpea seeds with small eyes and wrinkled or rough testa that can quickly imbibe water to facilitate the easy removal of the seed coat during food processing (Singh and Ntare, 1985).

The breeding strategy in the 1980s placed more emphasis on matching consumer preferences and on avoiding susceptibility to the extreme severity of insect damage, while retaining the superior traits identified in earlier initiatives. Cowpea improvement work in the subsequent decade also focused on developing extra-early

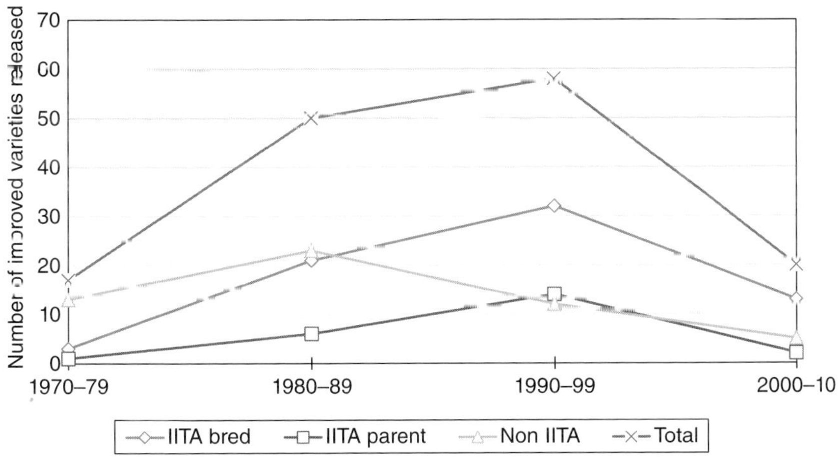

Fig. 6.2. Trends in cowpea varietal releases by germplasm content, 1970–2010.

maturing cowpea varieties with low pesticide requirement suitable for short rainy season areas or for relay cropping in the sub-humid and humid tropics. In the process, varieties IT82E-18, IT82E-16 and a series of similar varieties showed resistance to 11 major cowpea diseases including those transmitted by viruses. A backcrossing programme was undertaken that incorporated aphid resistance to existing lines and led to the isolation of one early maturing variety (IT82E-60) and to the development of medium maturing varieties TVx 3236, IT81D-1137, IT82D-699, IT81D-985, IT81D-994 and IT85D-3516-2, which were initially susceptible but were successfully upgraded to become resistant. Most of these varieties were as good as or better than other popular varieties, such as Ife Brown. For instance, demand for the variety IT82E-60 is strong among farmers in short rainy season regions like northern Kano and Bida in Nigeria, because it is white-seeded and has a maturity period of 60 days, which is about two weeks earlier than the standard varieties. Based on multi-location trials conducted by IITA and its partners, the varieties IT84S-2246-4, IT83S-742-11, IT82D-716 and IT81D-1020 have shown combined resistance to cowpea diseases and pests. These varieties have been tested and recommended for release by numerous national programmes across Africa and are still being grown in West, Central and Southern Africa (i.e. Benin, Burkina Faso, Cameroon, Malawi, Mozambique, Nigeria and Zambia).

Demand-driven breeding has continued. This process takes into account contemporary challenges as experienced by the end-users, while retaining the superior performance demonstrated by early varieties. The breeding programme has identified emerging themes and in the process rightfully incorporated desired traits such as resistance to parasitic weeds like *Striga* (e.g. VITA 3, IT81D-985, IT90K-76, IT90K-59, TN 121-80, IT81D-994, IT82D-849 and IT89KD-245), drought tolerance (e.g. IT96D-604), dual purpose cowpea with high grain yield and fodder quality for crop-livestock systems (e.g. IT89KD-288 and IT90K-277-2) and varieties suitable for cereal–cowpea intercropping (e.g. IT95K-193-12 and IT95K-222-3).

In spite of the progress made in cowpea breeding both nationally and internationally, varietal releases over the period between 1970 and 2010 have been episodic. Nigeria tops the table in terms of frequency of release with 14 released varieties in the 41 years of the review period. Most of the countries, especially Malawi, Mozambique, Uganda, Zimbabwe, Guinea and Togo are characterized by large coefficients of variation indicative of only a few release events since 1970.

Of the 18 countries surveyed, only six countries (DR Congo, Ghana, Malawi, Niger, Tanzania and Uganda) showed a positive trend in mean annual variety release and decreasing variability as evidenced by a decline in the coefficient of variation between 1970–1997 and 1998–2010. This trend in Ghana, Tanzania and Uganda – three of the region's 'Big Eight'[2] – is expected to continue as it coincides with the increase in research investment in these countries. Agricultural research and development spending more than doubled in Ghana during 2000 to 2008. Uganda experienced a threefold increase in agricultural research spending and Tanzania also registered a 1.4% annual gain over the same period. Despite plummeting agricultural research and development spending owing to sociopolitical tension in the 1990s, DR Congo maintained the same annual release rate and reduced variability in release over time. In Niger, cowpea is the second most researched crop. A small increase in annual varietal release from 0.32 to 0.46 between the two periods was anticipated. In these countries, change stems from increased donor support along with growth in government funding to support increased staffing and salary levels, as well as substantial investment in research infrastructure and equipment (Flaherty *et al.*, 2010; Stads *et al.*, 2010; Beintema and Rahija, 2011).

Since 1998, six of the 18 survey countries have not formally released new cowpea varieties. This apparent lack of progress is especially disappointing in a large producer such as Burkina Faso. Additionally, the mean annual release rate of 1.0 for Nigeria fell to 0.62 in the recent period from 1998 to 2010. Much of the decline in release activity is attributed to the closure of the aforementioned SADC-related cowpea project that sparked increased varietal output among several smaller-producing countries in Southern Africa in the 1990s. But several West African countries were also characterized by poor performance in the recent past. Senegal has released only one variety since 1997.

The incidence of spill-over varieties, i.e. those released in more than one country, is high in cowpea genetic improvement in SSA (Table 6.8). TV×3236 and IT82E-16 have been released for cultivation in Benin, Burkina Faso, Cameroon and Nigeria in West and Central Africa, and also in Malawi, Mozambique and Zambia in Southern Africa. IT84S-2246-4, VITA-5, Ife Brown, IT82E-32 (Asontem), IT 89KD-374 (Korobalen) and IT89KD-374-57 are found in at least three countries but are largely restricted in West and Central Africa.

Typical of several of these spill-over varieties, IT84S-2246-4 combines multiple sources of disease resistance with desirable agronomic and seed characteristics (IITA, 1987). Other important traits of IT84S-2246-4 are its short duration (65 days), rough medium-sized seeds, high protein content and relatively short cooking time (30–45 min).

Adoption of improved cowpea varieties

Adoption of improved cowpea cultivars is generally low in SSA, but the estimates in Table 6.9 display considerable variation across countries. In 2009, an estimated 12 million hectares were allocated to cowpea production; expert opinion estimates suggest that, of these, 2.7 million hectares were under improved cowpea varieties. This translates into an estimated adoption rate of 23% for SSA. The adoption of IITA-related cowpea varieties is estimated at 16% of the total cowpea area in the region. In West and Central Africa, some 25% of the cowpea area was under improved varieties.

The highest adoption rates are reported for DR Congo, Ghana, Cameroon, Guinea, Benin and Mali, where the share of improved cowpea area is between 50% and 87% of the total area under cowpea cultivation. Six countries (Zimbabwe, Togo, Nigeria, Tanzania, Senegal and Côte d'Ivoire) are classified in the middle category for improved cowpea adoption in Africa. These countries have reached adoption rates of at least 20% but fall below 50% of total cowpea area. Nigeria has the largest number of released varieties and relatively higher research expenditures committed towards cowpea improvement programmes in Africa but the country has a modest cowpea adoption rate estimated at 39% compared to other countries in the high-adoption category. Despite the high economic importance

Table 6.8. Varietal spill-over of improved cowpea varieties in sub-Saharan Africa, 1970–2010.

Name of variety	Release year	Countries releasing/adopting	Countries (number)
Ife brown (Irawo)	1970	Nigeria, Cameroon, Guinea	3
VITA-5	1974	Benin, Nigeria, Togo	3
IT81D-985 (VITOCO)	1978	Cameroon, Togo	2
Gorom locale	1979	Mali, Burkina Faso	2
VITA-7 (KN-1)	1982	Burkina Faso, Côte d'Ivoire, DR Congo, Guinea	4
IT82E-16	1984	Benin, Malawi, Mozambique, Zambia	4
TVx 3236	1984	Benin, Burkina Faso, Cameroon, Nigeria	4
IT81D-985 (BR1)	1985	Cameroon, Togo	2
IT81D-994 (BR2)	1985	Cameroon, Nigeria	2
IT82E-32	1985	Benin, Cameroon, Ghana	3
Vya	1986	Cameroon, Guinea	2
IT84S-2246-4	1988	Benin, Guinea, Nigeria	3
IT89KD-374-57	1991	Mali, Niger, Nigeria	3
IT84S-2246-4	1993	Benin, Guinea, Nigeria	3
IT89KD-374 (Korobalen)	1993	Mali, Niger, Nigeria	3
IT90K-277-2 (GLM-93)	1993	Cameroon, Nigeria	2
IT82E-18	1994	Mozambique, Zambia	2
IT90K-372-1-2 (Wilibali)	1996	Mali, Niger	2
IT90K-372-1-2	2001	Mali, Niger	2
IT97K-499-35 (Songotra)	2008	Ghana, Nigeria	2
IT90KD-277-2	2008	Cameroon, Nigeria	2

Table 6.9. Adoption of improved cowpea varieties in sub-Saharan Africa, 2009.

Country	2007–2009 cowpea area ('000 ha)	Proportion of total cowpea area planted to: Local varieties (%)	Improved varieties (%)	Materials containing IITA germplasm or directly related to IITA activities '000 ha	MV area (%)
Benin	68	49	51	31	45
Burkina Faso	623	91	10	33	5
Cameroon	111	30	71	55	50
DR Congo	96	13	87	49	51
Ghana	164	19	82	127	78
Guinea	4	38	62	2	60
Malawi	83	90	10	8	10
Mali	264	47	53	80	31
Mozambique	352	89	11	30	9
Niger	5,102	91	9	159	3
Nigeria	3,768	62	39	1,066	28
Senegal	219	73	27	42	19
Tanzania	148	69	31	46	31
Togo	179	60	40	71	40
Uganda	74	85	15	11	15
Zambia	74	83	17	8	11
All	11,328	76	24	1,820	16
Sub-Saharan Africa	11,504	77	23	1,820	16

of cowpea in Niger and Burkina Faso, the two countries reported very low levels of adoption of improved varieties.

Low adoption could be explained by a size of country production that dwarfs the levels of human and financial resource investments. Estimated research intensities per unit area of land (US$/ha and scientist/ha) are very low in Burkina Faso and Niger and could be inadequate to leverage meaningful adoption. The other factor responsible for low adoption could be limited access to improved cowpea seeds or lack of desirable quality traits that appeal to the users.

Adoption rates below 20% were also estimated for Malawi, Mozambique and Zambia. Except for Mozambique, these countries have released few improved cowpea varieties and farmers have limited opportunities for the uptake of suitable varieties. The pool of available improved varieties may not substantially address the needs of the cowpea consumers in the changing biotic and socio-cultural environment. Malawi, Mozambique and Zambia have low to medium research expenditures committed towards cowpea improvement activities (Table 6.6). It is noteworthy that apart from USAID's Bean/Cowpea CRSP, no other major collaborative research programme has been carried out in Southern Africa that is of a magnitude equal to those implemented in West and Central Africa.

Cowpea varieties that accounted for at least 1% of national area are considered to be successful and economically important. These varieties are presented in Table 6.10 for the 18 survey countries in sub-Saharan Africa. Popular varieties in terms of farmer adoption are IT82E-32 covering 23% of the total cowpea area in Ghana, 11% in Benin and 2% in Cameroon followed by VITA-7 accounting for 22% of total cowpea area in Guinea and 13% in DR Congo. The adoption rate for variety IT81D-1137 is estimated at 17% in DR Congo and 14% in Benin. These varieties are attractive to the farmers because they embrace multiple attributes such as high yield potential, disease tolerance and short maturity period. Other varieties with high single-country adoption rates are: IT81D-985 or BR1 (30%) and Lori Niebe (18%) in Cameroon; IT82E-16 (8%) in Benin; H36 (33%) and Diamant (11%) in DR Congo; IT87D-1951 or Asetenapa (19%), ITXP-148-1 (Apagbaala) (14%), IT97K-499-35 (Songtra) (10%) and IT95K-193-2 (Bawutawuta)

Table 6.10. Economically important improved cowpea varieties in sub-Saharan Africa, 2009.

Country	Variety	Release year	Adoption (% area)
Benin	IT 82 E- 32	1985	13
	IT 81D 1137	1990	11
	NI 86-650-3	Informal	6
	KV×396-18	1988	5
	IT99K- 494-6	Informal	3
	IT 97K-568-18	Informal	3
	TV×1850-01E	1987	3
	TV× 32-36	1988	3
	IT95K-193-12	Informal	2
	IT84D-513	Informal	2
	IT98D-1399	Informal	1
	KV×313-2	Informal	1
	All MVs (National)		51
Burkina Faso	Gorom locale	1982	4
	KVx 396-4-5-2D	1990	3
	KVx 61-1	Informal	2
	All MVs (National)		9
Cameroon	IT81D-985 (BR1)	1985	30
	Lori Niebe (C93W 24-130)	1999	18
	Asontem	1995	2
	All MVs (National)		71
Côte d'Ivoire	KN1	Informal	16
	IT86-D400	Informal	6
	IT88D-363	Informal	3
	All MVs (National)		27
DR Congo	DIAMANT	2000	18
	MUYAYA	Informal	17
	Limbimi (IT87D-1137)	2002	17
	VITA7 (TVx 289-46)	1988	13
	H4	1988	9
	H36	1988	7
	H204	Informal	3
	VIMPI	2003	3
	All MVs (National)		87
Ghana	IT82E-32 (Asontem)	1988	23
	IT87D-1951 (Asetenapa)	1992	19
	ITXP-148-1 (Apagbaala)	2003	14
	IT97K-499-35 (Songotra)	2008	10
	IT95K-193-2 (Bawutawuta)	1992	6
	SARC 3-122-2 (Padi-Tuya)	2005	5
	SARC 4-75 (Zaayura)	2005	3
	Bengpla (IT83S-818) & Marfo Tua (SUL 518-2)	1992/2003	1
	All MVs (National)		82
Guinea	VITA-7	1993	22
	IT83S-899	1992	11
	TV4-3000	1993	7
	Ife Brown	1993	5
	IT84S-888	1993	5
	IT84S-2246-4	1993	4
	IT86F-2014-1	1992	2
	IT84E-116	1992	2

Continued

Table 6.10. Continued.

Country	Variety	Release year	Adoption (% area)
	KV×414-22-21	1993	2
	IT81D-1228-14	1992	1
	IT84S-22	1993	1
	VYA	1993	1
	All MVs (National)		62
Malawi	Sudan-1	2003	7
	IT 82E-16	2003	3
	All MVs (National)		10
Mali	Korobalen (IT89KD-374)	1993	18
	Sangaraka (IT89KD-245)	1993	12
	Yere wolo (PRL 73)	1986	11
	Djemani (PBL 22)	1998	7
	Dounan fana	1986	5
	All MVs (National)		53
Mozambique	IT 18	1995	8
	INIA 36	Informal	2
	Timbawene	1995	1
	IT 16	2011	1
	others		0
	All MVs (National)		11
Niger	TN5-78	1984	2
	KV×30-309-6G	1994	1
	TN28-87	Informal	1
	IT89KD-374-57	2001	1
	IT90K372-1-2	2001	1
	TN27-80	1984	1
	TN121-80	1995	1
	All MVs (National)		9
Nigeria	IT90K-277-2	2008	11
	IT89KD-288	2009	6
	Ife brown	1970	5
	Sampea-7 (IAR 48)	1986	4
	IT97K-499-35	2008	4
	IT89KD-391	2009	3
	IT89KD-374-57	1991	2
	IT88D-867-11	Informal	2
	Ife Bimpe (Ife BPC)	1985	1
	IT84S-2246-4	1991	1
	All MVs (National)		39
Senegal	Mélakh	1995	18
	Yacine	2005	8
	Mouride	1991	1
	All MVs (National)		27
Tanzania	IT85F-2020 (VULI-2)	2003	11
	TUMAINI (TVU 410/TVU 2616/SVS 3)	1982	9
	TV×1948-01F (FAHARI)	1982	9
	IT82D-889 (VULI-1)	1987	3
	All MVs (National)		31
Togo	VITOCO	1978	27
	Vita-5	1974	13
	All MVs (National)		40

Continued

Table 6.10. Continued.

Country	Variety	Release year	Adoption (% area)
Uganda	SECOW1T (K21)	2002	8
	SECOW2W (IT81-D-985)	2002	8
	All MVs (National)		15
Zambia	Bubebe (IT82E-16)	1995	11
	Katete	2004	3
	Lutembwe	1993	3
	All MVs (National)		17
Zimbabwe	IT18	1994	45
	All MVs (National)		45

(6%) in Ghana; IT83S-899 (11%) in Guinea; and IT89KD-374 (Korobalen) (18%), IT89KD-245 (Sangaraka) (12%), PRL 73 (Yere Wolo) (11%) and PBL 22 (Djemani) (7%) in Mali.

Maize

Maize is a major food crop in West and Central Africa and the trends in maize production indicate a steady growth due to both area expansion as well as increased yields. For instance, the average maize yield in Africa during 1989–1991 of 1.2 tonnes per hectare was twice that estimated for the 1950s, before improved varieties were generally available (Byerlee and Heisey, 1996). Widespread adoption of improved maize varieties in the savannahs of West and Central Africa has changed the status of maize from a backyard crop to a major cereal grown for both cash and food (Smith *et al.*, 1994; Alene *et al.*, 2009).

IITA has a regional mandate for maize research in West and Central Africa. IITA works in partnership with international and national research and extension services to develop and disseminate improved maize varieties and hybrids that meet the requirements of smallholder farmers. The first scientific breakthrough in West Africa came in the 1970s with the release of the IITA-developed open-pollinated varieties (OPVs), TZB and TZPB. These varieties combined high yields with resistance to rust and blight (TZPB) and drought tolerance (TZB), spearheading the Nigerian maize revolution in the 1980s (Smith *et al.*, 1994). They also have been widely adopted elsewhere in West Africa. Later varieties focused on streak virus resistance and are the basis for currently grown varieties.

The 11 surveyed countries in the DIIVA Project accounted for about 10 million hectares of area in 2007–2009. The majority produced several hundred thousand hectares of maize. Nigeria was the leading producer in the region where the area under maize increased from nearly 4 million hectares in 2007–2009 to 6 million hectares in 2011. Currently Nigeria accounts for over 50% of the maize area and production in West and Central Africa.

Scientific staffing of maize improvement programmes

In general, the total number of FTE scientists in maize national programmes is about the same as in cassava. But, unlike the more diversified allocation in cassava, maize improvement programmes are more heavily concentrated in fewer disciplines, especially breeding, agronomy, seed systems and postharvest technologies (Table 6.11). Nigeria employed more than 77 FTE researchers, which represents a 4–5-fold increase from the 16 FTE researchers employed in 1998 (Alene *et al.*, 2011). Cameroon is the second highest employer in maize research with 17 FTE researchers. Allocations for five other sampled countries fell within the range of 5–10 FTE researchers: Ghana, Côte d'Ivoire, Senegal, Togo and DR Congo. The small maize research systems of Benin, Burkina Faso, Mali and Guinea devoted less than five FTE researchers each.

The large number of FTE researchers working in Nigeria is consistent with the importance of maize and also with recent evidence showing that maize accounts for about 4% of total FTE

Table 6.11. Full-time equivalent staff by major specialization working on maize improvement, 2009.

CGIAR/ NARS Programme	Major specialization											
	Germplasm conservation	Breeding	Pathology	Molecular biology	Entomology/ Nematology	Agronomy	Seed production	Tissue culture	Postharvest	Social science	Food science	Others	Total
CGIAR (IITA)	0.10	6.00	0.35	0.10	0.00	0.15	0.00	0.00	0.00	0.00	0.30	1.00	8.00
NARS	7.37	32.60	6.92	2.80	9.85	26.67	20.90	3.87	16.03	12.35	0.00	0.18	140
Benin	0.30	0.90	0.00	0.30	0.70	1.10	0.45	0.00	0.20	0.40	0.00	0.00	4.35
Burkina Faso	0.10	1.50	0.20	0.05	1.00	0.50	0.25	0.00	0.00	0.10	0.00	0.00	3.70
Cameroon	1.87	5.20	0.57	0.30	0.40	3.87	4.20	0.57	0.13	0.40	0.00	0.00	17.50
Côte d'Ivoire	1.30	1.60	0.00	0.00	1.00	1.00	1.50	0.00	0.00	0.00	0.00	0.00	6.40
DR Congo	0.00	3.00	1.00	0.00	0.00	1.00	0.00	0.00	0.00	0.00	0.00	0.00	5.00
Ghana	0.00	3.10	0.20	0.10	1.10	1.50	0.35	0.00	0.20	0.55	0.00	0.00	7.10
Guinea	0.00	1.10	0.00	0.00	0.00	0.70	0.80	0.00	0.00	0.15	0.00	0.18	2.93
Mali	0.00	1.50	0.05	0.05	0.05	1.50	0.20	0.00	0.00	0.05	0.00	0.00	3.40
Nigeria[a]	1.00	12.10	4.00	2.00	5.40	14.00	11.00	3.30	15.50	9.20	0.00	0.00	77.50
Senegal	1.50	2.00	0.30	0.00	0.20	0.30	1.25	0.00	0.00	0.30	0.00	0.00	5.85
Togo	1.30	0.60	0.60	0.00	0.00	1.20	0.90	0.00	0.00	1.20	0.00	0.00	5.80
Average	0.67	2.96	0.63	0.25	0.90	2.42	1.90	0.35	1.46	1.12	0.00	0.02	13

[a]Data for Nigeria include private sector (Premier Seed Ltd) as well as university researchers (Amhadu Bello University and Obafemi Awolowo University).

researchers (Flaherty *et al.*, 2010). On the other hand, the fivefold increase in maize research capacity could be a result of a 35% increase in total agricultural research capacity between 2000 and 2008. For maize and cowpea in Nigeria, the data on scientific staffing includes private-sector (Premier Seed Ltd) and university (Amhadu Bello University in the north and Obafemi Awolowo University in the south) researchers working on maize improvement. Expanded sector participation partly explains the greater concentration of scientific staff working on maize relative to other crops (e.g. cassava) in Nigeria as well as the greater concentration of scientific staff working on maize in Nigeria relative to other countries in West and Central Africa. In particular, the greater number of researchers carrying out research in the areas of postharvest (15 FTEs) and socioeconomics (9 FTEs) is attributable to university programmes that have a large number of faculty staff studying a range of crops with a practical focus on maize as part of larger collaborative efforts led by IITA and other international research and development agencies.

With a regional mandate for international maize research in West and Central Africa, IITA had about eight FTE researchers working on maize improvement in 2009. Unlike East and Southern Africa, West and Central Africa has not benefited from private sector investment in maize improvement. With the exception of Premier Seed Ltd in Nigeria, there is no private sector investment in maize research in West and Central Africa. Premier Seed had about 6 FTE researchers working on maize breeding and seed production in 2009.

The contrast between East and Southern Africa and West and Central Africa in terms of the relative importance of the private sector is largely related to the relative historical emphasis of the respective maize research programmes on OPVs and hybrids. The use of hybrids in East and Southern Africa reflects the fact that maize research programmes in this region were developed originally for the large-scale commercial sectors in Kenya, Zimbabwe and South Africa. Hybrids later spread to smallholder farmers in these and neighbouring countries, such as Malawi, Swaziland and Zambia. In most of these countries, the dominant maize-growing ecology is located in mid- and high-altitude areas. In West and Central Africa, on the other hand, improved OPVs are more important than hybrids. In these areas, the dominant ecology for maize is the tropical lowland. Hybrid materials were unavailable for this ecology until recently (Byerlee and Heisey, 1996).

Besides improved capacity in the implementation of the research programmes, PhD scientists are needed to facilitate internal capacity for transformation through in-service training where senior researchers offer training to junior researchers. In-service training through 'learning by doing' facilitates the smooth takeover and continuity of research programmes. Furthermore, senior scientists help in the organization of the breeding programme and offer strategic direction. In 2009, close to three-quarters of the total maize FTE researchers in the sample of 11 countries had postgraduate-level training. With 29 PhD and 29 MSc researchers, Nigeria had the most educated staff working on maize research, followed by Cameroon with 10.4 MSc and 4 PhD researchers involved in maize genetic improvement. Almost half of the postgraduate staff working on maize research had a PhD degree. High postgraduate qualifications are expected in Nigeria and Ghana because they are among the countries with the most complex agricultural research systems.

With almost all staff qualified to BSc level only, researchers in Guinea are the least highly qualified of those in the 11 countries surveyed, followed by DR Congo with three-fifths of total maize FTE researchers qualified to BSc level only. Both countries have experienced an overall reduction in research and development spending since 2000. According to Stads *et al.* (2010), low levels of postgraduate-qualified researchers in Guinea are due to the country's lack of agricultural development and to its isolation until the mid-1980s. Guinean universities do not currently offer PhD-level courses in agricultural and veterinary sciences. In DR Congo, the decline is largely attributed to the retirement of a large number of researchers employed by research centres placed under the General Delegation of Scientific and Technical Research (DGRST), exacerbated by a public-sector hiring freeze (Stads *et al.*, 2010).

Relative to size of production in the surveyed countries, human resource investment in maize research is substantially greater than for cassava and cowpea. In 2009, West and Central African countries invested 8 FTE researchers per million tonnes of maize production. Estimated research intensities in maize improvement programmes are also more tightly clustered than comparable ratios in cassava and cowpea programmes. Inter-country comparisons show that intensity ratios vary from 2.9 to 19.2 FTE researchers per million tonnes of production. This difference among crops suggests that countries have either prioritized maize investments or found ways to protect maize research investments to ensure a reasonable level of capacity of human resources.

Intensity ratios that stand out as high are those for Senegal (19.2), Cameroon (11.6), Nigeria (10.5), Côte d'Ivoire (9.8) and Togo (9.3). The West Africa Agricultural Productivity Program (WAAPP), which was first implemented in 2007, could be the reason behind high investment levels in maize research in Senegal. Financed through World Bank loans, WAAPP's objective is to develop and disseminate improved agricultural technologies in the participating countries and Senegal was put in charge of cereals. Benin, Burkina Faso, Ghana, DR Congo and Guinea allocated 3–5 FTE researchers per million tonnes of maize production. Relatively low investment in maize research in Burkina Faso and DR Congo could be attributed to diversity in their agricultural research systems, which translate into low budgetary shares of 30% in Burkina Faso and less than 35% in DR Congo for crop research (Beintema and Rahija, 2011). Mali was characterized by the lowest research intensity in maize research, employing only 2.9 FTEs for each million tonne of production. It is likely that this low estimate was precipitated by staff layoffs that followed the completion of large donor-funded projects in Mali between 2001 and 2008, financed by the World Bank, the Netherlands and the Syngenta Foundation (Stads and Maiga, 2010).

Varietal output of maize improvement

From 1970 to 2010, 327 improved maize varieties have been released by the national programmes or informally deployed for cultivation by farmers (Table 6.12). Nigeria accounts for about one-third of total releases but all surveyed countries have released at least ten varieties.

The use of IITA germplasm by maize breeding programmes has been extensive in West and Central Africa. More than half (54%) of the entries in Table 6.12 were IITA-bred and 16% were developed from IITA parents, indicating that about 70% of the maize varieties released have IITA germplasm. Indeed, more varieties have been developed from crossing with IITA parents than via selection of IITA-bred progenies by NARS.

Country differences are evident in the IITA content of the released varieties. Maize breeding in Togo, Nigeria, Benin, Guinea, Senegal and Cameroon relies heavily on IITA materials. In contrast, variety releases without IITA germplasm are dominant in Burkina Faso and Ghana. Nigeria, Cameroon and Benin have stronger capacity to develop and release varieties as evidenced by

Table 6.12. The IITA content of improved maize varieties in West and Central Africa, 1970–2010.

	Number of varieties released				Percentage of total release		
Country	IITA-bred	IITA-parent	Non- IITA	Total	IITA-bred	IITA-parent	Non-IITA
Benin	21	7	8	36	58	19	22
Burkina Faso	10	3	19	32	31	9	59
Ghana	12		15	27	44	0	56
Cameroon	11	19	14	44	25	43	32
Côte d'Ivoire	5	1	5	11	45	9	45
Guinea	8	1	3	12	67	8	25
Mali	10	2	9	21	48	10	43
Nigeria	75	16	20	111	68	14	18
Togo	12	1		13	92	8	0
Senegal	6	1	3	10	60	10	30
DR Congo	5		5	10	50	0	50
Total	175	51	101	327	54	16	31

their use of IITA parents or ancestors without direct IITA involvement.

About 13% of the released varieties in Table 6.12 are informally released varieties or varieties with unknown release years. Unlike other crops reported in this chapter, most of the surveyed countries have one or more unofficial maize releases.

Of the 284 dated releases, 184 (65%) were released between 1970 and 1997, and 100 (or 35%) over the recent period between 1998 and 2010. A regional average release rate of eight varieties per year in the recent period indicates sustained progress in public sector maize breeding in most of the surveyed countries. The exceptions are Côte d'Ivoire and DR Congo with no releases in the recent period.

Compared to the 1970s, maize releases doubled in the 1980s and 1990s and trended further upward in the 2000s (Fig. 6.3). Much of this trend is driven by the increase in the use of IARC-related materials, especially those from IITA. IITA-bred varieties increased from three in the 1970s to 46 releases in the 1980s. Over the same period, total releases in the region also doubled from 40 to 80 varieties. The sharp rise is varietal output reflects the activities of Semi-Arid Food Grain Research and Development (SAFGRAD) project mainly funded by USAID with a vision to reinforce and coordinate agricultural research and develop suitable farming systems for increased productivity of major food staples including maize. Collaborative efforts of SAFGRAD and IITA were the main mover of maize germplasm improvement in West and Central Africa (Sanders *et al.*, 1993). SAFGRAD worked as an intermediary between IARCs (IITA and CIMMYT) and NARs by facilitating the movement of new germplasm and new technology concepts. The phasing out of the SAFGRAD project coincided with a slight fall in number of IITA-bred maize variety releases from 46 to 35 in the 1990s, and an increase in the release of varieties with IITA parents from nine to 17. Increased usage of IITA germplasm as parents in maize breeding implies growth and development in maize breeding systems in the region. Several NARs are capable of carrying out crossing programmes and producing improved varieties on their own.

All the countries surveyed experienced periods of inactivity where no varieties were released for several consecutive years. Nigeria has the highest average release rate of three varieties per year. Six of the 11 countries, namely Côte d'Ivoire, DR Congo, Guinea, Mali, Senegal and Togo, are characterized by a release rate of below one variety per year.

Years with zero releases are strikingly high: seven of the 11 surveyed countries did not release maize varieties in 40 or more years from 1960–2010. At the opposite end of the output spectrum, Nigeria registered a positive outcome in varietal release in 25 years from 1960. Releasing maize varieties in roughly half of the years of a 50-year period is truly a remarkable achievement for a developing country.

For the majority of the surveyed countries, the abundance of non-release years signifies inactivity or lack of progress in a plant-breeding

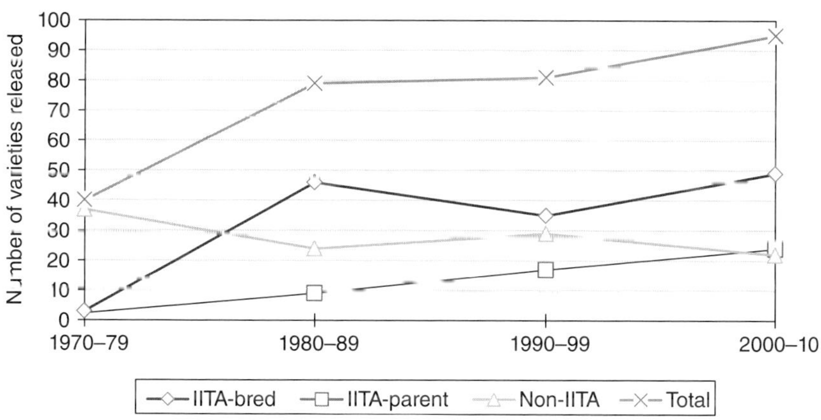

Fig. 6.3. Trends in maize varietal releases by germplasm source, 1970–2010.

programme. Lack of continuity, in turn, could be explained by several constraints, such as weak scientific capacity, funding scarcities, inadequate linkages with research systems outside the region and non-adherence to a research master plan (Venkatesan and Kampen, 1998). A comparison of decadal intervals shows that highest incidences of variety releases were recorded in the period starting from the 1990s extending into 2000s for most of the countries in the region.

Average annual releases for West and Central Africa between 1970–1997 and 1998–2010 changed remarkably from 0.43 to 0.77, up by 80%. Nigeria and Togo registered very high increases in mean releases from 1.61 and 0.05 in 1970–1997 to 2.92 and 0.77 in 1997–2010. The presence of IITA headquarters in Nigeria together with the size of NARs explains the success of the maize breeding system in Nigeria. In addition to providing financial assistance, breeders at IITA backstop research activities by providing material as well as technical assistance and capacity building to Nigeria's maize improvement programme.

In Togo, maize has become the most researched crop, accounting for 14% of total FTE researchers. Variability in annual releases as measured by coefficient of variation for varietal releases has also declined in Benin, Burkina Faso, Cameroon, Côte d'Ivoire, Guinea, Senegal and Togo between the two periods, signifying positive progress in productivity of maize research in SSA.

Adoption of improved maize varieties

During the period between 2007 and 2009, the total area allocated to maize production in the region was estimated at 12 million hectares, of which 5.3 million hectares were under improved maize varieties in the 11 country observations in the DIIVA Project (Table 6.13). This represents an adoption level of 67% of the total maize area in the countries surveyed and 57% of the maize area in West and Central Africa (assuming no adoption of improved varieties in the non-surveyed countries).

Both Nigeria and Senegal are approaching full adoption of improved maize varieties. Adoption of improved varieties in Cameroon, Burkina Faso and Mali was also high, occupying 82%, 72% and 71% of maize harvested area, respectively. Benin, Guinea, Côte d'Ivoire and Ghana allocated more than 50% but less than 70% of their maize area to modern varieties and

Table 6.13. Adoption of improved maize varieties in West and Central Africa, 2009.

		Proportion of total maize area planted to:		Materials containing IITA germplasm or directly related to IITA activities	
Country	2007–2009 maize area ('000 ha)	Local varieties (%)	Improved varieties (%)	'000 ha	MV area (%)
Benin	830	46	54	291	35
Burkina Faso	555	28	72	290	52
Cameroon	686	18	82	444	65
Côte d'Ivoire	302	46	54	51	17
DR Congo	1,489	85	15	1	0.1
Ghana	864	43	57	294	34
Guinea	440	33	67	64	15
Mali	409	29	71	190	47
Nigeria	3,708	3	97	3,586	97
Senegal	191	3	97	139	73
Togo	445	95	5	20	4
All	9,919	33	67	5,369	54
West and Central Africa	11,702	43	57	5,369	46

hybrids. With only 15% and 5% of area under improved varieties, the DR Congo and Togo lagged far behind the other surveyed countries in the uptake of modern varieties.

How do the findings in Table 6.13 compare with those of the 1998 initiative by Manyong *et al.* (2003)? Area planted to improved varieties in West and Central Africa increased from 37% in 1998 to 57% in 2009. In Nigeria, adoption of improved varieties was very rapid, reaching 97% by 2010 compared to 40% recorded in 1998. All other surveyed countries except DR Congo and Ghana more than doubled the area under improved varieties between 1998 and 2009. Area under improved varieties in Ghana increased only by 4 percentage points, from 53% in 1998 to 57% in 2010.

Of the total 11.7 million hectares planted to maize in West and Central Africa, about 5.3 million hectares (46%) were planted to varieties containing IITA germplasm (Table 6.13). However, the use of IITA-derived varieties varied widely by country. The whole area under improved varieties in Nigeria was under IITA-related varieties. More than 50% of maize area was planted to IITA-related varieties in Senegal, Cameroon and Burkina Faso. Conversely, in DR Congo and Togo, only 0.1% and 4% of maize area was devoted to IITA-derived varieties. Togo spends less on research and development than most other countries in the region; worse still, expenditure on agricultural research has been declining at a rate of 4% over the past 10 years. Recent releases have not yet translated into adoption.

Important varieties, estimated to contribute at least 1% to national adoption, are listed in Table 6.14. An IITA high-yielding dent variety called DMR ESR W, which was released in 1987, is the leading modern maize cultivar in Benin, occupying about 23% of total maize area. EVDT 97 STR C1 is another important IITA-derived variety prized for its drought tolerance and high yield. It was formally released in 1999, first adopted in 2002 and now accounts for about 7% of total maize area in Benin.

Six improved varieties occupy at least 3% of total maize area in Burkina Faso. The most important varieties are SR21 (EV8421 SR) with about 20% coverage and Obatanpa (Pop 62) with 19%. SR21 (released in 2001) is an IITA-bred variety with Maize Streak Virus resistance and average grain yield of 5 tonnes/ha. Released in 1998, Obatanpa is a protein-rich variety that was derived from IITA and CIMMYT genetic materials.

In Cameroon, Kassai (CHC 201) is the leading variety. It is a 22-year old variety derived from an IITA parent. About one quarter of the area in Côte d'Ivoire is occupied by a non-IITA variety called F 7928, which was released in 1992. It is high yielding as well as tolerant to streak virus. In DR Congo, almost all the maize area under improved varieties is under a MUS-1 (MUS-1 is a NARS-bred variety released in 1996), accounting for nearly 15% of area. In Ghana, Obatanpa was released in 1992. It occupies 26% of the maize area under improved varieties. Etubi is another important variety in Ghana. It is a variety released in 2007 and it now occupies about 11% of total maize area. It is a hybrid of medium maturity and can yield as much as 6.5 tonnes per hectare. Oba 98, Suwan 1-SR, TZE-Y, Sammaz 11 (Acr 97 TZL Com. 1-W), NARZH 1 (Oba Super1) and TZEE-W are the most important varieties and hybrids with adoption rates ranging from 10 to 17% of total maize area in Nigeria. Suwan 1-SR (released in 1988) and NARZH 1 (Oba Super1) (released in 1984) are the oldest varieties under wide cultivation. Suwan 1-SR is a late-maturing variety with resistance to streak and downy mildew, whereas NARZH 1 (Oba Super1) is a semi-flint hybrid with resistance to *Striga*, streak, and weevils. Oba 98 and Sammaz 11 (Acr 97 TZL Com. 1-W) were both released in 2001 from IITA materials. TZEE-Y and Early Thai are the dominant improved varieties in Senegal with TZEE-Y occupying 47% and Early Thai 24% of the total maize area. TZEE-Y is an IITA-bred variety, whereas Early Thai is a CIMMYT-bred variety. Both are high-yielding varieties with resistance to lodging and are insensitive to blight. Nieleni (DMR ESR Y) is a multi-disease resistant IITA-bred variety occupying 17% of the area in Mali and K 9101 (non-IITA) is the most adopted improved variety in Guinea accounting for 39% of the maize area.

The release dates combined with the adoption levels in Table 6.14 suggest that the age of modern varieties in farmers' fields exceeds 20 years for Côte d'Ivoire, Cameroon and Mali. Slow varietal turnover is a cause for concern in these countries that have released few if any varieties in the recent past.

Table 6.14. Economically important improved maize varieties in West and Central Africa, 2009.

Country	Variety	Release year	Adoption (% area)
Benin	DMR ESR W	1987	23
	EVDT 97 STR C1	1999	7
	DMR	2005	7
	TZPB SR W	1987	5
	QPM	1994	4
	FAABA (QPM)	1987	4
	EVDT 97 STR W	1989	2
	TZPB	1999	1
	2000 SYN EE W	2005	1
	All MVs (national)		54
Burkina Faso	SR21	2001	20
	OBATANPA	1998	19
	FBC6	1999	11
	WARI	2007	9
	ESPOIR	2004	8
	BARKA	2007	3
	All MVs (national)		72
Cameroon	Kassai (CHC 201)	1987	13
	Shaba	1991	11
	CMS 8704	1987	9
	CMS 8501	1985	8
	ATP SR Y (CHC 202)	1990	7
	CLH 103	2000	6
	TZEE-W	2004	5
	Coca	1991	4
	CMS 8806	1988	4
	Obatanpa	2003	4
	CMS 9015	1990	3
	BSR 81	1987	1
	TZEE	Informal	1
	CHH 105	1993	1
	CHC 101	Informal	1
	All MVs (national)		82
Côte d'Ivoire	F 7928	1992	23
	EV99-QPM	Informal	6
	MTS (Maïs témoin station)	1990	5
	EV8766-SR-QPM	Informal	5
	TZE Comp 4	Informal	4
	TZL Composite 3	Informal	3
	CJB	1990	2
	All MVs (national)		54
DR Congo	MUS-1	1996	14.85
	AK 9331 DMR-ESR-Y	1996	0.075
	Salongo-2	Informal	0.045
	Kasai-1	1976	0.03
	All MVs (national)		15
Ghana	Obatanpa	1992	26
	Etubi	2007	11
	Mamaba (GH 110)	1996	11
	Okomasa	1988	5
	Dorke-SR	1992	3
	Kawandzie	1984	1
	All MVs (national)		57

Continued

Table 6.14. Continued.

Country	Variety	Release year	Adoption (% area)
Guinea	K 9101	1992	39
	DMR ESR Y	1996	12
	CMS 475	1999	11
	Obatanpa	1997	3
	Perta	1969	2
	EV28SR	1989	1
	All MVs (national)		67
Mali	Nieleni	1992	17
	Sotubaka	1992	12
	Molobala2	1983	10
	Kogoni B	1970	10
	Dembanyuman	1995	7
	EV8422 SR	1984	6
	Sataba	Informal	2
	Kababa	Informal	2
	Znfiè	Informal	2
	Apollo	1996	2
	TZE SR W	1983	1
	Zanguereni	1972	1
	All MVs (national)		71
Nigeria	Oba 98[a]	2001	17
	Suwan 1-SR	1988	15
	TZE-Y	Informal	13
	Sammaz 11 (ACR 97 TZL Com. 1-W)	2001	12
	NARZH 1 (Oba Super 1)[a]	1984	10
	TZEE-W		10
	NARZH 15 (Oba Super 2)[a]	1996	6
	NARZO 28 (TZMSR-W)	1985	4
	8535-23	Informal	1
	NARZO 24 (DMR-LSR-W)	1984	1
	All MVs (national)		97
Senegal	TZEE-Y	1998	47
	Early Thai	1998	24
	Suwan 1 SR	1999	15
	Obatanpa	2000	9
	Pool 16 DR	1997	1
	DMR ESR W	1998	1
	All MVs (national)		97
Togo	Ikenne 9449 SR	2003	3
	Obatanpa	2003	2
	AB11	1985	1
	All MVs (national)		6

[a]Hybrids

Soybean

Soybean is a relatively new but expanding food crop in SSA and constitutes an important component of the smallholder cropping systems with considerable potential for arresting soil fertility decline, raising household incomes, and enhancing household food and nutrition security. With average yields of about 1.3 tonnes per hectare and total production of 1.8 million tonnes, however, soybean production in Africa is characterized by low production and productivity. Nigeria and the Republic of South Africa are the largest producers and together account for more than 50% of the total soybean production in Africa. IITA embarked on soybean improvement around 1974 with the goal of improving soybean yields in tropical Africa, which at the time were less than half a tonne per hectare. The main goal of soybean improvement at IITA has been to develop high-yielding and stable soybean varieties that are tolerant to biotic and abiotic constraints and to promote soybean processing and utilization suitable for smallholder farmers in tropical Africa. Given the low soybean yields in Africa relative to other continents, high and stable grain yield was the main objective from the outset and this is still a top priority in soybean breeding and crop management. Other constraints of importance to soybean research include seed viability, nodulation with *Rhizobium* available in the soil and pod shattering.

Scientific staffing of soybean improvement programmes

With a regional mandate for international soybean research in the CGIAR since the 1970s, IITA has contributed to soybean improvement efforts as well as capacity strengthening of several national soybean research programmes in SSA. In 2009, IITA had 6 FTE researchers working on soybean improvement (see Table 6.15).

Despite isolated cases of strong and concerted effort on varietal improvement, regional human resource capacity remains limited. Scientific staff time allocated to soybean research in the region ranged between 0.8 FTE in DR Congo to 14.6 FTE in Nigeria, with most countries clustered around 2.0 FTE researchers.

This distribution of researchers was expected because almost all countries are still only very small producers of soybean. Nigeria is by far the largest producer with about 650,000 hectares under cultivation. Uganda is a distant second to Nigeria in size of production, with an area approaching 150,000 hectares. Indeed, 5 of the 14 surveyed countries had less than 10,000 hectares in production in 2007–2009.

Nigeria with 14.6 FTE Researchers and Zambia with 10 are the only countries that allocated scientific resources to soybean above the regional average. Nigeria's high investment level could be explained by the doubling of agricultural research and development spending between 2000 and 2008, which saw a rise in research capacity by a factor of 0.54 (Flaherty *et al.*, 2010). Nigeria's relatively greater investment in soybean improvement is also consistent with the fact that it is the largest producer of soybean in SSA. In 2007–2009, about 1.3 million tonnes of soybean was produced in SSA, with Nigeria alone accounting for nearly half (over 600,000 tonnes). During the period 1970–2008, soybean production in Nigeria grew by about 7.5% per year, with area and yield each accounting for nearly half of this growth.

In Zambia, lifting of a hiring freeze in 2006 resumed growth in research capacity. With only about 50,000 hectares of area under soybean, however, Zambia's investment in soybean research is among the highest estimated research intensity of any country-by-crop observation in the DIIVA Project. This commitment to soybean research is fuelled by private sector participation and rising expectations for the commercial production of a commodity that is rapidly expanding from a very small base.

The total number and the disciplinary composition of researchers in Table 6.15 warrant discussion. Although a team of 52 FTE scientists seems small in absolute terms, it is large in relative terms because the total area of soybeans only slightly exceeds 1.1 million hectares in SSA. For comparison, cowpea genetic improvement programmes were characterized by about 2.7 times as many researchers but with about 10.9 times the planted area.

The disciplinary allocation shows a greater emphasis on germplasm conservation than any of the other four food crops described in this paper. A 20% share in breeding is considerably smaller than what is reported in other food crops that are sexually propagated. Taken together, a higher emphasis on germplasm conservation and a lower emphasis on breeding suggest that soybean improvement programmes in SSA rely heavily on borrowing elite lines and finished varieties from mature country programmes in Asia, Latin America and North America, and from international providers such as IITA and INTSOY than similar programmes in other food crops. Only four of the countries have the equivalent of at least one FTE breeder who is focused wholly on soybean breeding.

Table 6.15. Full-time equivalent staff by major specialization working on soybean improvement, 2009

CGIAR/ NARS programme	Major specialization												
	Germplasm conservation	Breeding	Pathology	Molecular biology	Entomology/ Nematology	Agronomy	Seed production	Tissue culture	Postharvest	Social science	Food science	Others	Total
CGIAR (IITA)	0.10	3.70	0.15	0.00	0.10	1.20	0.00	0.00	0.00	0.40	0.15	0.00	6
NARS	0.15	9.65	5.50	0.10	7.45	9.65	7.60	0.00	0.90	1.10	0.00	0.05	52
Benin	0.10	0.30	0.30	0.00	0.30	0.20	0.10	0.00	0.05	0.15	0.00	0.05	1.55
Burundi	0.00	1.10	0.00	0.00	0.00	0.90	0.90	0.00	0.00	0.30	0.00	0.00	3.20
Cameroon	0.80	0.05	0.00	0.00	0.00	0.50	1.25	0.00	0.00	0.00	0.00	0.00	2.60
Côte d'Ivoire	2.70	0.40	0.10	0.00	0.10	0.40	0.40	0.00	0.40	0.00	0.00	0.00	4.50
DR Congo	0.25	0.20	0.00	0.00	0.00	0.15	0.20	0.00	0.00	0.00	0.00	0.00	0.80
Ghana	0.00	0.50	0.00	0.00	0.30	0.50	0.40	0.00	0.00	0.20	0.00	0.00	1.90
Kenya	0.00	0.25	0.00	0.00	0.05	0.60	0.05	0.00	0.05	0.00	0.00	0.00	1.00
Malawi	0.00	0.80	0.20	0.00	0.20	0.00	0.70	0.00	0.00	0.20	0.00	0.00	2.10
Nigeria	2.00	1.25	2.30	0.00	2.00	3.50	2.90	0.00	0.40	0.25	0.00	0.00	14.60
Tanzania	0.00	0.40	0.50	0.00	0.10	0.00	0.00	0.00	0.00	0.00	0.00	0.00	1.00
Togo	1.40	0.00	0.60	0.00	0.00	0.70	0.50	0.00	0.00	0.00	0.00	0.00	3.20
Uganda	0.00	0.80	0.00	0.00	1.00	1.00	0.00	0.00	0.00	0.00	0.00	0.00	2.80
Zambia	2.80	1.70	1.50	0.10	3.40	1.20	0.20	0.00	0.00	0.00	0.00	0.00	10.90
Zimbabwe	0.10	1.90	0.00	0.00	0.00	0.00	0.00	0.00	0.00	0.00	0.00	0.00	2.00
Average	0.73	0.69	0.39	0.01	0.53	0.69	0.54	0.00	0.06	0.08	0.00	0.00	3.73

The small-size distribution of production also has implications for differences in the educational profile of soybean researchers and those in other IITA-mandated crops. In particular, small size combined with less economic importance may result in a higher allocation of researchers with lower educational attainment. That expectation is confirmed because soybean is the only crop for which BSc researchers are more numerous than their colleagues with PhDs and MScs. On average, of the soybean FTE researchers in the countries surveyed in SSA 28% held PhD, 34% held MSc and 38% held BSc degrees.

These mean estimates mask widespread variation in educational attainment among NARS programmes. Soybean improvement in Nigeria is primarily carried out by a team of researchers with PhDs (8.35) and MScs (3.85). All researchers are educated at the PhD level in Ghana's soybean crop improvement programme. In Kenya and Benin, the soybean programme depends mainly on MSc level staff. Relatively equal proportions of researchers holding PhD, MSc and BSc degrees were observed in Malawi and Uganda. Despite registering the second highest scientific staffing of more than 10 FTE researchers in the region, Zambia's public sector scientific staffing for soybean is dominated by BSc-level staff. All the FTE researchers (2.0) in Zimbabwe were trained at BSc level.

Varietal output of soybean improvement

Between 1970 and 2011, 195 soybean varieties have been released in SSA (Table 6.16). Most (51%) were produced in Eastern and Southern Africa, with Zimbabwe (31) and Zambia (30) dominating the number of releases in the region. This high level of varietal output is attributed to the involvement of private companies in soybean genetic improvement, such as SeedCo in Zimbabwe and ZamSeed in Zambia (Beintema *et al.*, 2004). In West and Central Africa, Nigeria has the highest number of releases (20), followed by Côte d'Ivoire (18), Benin (12) and Cameroon (10).

Of the 195 releases between 1970–2010, 119 (or 61%) varieties were private-sector or NARS bred and selected in country (Table 6.16). Of these, 104 (or 87%) were released in East and Southern Africa. Private sector and NARS efforts in East and Southern Africa are evident in six of the surveyed countries. In Zambia, Zimbabwe, Malawi, Tanzania, Uganda and Burundi, NARS contributed more than 70% of the releases. This shows that the private seed companies and NARS in these countries are taking more responsibility for soybean variety development.

IITA has made considerable progress in soybean improvement, especially in West and

Table 6.16. The IITA content of improved soybean varieties in sub-Saharan Africa, 1970–2011.

	Number of varieties released				Percentage of total release		
Country	IITA-bred	IITA-Parent	Non-IITA	Total	IITA-bred	IITA-Parent	Non-IITA
Benin	9	0	3	12	75	0	25
Cameroon	7	1	2	10	70	10	20
DR Congo	5	0	0	5	100	0	0
Côte d'Ivoire	15	0	3	18	83	0	17
Ghana	5	0	2	7	71	0	29
Nigeria	16	0	4	20	80	0	20
Togo	6	0	1	7	86	0	14
Burundi	2	1	8	11	18	9	73
Kenya	4	0	6	10	40	0	60
Malawi	0	0	14	14	0	0	100
Tanzania	0	2	7	9	0	22	78
Uganda	2	1	8	11	18	9	73
Zambia	0	0	30	30	0	0	100
Zimbabwe	0	0	31	31	0	0	100
Sub-Saharan Africa	71	5	119	195	36	3	61

Central Africa. Of the 195 improved varieties, 71 (or 36%) were IITA elite lines and finished materials designated as TGx, which stands for tropical glycine cross. Only five of the varieties were NARS selections from introduced IITA crosses and progenies. About 63 of the 79 varieties released in West and Central Africa are a direct result of IITA germplasm enhancement, reflecting the important role that IITA has played in soybean improvement in the region. All improved varieties released in DR Congo were IITA lines. More than 70% of the improved varieties released in Benin, Nigeria, Cameroon, Ghana, Côte d'Ivoire and Togo were from IITA materials. In contrast, IITA elite lines and finished materials account for only 12 (or 10%) of the 116 varieties released in East and Southern Africa, where the private-sector is a major player in breeding and seed production.

In West and Central Africa, soybean improvement started in Nigeria in 1974 and it spilled over to Côte d'Ivoire, Ghana and Cameroon. With a base in Malawi, IITA only recently established soybean improvement in Southern Africa in 2007. Several IITA-related varieties have been released by national programmes in East and Southern Africa: two varieties in Burundi (TG× 1485-1D and TG× 1019-2EB); three in Kenya (TG× 1740-2F, TG× 1835-10E, and TG× 1895-33F); two in Uganda (Namsoy 1 and TG× 1835-10E); and one in Malawi (TG× 1740-2F).

Largely because of private-sector participation, only Zambia, Zimbabwe and Malawi have demonstrated significant ability to develop varieties from their own crosses in the region. The other 11 countries are characterized by a weak capacity to undertake crossing where access to IITA elite materials is of paramount importance to soybean improvement in these countries. As yet, NARS in West and Central Africa have released only one variety that they crossed using one or more IITA parents. The total dependence on finished materials is expected in these very small crop improvement programmes but it is surprising in Nigeria with 14.6 FTE scientists.

About 45 of the 195 improved varieties described as released are either not formally cleared by variety release committees or their dates of release are unknown. Release trends for the remaining 150 varieties are documented in Fig. 6.4 by germplasm source. IITA-related soybean varieties increased during the period 1970–1999 but declined sharply in the 2000s from their peak level in the 1990s. This trend is also reflected in the increase in the total number of varieties released in SSA. Total releases peaked between 1990 and 1999.

The share of non-IITA released varieties shows a gradual rise over the same period. During the period 2000–2010, non-IITA varieties eclipsed IITA-bred releases. The increase in non-IITA related releases is mostly due to the release of 16 varieties by the private sector in Zambia during the past decade.

Given soybean's promise as an emerging food and cash crop in SSA, the decline in total

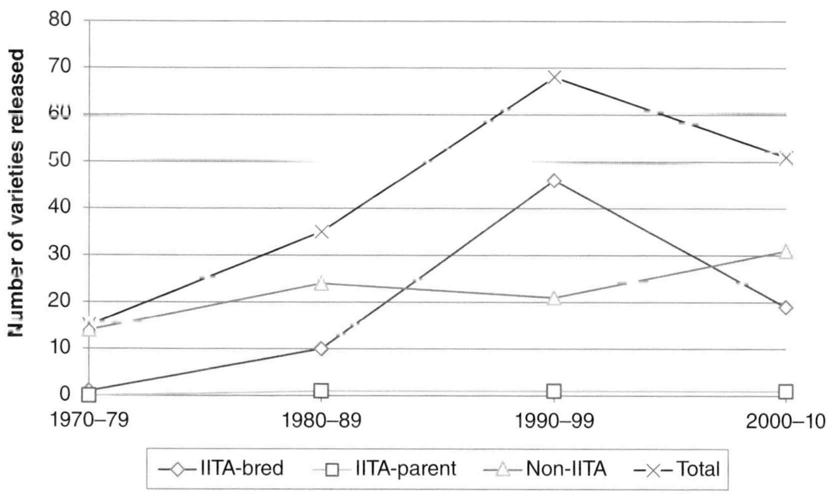

Fig. 6.4. Trends in soybean varietal releases by germplasm source, 1970–2010.

releases in the 2000s was unexpected. Tightening international resources for crop improvement research in the 1990s and early 2000s seems to have resulted in a restricted flow of improved soybean varieties to farmers during the 2000s in West and Central Africa.

Of the 14 countries surveyed, Zambia, Nigeria, Côte d'Ivoire, Zimbabwe and Malawi had the highest rate of varietal release, ranging from 0.37 to 0.71 annually. These countries also are the highest adopters of improved soybean varieties with more than 95% of soybean area planted to improved cultivars.

During the period 1998–2010, Kenya, Benin, Tanzania, Burundi, Ghana and Uganda more than doubled the average varietal releases observed between 1970 and 1997. In all these countries, the advancement in varietal release rates corresponded to increases in agricultural research investment, as well as to increased donor support in terms of funding and human capacity development (Stads and Hinvi, 2010). During the same period, the other countries, including the highest soybean producer, Nigeria, experienced a slowdown in varietal releases. This decline in varietal output could be attributed to declining investments in soybean improvement at IITA since the mid-1990s.

Many of the smaller producers released varieties in five or fewer years from 1970 to 2010. In terms of release years, Kenya and Togo were characterized by only one release event in the 41-year period. With more than 10-release year events, Zambia and Zimbabwe had the most stable release patterns over time. The annual incidence of releases is increasing in Zambia but decreasing in Zimbabwe where economic decline has constrained the government's ability to provide adequate funding for agricultural research. This situation was exacerbated by the withdrawal of donor funding in 2003. In addition, private companies affected by the economy were unable to contract research services as they had done in the past (Flaherty and Mwala, 2010).

Adoption of improved soybean varieties

With limited competition from traditional varieties, the uptake of modern varieties is higher in soybean than in any other crop in the DIIVA Project. Largely because soybean itself is new as a crop and its cultivation is made possible through improved varieties, the total soybean area in Côte d'Ivoire, Malawi, Zambia and Zimbabwe is under improved varieties (Table 6.17). Adoption levels are 97% in Uganda, 96% in Nigeria and 94% in Ghana. In contrast, modern varieties

Table 6.17. Adoption of improved soybean varieties in sub-Saharan Africa, 2009.

Country	2007–2009 soybean area ('000 ha)	Proportion of total soybean area planted to:		Materials containing IITA germplasm or directly related to IITA activities	
		Local varieties (% area)	Improved varieties (% area)	'000 ha	Area (%)
Benin	38	50	50	0.9	2
Burundi	4	96	4	0	0
Cameroon	13	25	75	10	75
Côte d'Ivoire	1	0	100	0.6	78
DR Congo	35	0	100	35	100
Ghana	70	6	94	42	61
Kenya	3	26	74	1.1	42
Malawi	80	0	100	0	0
Nigeria	613	4	96	503	82
Tanzania	9	21	79	1.2	13
Uganda	148	3	97	79	53
Zambia	45	0	100	0	0
Zimbabwe	67.3	0	100	0	0
All	1113	9	95	673	60
Sub-Saharan Africa	1295	21	82	673	52

have yet to make much headway in Burundi (4%) and Togo (30%). Of the 1.3 million hectares planted to soybean in SSA, about 0.7 million hectares (52%) were planted to IITA-related varieties, reflecting the important role that IITA plays in soybean genetic improvement in SSA.

Of the total area planted to soybean, the share of IITA-related varieties ranged from 2% in Benin to 75% and above in Cameroon, Côte d'Ivoire, Nigeria and the DR Congo. The high relative importance of IITA-related varieties stems from their popularity in Nigeria, where 82% of the area was planted to IITA-related varieties. More than half of the soybean area in Uganda is planted to IITA-developed varieties.

Four countries from Eastern and Southern Africa (Burundi, Malawi, Zambia and Zimbabwe) did not grow any IITA crosses. On the basis of these results, IITA has made greater contribution to soybean improvement in West and Central Africa, relative to Eastern and Southern Africa. IITA's soybean breeding activities had in the past concentrated in West Africa and its influence in Southern Africa was minimal.

Two improved soybean varieties are widely cultivated in the region (Table 6.18). First, TG× 1448-2E, a shattering and frog-eye, leaf-spot resistant IITA-bred variety, occupies slightly more than 60% of soybean area in Nigeria, more than 20% in Ghana and 15% in Cameroon. Second, TGx 1835-10E, another IITA-developed variety that is desired for its early maturity and resistance to soybean rust, pod shattering and lodging, dominates soybean area in Uganda (50%) and covers 26% of total land area in Cameroon and 6% in Kenya.

Other modern varieties have been adopted in more specific circumstances. The variety Anidaso (TG× 813-6D), which was released in 1992 in Ghana, covers 56% of the area planted to soybean. The varieties Vuangi with 55% of area and Munanga with 35% are the most popular IITA-bred varieties in DR Congo with resistance to pests and diseases, and pod shattering. The variety Bossier, a landrace with wide agroecological adaptation and intermediate maturity that was introduced in Tanzania in 1978, accounts for nearly 50% of soybean area. Other leading national varieties enjoying high adoption rates and occupying over 44% of soybean area are Safari in Zimbabwe and Ocepara-4 in Malawi.

The variety-specific adoption estimates in Table 6.18 also suggest that slow varietal turnover is becoming a problem in several countries. In the DR Congo, for example, not a single improved variety has been released since 1998. Negligible varietal output has limited farmers' choices and most continue cultivating older improved varieties. The durability of first-generation releases, such as Bossier and H3 in Tanzania, also points to slow varietal turnover and shows a lack of progress in plant breeding and technology transfer.

Slow varietal turnover despite a steady release of improved varieties could be due to constraints outside the research system, such as weak extension and seed delivery systems. Given weak extension systems, a large number of farmers are unaware of the availability and value of a number of newer varieties. Many farmers also lack the physical and economic access to seed of improved varieties due to a weak seed and credit delivery system.

In contrast, slow varietal turnover in the face of slow release of improved varieties points to constraints within the research system. In Nigeria, for example, one of the most popular IITA-developed improved soybean varieties, TG× 1448-2E, accounts for almost 60% of the current soybean area because it was disseminated widely at a time when there were several aggressive soybean popularization campaigns and technology dissemination programmes in the early 1990s. The dominance of this variety about 20 years after its release and first adoption by farmers is, however, partly because the research system has not produced newer varieties since then until 2008 when varieties TG×1835-10E and TG× 1740-2F were released, followed by TG× 1904-6F in 2009, and TG× 1987-10F and TG× 1987-62F in 2010. Owing to a strong seed system component of the Bill & Melinda Gates Foundation-funded Tropical Legumes II project, these newer varieties are now being disseminated widely in the pilot communities in northern Nigeria.

Yam

Although the value of production of yams is higher than that of any other food crop in the FAO production data for SSA, African yam species have not commanded much research attention. Heavy yields from a long growing season of

Table 6.18. Economically important improved soybean varieties in sub-Saharan Africa, 2009.

Country	Variety	Release year	Adoption (% area)
Benin	Jupiter	Informal	29
	ISRA 44A/73	Informal	11
	ISRA 29/72	Informal	6
	TGX 536-02D	1989	2
	Others		1
	All MVs (national)		50
Burundi	Ogden	Informal	3
	Namsoy	Informal	1
	All MVs (national)		4
Cameroon	TGx1835-10E	2003	28
	TGX 1448-2E	1994	15
	TGX 1660-19F	1994	13
	TGX 1440-1E	1994	8
	SG 299	Informal	8
	SG 320	Informal	2
	Others		1
	All MVs (national)		75
Côte d'Ivoire	32-R2-231	1997	39
	33-R8-271	1997	22
	TGx 1740-2F	1997	18
	Canarana	1983	9
	Doko	1983	9
	Emgopa 308	1983	5
	All MVs (national)		100
DR Congo	Vuangi	1989	55
	Munanga	1997	35
	Kitoko	1997	10
	All MVs (National)		100
Ghana	Anidaso	1992	56
	Jenguma (TGx 1448-2E)	2003	22
	Quarshie (TGx 1445-2E)	2003	11
	Salintuya 1	1992	5
	All MVs (National)		94
Kenya	SB 19 (TGx 1740-2F)	2010	34
	SB 25 (Namsoy 4M)	Informal	25
	Gazelle	2009	6
	SB 3 (TGx 1835-10E)	2009	6
	SB 8 (TGx 1895-33F)	2010	2
	All MVs (national)		74
Malawi	Ocepara-4	1993	47
	Makwacha	2008	27
	427/5/7	1993	16
	Magoye	1985	11
	All MVs (national)		100
Nigeria	TGx-1448-2E	1992	60
	Samsoy-2 (M-216)	1983	14
	TGx-1440-1E	1990	10
	TGx-1485-1D	1990	7
	TGx-1019-2EB	1990	2
	TGx-1019-2EN	1990	2
	TGx-923-2E	1990	0.9
	Samsoy-1(M-79)	1983	0.4
	TGx-536-02D	1985	0.4

Continued

Table 6.18. Continued.

Country	Variety	Release year	Adoption (% area)
	All MVs (National)		96
Tanzania	Bossier	1978	48
	Uyole Soya-1	2002	13
	H3	1974	5
	MAL	Informal	4
	ZAM	Informal	4
	ZIM	Informal	4
	Others		0
	Uyole Soya-2	Informal	0
	All MVs (national)		79
Togo	TG× 1910-14F	1997	39
	All MVs (national)		39
Uganda	Maksoy 1N (TGx 1835-10E)	2004	53
	Namsoy 4N	Informal	32
	Maksoy 2N	2008	10
	Namsoy 4M	2004	2
	All MVs (national)		97
Zambia	SC satellite	2008	23
	PAN 1856	2008	20
	Lukanga	2004	13
	SC Sirocco	2007	12
	Mulungushi	2005	1
	Others		32
	All MVs (national)		100
Zimbabwe	Safari	2001	44
	Serenade	2008	31
	Siesta	Informal	16
	Santa	Informal	9
	All MVs (national)		100

about 10 months coupled with attractive market prices result in high levels of value of production per hectare.

Research neglect is largely attributed to the spatial concentration of production in the humid tropics of West and Central Africa. With an area of about 3.0 of the 4.5 million hectares grown in West and Central Africa, Nigeria is by far the largest producer. According to FAOSTAT, the value of yam production in Nigeria was nearly US$6 billion in 2009. Ghana and Côte d'Ivoire are the second and third most important producers. On a per capita basis, yams are also economically important in Benin where they are regarded as a staple food and are the second most researched commodity after cassava (Stads and Hinvi, 2010). Benin has the highest yam consumption per capita in the world at 395 kcal per day (Gedil and Sartie, 2010). Cameroon and Togo also produce more than 0.5 million tonnes annually. Among the surveyed countries, Guinea and Uganda are very small producers.

With a global mandate for international yam research in the CGIAR since the 1970s, IITA has contributed to yam improvement efforts as well as capacity strengthening of several national yam research programmes. The most popular yam species in the region are *Dioscorea rotundata* (TDr) and *Dioscorea alata* (TDa). The two species are IITA-bred and are high yielding and resistant to anthracnose and virus diseases.

Scientific staffing of yam improvement programmes

In 2009, IITA employed 7 FTE researchers working on yam improvement research, whereas national programmes in the eight surveyed countries employed only 49 FTE researchers (Table 6.19).

Table 6.19. Full time equivalent staff by major specialization working on yam improvement in sub-Saharan Africa, 2009.

CGIAR/ NARS programme	Major specialization												
	Germplasm conservation	Breeding	Pathology	Molecular biology	Entomology/ Virology	Agronomy	Seed production	Tissue culture	Postharvest	Social science	Food science	Others	Total
CGIAR (IITA)	0.4	1.2	0.4	0.1	0.2	0.1	0.0	1.0	0.0	3.3	0.2	0.2	7.0
NARS	8.7	7.7	1.6	2.5	2.7	6.7	5.5	1.9	2.5	8.5	0.0	1.3	49.3
Benin	2.9	1.1	0.1	1.0	0.2	0.6	2.7	0.2	1.0	2.0	0.0	0.4	12.1
Cameroon	0.0	0.3	0.1	0.0	0.0	0.5	0.8	1.0	0.3	0.5	0.0	0.0	3.3
Côte d'Ivoire	3.2	1.0	0.5	0.3	0.0	0.5	0.2	0.2	0.6	0.6	0.0	0.0	7.1
Ghana	0.4	1.0	0.2	0.2	0.2	0.2	0.2	0.0	0.2	0.4	0.0	0.0	3.0
Guinea	0.3	1.1	0.2	0.0	0.0	1.0	0.2	0.0	0.0	0.2	0.0	0.2	3.1
Nigeria	0.3	2.0	0.5	0.0	1.0	2.0	0.0	0.5	0.3	4.8	0.0	0.5	11.8
Togo	0.7	1.3	0.1	0.5	0.0	0.9	0.6	0.0	0.2	0.1	0.0	0.2	4.5
Uganda	1.0	0.0	0.0	0.5	1.3	1.0	0.8	0.0	0.0	0.0	0.0	0.0	4.6
Average	1.1	1.0	0.2	0.3	0.3	0.8	0.7	0.2	0.3	1.1	0.0	0.2	6.2

Although there have been concerted efforts by IITA and NARS on yam improvement, scientific capacity remains inadequate relative to the perceived economic importance of the crop. Only Nigeria and Benin have invested in programmes that exceed 10 FTE researchers in size (Table 6.19). Less than 10 of the 56 international and national FTE researchers are allocated exclusively to plant breeding. Indeed, there are more FTE researchers working in social sciences than in yam breeding.

The evidence suggests that yam improvement research probably has one of the lowest human resource investments among the major food crops in SSA. The surveyed countries in SSA, which together account for 93% of world's annual yam production, employed a little over one FTE researcher per million tonnes of production. Yams in Nigeria were characterized by the lowest estimated research intensity at 0.3 of any crop by country observation in the DIIVA Project. Scientific investment in yams in Ghana also fell below a threshold of one FTE researcher per million tonnes of production.

On a more positive note, the educational status of human resources seems to be high in yam improvement programmes. On average in the countries surveyed in SSA, 2.5 (or 40%) of these researchers held PhD degrees, 2.2 (or 35%) held MSc degrees and 1.6 (or 25%) held BSc degrees. Yam improvement in Benin is carried out by a group of researchers with PhDs (2.9) and MScs (5.2), whereas in Nigeria the level of staffing is skewed toward PhD holders (7.5) and a few have MScs (3.3). Overall, yam improvement programmes in Nigeria, Côte d'Ivoire, Uganda and Ghana rely heavily on PhD level staff. By contrast, most of the researchers working on yam improvement in Benin, Cameroon and Togo only have MSc-level qualifications.

Guinea is the only country that has no PhDs involved in yam improvement. Reasons for their absence are the same as those discussed above on scientific staffing in maize. Moreover, very low levels of production argue against an investment in postgraduate scientists for yam improvement. A volatile political climate in Guinea has also hindered the development of agricultural research and development, and continues to deter investment by foreign donors.

Varietal output of yam improvement

During the period 1970–2010, international and national yam research programmes have developed and/or identified 78 improved yam varieties for release (Table 6.20). Almost all (96%) of the varieties have been released in West and Central Africa. As the world's leading producer of yam, Nigeria has the highest (24) number of varietal releases, followed by Benin (15) and Togo (14). Most of these released improved varieties are high yielding and combine major disease resistance with good tuber qualities. IITA-related varieties account for nearly 80% of the 78 yam varieties developed and released in SSA during the last four decades. About 72% of

Table 6.20. The IITA content of improved yam varieties in sub-Saharan Africa, 1970–2010.

	Number of varieties released				Percentage of total release		
Country	IITA-bred	IITA-parent	Non IITA	Total	IITA-bred	IITA-parent	Non IITA
Benin	6	5	4	15	40	33	27
Cameroon	10	0	0	10	100	0	0
Côte d'Ivoire	3	0	2	5	60	0	40
Ghana	1	0	2	3	33	0	67
Guinea	4	0	0	4	100	0	0
Nigeria	20	0	4	24	83	0	17
Togo	9	0	5	14	64	0	36
West and Central Africa	53	5	17	75	71	7	23
Uganda	3	0	0	3	100	0	0
East and Southern Africa	3	0	0	3	100	0	0
Sub-Saharan Africa	56	5	17	78	72	6	22

the IITA-related varieties were bred by IITA; the others featured IITA germplasm as a parent.

Although only IITA-bred varieties (i.e. finished materials) have been released in Uganda, Guinea and Cameroon, both IITA-bred and NARS-bred varieties (i.e. using IITA parent materials) have been released in Benin, Nigeria, Ghana, Côte d'Ivoire and Togo. This is consistent with the scientific strength of the various national programmes in terms of FTE researchers working on yam improvement (Table 6.19) where weaker national yam programmes such as Cameroon and Guinea, each having not more than three researchers, rely heavily on IITA for finished materials for direct release to farmers with little or no adaptation.

The data also show that 31 of the 78 improved yam varieties were either not officially released or their dates of release are not known. These 'informal' releases mainly come from Benin and Cameroon where no official releases have taken place.

The novelty or immaturity of yam improvement research is reflected by the fact that only 6 of the 47 dated improved varieties were released before 1998. The incidence of release has markedly increased over time in most of the countries in Table 6.20 with the exception of Ghana and Guinea. During the period 1998–2010, Benin, Côte d'Ivoire, Nigeria and Togo more than doubled the average varietal releases compared with the number of varieties released during the period 1970–1997. Recent improvements in varietal release are attributed to the fact that most CGIAR centres and NARS have received substantial donor funding for agricultural research programmes (Stads, 2011).

Notwithstanding recent improvements, release events are still rare outcomes in yam genetic improvement in West and Central Africa. During the period 1970–2010, years with zero releases were common, ranging from 32 in Nigeria to 39 in Guinea and Togo. Ghana and Côte d'Ivoire only released varieties on two occasions in the 41-year period of analysis. Absence of release events again points to low investments in yam improvement in most of the countries in West and Central Africa.

Adoption of improved yam varieties

Improved varieties account for about 28% of the yam area in the surveyed countries and 26% in West and Central Africa as a whole (Table 6.21). Inter-country comparisons show that adoption rates of improved varieties vary from 4% to 75%. Côte d'Ivoire has the highest adoption rate of 75%; most of the countries have adoption rates between 4% and 28%. However, only 5% of the yam area in Côte d'Ivoire is planted to IITA-bred varieties or varieties containing IITA germplasm. Uptake of improved varieties in Benin (4%) and Guinea (5%) is negligible. Adoption levels are low in countries like Cameroon because of the

Table 6.21. Adoption of improved yam varieties in sub-Saharan Africa, 2009.[a]

		Proportion of total yam area planted to:		Materials containing IITA germplasm or directly related to CG Center activities	
Country	2007–2009 yam area ('000 ha)	Local varieties(%)	Improved varieties (%)	'000 ha	Area under MVs (%)
Benin	171	96	4	0	0
Cameroon	33	91	9	3	9
Côte d'Ivoire	758	25	75	40	5
Ghana	350	90	10	0	7
Guinea	2	95	5	17	5
Nigeria	2981	80	20	551	18
Togo	60	72	28	3	5
Surveyed countries	4355	72	28	614	14
West & Central Africa	4567	74	26	614	13

[a]Adoption estimates are not presented for Malawi and Uganda because yam area is not reported by FAO.

scarcity of high-yielding seedlings resulting in low yields (Nchinda *et al.*, 2010).

IITA-related varieties account for 13% (over half a million hectares) of the yam area. Nigeria has the highest area (18%) under improved yam varieties from IITA. The estimates in Table 6.21 illustrate that adoption of improved yam varieties is low and uptake is slow, with much of the area still planted to local landraces.

In spite of low levels of adoption in all countries except Côte d'Ivoire, it is possible to find improved varieties that are commonly grown in at least two countries in the region with positive spill-over effects (Table 6.22). Varieties such as

Table 6.22. Economically important improved yam varieties in sub-Saharan Africa, 2009.

Country	Variety	Release year	Adoption (% area)
Benin	Florido	Informal	4
	All MVs (national)		4
Cameroon	TDr 95/19127	Informal	5.0
	TDr 84/02461	Informal	2.4
	TDr 89/02677	Informal	1.2
	TDr 95/19158	Informal	0.4
	Others		0.1
	All MVs (national)		9
Côte d'Ivoire	C18	1998	61
	C20	2000	9
	TDR205	2000	2
	TDR608	2000	2
	NDRBD10	2000	1
	All MVs (national)		75
Ghana	TDr 89/02665	1992	7.34
	CRIPona	1992	1.49
	CRIKukrupa	2003	0.87
	All MVs (national)		10
Guinea	TDr 131	1997	3.7
	TDr 608	1990	0.8
	TDr 745	1996	0.2
	TDr 205/TDr 179	1994	0.2
	All MVs (national)		5
Nigeria	TDr 89/02660	2009	6.0
	TDr 89/02665	2003	4.0
	TDr 89/02602	2009	4.0
	DRN 200/4/2	2008	1.4
	TDa 98/01176	2008	0.7
	TDa 98/01168	2008	0.7
	TDa 98/01166	2008	0.7
	TDr 89/02672	2008	0.3
	TDr 95/19158	2009	2.1
	All MVs (national)		20
Togo	Florido	1996	22.9
	TDr 747	1999	3.2
	TDr 89/02665	2006	1.0
	TDr 95/19156	2006	0.2
	TDr 97/00903	2009	0.1
	TDr 89/02475	Informal	0.1
	TDa 98/1166	2005	0.1
	TDa 98/01169	2005	0.1
	TDa 99/01176	Informal	0.1
	All MVs (national)		28

TDr 89/02665 are common in Nigeria (4% of total yam land area) and Ghana (7%). Florido is the most common variety of yam, and it accounts for some 23% of total yam area in Togo and 4% in Benin. Varieties TDr 95/19127 and TDr 131 are accepted by a sizeable minority of farmers in Cameroon and Guinea as they account for about 5% of the yam area.

Characteristic of a crop with a low multiplication ratio in terms of output per unit of planting material, diffusion of improved yam cultivars is usually a slow process, but one variety's uptake (Table 6.22) has been as rapid and extensive as to warrant mention in CABI's invasive species compendium. The most popular yam variety in Côte d'Ivoire is C18, a *Dioscorea alata* variety introduced in 1992 and subsequently released in 1998. Its quick and widespread diffusion is attributed to its superior cooking quality (Kouakou *et al.*, 2012).

Summary

Several findings emerge from a comparative examination of the evidence discussed for the five crops in this chapter. Perhaps most impressive is the significant uptake of improved maize and cassava varieties between 1998 and 2009. The level of adoption of modern cultivars more than doubled in many of the project countries in both crops. Fueled by the rapid diffusion of mainly improved OPVs in Nigeria, adoption of modern varieties increased from 37% of the maize area in 1998 to 57% in 2009. The uptake of improved varieties doubled in cassava from 18% of the cassava area in 1998 to 36% in 2009.

Equally impressive is the critical role that IITA has played in providing finished varieties and germplasm. With the exception of soybean, the majority of releases in the surveyed countries during the period 1970–2010 contained IITA-related materials, usually in the form of elite varieties. With the exception of yam, releases related to IITA accounted for a larger share in adopted area than their share in varieties released. Slower than expected private sector participation in the genetic improvement of maize and soybean is one reason for IITA's seemingly unchanging key role as the supplier of much needed genetic materials. Slower than expected NARS development in crossing parental materials and selecting from them, in lieu of a marked and continuing tendency to focus on the selection of finished varieties, is another potential reason why IARC input in the improvement of these five crops is still essential. In principle, one to three stronger national programmes in each crop have the capacity to adopt a more mature role in genetic improvement for themselves and for other smaller programmes in the region. In practice, few have done so in a manner that attests to the logical evolution and maturation of a plant-breeding programme.

The transparent documentation in all five crops of spill-over varieties that are adopted by many farmers in multiple countries reinforces the observation that wide adaptation and high returns can be achieved from a regional plant-breeding perspective. The modern quality protein maize OPV Obatanpa, which has been adopted by farmers in nine of the 11 surveyed countries, epitomizes this type of success.

The discussion of results in this chapter also sheds light on areas of concern and future challenges. The frequency of varietal releases has declined in the recent past for both cowpea and soybean in West and Central Africa. In soybean, a downturn in releases reflects decreased funding for international genetic research in the 1990s. Unavailability of new adapted materials may dampen and even stall the rapid expansion of soybean in SSA. More than any other crop, soybean, arguably Africa's newest food crop, is paradoxically characterized by some of the oldest modern varieties in farmers' fields. The durability of first-generation modern varieties is a cause for concern in several soybean-growing countries, especially Nigeria.

In all five crops, one can cite countries where lack of recent varietal output has resulted in ageing varieties in farmers' fields. Varieties that date from the 1980s should have been replaced by now if genetic improvement was efficient and technology transfer was effective. There is a need to increase the frequency of varietal releases and again put in place good seed and planting material production and distribution schemes that require public–private partnership as well as efficient extension systems to popularize new varieties.

Although inputs were not explicitly related to outputs, outcomes and impacts in this chapter, examples were given where low levels of research intensities were associated with low output in the form of released varieties, which in turn were associated with low adoption. The problem of low research intensities seems chronic in yam and also remains a concern in cassava and cowpea. Policy makers and donors need to take the potential for genetic improvement in these crops more seriously in the allocation of research resources.

The results also carry some good news about varietal releases. Releases are trending upwards in cassava, maize and yam. The variability in releases over time is also declining in many countries characterized by a steadier stream of varietal output. Nigeria has imparted regional stability to releases in all five crops. Having a national programme that displays a steady pattern of releases over time is highly advantageous from the perspective of regional genetic improvement. Although many NARS have experienced long periods without varietal releases, Nigeria has managed to release varieties in every decade and in most 5-year periods in maize, cassava, cowpea and soybean since 1970.

The problem of countries lagging in adoption is a challenge that applies to four of the five crops. In cassava, Angola and Mozambique are two of several countries where the crop is the staple food and where adoption of improved varieties is less than 20%. In maize, the DR Congo, the second largest maize-producing country in cultivated area, is an important laggard. In cowpea, Niger and Burkina Faso rank first and third in growing area and also rank in the bottom quartile of countries in adoption of modern varieties. In yam, low adoption is not associated with one or two laggards but is a problem that is shared by most of the surveyed countries with the exception of Côte d'Ivoire where the rapid diffusion of C18 in the 1990s and 2000s suggests that speedy diffusion of highly acceptable material can be achieved even in a crop with a sparse research tradition and a low multiplication ratio. Unless adoption picks up these lagging countries, it will be difficult to sustain a pace of varietal change equivalent to a linear gain of 1% per annum from now until 2020 or 2025.

The results also furnish sufficient information on varietal characteristics to form a generalized picture of what worked and what did not in terms of breeding strategy. In all five crops, a high-yielding agronomic background was portrayed as essential for adoption. Disease resistance or tolerance was highly desirable and found in some materials in most crops. Early maturity and short duration featured in crops such as cowpea. And recently, breeders seem to be paying considerably more attention to traits related to consumer acceptance.

Digesting the results in this chapter provides a basis for arriving at a consensus on areas for future research that apply to most if not all five crops. Diagnostic research is needed on constraints to varietal adoption in the lagging countries. Specific issues to address include: To what extent are low adoption levels primarily the result of released varieties that are not preferred by producers or consumers or that do not yield well in their circumstances? Or, is negligible adoption the result of inadequate technology transfer that is conditioned by poor availability of seed and planting materials or lack of information that determines awareness? Searching for plausible explanations to these questions requires more systematic testing of improved materials in typical farmer circumstances.

Validation research that is synonymous with ground truthing of the expert-opinion estimates is also needed. Two thorough adoption studies that also supplied information for validation of expert estimates have given mixed results. In six states in south-western Nigeria, the expert estimates were about 15% higher than the adoption results based on household surveys with varietal identification in farmers' cassava fields (Alene *et al.*, 2012). In Ghana, a recent IFPRI study generated an aggregate level of adoption of improved maize varieties that were almost identical to the estimate from expert opinion (Ragasa *et al.*, 2013). More of these validation exercises should be conducted especially for crop and country observations, such as for cassava in the DR Congo, where expert opinion provides seemingly optimistic estimates and where adoption studies are lacking even for smaller areas and subregions.

Providing feedback information for breeders on the adoption estimates to determine what has worked and what has not worked is also a priority for research. Such research is highly crop specific and requires familiarity with the crop to draw inferences about farmers' demand for characteristics in well-defined production and consumption contexts. The baseline generated by the DIIVA Project on the performance of crop improvement should have the capability to generate lessons for multiple audiences, including plant breeders.

Notes

[1] This paper is a revised and abridged version of Alene and Mwalughali, 2012.
[2] Beintema and Stads (2011) describe Ethiopia, Kenya, Nigeria, Sudan, Ghana, South Africa, Tanzania and Uganda as the 'Big Eight,' because they dominate the levels of investment in research and development in Africa.

References

Alene, A.D. and Mwalughali, J. (2012) *The Effectiveness of Crop Improvement Programs in Sub-Saharan Africa from the Perspectives of Varietal Output and Adoption: The Case of Cassava, Cowpea, Maize, Soybean and Yam.* Diffusion and impact of improved varieties in Africa (DIIVA), Objective 1 Report. International Institute of Tropical Agriculture, Lilongwe, Malawi.

Alene, A.D., Menkir, A., Ajala, S.O., Badu-Apraku, B., Olanrewaju, A.S. *et al.* (2009) The economic and poverty impacts of maize research in West and Central Africa. *Agricultural Economics* 40(5), 535–550.

Alene, A.D., Yigezu, Y., Ndjeunga, J., Labarta, R., Andrade, R. *et al.* (2011) *Measuring Effectiveness of Agricultural R&D in Sub-Saharan Africa from the Perspective of Varietal Output and Adoption. Initial Results from the Diffusion of Improved Varieties in Africa (DIVA) Project*. Paper prepared for the ASTI/IFPRI-FARA Conference, Accra, Ghana.

Alene, A.D., Khonje, M., Abdoulaye, T., Kulakow, P. and Manyong, V.M. (2012) *Adoption of Improved Cassava Varieties in Southwestern Nigeria, Objective 2 Technical Report.* International Institute of Tropical Agriculture (IITA), Lilongwe, Malawi.

Beintema, N. and Rahija, M. (2011) Human resource allocation in African agricultural research: revealing more of a story behind the regional trends. *Conference Working Paper 12*. Presented at the ASTI/IFPRI-FARA Conference, Accra, Ghana.

Beintema, N. and Stads, G. (2011) *African Agricultural R&D in the New Millennium: Progress for Some, Challenges for Many*. IFPRI, ASTI, Washington, DC.

Beintema, N.M., Castelo-Magalhaes, E., Howard, E.H. and Mwala, M. (2004) *Agricultural Science and Technology Indicators. Zambia*. ASTI Country Brief No. 18.

Bohn, A. and Byerlee, D. (1993) *Wheat Breeding Research in Developing Countries: An Analysis of Investments and Impacts.* Part I of 1992/93 CIMMYT World Wheat facts and trends. The wheat breeding industry in developing countries: an analysis of investments and impacts. CIMMYT, Singapore.

Bohn, A., Byerlee, D. and Maredia, M.K. (1999) Investment in wheat improvement research in developing countries. In: Maredia, M.K. and Byerlee, D. (eds) *The Global Wheat Improvement System: Prospects for Enhancing Efficiency in the Presence of Spillovers*. CIMMYT Research Report No. 5. CIMMYT, Mexico (DF).

Byerlee, D. and Heisey, P. (1996) Past and potential impacts of maize research in sub-Saharan Africa: a critical assessment. *Food Policy* 21(3), 255–277.

Flaherty, K. and Mwala, M. (2010) Recent developments in agricultural research. Zambia ASTI-Country note, July. Available online: http://www.asti.cgiar.org/pdf/Zambia-Note.pdf (Accessed 30 April 2015)

Flaherty, K., Ayoola, G., Ogbobo, J. and Beintema, N.M. (2010) Nigeria. ASTI country note. Washington, DC, and Abuja, Nigeria. International Food Policy Research Institute, Agricultural Research Council of Nigeria, and the Farm and Infrastructure Foundation.

Gedil, M. and Sartie, A.M. (2010) Perspectives on molecular breeding of Africa's main staple food crops: cassava and yam. *Aspects of Applied Biology* 96, 123–136.

IITA (1987) IITA Annual Report 1987. Ibadan, Nigeria.

IITA (1992) IITA Annual Report 1992. Ibadan, Nigeria.

Johnson, N.L., Manyong, V.M., Dixon, A.G.O. and Pachico, D. (2003) The impact of IITA genetic improvement programs on cassava. In: Evenson, R.E. and Gollin, D. (eds) *Crop Variety Improvement and Its Effect on Productivity: The Impact of International Agricultural Research*. CAB International, Wallingford, UK.

Kavia, F.Y., Mushingo, C.C. and Sonda, G.B. (2007) Factors affecting adoption of cassava varieties: a case of cassava mosaic disease tolerant varieties in Lake Zone Regions Tanzania. *African Crop Science Conference Proceedings*, Vol. 8. African Crop Science Society, El-Minia, Egypt, pp. 1875–1878.

Kouakou, A.M., Zohouri, G.P., Dibi, K.E., N'zué, B. and Foua-Bi (2012) The emergence of a new yam variety of Dioscorea Alata species, C18, in Côte d'Ivoire. *Journal of Applied Biosciences* 57, 4151–4158.

Manyong, V.M., Kling, J.G., Makinde, K.O., Ajala S.O. and Menkir, A. (2003) Impact of IITA germplasm improvement on maize production in West and Central Africa. In: Evenson, R.E. and Gollin, D. (eds) *Crop Variety Improvement and Its Effect on Productivity: The Impact of International Agricultural Research*. CAB International, Wallingford, UK.

Nchinda, V.P., Ambe, T.E., Nathalie, H., Leke, W., Che, M.A. *et al.* (2010) Factors influencing the adoption intensity of improved yam (*Dioscorea* spp.) seed technology in the western highlands and high guinea savannah zones of Cameroon. *Journal of Applied Biosciences* 36, 2389–2402.

Nweke, F. (2009) *Controlling Cassava Mosaic Virus and Cassava Mealybug in Sub-Saharan Africa.* IFPRI Discussion Paper, Washington, DC.

Nweke, F., Spencer, D.S.C. and Lynam, J.K. (2002) *The Cassava Transformation: Africa's Best-Kept Secret*. Michigan State University Press, East Lansing, Michigan.

Pardey, G.P., Roseboom, J. and Anderson, J.R. (1991) Topical issues on NRS. In: Pardey G.P., Roseboom, J. and Anderson, J.R. (eds) *Agricultural Research and Policy. International Quantitative Perspective*. Cambridge University Press, Cambridge.

Ragasa, C., Dankyi, A., Acheampong, P., Wiredu, A.N., Capoto, A. *et al.* (2013) *Patterns of Adoption of Improved Maize Technologies in Ghana*. GSSP Working Paper 36. IFPRI, Washington, DC, p. 28.

Sanders, J.H., Bezuneh, T. and Schroeder, A.C. (1993) *Impact Assessment of the SAFGRAD Commodity Networks*. USAID/AFR OAU/STRC-SAFGRAD. PN-ABP-430.

Singh, B.B. and Ntare, B.R. (1985) Development of improved cowpea varieties in Africa. In: Singh, S.R. and Rachie, K.O. (eds) *Cowpea Research, Production and Utilization*. John Wiley & Sons Ltd., London.

Smith, J., Barau, A.D., Goldman, A. and Mareck, J.H. (1994) The role of technology in agricultural intensification: the evolution of maize production in the northern Guinea savanna of Nigeria. *Economic Development and Cultural Change* 42(3), 537–554.

Stads, G.J. (2011) *Benchmarking Agricultural Research Investment and Capacity Indicators in West and Central Africa.* ASTI Background Note. International Food Policy Research Institute, Washington, DC.

Stads, G. and Hinvi, J. (2010) *Recent Developments in Agricultural Research*. Benin ASTI Country note, July. Available online: http://www.asti.cgiar.org/pdf/Benin-Note.pdf (Accessed 30 April 2015)

Stads, G.J. and Maiga, A. (2010) *Mali. ASTI Country Note*. International Food Policy Research Institute (IFPRI), Agricultural Science and Technology Indicators (ASTI) and Rural Economy Institute, Washington, DC and Bamako, Mali.

Stads, G.J., Beavogui, F. and Diawara, S. (2010) *Guinea. ASTI Country Note*. International Food Policy Research Institute (IFPRI) and Guinean Institute of Agricultural Research. Washington, DC and Conakry, Guinea.

Venkatesan, V. and Kampen, J. (1998) *Evolution of Agricultural Services in Sub-Saharan Africa*. World Bank, Washington, DC.

Walker, T.S. (2010) *Guidelines for Data Collection for Objective 1 of the DIIVA Project*. Fletcher, North Carolina.

Appendix 6.1

Table 6.A1. Country and crop coverage in the DIIVA survey in sub-Saharan Africa.

		Commodity				
No.	Country	Cassava	Cowpea	Maize	Soybean	Yam
1	Angola	√				
2	Benin	√	√	√	√	√
3	Burkina Faso		√	√		
4	Burundi	√			√	
5	Cameroon	√	√	√	√	√
6	Côte d'Ivoire	√	√	√	√	√
7	DR Congo	√	√	√	√	
8	Ghana	√	√	√	√	√
9	Guinea	√	√	√		√
10	Kenya	√			√	
11	Malawi	√	√		√	
12	Mali		√	√		
13	Mozambique	√	√			
14	Niger		√			
15	Nigeria	√	√	√	√	√
16	Senegal		√	√		
17	Tanzania	√	√		√	
18	Togo	√	√	√	√	√
19	Uganda	√	√		√	√
20	Zambia	√	√		√	
21	Zimbabwe	√	√		√	
Total		17	18	11	14	8
Grand Total				68		

7 Assessing the Effectiveness of Agricultural R&D for Groundnut, Pearl Millet, Pigeonpea and Sorghum in West and Central Africa and East and Southern Africa

J. Ndjeunga,[1]* K. Mausch[2] and F. Simtowe[3]

[1]*International Crops Research Institute for the Semi-Arid Tropics (ICRISAT) West and Central Africa, Niamey Niger;* [2]*ICRISAT, Nairobi;* [3]*CIMMYT, Nairobi (formerly ICRISAT, Nairobi), Kenya*

Introduction[1]

Arable land in sub-Saharan Africa is often cultivated during seasonal rains in regions where the supply of rainfall exceeds the demand for rainfall for only 2–7 months of the year. These rainfall supply and demand conditions define rainfed agriculture in the semi-arid tropics (SAT). In 1972, the International Crops Research Institute for the Semi-Arid Tropics (ICRISAT) was established in India with a global mandate to increase agricultural production in the SAT, thereby enhancing poor people's welfare in these rainfall-unassured production environments.

Technically, the SAT encompassed large areas of Australia, Latin America and Asia, but the geographic focus at ICRISAT was always on peninsular India and sub-Saharan Africa where most rural and urban poor lived. By 2020, the total population of people in Asia's and Africa's SAT is projected to be about 850 million, comprising a 70% share for Asia and a 30% share evenly split between West and Central Africa and East and Southern Africa (Walker, 2009). When ICRISAT was founded in 1972, the relative importance of the two continental populations was about 80% for Asia's SAT and 20% for Africa's SAT. The total SAT population in 1972 was only about 35% of the projected population in 2020.

Sorghum, pearl millet, groundnut, chickpea and pigeonpea are cultivated wholly or mostly in India's SAT. Sorghum, pearl millet and groundnut production also define West and Central Africa's SAT, where they account for 40% of arable cultivation. Sorghum, millet and groundnut are also cultivated on about 15% of arable land in the SAT of East and Southern Africa. ICRISAT's area mandate of the SAT provided the rationale for its crop mandate of sorghum, pearl millet, groundnut, chickpea and pigeonpea.

From its headquarters in Central India, ICRISAT began to invest in international agricultural research infrastructure and programmes in sub-Saharan Africa. Highlights include the following:

- Establishment of a research centre on pearl millet, groundnut and resource management in Niamey, Niger, in response to the needs of the dry SAT in the Sahel in 1983;
- Posting of ten internationally recruited plant breeders in country national programmes via a long-term grant from the

* E-mail: n.jupiter@cgiar.org

 Crop Improvement, Adoption and Impact of Improved Varieties in Food Crops in Sub-Saharan Africa (eds T.S. Walker and J. Alwang)

United Nations Development Program (UNDP) starting in the late 1970s and ending in the early 1990s;
- Establishment of a regional office in Nairobi, Kenya, to attend to its crop and natural resource management mandate in East Africa;
- Initiation of the long-term Southern African Development Community (SADC)-ICRISAT groundnut project in 1982 at the Chitedze Research Station in Malawi for Southern Africa;
- Establishment of a regional research centre in Bulawayo, Zimbabwe, in the early 1990s to respond to its crop mandate in Southern Africa;
- Founding of a long-term partnership with the national programme in Mali in the late 1990s to conduct regional research on sorghum and groundnut;
- Establishment of the SADC/ICRISAT Sorghum and Millet Improvement Program in Southern Africa with major funding from USAID from 1983 to 2003.
- Longer term investment in 6-month applied training at its main station in Patancheru, India, of about 120 African crop improvement scientists and technicians annually from the late 1970s to the early 1990s.

The above is a non-exhaustive list of discrete resource investments that ICRISAT has made to improve productivity in sorghum, pearl millet, groundnut, and, to a lesser extent, pigeonpea and chickpea in sub-Saharan Africa. Institutionally and internationally, ICRISAT has not acted alone. Prior to and since independence, the French research institute Institut de Recherche pour Les Huiles et Oleagineux (IRHO) invested in genetic improvement, especially of high value and export crops, such as cotton and groundnut, in West Africa with stations in Bambey, Senegal, and sub-stations in Niger. Investments by IRAT (Institut de Recherches Agronomiques Tropicales) in the improvement of food-security crops, sorghum and pearl millet followed in the early 1960s. Since the 1960s and 1970s, a foundation for modern varietal change was laid by public-sector national agricultural research programmes (NARS) in the countries of sub-Saharan Africa. Starting in the late 1970s, in addition to the above players, USAID's Collaborative Research Support Projects (CRSPs) invested in crop improvement in sorghum, pearl millet and groundnut in many countries in West Africa.

Summing up, NARS, ICRISAT, the International Sorghum and Millet Innovation Laboratories of the United States Agency for International Development (USAID) – INTSORMIL – and the Peanut Collaborative Research Support Program of USAID (CRSP), and CIRAD (Centre de Coopération Internationale en Recherche Agronomique pour le Développement; formerly IRAT and IRHO) have invested in sorghum, pearl millet and groundnut crop improvement programmes in sub-Saharan Africa. An understanding of the performance of sorghum, pearl millet, and groundnut programmes from the perspectives of scientific capacity, output of released varieties and the level of adoption of these improved varieties is not, however, as comprehensive as in other staple food crops such as maize. This chapter attempts to systematically improve coverage on these aspects and thus inform on the performance of research on crop improvement for these staple food and cash crops in Africa's SAT.

This chapter is organized along regional lines. In presenting and discussing research results in each of the sections that follow, findings are reported separately for West and Central Africa's SAT and for the SAT of East and Southern Africa. This regional orientation is preferred to a crop-wise organization because several of the major findings are regional and apply to the same crops within a region of Africa's SAT. Teasing out lessons from a comparison of regional findings leads to a more fundamental definition of the problems, opportunities and successes of and from crop improvement than a comparison of results across crops. These lessons are described in the concluding section of the chapter.

Country Coverage, Methods and Data Collection

West and Central Africa

In West and Central Africa, six countries were originally targeted for assessing the effectiveness of crop improvement in groundnut, pearl millet and sorghum, which are the most important and extensively grown ICRISAT-mandated crops in the region. Country selection for the assessment was based on national production areas. Initially, Burkina Faso, Chad, Mali, Niger, Nigeria and Senegal were proposed for study. Although adoption inquiries were available in Chad at the project level on sorghum modern varieties (MVs), Chad, the only country from Central Africa originally targeted, was dropped because of logistical issues and resource limitations.

Fifteen crop-by-country observations are reported on in this chapter. Information on sorghum, pearl millet and groundnut were gathered in Burkina Faso, Mali, Niger, Nigeria and Senegal. The five countries account for about 83% of sorghum area in West and Central Africa, 87% of the area cultivated in pearl millet, and 62% of area cultivated in groundnut (FAO, 2011).

Representation is more than adequate because the largest producers in the region are included in the assessment. With a summed area approaching 4 million hectares, Nigeria and Senegal are the largest producers of groundnut. By the same measure, Niger and Nigeria are the heaviest producers of pearl millet in the region; all five countries harvest more than 1 million hectares of pearl millet annually. Sorghum production in the study countries exceeds 10 million hectares.

Data collection focused on varietal output (release), the strength of NARS and IARCs in food-crop commodity improvement, the level of adoption in important countries by food crop, and the contribution of genetic materials from different institutional sources in national varietal output and adoption.

Because groundnut, pearl millet and sorghum were in the 1998 global initiative, it was proposed that data collection be restricted to an update of information gathered in 1998. However, the 1998 global database did not cover West Africa adequately (Bantilan and Deb, 2003; Bantilan *et al.*, 2003; Deb and Bantilan, 2003). Few countries overlapped for a before-and-after comparison, and ICRISAT related materials were emphasized in 1998. Therefore, the release database had to be constructed from scratch for each of the three crops in the five countries from 1970 to 2010. The database contains information on the following variety characteristics: official name of the release (and other local names if any), the year of release, the origin of the germplasm, the breeding scheme, genetic background (parentage, genetic ancestry, pedigree), institutional source of the material, the variety maintainer, the country of origin, the relevant agroecological zone (in terms of length of growing period in days and rainfall in mm), the genetic background and the release classification (type of material, NARS input). Varietal characteristics including the average yield potential (tonnes/ha, on-station and on-farm), plant height (cm), tillering, panicle length, weight of 1000 grains (g), panicle compactness, grain colour, plant type and other selected varietal traits were gathered where such data were available.

Data on scientific capacity were collected on the personnel by crop and institution in 2009 on the strength of sorghum, groundnut and pearl millet improvement programmes. The variables include the name of scientist, gender (male or female), age group, function, specialization, level of education and scientist (full-time equivalent (FTE)) status. Data on research investment in monetary terms were difficult to obtain due to attribution issues and time needs for accountants to gather information.

To collect the data on adoption of released varieties, focus and individual meetings with key partners, including breeders, agronomists, technicians, managers of seed companies, farmers' organizations, seed producers, etc., were to be conducted. After the first meeting in Niger, it was evident that stakeholders had difficulties estimating the potential area occupied by released varieties nationwide. It was easy, however, for scientists to point to individual locations at districts or regional levels and to assert the percentage adoption in those locations. Thus, individual interviews were carried out with scientists and selected technicians in the crop improvement programmes in the selected countries. Locations were geo-referenced and spatial areas were computed using geographic information system (GIS) tools. These spatial areas were aggregated to the national level and the proportion of area occupied by the variety relative to total cultivated area was taken as the estimate of adoption rate. For Nigeria, results from large nationally representative adoption surveys of households undertaken in northern Nigeria in 2009 and 2010 were used to estimate national adoption.

East and Southern Africa

Coverage in East and Southern Africa (ESA) also focused on five countries: Kenya, Malawi, Tanzania, Uganda and Zambia. Of the ICRISAT-mandate crops, commodity coverage was restricted to groundnut, sorghum and pigeonpea. In general, pearl millet is not cultivated much in this region. Finger millet is the dominant millet species. The area under chickpea is expanding in some countries from a very low base. Its area and production are heavily concentrated in Ethiopia. Chickpea in Ethiopia is reported on in Chapter 12, this volume.

Unlike West Africa, coverage in East and Southern Africa is unbalanced because some crops are of minor importance in one or more of the five countries. Groundnut's coverage extends to the five countries. The area of pigeonpea includes Kenya, Malawi and Tanzania. Sorghum's coverage is restricted to Kenya and Tanzania. Hence, ten crop-by-country observations are available for analysis in the East and Southern Africa region for groundnut, pigeonpea and sorghum crop improvement.

This coverage was strengthened by complementary research by INTSORMIL in the Sudan, which has more land under sorghum cultivation than any other country in Africa (Zereyesus and Dalton, 2012). That research has used the same protocols and has stored their information in the same database as those discussed in Chapters 3 and 4 of this volume.

Including sorghum in the Sudan, the 11 crop improvement observations comprise three observations all in Kenya with less than 100,000 tonnes of production, six observations between 100,000 and 350,000 tonnes of production, sorghum in Tanzania with about 750,000 tonnes of production, and sorghum in the Sudan with 4.2 million tonnes of production in 2009. Total production in 2009 summed to about 6.4 million tonnes across all study countries and crops in ESA, which is decidedly inferior to West Africa's total that approached 30 million tonnes for the 15 study countries in 2009.

Given the small and modest levels of production in nine of the eleven observations, it is reasonable to expect less representative coverage than in West and Central Africa (WCA) unless crop production is concentrated in very few countries. That is the case for pigeonpea because the three country observations account for 75% of production in all of sub-Saharan Africa. In contrast, coverage in groundnut is substantially lower at 45% of production. Production of sorghum in the Sudan and Tanzania represents about half of the regional production. There are 15 excluded groundnut-growing countries and 16 excluded sorghum-producing countries. Usually, these are characterized by small amounts of production but there are some major omissions. In groundnut, the Sudan with 37% of regional production is a large producer with an area that exceeds 1 million hectares. In sorghum, omitted countries that should be considered in the next baseline are Ethiopia, the centre of domestication of the crop, and Uganda where sorghum is used extensively in beer making. Production in these two countries adds up to more than one-third of the production in the region.

The data collected in East and Southern Africa are not as comprehensive as the data gathered in West Africa but they are very similar. An important difference is the manner in which the adoption estimates were deduced. The estimates in East and Southern Africa were based on a combination of sources that included expert opinion as well as secondary literature. In each country, discussions and workshops were held with breeders, seed producers and a variety of key stakeholders including those in the private sector involving crop improvement programmes or seed production for the specified crops. The results obtained from such expert opinions at the country level were shared with ICRISAT scientists before arriving at the final estimates. In more than 80% of the cases, the ICRISAT scientists' estimates were consistent with estimates from experts in each of the countries. Although it was easy to identify agroecological zones and districts where each of the varieties is grown, adoption estimates by district were difficult to derive with the desired degree of certainty and confidence.

Scientific Capacity of NARS

Although commercial hybrids of sorghum, pearl millet and even pigeonpea are readily available in other parts of the world, the private sector is still an insignificant player in genetic research in these crops in sub-Saharan Africa. For this reason, the scientific capacity of NARS in this section focuses exclusively on research in the public sector.

West Africa

This section reports on the strength of NARS proxied by the number of FTE scientists in each institution and programme, and the research intensities defined as the number of scientists per million tonnes of production in 2009/10. The number of scientists by crop and country programme is presented in Table 7.1, which contains three findings. First, a total of only 50–60 FTE scientists are working in groundnut, pearl millet and sorghum improvement programmes in the five countries in West Africa. Secondly, a measure of parity exists across all country programmes. All programmes have more than two FTE scientists and no programmes have more than ten scientists. Parity in the number of scientists is notable in the groundnut programmes that are staffed in the narrow range of 2–4 FTE scientists. With the exception of Senegal, which traditionally is an exporter of groundnut, sorghum and pearl millet, each command a greater allocation of research resources compared to groundnut. Thirdly, inter-country disparities in staffing are substantial. Mali accounts for more than one-third of scientists in pearl millet and sorghum crop improvement among the five countries studied from this region. Given the breadth and depth of Nigeria's investment in education relative to other countries in West Africa, its total of 5.5 FTE scientists in three economically important crops is unexpectedly low.

The estimated research intensities in Table 7.2 reinforce one of the story lines introduced in Table 7.1. Nigeria is characterized by very low research intensities across the three crops. The estimate of 0.3 for sorghum and 0.4 for pearl millet are two of the lowest research intensities found in the literature. Nigeria's apparent lack of commitment to research in groundnut, pearl millet and sorghum is strongly felt at the regional level. The weighted average research intensities range from 1.7 to slightly over 3.0 for the three crops. Per million tonnes of production summed across the three crops, Mali has allocated 17 times more scientists to genetic improvement research than has Nigeria. These low values for the coarse cereals and groundnut stand in sharp contrast to the estimate of 77.5 FTE scientists working on maize in Nigeria that resulted in an estimated research intensity of 10.3 (Chapter 6, this volume).

The other recurring theme centres on the parity in research attention across groundnut, pearl millet and sorghum for the five programmes. The estimated weighted averages in the last column of Table 7.2 range from 2.0 to slightly over 3.0 by crop. Although pearl millet is somewhat discriminated against in terms of its resource allocation of FTE scientists relative to its economic importance, estimated research intensities are not systematically higher or lower by crop across the five countries.

The country crop improvement programmes are heavily concentrated on breeding, with slightly over half of the total of 60 FTE scientists involved in this area (Table 7.3). Only Mali shows a relatively diversified allocation of scientific resources across disciplinary specializations. Entomology, pathology, agronomy and weed science also play supportive roles in several of the programmes. Mali, Niger and Senegal have invested in about 1 FTE scientist in postharvest technology. The low representation of biotechnologists and social scientists, including agricultural economists, is notable in Table 7.3. This could partially explain the low adoption of most of the varieties bred without involving farmers through participatory processes.

The scientists are well educated in terms of their qualifications. More than half of the scientists in Table 7.4 have PhDs. But what is most striking about the information presented in Table 7.4 is the low frequency of presumably younger scientists who only have BScs. They comprise only about one scientist in six.

Scientists in these crop improvement programmes in West Africa are few in number, well educated and old. The highest frequency 5-year age cohorts are 50–55 and 55–60 (Table 7.5). The concern for ageing scientists is especially relevant in Mali, Nigeria and Senegal. Only Niger has an age profile that would seem to facilitate on-the-job learning-by-doing that is essential to sustaining an effective crop improvement programme. The problem of ageing in key programmes poses a threat to national research capacity to undertake the crucial research needs

Table 7.1. Scientific strength of groundnut, pearl millet and sorghum crop improvement programmes by disciplinary area in West Africa, 2009.

Crop	Country	Breeding	Pathology	Molecular biology	Entomology	Agronomy	Seed production	Social science	Food science	Soil science	Total
Groundnut	Burkina Faso	1.2	1.0	0	0	0.2	0	0	0	0	2.4
Groundnut	Mali	0	0	0	0	2.0	0	0	0	0	2.0
Groundnut	Niger	2.2	0	0	0	0	0	0	0	0	2.2
Groundnut	Nigeria	1.2	0.3	0	0	0	0	0	0	0	1.5
Groundnut	Senegal	1.3	0	0	0	0.3	0	0	0	0	1.6
Total		5.9	1.3	0	0	2.5	0	0	0	0	9.7
Pearl Millet	Burkina Faso	3.0	0.3	0	0.5	0.3	0.2	0.1	0	0	4.4
Pearl Millet	Mali	1.5	0.6	0	1.0	1.5	1.2	0	0.3	0.3	6.4
Pearl Millet	Niger	3.0	0.5	0	1.0	0	0	0	0.5	0	5.0
Pearl Millet	Nigeria	1.5	0	0	0	0	0	0	0	0	1.5
Pearl Millet	Senegal	1.6	0	0	0.2	0.6	0.2	0	0.6	0	3.2
Total		10.6	1.4	0	2.7	2.4	1.6	0.1	1.4	0.3	20.5
Sorghum	Burkina Faso	1.5	0.3	0	0.5	0.3	0.2	0.1	0	0	2.9
Sorghum	Mali	3.2	1.0	1.0	0.2	2.0	0	0	0.3	0	7.7
Sorghum	Niger	3.0	0	0	1.0	0	0	0	0.5	0	4.5
Sorghum	Nigeria	2.0	0.3	0	0.2	0	0	0	0	0	2.5
Sorghum	Senegal	0.8	0	0	0.2	0.6	0.2	0	0.6	0	2.4
Total		10.5	1.6	1.0	2.1	2.9	0.4	0.1	1.4	0	20.0

Table 7.2. Estimated research intensities expressed in FTE scientists per million tonnes of production in 2009–2010 by crop and country.

	Research intensity by country					
Crop	Burkina Faso	Mali	Niger	Nigeria	Senegal	Weighted average
Groundnut	9.3	8.2	12.6	0.8	4.9	3.1
Pearl millet	4.3	5.2	1.6	0.4	5.3	1.7
Sorghum	1.9	8.3	5.3	0.3	14.8	2.0
Weighted average	3.5	6.8	3.1	0.4	6.2	2.0

Table 7.3. Number of FTE scientists by specialization and country in 2009–2010.

	Country					
Specialization	Burkina Faso	Mali	Niger	Nigeria	Senegal	Total
Agricultural economist	0.3	0.25	0.3	0.3	0.05	1.2
Agronomist	0.4	2.3	0.55	0.4	0.0	3.65
Soil scientist	0.0	1.5	0.0	0.0	0.0	1.5
Biotechnologist	0.0	1	0.0	0.0	0.0	1
Breeder	6.1	5.65	9.15	5.0	6.15	32.05
Entomologist	1	1.2	1.75	0.85	0.4	5.2
Food technologist	0.0	1.3	1	0.0	1.2	3.5
Genetic resources	0.25	0.0	0.0	0.0	0.0	0.25
Weed scientist	0.53	2	0.0	0.0	0.6	3.13
NRM expert	0.0	0.33	0.0	0.0	0.0	0.33
Nematologist	0.0	0.0	0.0	0.3	0.0	0.3
Pathologist	0.71	1.65	0.55	0.65	0.05	3.61
Physiologist	0.0	0.0	0.0	0.0	0.6	0.6
Seed technologist	0.4	1.2	0.3	0.0	0.6	2.5
Virologist	1		0.0	0.2	0.0	1
Total	10.69	18.38	13.6	7.70	9.65	60.02

NRM, natural resources management.

Table 7.4. Distribution of scientists by level of education and country in 2009–2010.

	Country					
Level of education	Burkina Faso	Mali	Niger	Nigeria	Senegal	Total
BSc	1.2	4.0	4	0.0	0.6	9.8
MSc	1.4	6.5	6.6	0.2	3.7	18.3
PhD	8.1	8.0	2.6	7.5	5.2	31.3
Total	10.7	18.5	13.1	7.7	9.5	59.4

of adaptation, identification and release of varieties. NARS have no succession plans to renew ageing staff.

East and Southern Africa

There are about 40 FTE scientists working in the ten national groundnut, pigeonpea and sorghum crop improvement programmes in East and Southern Africa (Table 7.6). With the exception of sorghum in the Sudan, the scientists work on several crops in larger cereal, grain legume, pulse or oilseed improvement programmes. Therefore, the actual number of scientists is 2–4 times larger than the FTE estimate that reflects the sum of the percentage allocations across all scientists working on the crop.

Table 7.5. Distribution of scientists by age cohort and country in 2009–2010.

	Country					
Age group	Burkina Faso	Mali	Niger	Nigeria	Senegal	Total
30–35	0.0	0.0	3	0.0	0.0	3
35–40	2.55	0.3	1.3	0.5	2.0	6.65
40–45	0.0	1.15	2	0.0	1	4.15
45–50	1.66	3.5	3.5	0.9	0.7	10.26
50–55	2.58	2.85	1.5	2.7	4.55	14.18
55–60	3.75	10.23	0.75	3.1	1.25	19.08
60–65	0.15	0.35	1.55	0.5	0.15	2.7
Total	10.69	18.38	13.6	7.7	9.65	60.02

With about 18 FTE scientists, Sudan is the only large national programme among the ten crop-by-country observations. Sudan is also the only programme that has invested in biotechnology that is beginning to pay dividends (ICRISAT, 2013). Four of the programmes, two in groundnut and two in pigeonpea, are very small with less than 2.0 FTE scientists in total. All of the programmes have invested at least 0.5 FTE scientists in plant breeding, the area of specialization of about 55% of the scientists in Table 7.6. About 20% of the scientific staff is agronomists. The other disciplinary areas are sparsely represented, although a few of the programmes have made a commitment to social science to provide research support.

The very high estimated research intensities in the last column of Table 7.6 for groundnut and pigeonpea in Kenya are typical of small crop improvement programmes with small quantities of national production. Estimated research intensities below 4.0 for groundnut and pigeonpea in Malawi are very low for grain legume improvement programmes. These estimates suggest that scientific staff strength is inadequate in Malawi or relies heavily on ICRISAT in-country support to address crop improvement in groundnut and pigeonpea. One or fewer FTE scientists in total for groundnut and pigeonpea in Malawi would seem to be unable to respond to research requirements of crops characterized by production levels approaching 200,000–300,000 tonnes per annum.

The estimated research intensity for the large sorghum improvement programme in the Sudan exceeds 4.0, which is high for a cereal that is planted on more than 7 million hectares with more than 4 million tonnes of production. In contrast, more rainfall-assured Nigeria with considerably less area and somewhat more production has an estimated research intensity that is only one-tenth of the estimate for rainfall-unassured Sudan. Compared to Nigeria, the sorghum programme in the Sudan is well staffed scientifically and also features a diversified allocation across several research support disciplines with the exception of social science.

Varietal Output

West and Central Africa

Between 1970 and 2010, Burkina Faso, Mali, Niger, Nigeria and Senegal released a total of 326 groundnut, pearl millet and sorghum varieties. Of these improved genotypes, 313 have information on year of release. Although more sorghum varieties (131) have been released than groundnut (87) and pearl millet (95) improved cultivars, the incidence of release follows the same temporal pattern in the three crops: total releases started from a relatively low but firm base in the 1970s, peaked in the 1980s and 1990s, and tapered off in the 2000s (Table 7.7). This increasing–decreasing pattern was more marked in pearl millet than in groundnut and sorghum as pearl millet releases peaked earlier in the 1980s and fell off more sharply in the 2000s.

Declining varietal releases in the recent past reflects decreasing efforts in genetic improvement

Table 7.6. Scientific strength of groundnut, pigeonpea and sorghum crop improvement programmes by disciplinary area in East and Southern Africa, 2009.

Crop	Country	Breeding	Pathology	Molecular biology	Entomology	Agronomy	Seed production	Social science	Food science	Soil science	Total	FTE scientists per million tonnes production
Groundnut	Kenya	1.0	0	0	0	1.0	1.0	0	0	0	3.0	139.8
Groundnut	Malawi	1.0	0	0	0	0	0	0	0	0	1.0	3.6
Groundnut	Tanzania	1.5	1.0	0	0	1.0	0	0.3	0	0	3.8	10.9
Groundnut	Uganda	1.0	1.0	0	1.0	1.0	0	1.0	0	0	5.0	19.4
Groundnut	Zambia	0.5	0	0	0	0.4	0.2	0	0	0	1.1	9.1
Total		5.0	2.0	0	1.0	3.4	1.2	1.3	0	0	13.9	13.6
Pigeonpea	Kenya	2.0	0	0	0	1.0	0	2.0	0	0	5.0	107.6
Pigeonpea	Malawi	0.7	0	0	0	0	0	0	0	0	0.7	3.8
Pigeonpea	Tanzania	0.5	0	0	0	0.7	0	0	0	0	1.2	9.9
Total		3.2	0	0	0	1.7	0	2.0	0	0	6.8	19.3
Sorghum	Kenya	1.5	0.5	0	0	1.0	0	0	0	0	3.0	30.3
Sorghum	Sudan	11.6	0.5	0.7	0.7	2.2	0	0	0.9	0.5	17.1	4.1
Total		13.1	1.0	0.7	0.7	3.2	0	0	0.9	0.5	20.1	4.7

Estimates on scientific strength for sorghum in Tanzania were not presented; therefore, this table refers to only ten crop-by-country observations.

caused by reductions in funding. Many countries in West Africa received World-Bank-funded loans to strengthen research and extension services during the 1980s to 2000. This is probably when most output was generated. Varietal output fell as the projects ended.

More releases in sorghum and pearl millet are a reflection of larger investments in those crops than in groundnut in West and Central Africa. INTSORMIL and ICRISAT were additional partners who made substantial investments in strengthening NARS in the sorghum and pearl millet crop improvement programmes.

The total number of releases is substantially higher in Mali than in the other four study countries of the WCA region (Table 7.8). Mali accounts for more than one-third of the total releases. Cropwise, Mali and Senegal have more groundnut releases; Mali and Niger have, by far, the highest number of pearl millet releases; and Mali, Nigeria and Burkina Faso rank first, second and third in the number of sorghum releases. At the other extreme, Burkina Faso, Niger and Senegal have released fewer than ten varieties in at least one of the three crops. These differences in total varietal output to some extent mirror the differences in scientific capacity discussed in the previous section.

Dividing the data in Tables 7.7 and 7.8 into two periods (1970–1990 and 1990–2010) shows that about half of the varieties were released before 1990 and half after 1990. However, this 50:50 split does not apply to each of the five countries. Nigeria with 45 releases had more varietal output from 1970 to 1990 than any other country. After 1990, Nigeria has only released a total of 14 varieties of groundnut, pearl millet and sorghum. Senegal and Niger also were characterized by fewer releases since 1990. In contrast, Mali has released 88 genotypes in the latter period and has made available more varieties in each crop than any other country since 1990.

In numerical terms, the steepest drop in releases between the two periods for any of the 15 crop-by-country observations occurred in sorghum in Nigeria. Prior to 1990, 27 sorghum varieties were released in Nigeria; post-1990 only five varieties have been approved for release.

Recent dry spells in generating varietal output are also evident in the release database. Senegal did not release an improved pearl millet or sorghum variety between 1990 and 2010. In the same period, Nigeria has released only three pearl millet varieties, Niger has released only three sorghum varieties, and Burkina Faso

Table 7.7. Number of varieties released by year range and crop from 1970 to 2010.

	Crop			
Decade	Groundnut	Pearl millet	Sorghum	Total
1970s	16	15	13	44
1980s	33	45	40	118
1990s	22	28	46	96
2000s	16	7	32	55
Total	87	95	131	313

Table 7.8. Number of varieties released by country and crop from 1970 to 2010.

	Crop			
Country	Groundnut	Pearl millet	Sorghum	Total
Burkina Faso	8	9	26	43
Mali	26	33	60	119
Niger	13	37	7	57
Nigeria	17	10	32	59
Senegal	23	6	6	35
Total	87	95	131	313

has released only one groundnut and two pearl millet varieties.

Since 1990 varietal output by crop is concentrated in two countries of the region. Mali and Senegal account for 31 of the 44 groundnut releases, Mali and Niger account for 32 of the 36 pearl millet releases, and Mali and Burkina Faso account for 75 of the 83 sorghum releases.

From the perspective of the development of crop improvement programmes, sufficient information was available on 172 entries in the release database to distinguish among different types of products from crop improvement. 'Purified varieties' refer to local landraces that were made as genetically homogeneous as possible and were subsequently released. 'Adaptation' refers to selection of finished elite varieties in multi-locational trials. Usually, these varieties are imported directly from germplasm distribution networks. The descriptor 'Crossing' in the following tables indexes varieties that were selected from crossed materials. The selected progenies could come from populations provided by IARCs or from crosses made by the NARS themselves. Progeny selection from segregating populations requires more applied plant breeding effort than selection of finished varieties, which, in turn, is more technically demanding than purification of landraces.

About 45% of the varieties were released following adaptation trials; 24% resulted from variety purification. Only about 30% of varieties released were developed from crossing (Table 7.9). These frequencies vary crop-wise where 81% of the groundnut releases resulted from adaptation trials against 15% resulting from crossing compared to sorghum or pearl millet where 35–40% of the releases were derived from crossing. In fact, there have been few mature breeding efforts in the region in groundnut except for ICRISAT and the University of Georgia's investment in Nigeria in the 1990s. The data in Table 7.9 also suggest that the release of purified landraces in this region was rare in groundnut but was common in the two cereals.

Across countries, the differences in the use of different sources and procedures for the release of varieties is not as marked across crops because adaptation trials and progeny selection are common to the five country programmes (Table 7.10). However, the release of purified landraces is mainly confined to Mali and Niger. The low incidence of crossing is surprising in a large NARS like Nigeria where breeding efforts on groundnut, pearl millet and sorghum seem to have declined over time.

The contribution of ICRISAT to the total releases since 1970 seems modest (Table 7.11). Only about 24% of the total releases had ICRISAT

Table 7.9. Distribution of varieties released by breeding scheme and crop in the five countries.

	Crop			
Breeding scheme	Groundnut	Pearl millet	Sorghum	Total
Adaptation	44	17	17	78
Purification	2	28	12	42
Crossing	8	25	19	52
Total	54	70	48	172

Table 7.10. Distribution of varieties released by breeding scheme and country.

	Country					
Breeding scheme	Burkina Faso	Mali	Niger	Nigeria	Senegal	Total
Adaptation	5	43	13	9	8	78
Purification	2	17	21	2	0	42
Crossing	6	23	11	5	7	52
Total	13	83	45	16	15	172

parents or were selected from ICRISAT progenies, lines or elite varieties. Attribution to ICRISAT is the lowest for sorghum (8% of varieties with an ICRISAT parent or crosses). In terms of number of ICRISAT-related varieties, the institute's presence has been more pronounced in pearl millet than in groundnut and sorghum. Only one groundnut variety and only eight sorghum varieties were selected from ICRISAT materials that were not ICRISAT-bred varieties or lines.

Low attribution was unexpected given ICRISAT's historical investment in crop improvement in West and Central Africa. A paucity of ICRISAT-related released materials in sorghum is partially explained by the lack of adaptability of some ICRISAT germplasm in the late 1970s and early 1980s when the breeding work started in the region. At the early stages, the breeding scheme was oriented towards the *Caudatum* race of material (popular in India and the USA), whereas the *Guinea* race of material was predominant in Burkina Faso and Mali. Poorly accepted releases of the *Caudatum* types were largely explained by preferences of farmers for tall *Guinea* types that produce plenty of stalks for use as construction material, fuel and fodder for livestock, and non-preferred cooking quality of the large white-seeded and chalkier grain of the *Caudatum* types that were more susceptible to diseases (particularly grain mould) and pests (including head bugs and grain-feeding birds) than the locally preferred *Guinea* race types.

Another reason for the low institutional attribution to ICRISAT that became apparent in consultations with stakeholders was the imprecise knowledge of variety information with regard to parents. There may be many more ICRISAT parents involved than was reported.

Through time, the relative importance of ICRISAT-related materials has increased from 11% of varieties with an ICRISAT parent or crosses during 1970–1980 to about 24% during 2000–2010 (Table 7.12). However, the number of ICRISAT-related releases mirrors the trend of total releases described in Table 7.12: it peaked in the 1980s and 1990s and has since declined.

Data on the contribution from other institutions such as INTSORMIL, CIRAD and the Peanut CRSP were not clearly elicited when the variety release database was assembled. At the stakeholder meeting that was held in Niamey with breeders and agronomists on 6–7 August 2012, partners were asked about pedigree information and institutional attribution. Results are presented in Table 7.13 by crop and Table 7.14 by country.

Table 7.11. Distribution of released varieties related to ICRISAT by crop from 1970 to 2010.

	Crop			
Germplasm origin	Groundnut	Pearl millet	Sorghum	Total
Not ICRISAT Germplasm	43	53	107	203
Parent ICRISAT/Cross NARS	1	0	4	5
Cross ICRISAT/Selection NARS	0	14	4	18
Cross ICRISAT/Selection ICRISAT	11	23	6	40
Total	55	90	121	266

Table 7.12. Distribution of released varieties related to ICRISAT by decade from 1970 to 2010.

	Year range				
Germplasm origin	1970–1980	1980–1990	1990–2000	2000–2010	Total
Not ICRISAT Germplasm	34	70	61	38	203
Parent ICRISAT/Cross NARS	0	0	0	5	5
Cross ICRISAT/Selection NARS	0	5	11	2	18
Cross ICRISAT/Selection ICRISAT	4	17	14	5	40
Total	38	92	86	50	266

Table 7.13. Distribution of released varieties by crop and institution that provided at least one of the parents.

Whose baby is it?	Crop: Groundnut	Pearl millet	Sorghum	Total
1970–1990				
CIRAD	2	0	0	2
ICRISAT	0	19	1	20
ICRISAT–INTSORMIL	0	0	1	1
IRAT	0	13	1	14
IRHO	11	0	0	11
Local and NARS varieties	19	14	40	73
CRSP, USA	2	0	0	2
Missing	10	7	9	26
Sub-total	44	53	52	149
1970–2010				
AMU	1	0	0	1
CIRAD	3	0	5	8
CRSP, USA	5	0	0	5
Taiwan	1	0	0	1
ICRISAT	5	37	8	50
ICRISAT–INTSORMIL	0	0	1	1
ICRISAT–Purdue	0	0	1	1
INTSORMIL	0	0	6	6
IRAT	0	13	3	16
IRHO	11	0	0	11
Local and NARS varieties	32	44	77	153
Tifton	1	0	0	1
USA	3	0	0	3
Missing	15	13	34	62
Sub-total	77	107	135	319

AMU, Texas A&M University; CIRAD, Centre de Coopération Internationale en Recherche Agronomique pour le Développement, France; CRSP, Collaborative Research Support Project; CRSP, USA, University of Georgia, USA; ICRISAT, International Crop Research Institute for Semi-Arid Tropics; INTSORMIL, International Sorghum and Millet CRSP; IRAT, Institut de Recherches Agronomiques Tropicales; IRHO, Institut de Recherche pour Les Huiles et Oleagineux (France).

Of the 257 varieties with parental information, 104 were linked to institutes external to the countries where the varieties were released. The other 153 varieties contained materials internal to the countries where they were released. Overall, 16% of varieties bred had ICRISAT germplasm, followed by 13% from CIRAD (includes IRHO, IRAT and CIRAD), 4% from INTSORMIL and 2% from universities (including Purdue University; Tifton Coastal Plain Experiment Station, USDA/University of Georgia; Texas A&M University; North Carolina State; Florida A&M; and Gujarat Agricultural University (Junagadh, India)). Crop-wise before 1990, the contribution of international institutions was high for pearl millet where about 60% of the varieties had a parent from an international organization, followed by groundnut with 34%. The estimate for sorghum was a very low 6%. The same pattern is observed during 1990–2010 by crop. The contribution of international organizations to pearl millet had, however, decreased somewhat and had risen for groundnut.

By country, the contribution of international organization averaged about 31% and has not changed much over time (Table 7.14). However, it is estimated that the contribution of international organizations decreased significantly in Burkina Faso and fell slightly in Mali, reflecting the increasing strength of INERA (Institut de l'Environnement et de Recherches

Table 7.14. Distribution of released varieties by country and institution that provided at least one of the parents.

	Country					
Whose baby is it?	Burkina Faso	Mali	Niger	Nigeria	Senegal	Total
1970–1990						
CIRAD	0	0	0	0	2	2
ICRISAT	6	3	7	0	4	20
ICRISAT-INTSORMIL	0	1	0	0	0	1
IRAT	5	2	5	0	2	14
IRHO	7	1	0	1	2	11
Local	1	15	13	43	1	73
Missing	2	1	8	9	6	26
Tifton	0	1	0	0	0	1
USA	0	0	0	1	0	1
Sub-total	21	24	33	54	17	149
1970–2010						
AMU	0	1	0	0	0	1
CIRAD	4	1	0	0	3	8
CIRAD-CRSP	0	0	0	0	5	5
Taiwan	0	1	0	0	0	1
ICRISAT	12	14	16	4	4	50
ICRISAT-INTSORMIL	0	1	0	0	0	1
ICRISAT-Purdue	0	0	1	0	0	1
INTSORMIL	0	5	1	0	0	6
IRAT	5	4	5	0	2	16
IRHO	7	1	0	1	2	11
Local	14	71	17	49	2	153
Missing	22	6	16	12	6	62
Tifton	0	1	0	0	0	1
USA	0	1	0	2	0	3
Sub-total	64	107	56	68	24	319

AMU, Texas A&M University; CIRAD, Centre de Coopération Internationale en Recherche Agronomique pour le Développement. France; CRSP, Collaborative Research Support Project; ICRISAT, International Crop Research Institute for Semi-arid Tropics; INTSORMIL, International Sorghum and Millet CRSP; IRAT, Institut de Recherches Agronomiques Tropicales; IRHO, Institut de Recherche pour Les Huiles et Oleagineux (France); Tifton, University of Georgia, USA.

Agricoles) of Burkina Faso and IER (Institut d'Economie Rurale) in Mali. This decline was compensated by small increases in external institutions' contributions in Niger, Nigeria and Senegal.

Nigeria's self-sufficiency in material related to released varieties is the most noteworthy finding in Table 7.14. Of 54 released varieties with pedigree information, only five were identified as coming from international sources. The absence of collaboration with French crop improvement research institutes was expected for Nigeria but the apparently low level of effective interaction with other international organizations, including ICRISAT, was surprising.

East and Southern Africa

In this section, we start by providing an overview of variety releases derived from ICRISAT-supplied germplasm for three of ICRISAT's mandate crops (groundnut, pigeonpea and sorghum) in 18 countries of ESA. We then move on to discuss variety releases in just the five countries within the region that were selected for the in-depth study. Again, these data are complemented by INTSORMIL's research on crop improvement in sorghum in the Sudan.

Table 7.15 shows the distribution of variety releases of groundnut, pigeonpea and sorghum across all 18 ESA countries over the period

1975–2010, which is further disaggregated into two time periods (before and after 1998). The results indicate contrasting trends in outputs between crops over the years. Variety releases for legumes increased substantially after 1998, reflecting maintained efforts by and funding for genetic improvement. In contrast, variety releases for sorghum have been decreasing since 1999, reflecting the closure of a large USAID-funded programme on sorghum and millet improvement in Southern Africa under the auspices of SADCC (Southern African Development Coordination Conference).

Within the study countries of Malawi, Tanzania, Uganda, Zambia, Kenya and the Sudan, the trend in varietal release for groundnuts and pigeonpea is consistent with the trend in the ESA region. As depicted in Table 7.16, the incidence of varietal release was at a low-level equilibrium in the 1960s, in the 1970s, and the 1980s of about 10 varieties released per decade. Releases gathered momentum in the 1990s and surged upward in the 2000s.

The character of the releases also changed over time. Before the 1970s, releases consisted mainly of shared landraces in the East African community and landraces from South Africa such as the popular groundnut cultivar Natal Common. In the past two decades, the frequency of landraces in releases has declined and the incidence of bred varieties has increased.

The character of releases also varies by strength of NARS and crop type. Sorghum breeders in the Sudan and Kenya have selected materials from their own parental crosses using NARS or IARC materials. In contrast, pigeonpea and groundnut breeders have released varieties mainly from adaptation trials of ICRISAT-bred materials.

Of the 105 releases in Table 7.16, 57 are related to ICRISAT. A few were distributed via ICRISAT from the Indian Council of Agricultural

Table 7.15. Number of varieties released for the period 1975–2010 released using ICRISAT-supplied germplasm in 18 countries in East and Southern Africa.

	Year				
	1975–1998		1999–2010		
Crop	Number of varieties	Percentage of total	Number of varieties	Percentage of total	Total
Groundnut	17	30.4	39	69.6	56
Pigeonpea	6	28.6	15	71.4	21
Sorghum	56	67.5	27	32.5	83
Total	79		81		160

Source: ICRISAT's variety release database.

Table 7.16. Varietal releases by crop and country across five time periods in East and Southern Africa.

Crop	Country	Pre-1970	1970s	1980s	1990s	2000–2011	Total
Groundnut	Kenya	5	0	0	0	0	5
Groundnut	Malawi	0	0	0	1	5	6
Groundnut	Tanzania	0	1	2	0	7	10
Groundnut	Uganda	0	0	0	1	5	6
Groundnut	Zambia	4	0	1	5	7	17
Pigeonpea	Kenya	0	0	2	1	4	7
Pigeonpea	Malawi	0	0	1	0	6	7
Pigeonpea	Tanzania	0	0	0	1	2	3
Sorghum	Kenya	0	2	2	3	7	14
Sorghum	Sudan	1	5	1	8	8	23
Sorghum	Tanzania	1	1	0	2	3	7
Total		11	9	9	22	54	105

Research (ICAR) to national groundnut programmes in the region. JL 24 and Robut 33-1 are two of the three old and still very popular improved cultivars that dominate groundnut production in India. Most others were ICRISAT-bred varieties. A few were derived from the use of ICRISAT parental materials. Two were efforts partnered by INTSORMIL and the Sudanese national programme, the Agricultural Research Corporation (ARC).

The majority of these ICRISAT-affiliated releases occurred in the 2000s, especially in groundnut and pigeonpea. However, ICRISAT-related release activity was higher in the Sudan in the late 1980s and early 1990s when a joint ICRISAT–INTSORMIL breeder was posted in the country at ARC.

Other institutions have also contributed to releases. INTSORMIL was associated with several releases in the Sudan. The private sector has also been involved in the production of several groundnut varieties in Zambia, a sorghum hybrid in Sudan and a sorghum OPV in Tanzania.

Although the data highlight many positive aspects of varietal output in ESA, three concerns are worth mentioning. First, the fact that Kenya has not released any groundnut cultivars since the 1960s is puzzling. Secondly, long dry spells in release behaviour were documented in a crop improvement programmes as mature as that of sorghum in the Sudan, which has recently released cultivars targeted for its smallholder traditional sorghum-production sector. Sudan only released one improved cultivar, albeit a very important one[2], between 1978 and 1991. Lastly, sorghum varietal output is declining in the region. This may reflect the substitution of maize for sorghum or the decline in sorghum consumption with urbanization. The interest in sorghum seems to be waning in several of the smaller producers in the region.

Before moving to the next section on adoption, a brief comparison in varietal output between West Africa and East and Southern Africa is timely. The two regions are characterized by two very different varietal release profiles over time, by the differences in importance of landrace releases vis-à-vis bred varieties, and by the level of influence of IARC materials in varietal output. All three aspects suggest a more favourable experience in East and Southern Africa than in West Africa.

Adoption of Improved Varieties

West Africa

The adoption estimates are derived from expert opinion and GIS information as described in the methods section of this chapter for groundnut, pearl millet and sorghum in Burkina Faso, Mali, Niger and Senegal. In Nigeria, estimates are taken from nationally representative surveys of adoption and impact of improved pearl millet and sorghum varieties (Ndjeunga *et al.*, 2011) and modern groundnut cultivars (Ndjeunga *et al.*, 2012b). The cereal survey was conducted in 2009; the groundnut survey was undertaken in 2011.

It is important to point out that several popular improved varieties are not considered as modern varieties for the purpose of this investigation. The groundnut variety 55-347 is the dominant variety in Senegal with an area share of 15%. It is also the leading variety in Nigeria with an area share exceeding 40%. Likewise, the variety 47-10 accounts for over 40% of groundnut growing area in Mali. It was released in Mali in 1957; therefore, it does not qualify on the age criterion of having a release date after 1970. Variety 55-347 and several kindred cultivars do qualify in principle because they were re-released in the late 1980s in several groundnut-growing countries in West Africa. But they do not qualify in practice because they were bred during the colonial era in the 1950s and were initially released during the 1960s. These varieties have been around for a long time; 55-347 is still expanding in area in Nigeria.

The pearl millet open-pollinated variety HKP addresses a different aspect in the definition of a modern variety. It was released in 1977. HKP was derived by selection from a local landrace and is still the most widely multiplied pearl millet in Niger's certified seed programme. Pearl millet is a highly cross-pollinated crop that is very prone to outcrossing so the HKP seed that is marketed now is different from the original. It should, however, still offer some advantage compared to the local landrace from which it was developed because it has a lower frequency of 'shibras', which are weedy intermediates between its cultivated form and its wild progenitor. Nonetheless, HKP was not regarded as an improved variety in this study because it is derived wholly from local landrace materials. Being more inclusive in the definition of a modern variety results in a doubling of the adoption levels of improved varieties in groundnut in Mali, Nigeria and Senegal (Ndjeunga *et al.*, 2012a). Including HKP leads to a tripling of the level of pearl millet improved-variety adoption in Niger.

National adoption estimates of improved varieties are presented for the three crops and five countries in Table 7.17. The area weighted mean adoption estimates for improved cultivars of each crop across the five countries varied from about 18% for pearl millet and sorghum to 25% for groundnut. Each crop was characterized by one or more lagging countries where uptake of modern varieties was substantially below this mean level. Niger lagged behind in groundnut. Burkina Faso and Niger were slow to adopt improved varieties of pearl millet. Adoption of modern varieties of sorghum was negligible in Burkina Faso.

Low adoption is partly explained by the slow release of modern varieties, therefore limiting the availability of higher performing varieties that can readily attract smallholder farmers. In effect, countries have historically had weak pearl millet and sorghum breeding programmes. Since 1990, countries in WCA have released on average less than one pearl millet variety per year and less than one sorghum variety per year.

Table 7.17. Adoption of improved varieties of groundnut, pearl millet and sorghum in West and Central Africa, 2009.

Country	National/ agroecology	Area (million ha)	Area MVs (%)
Groundnut			
Burkina Faso	National	0.5	24.8
Mali	National	0.3	19.6
Niger	National	0.6	11.9
Nigeria	National	2.6	19.4
Senegal	National	1.0	47.4
Pearl millet			
Burkina Faso	National	1.2	2.6
Mali	National	1.5	31.1
Niger	National	6.5	11.5
Nigeria	National	3.7	25
Senegal	National	1.0	34.5
Sorghum			
Burkina Faso	National	1.8	3.3
Mali	National	1.0	32.6
Niger	National	2.5	15.1
Nigeria	National	4.7	20
Senegal	National	0.2	41.2

There are, however, disparities between countries. The correlation between the number of releases and adoption is positive and significant. In countries such as Mali, where the number of releases is high, the adoption rate is also relatively high compared to countries like Senegal that have released few if any improved varieties in 20 years. The adoption rate is partly explained by the strength of the breeding programmes.

Low adoption is also attributed to a lack of promotion of released varieties. For example, the hybrid sorghum variety NAD1 developed by both INTSORMIL and the Institut National de la Recherche Agronomique du Niger (INRAN), Niger's national programme, has been largely constrained by incomplete knowledge of hybrid production by 27 seed producers mainly organized in farmer associations. One of their pervasive problems in producing viable hybrid seed is timing the crossing of in-bred lines.

Few if any of the adopted improved groundnut varieties in Table 7.18 could be called dominant or even leading varieties. Only two varieties in Senegal account for more than 10% of cultivated area (Table 7.18). And neither 73-33 or Boulkoss are adopted by other countries in the region. Nevertheless, the incidence of spill-over varieties released and adopted in more than one country in West Africa is quite high. Using an expanded, more inclusive definition of a modern variety, Ndjeunga *et al.* (2012a) found that 12 varieties were sown in two or more countries. Fleur 11, a recent introduction from Asia, is sown in three countries. The old, extensively cultivated bred short-duration groundnut variety 55-437 is released and grown in every West African country included in this study, except Burkina Faso.

In contrast to groundnut, pearl millet is characterized by several recent releases that satisfy the criterion of a leading variety tending towards dominance. SOSAT-C88 in Nigeria and Toronion C1 in Mali are potential members of a set of leading varieties with appreciable levels of adoption (Table 7.19). Similar to groundnut,

Table 7.18. Economically important improved groundnut varieties in West and Central Africa, 2009, by area adopted.

Country	Variety	Area (%)
Burkina Faso	TS32-1	5.98
Burkina Faso	E(104)	5.74
Burkina Faso	CN94_C	4.21
Burkina Faso	SH470-P	3.42
Burkina Faso	QH243-C	2.01
Burkina Faso	SH67-A	0.69
Burkina Faso	RMP12	0.69
Burkina Faso	RMP91	0.69
Burkina Faso	KH149-A	0.69
Burkina Faso	Fleur 11	0.69
Mali	JL 24	4.44
Mali	Fleur 11	3.04
Mali	ICGS(34)E	2.64
Mali	28-206	2.50
Mali	ICG(FRDS)4	2.26
Mali	ICGV 86124	2.01
Mali	CN94C	1.55
Mali	ICGV86015	1.16
Niger	RRB	8.54
Niger	JL 24	1.22
Niger	TS 32-1	1.02
Niger	J11	0.51
Niger	Fleur 11	0.51
Niger	ICG 9346	0.10
Niger	O-20	0.00
Nigeria	SAMNUT 23 (ICGV-IS 96894)	4.21
Nigeria	SAMNUT22 (M572.80I)	3.21
Nigeria	SAMNUT21 (UGA 2)	3.2
Nigeria	RMP 91	2.09
Nigeria	SAMARU	2.09
Nigeria	MK 374	1.24
Nigeria	RRB	1.24
Nigeria	RMP 12	1.14
Nigeria	ICIAR 19bt	0.45
Nigeria	M 25.68	0.28
Nigeria	ICIAR 6at	0.11
Nigeria	F452.2	0.04
Nigeria	M 412.801	0.04
Nigeria	M 318.7	0.02
Nigeria	ICIAR 7b	0.01
Senegal	73-33	12.26
Senegal	Boulkouss	11.31
Senegal	H75-0	6.47
Senegal	28-206	3.93
Senegal	PC7979	3.62
Senegal	Fleur 11	3.55
Senegal	78-936	3.15
Senegal	73-911	3.15

Table 7.19. Economically important improved pearl millet varieties in West and Central Africa, 2009.

Country	Variety	Area (%)
Burkina Faso	IKMP 5	1.11
Burkina Faso	IMKV 8201	0.62
Burkina Faso	IKMP 1	0.60
Burkina Faso	SOSAT-C88	0.27
Mali	Toronion C1	16.61
Mali	SOSAT-C88	5.23
Mali	Sanioba 03	3.15
Mali	Djuiguifa	2.38
Mali	Indiana	1.99
Mali	Benkadinion	1.38
Mali	Sanioteli53	0.35
Mali	Amel.M01	0.00
Mali	IKMV 82-01	0.00
Mali	Pool 9	0.00
Mali	PN4 C1	0.00
Nigeria	SOSAT C88	23.95
Nigeria	GB 8735	0.30
Nigeria	ICMV-IS 89305	0.22
Nigeria	LCIC 9703	0.18
Nigeria	LCIC 9702	0.05
Niger	HKP	5.00
Niger	Moro	2.21
Niger	SOSAT-C88	1.29
Niger	Zatib	1.23
Niger	ANK P1 (Ankoutess)	0.38
Niger	ICMV-IS 89305	0.31
Niger	GB8735	0.26
Niger	H80 10 GR	0.19
Niger	Souna3	0.19
Niger	BAP1	0.16
Niger	ICMV-IS 99001	0.06
Niger	CT 6	0.05
Niger	HKB0P1	0.04
Niger	MTDO	0.04
Niger	CEY	0.04
Niger	Zongo Kollo	0.01
Senegal	Thialack2	16.50
Senegal	Souna3	14.00
Senegal	IBMV8402	4.00

Ndjeunga *et al.* (2012a) also found a high incidence of spill-over varieties in pearl millet. Thirteen cultivars were released and partially adopted in two countries. SOSAT-C88 is released and adopted in four of the five countries included in this study (all except Senegal), and at least three additional countries in the region (Cameroon, Chad and Mauritania). Its cultivated area in West Africa approaches 1 million hectares.

Sorghum seems to be a different case than either groundnut or pearl millet in terms of leading and spill-over varieties. There are no persuasive candidates for leading varieties in Table 7.20. Moreover, only five improved varieties were released and partially adopted in only two countries (Ndjeunga *et al.*, 2012a). None was cultivated in more than two countries of the region.

In all the countries, the turn-over of varieties is low. In Niger, the pearl millet variety HKP released in 1975 is still the most popular despite the release of more modern varieties (including ICMV-IS 99001, which is itself a higher-yielding re-selected version of HKP). Similarly the groundnut variety 55-437 released in Senegal some 50 years ago is still the dominant variety in Niger. In Mali, the groundnut variety 47-10 released in the 1950s is still dominant. In Nigeria, the variety 55-437 continues as the leading variety. These varieties are still difficult to replace. Existing groundnut cultivars have staying power in farmers' fields because the crop is highly self-pollinated, the multiplication rate of seed is low, seeding rate is high, and the rate of degeneration from outcrosses, mutation or viruses is negligible.

Table 7.20. Economically important improved sorghum varieties in West and Central Africa, 2009.

Country	Improved variety	Area (%)
Mali	Seguifa	6.97
Mali	Tiandougou coura	4.36
Mali	Grinkan	4.26
Mali	Sewa (hybrid)	4.02
Mali	Jacunbe	2.68
Mali	Unnamed hybrid	2.01
Mali	Darrelken	1.48
Mali	Djiguisene	1.46
Mali	Niatitiama	1.14
Mali	Wassa	1.11
Mali	97-SBF5DT-150	0.79
Mali	Kalaban	0.68
Mali	Marakanio	0.68
Mali	ICSV401	0.65
Mali	Tiandougou	0.12
Mali	98-SB-F2-78	0.11
Mali	97-SB-F5DT-63	0.02
Mali	IS15401	0.00
Niger	Sepon 82	4.95
Niger	MM (Mota Maradi)	3.63
Niger	SSD35	2.91
Niger	IRAT204	2.74
Niger	NAD-1 (hybrid)	0.45
Niger	90SN7	0.14
Niger	S35	0.13
Niger	SRN39	0.07
Niger	MAR	0.02
Niger	90SN1	0.01
Nigeria	ICSV 111	8.65
Nigeria	ICSV 400	8.35
Nigeria	SK 5912	2.76
Nigeria	ICRISAT Hybrid Sorghum	0.27
Senegal	F2-20	12.03
Senegal	CE151	10.57
Senegal	CE145-66	9.30
Senegal	CE181	9.30

On average, during the period 1970–1990 and equating adoption to all released varieties, the age of groundnut varieties is estimated at 20 years, pearl millet at 21 years and sorghum at 19 years (Table 7.21). During the period 1990–2010, the age of varieties was estimated at 12 years for groundnut, 14 years for pearl millet and 12 years for sorghum, signalling low turnover even for the recent time period of the last two decades. There are, however, three cases where varietal age is less than 10 years. In Mali, the turnover for sorghum varieties is relatively higher, i.e. 9 years, and in Niger, the turnover of modern groundnut varieties is fast, estimated at 6 years. In both countries the presence of ICRISAT has played a significant role. Many groundnut varieties released in Niger during the last 10 years are from ICRISAT parents or crosses. Similarly, in Nigeria, pearl millet varieties released during the last 10 years have ICRISAT parents or crosses. In Senegal, the results in Table 7.21 are meaningless because no sorghum or pearl millet varieties were officially released in the recent period.

East and Southern Africa

The three crop estimates for adoption of improved varieties in Tanzania are derived from a nationally representative adoption survey in 2011 (Mausch and Simtowe, 2012). The other eight crop-by-country estimates come from a mixture of expert opinion reinforced by the existing literature on adoption as described earlier in this chapter.

Table 7.21. Average age of varieties (years) released during the entire period 1970–2010 and 1990–2010.

Country	Period	Crop			Total
		Groundnut	Pearl millet	Sorghum	
Burkina Faso	1970–2010	23.25	24.11	17.74	20.04
	1990–2010	16.00	16.00	16.36	16.32
Mali	1970–2010	16.38	17.69	11.61	14.34
	1990–2010	12.05	13.37	9.34	10.79
Niger	1970–2010	19.62	21.75	22.57	21.37
	1990–2010	6.20	15.00	19.33	13.52
Nigeria	1970–2010	25.27	23.5	28.87	27.00
	1990–2010	12.33	7.33	11.00	10.78
Senegal	1970–2010	19.57	28.50	28.00	22.54
	1990–2010	12.5	0.00	0.00	12.50
Total for all five countries	1970–2010	20.36	21.18	18.91	19.98
(weighted average)	1990–2010	11.72	13.53	11.92	12.22

The adoption levels of improved varieties range from 30 to 60% for all 11 of the crop-by-country observations (Table 7.22). In Sudan, the estimates were elicited for the three major sorghum-producing agroecologies. They reflect a wide range in production and economic conditions varying from almost full adoption in irrigated production to very partial adoption of less than 20% in the smallholder rainfed sector that is largely unmechanized.

Adoption estimates by cultivar are presented for groundnut in Table 7.23, for pigeonpea in Table 7.24 and for sorghum in Table 7.25. The rapid uptake of recently released groundnut varieties in Uganda is an emerging success story. The diffusion of these ICRISAT-bred and NARO-selected varieties has been aided and abetted by inputs from USAID's Peanut CRSP (John Williams, 2012, personal communication).

Several of the leading groundnut varieties in Table 7.23 warrant comment. CG 7 (ICGV 83708) was released in Malawi in 1990. CG 7 is a high-yielding, red-seeded Virginia bunch variety that is known for its drought tolerance. In 1997, famine-monitoring survey data suggested that CG7 was planted on about 10% of groundnut area and was replacing Chalimbana, the dominant landrace variety introduced by the EAC (East African Community) in the 1960s (Subrahmanyam *et al.*, 2000). ICGV 83708 is also the leading improved variety in Zambia where it was released in 1990 as MGV 4. ICGV 83708 was released in Uganda as Serenut 1R.

Table 7.22. Adoption of improved varieties of groundnut, pigeonpea and sorghum in East and Southern Africa, 2009.

Country	National/ agroecology	Area (ha)	Area MVs (%)
Groundnut			
Kenya	National	20,640	47
Malawi	National	266,946	58
Tanzania	National	535,000	32
Uganda	National	253,000	55
Zambia	National	204,073	57
Pigeonpea			
Kenya	National	118,167	49.7
Malawi	National	175,734	50
Tanzania	National	72,000	49.8
Sorghum			
Kenya	National	173,172	40
Tanzania	National	874,219	37.7
Sudan	Irrigated	465,675	97.4
Sudan	Mechanized/ rainfed	3,991,500	45.8
Sudan	Traditional/ rainfed	2,195,325	18.5
Sudan	National	6,652,500	40.4

ICGV-SM 90704 is a rosette-resistant cultivar that was released in Malawi in 2000. ICGV-SM 90704 was generated by ICRISAT in 1983 from a cross of two varieties, one of which was developed in Malawi (Chiyembekeza *et al.*, 2000). ICGV-SM 90704 was extensively tested

Table 7.23. Economically important improved groundnut varieties in East and Southern Africa, 2009, by area adopted.

Country	Variety	Area (%)
Kenya	ICGV-SM 90704	22.00
Kenya	ICGV-SM 99568	16.00
Kenya	ICGV-SM 9991	6.00
Kenya	ICGV-SM 12991	3.00
Malawi	ICGV-83708	30.00
Malawi	ICGV-SM 90704	20.00
Malawi	JL 24	7.00
Malawi	ICG 12991	0.50
Malawi	ICGV-SM 99568	0.20
Malawi	C851/7	0.10
Tanzania	Pendo	18.4
Tanzania	Other improved	9.3
Tanzania	Sawia	3.7
Tanzania	Naliendele	0.5
Tanzania	Nachingwea	0.1
Tanzania	Mnanje	0.1
Uganda	Serenut 2T	16.80
Uganda	Serenut 3R	14.20
Uganda	Serenut 1R	12.70
Uganda	Serenut 4T	11.90
Zambia	MGV 4	23.00
Zambia	Chishango	10.00
Zambia	Kadononga	8.00
Zambia	MGV 5	6.00
Zambia	Natal Common	2.00
Zambia	Chipego	2.00
Zambia	Makulu red	2.00
Zambia	Kamulomo	2.00
Zambia	Luena	2.00
Zambia	Katete	0.10

in both on-station and on-farm trials in Malawi. ICGV-SM 90704 is a medium-duration, high-yielding cultivar that gives markedly heavier yields than susceptible varieties in years when rosette epidemics occur. Its incidence of infection with the rosette virus is significantly lower than susceptible varieties. ICGV-SM 90704 was also extensively tested in the region and was released as Serenut 2T in Uganda in 1999 and as Chishango in Zambia in 2001. Although it is not officially released in Kenya, it is believed to be the most widely adopted improved variety (Table 7.23). Among improved varieties of groundnut in ESA, ICGV-SM 90704 ranks first in area in Kenya and Uganda and is second in area in Malawi and Zambia. Consequently, ICGV-SM 90704 scores high marks on wide

Table 7.24. Economically important improved pigeonpea varieties in East and Southern Africa, 2009.

Country	Variety	Area (%)
Kenya	KAT 777	16.00
Kenya	ICEAP 00557	12.00
Kenya	ICPL 87091	10.00
Kenya	KAT 60/8	4.00
Kenya	ICEAP 00554	4.00
Kenya	KAT 81/3/3	0.08
Kenya	ICEAP 00040	0.06
Malawi	ICP 9145	25
Malawi	ICEAP 00040	20
Malawi	ICEAP 00557	5
Tanzania	ICEAP 00040	30.60
Tanzania	ICEAP 00053	12.80
Tanzania	ICEAP 00554	2.20
Tanzania	ICPL 87091	1.60
Tanzania	ICEAP 00557	0.80
Tanzania	ICEAP 00020	0.80
Tanzania	Other improved	0.70
Tanzania	ICEAP 00068	0.30

Table 7.25. Economically important improved sorghum varieties in East and Southern Africa, 2009.

Country	Improved variety	Area (%)
Kenya	Seredo	9.00
Kenya	KARI MTAMA-1	8.00
Kenya	IS21055	6.00
Kenya	KARI MTAMA-3	4.00
Kenya	KARI Mtama 2	4.00
Kenya	Serena	3.00
Kenya	IS8193	3.00
Kenya	GADAM	2.00
Sudan	Wad Ahmed	12.02
Sudan	Tabat	7.82
Sudan	Dabar	7.35
Sudan	Gadam Alhamam	4.45
Sudan	Arfaa gadamak 8	2.41
Sudan	PAN 606	2.05
Sudan	Yarwasha	1.75
Sudan	Arose el rimal	1.30
Sudan	Butana	0.66
Sudan	PAC 501	0.65
Sudan	Ingaz	0.06
Sudan	Hageen Dura 1	0.05
Tanzania	Macia(SDS 3220)	20.8
Tanzania	Tegemeo (ZK×17/B/1)	8.1
Tanzania	Wahi (P9406)	7.1
Tanzania	Hakika (P9405)	6.2

adaptation in ESA and is a prime example of a spill-over variety in the region.

Pendo was released in Tanzania in 1998. It is a short-duration, Spanish-type variety with negligible seed dormancy, has good market acceptance and is easy to hand shell. Pendo is a favourite choice in participatory varietal selection and has received support in seed production from the McKnight Foundation.

ICEAP 00040 is the improved pigeonpea cultivar that has been most widely disseminated in ESA (Table 7.24). It is a long-duration, *Fusarium*-wilt resistant variety with preferred market traits. Its long duration exploits the bimodal rainfall pattern prevailing in large parts of East Africa. The diffusion of ICEAP 00040 in northern and central Tanzania, Kenya and Malawi has resulted in increased grain yields and lowered production costs in comparison to local genotypes. Several medium-duration types, such as ICEAP 00557, are also gaining in popularity (Table 7.24).

According to sorghum experts in the Sudan, three varieties, Wad Ahmed, Tabat and Dabar, account for more than half of the area under improved varieties, which is equivalent to over 1.5 million hectares (Table 7.25). Because of sorghum's economic importance in the Sudan, Wad Ahmed is not only the most widely diffused improved variety in the country, but it is also the leading sorghum variety in sub-Saharan Africa. Wad Ahmed was selected from a NARS cross from NARS parents and was released in 1992. Tabat was selected from an ARC cross with ICRISAT parents and was released in 1996. Dabar was released in 1978 following selection from an introduced landrace. Wad Ahmed is the leading variety in both the mechanized rainfed and irrigated sectors. Tabat, like Wad Ahmed, is produced mainly in the mechanized rainfed and irrigated sectors. Dabar's cultivation is concentrated in the mechanized rainfed sector. None of these commercial varieties that are suitable for large-scale mechanization with an emphasis on grain production are popular in the smallholder traditional rainfed sector. However, with recently enhanced *Striga* resistance through marker-assisted selection, newly released, improved versions of Wad Ahmed and Tabat may become more relevant to small-scale production conditions than the original versions were in the past (ICRISAT, 2013).

Macia is a short-statured, early-maturing, white-grained sorghum line that was selected by breeders in the SADC/ICRISAT Sorghum and Millet Improvement Programme (SMIP) in 1984/85. In 1998, Tanzania became the fifth country in SADC to release this variety.

The 1998 Initiative did contain references to national adoption levels for five crop-by-country observations in ESA that overlap with the present set of observations. The estimates of adoption of improved varieties for sorghum were 20% of area for the Sudan and only 2% in Tanzania (Deb and Bantilan, 2003). Comparable estimates for groundnut in Malawi, Uganda and in Zambia were 10%, 10% and 5% (Bantilan *et al.*, 2003). The national adoption estimates in Table 7.22 are substantially higher than these 1998 estimates for all five crop-by-country observations suggesting considerable progress in the diffusion of these improved varieties since 1998.

Slow varietal turnover characterized by old improved varieties in farmers' fields was not a problem in groundnut and pigeonpea in ESA in 2010. High varietal age indicating slow turnover were most evident in sorghum in Kenya and Sudan where the weighted average age of improved varieties was about 19 years. However, recent releases in the 2000s and the early acceptance of those releases suggest that rising varietal age and slowing varietal turnover may not be that problematic in the future, especially in the Sudan.

Summary

Assessing the effectiveness of genetic improvement of the ICRISAT-mandate crops in West and Central Africa and in East and Southern Africa is the focus of this chapter. The assessment centres on three aspects of performance: (i) scientific capacity of NARS; (ii) varietal output in the form of releases; and (iii) adoption of improved varieties and hybrids.

Sorghum, pearl millet and groundnut are the major staples and cash crops for millions of farmers in the Semi-Arid Tropics of West and Central Africa. ICRISAT, the sorghum and millet (INTSORMIL) CRSP, the Peanut CRSP, the University of Georgia, CIRAD (former IRAT and IRHO) and NARS have invested in the development of more than 300 varieties of sorghum, pearl millet and groundnut in West Africa since the 1970s. Results showed that overall more varieties were released prior to the 1990s, less than one quarter of the releases was bred with IARCs' germplasm, and Mali stood out as a high performer in terms of releases after the 1990s. Research intensities (expressed in terms of FTE per million tonnes production) were low. Nigeria stood out as having the lowest apparent research intensity. About 30% of released varieties were bred from direct crosses or from populations; the remaining releases were selected from elite materials in adaptation trials or from the simple purification of local landraces.

The total number of FTE scientists working on sorghum, pearl millet and groundnut crop improvement programmes in five major producing countries was estimated at 60, which is substantially lower than the estimate in Chapter 6 for only one crop improvement programme in one country in the region: maize in Nigeria. In sorghum, pearl millet and groundnut, Mali had more scientists than the other country programmes. About half of the FTE scientists were breeders, and with the exception of IER in Mali, the programmes were not that diversified and supported by research in other relevant disciplines. Half of the scientists had PhDs, but more than two-thirds of the 60 FTE scientists are now more than 50 years old. Estimated research intensities in Nigeria of between 0.3 and 0.5 FTE scientists per million tonnes of production for sorghum, pearl millet and groundnut were among the lowest in the DIIVA Project.

Using a mixture of methods, from expert opinion at the national and regional level to interviews complemented by GIS tools at a higher spatial level of resolution to nationally representative surveys in Nigeria, adoption of improved varieties was estimated for each of the 15 crop-by-country observations. On the basis of a definition of improved varieties as being released in 1985 or later, adoption is estimated to range from 0% to 30% for pearl millet, 0% to 21% for sorghum and from 0% to 27% for groundnut. For groundnut, the dominant varieties in farmers' fields are 55-347 in Nigeria and Senegal and 47-10 in Mali. These varieties were bred during colonial rule and released in the 1950s and 1960s.

Not all is gloom and doom. SOSAT-C88 has emerged as an important cultivar with wide adaptation across several pearl-millet-producing countries in West Africa. Its area approached 1 million hectares in 2009. The high incidence of spill-over varieties in release and adoption from groundnut and pearl millet improvement is also impressive and speaks to the high potential for wide adaptability of these crops in the Sahelian, Sudanian and Guinean agroclimatic zones of West Africa.

In general, the findings support a positive correlation among inputs (FTE scientists), outputs (released varieties) and outcomes (adoption levels). Low numbers of FTE scientists are associated with few if any released varieties, which are associated with low levels of uptake. Relative to the size of its production, Mali has invested more in agricultural research, has released more varieties, and has had more favourable adoption outcomes in these three crops than the other four study countries in West Africa. A viable extension service, such as Nigeria's, and a well-targeted effort, such as ICRISAT's promotion of the diffusion of improved sorghum and pearl millet cultivars in northern Nigeria in the 1990s, can temporarily break these underlying associations, but such one-off initiatives have yet to lead to sustainable outcomes.

The ICRISAT-mandated crops are not nearly as economically important in East and Southern Africa as they are in West and Central Africa; however, groundnut, sorghum and millets still account for about 15% of arable land that is cultivated. Moreover, pigeonpea is an expanding export crop in East Africa.

The assessment in ESA drew on ten crop-by-country observations. They were a mixture of relatively small country programmes in groundnut, pigeonpea and sorghum with one very large country programme, sorghum in the Sudan. The results in ESA contrast sharply with those in West Africa. Releases have increased markedly across almost all programmes since the 2000s, continuing an upward trajectory that became apparent in the 1990s. ICRISAT has contributed to more than 70% of the total of 105 released varieties. With more sorghum-growing area but somewhat less production, sorghum in the Sudan is characterized by an estimated research intensity about 10 times greater than the figure for sorghum in Nigeria.

The adoption estimates for crops and countries in ESA were about twice as large as comparable figures for crops and countries in West Africa.

Similarities in results were not as numerous as contrasts between the two regions. One important similarity was the solid evidence for adaptability across countries in the analysis. In four of the five groundnut study countries, rosette-resistant ICGV-SM 90704 and drought-tolerant ICGV-83708, ranked first or second in the adoption of improved varieties.

Funding in the 1990s and early 2000s drove these regional contrasts. In particular, funding for crop improvement in groundnut, sorghum and pearl millet dried up at both the national and international levels during this period. The key issue has not changed in the recent past: ageing human resources and inability of research managers to replace scientists, as well as ageing research infrastructure and inadequate operational budgets for varietal development and testing. There is an urgent need for governments and donors to renew their investments in strengthening human capacity with adequate funding for operating budgets. Flexibility in varietal release procedures and more operational funding is also needed to stimulate variety release in several countries, such as Nigeria and Senegal, as many varieties are in the pre-release stage. The role of IARCs and CRSPs remains critical in an environment with decreasing human capital and funding, but the role of these international research organizations is largely contingent on the strength of partners. This calls for a collective effort (government and donors) to reinforce national research and extension services as a whole. Thinking outside the box may also be required to break some of the linkages among low input, low output and low adoption characteristics of most groundnut, pearl millet and sorghum improvement programmes in West Africa.

Notes

[1] This chapter is a revised and abridged version of Ndjeunga *et al.* (2012a) and Simtowe and Mausch (2012). It also draws heavily on Zereyesus and Dalton (2012).

[2] The research on and the impact of the hybrid Hageen Dura 1 won the World Food Prize for Gebisa Ejeta in 2009.

References

Bantilan, M.C.S. and Deb, U.K. (2003) Impacts of genetic improvement in pearl millet. In: Evenson, R.E. and Gollin, D. (eds) *Crop Variety Improvement and Its Effect on Productivity: The Impact of International Agricultural Research*. CAB International, Wallingford, UK, pp. 216–240.

Bantilan, M.C.S., Deb, U.K. and Nigam, S.N. (2003) Impacts of genetic improvement in groundnut. In: Evenson, R.E. and Gollin, D. (eds) *Crop Variety Improvement and Its Effect on Productivity: The Impact of International Agricultural Research.* CAB International, Wallingford, UK, pp. 293–314.

Chiyembekeza, A.J., Subrahmanyam, P., van der Merwe, P.J.A. and Kapewa, T. (2000) *A Proposal to Release ICGV – S M 90704 Rosette Resistant Groundnut Variety for Production in Malawi*. Department of Agricultural Research and Technical Services, Ministry of Agriculture and Irrigation, Malawi.

Deb, U.K. and Bantilan, M.C.S. (2003) Impacts of genetic improvement in sorghum. In: Evenson, R.E. and Gollin, D. (eds) *Crop Variety Improvement and Its Effect on Productivity: The Impact of International Agricultural Research*. CAB International, Wallingford, UK, pp. 183–214.

FAO (2011) FAOSTAT. FAO, Rome.

ICRISAT (2013) *First Marker-Assisted Bred Sorghum Varieties Released for Cultivation by Farmers in Sub-Saharan Africa*. ICRISAT, Patancheru, India.

Mausch, K. and Simtowe, F. (2012) *Adoption Rates and their Reliability Evidence Based on Expert Panel Discussions, Community Focus Groups and Household Survey in Tanzania – Groundnuts, Pigeonpea and Sorghum.* Diffusion and Impact of Improved Varieties in Africa (DIIVA) Project – Objective 2 Report. ICRISAT, Nairobi.

Ndjeunga, J. Umar, J., Ahmed, B., Aba Aba, Ibro, A., Abdoulaye, A. and Gwadi, K. (2011) *Adoption and Impacts of Modern Sorghum and Pearl Millet Varieties in Northern Nigeria*. ICRISAT, Niamey, Niger.

Ndjeunga, J., Hash, C.T., Faye, I., Sanogo, M., Echekwu, C.A. *et al.* (2012a) *Assessing the Effectiveness of Agricultural R&D in West Africa: Cases of Pearl Millet, Sorghum and Groundnut Crop Improvement Programs.* Diffusion and Impact of Improved Varieties in Africa (DIIVA) Project, Objective 1 Report. ICRISAT, Niamey, Niger.

Ndjeunga, J., Ntare, B.R., Ajeigbe, H., Echekwu, C.A., Ibro, A. and Amadou, A. (2012b) *Adoption and Impacts of Modern Groundnut Varieties in Nigeria.* Diffusion and Impact of Improved Varieties in Africa (DIIVA) Project – Objective 2 Report. ICRISAT, Niamey, Niger.

Simtowe, F. and Mausch, K. (2012) *Assessing the Effectiveness of Agricultural R&D in Eastern and Southern Africa: Cases of Groundnuts, Pigeonpea and Sorghum Crop Improvement Programs.* Diffusion and Impact of Improved Varieties in Africa (DIIVA) Project – Objective 1 Report. ICRISAT, Nairobi.

Subrahmanyam, P., Merwe, P.J.A. van der, Chiyembekeza, A.J., Ngulube, S. and Freeman, H.A. (2000) Groundnut variety CG 7: a boost to Malawian agriculture. *International Arachis Newsletter* 20, 33–35.

Walker, T.S. (2009) *Updating and Reviewing Future Challenges and Opportunities for Agricultural R&D in the Semi-Arid Tropics for ICRISAT Strategic Planning to 2020*. ICRISAT, Patancheru, India.

Zereyesus, Y.A. and Dalton, T.J. (2012) *The Impact of Sorghum Varietal Improvement Research in Sudan.* Diffusion and Impact of Improved Varieties in Africa (DIIVA) Project – Objective 3 Report. Draft. Kansas State University, Manhattan, Kansas.

8 The Performance of Bean Improvement Programmes in Sub-Saharan Africa from the Perspectives of Varietal Output and Adoption

R.A. Muthoni[1]* and R. Andrade[2]
[1]*CIAT, Uganda;* [2]*Department of Applied Economics (formerly of CIAT), University of Minnesota, USA*

Introduction[1]

The International Center for Tropical Agriculture (CIAT) was established in 1968 outside Cali, Colombia. The Center was entrusted with a global mandate for the genetic improvement of the common bean (*Phaseolus vulgaris L.*) that was domesticated in MesoAmerica and the Andes of Latin America. *Phaseolus* was viewed as the archetypal small-farm household crop in Latin America where it was often intercropped with maize with the intensive use of family labour and light-to-minimal use of purchased inputs such as inorganic fertilizer, insecticide and fungicide.

Beans are also very important in concentrated regions of sub-Saharan Africa (SSA) where 14 bean-growing agroecologies have been described (Wortmann *et al.*, 1998). Small-scale farmers in SSA, in cropping systems similar to those in Latin America, produce about 2.5 million metric tonnes, with Uganda, Kenya, Rwanda, Burundi, Tanzania and the DR Congo playing major roles. The common bean is a versatile, short-duration, self-pollinated crop that provides both food and cash to small-farm households in central, eastern and southern Africa (Katungi *et al.*, 2009).

CIAT and other institutions, notably the US-led Collaborative Research Support Program (CRSP), both invested in genetic improvement and sought to build research programmes in Africa. In 1984, a regional breeding programme was established in the Great Lakes Region of SSA. It focused on breeding for resistance to bean pests and diseases in conditions of low and declining soil fertility typical of small rural household production. To meet this challenge, the Pan-African Bean Research Alliance (PABRA) was launched as a CIAT project in 1996. It now consists of three regional networks, ECABREN, SABRN and WECABREN, and encompasses 29 countries in SSA. PABRA has a record of sustainability and growth that is unmatched by few other regional IARC-related crop improvement networks.

Beans are known not only for their location specificity in production constraints but also for their heterogeneity in market preferences. Preferences for traits in bean varieties are extremely diverse among farmers, traders, processors and consumers. Traits range from colour (red, white, black, red-mottled, cream, cream-mottled, yellow and others) through grain size (small- or large-seeded), growth habit (bush or climbing bean)

* E-mail: r.muthoni@cgiar.org

and use (dry bean, canning bean or green/snap bean pods). Very often, each country has germplasm preferences for more than one of these market types. Moreover, beans are usually grown in diverse environments that expose the crop to different stresses including drought, low soil fertility, pests and diseases. Noting the complexity of developing bean germplasm for different market classes with multiple attributes to overcome varying production constraints, PABRA scientists in 2000 developed a new bean breeding strategy that links the breeding programmes at CIAT headquarters in Colombia, the subregional breeding programmes in Africa and the NARS breeding programmes at the country level.[2] This strategy is presented graphically in Appendix Fig. 8.A1.

The rationale behind the strategy is that, in the past, bean breeding in Africa followed a monolithic approach where varieties were identified for their superiority in yield or resistance to a single biotic or abiotic stress with little or no consideration of the grain size, colour and other important socioeconomic factors. The current breeding strategy is grain-type led and is built on the premise that farmers produce beans for food and sale in their localities (neighbours, retail traders, schools and other institutions) and for larger domestic markets in urban centres and in regional and international markets.

To develop the breeding strategy, bean breeders from different PABRA countries as well as other breeding-related experts (pathologists, entomologists, market and agro-enterprise specialists, policy analysts) and stakeholders (e.g. representatives of seed companies, farmer associations, etc.) reviewed the major thrusts or objectives to be addressed in the breeding strategy. For each market class, the breeding objectives, methodology and germplasm requirements to meet breeding goals were clearly defined. Since no one bean variety or market class can meet the diversity of market needs in a country or region and because most national bean breeding programmes do not have the capacity to address all researchable issues and provide all the various bean product types the market demands, multiple breeding programmes that respond to food and market needs across various countries under PABRA were developed (Appendix Fig. 8.A1).

When CIAT scientists started their research on bean improvement, the cupboard was essentially bare of research findings and genetic materials that were conducted or generated in SSA from the colonial period. Fortunately, they could draw on research conducted by several national and regional institutes in Latin America, especially the larger NARS of Brazil and Mexico because Latin America is the center of genetic diversity for beans (S. Beebe, CIAT, Colombia, 2010, personal communication). Parallel to the growth and maturation of the bean improvement programme at CIAT, several smaller, largely university initiatives in developed countries played an important but secondary role of an alternative supplier of genetic materials. The dearth of any adapted research findings and materials to SSA and the existence of small alternative research options in the public sector distinguish beans from most of the other crops in this volume.

Country Coverage, Methods and Data Collection

Initially, ten countries were selected for study: Burundi, Democratic Republic of Congo (DR Congo), Ethiopia, Kenya, Malawi, Mozambique, Rwanda, Tanzania, Uganda and Zambia. Later, activities in Kenya were discontinued because of difficulties observed in the initial stages of project execution. Kenya was also characterized by incomplete data in the earlier initiative in 1998 described in Chapter 4 (Johnson *et al.*, 2003). Zimbabwe was added to the study at a later stage. Data from that country appear only in the next section on scientific capacity.

The importance of beans in area and production was the foremost criterion used in country selection. Data on area under beans was obtained from triangulation of data obtained from recent adoption studies, from local experts, from the central and national statistics bureaus, and from FAOSTAT. Data from recent adoption studies in east Kivu and South Congo were useful in DR Congo where this data was identified as the most reliable information available. Data for Zambia, Ethiopia, Uganda and Mozambique were obtained from their respective national statistics agencies. Data for Tanzania, Malawi and Rwanda were sourced from FAOSTAT. Data from the Zambian Central Statistics Office were informative about bean area in Zambia. In Ethiopia, data were obtained from the Ethiopian Central Statistics Authority (CSA). In Uganda, the data were procured from the Uganda Bureau of Statistics (UBOS). Similar data from the Trabalho de Inquerito (TIA) were useful in Mozambique.

The greatest variations in the data sources were observed in Burundi and Uganda. In Burundi, bean area was estimated at 405,715 hectares from expert sources, whereas the FAOSTAT estimate was 203,367 hectares. The results from the expert panel translate to 19% of the total agricultural land in Burundi under beans, whereas FAO results imply that only 9% of area is cultivated in beans. A large discrepancy was also observed in the data from Uganda: UBOS identified 532,883 hectares under beans, whereas FAOSTAT reported 917,000 hectares.

The nine countries can be divided into three groups: (i) larger producers represented by Burundi, Uganda and Tanzania with areas of more than 400,000 hectares; (ii) medium-sized producers, such as Rwanda, Malawi, the DR Congo and Ethiopia, with cultivated area of 200,000–300,000 hectares; and (iii) smaller producers in Southern Africa with national area approaching or slightly exceeding 100,000 hectares. With a total of about 3.3 million hectares of cultivated area, the nine study countries accounted for 61% of the total bean production and area in sub-Saharan Africa in 2010.

Consultative processes with experts in CIAT and national programme researchers also confirmed these countries as locations with the most representative information on adoption of improved varieties. Of the total varieties released in Africa, 80% were released in the nine study countries.

Having decided on the countries, the process of data collection featured a survey, an expert panel workshop and a review of locally available secondary data. The survey enabled collection of information on the list of improved varieties, their main characteristics (genetic background, source of the variety and germplasm, and role of CIAT), national research programme strength (scientific composition, degrees and disciplines) and the cost of specific lines of research where such information was available. A total of 13 expert panel meetings were held. One meeting was convened in all countries except in the DR Congo and Tanzania where two regional meetings were held. A CIAT staff person travelled to each of the countries and worked with the national programme to carry out the expert panels.

Each expert panel meeting had three sessions. National researchers responsible for providing their full-time equivalent (FTE) scientist estimates and the baseline on released varieties were the audience for the first session. Specialists active in the extension and delivery of bean varieties attended the second session. They provided information on adoption for popular varieties by ranking them. The third session pooled participants of both groups, who provided estimated area under beans in hectares and supportive secondary information for review and triangulation. Print copies of spatial maps of each country were made available to participants, and a session was facilitated to elicit bean area estimates in hectares. A generic protocol provided by the Diffusion and Impact of Improved Varieties in Africa (DIIVA) Project guided expert panel team exercises (Walker, 2010).

A total of 210 persons representing more than 150 different institutions were involved. Around 15% of the participants were women. Almost half of the participants were technicians and extensionists from non-governmental organizations (NGOs) and NARS who are dedicated to outreach and dissemination of beans. About one-quarter were researchers from NARS from different disciplines, mostly breeding and social science. The remaining one-quarter were from the private sector in the form of Community Based Organizations (CBOs), mainly seed producers and farmer associations. The participation of the CBOs was essential to avoid bias in the estimation of adoption. A social scientist and a spatial analyst from PABRA and the CIAT African bean programme supervised the data-collection process.

Scientific Capacity

NARS bean researchers as used here refers to scientists working in government programmes. In some instances, these staff are stationed at universities. Because of their rarity, the private sector was not included in assessing scientific capacity, even though there is an emerging group of private-sector bean researchers in Ethiopia and Zimbabwe.

Bean capacity in CIAT

CIAT's investment in bean improvement grew from about US$350,000 with two principal scientists in 1970 to about US$8.0 million in 1985 with 20 principal scientists. Expenditures peaked around US$13.8 million in 1990 with 26 principal scientists and 7 internationally recruited breeders. By 1997/98, investment in bean improvement by CIAT had declined to about US$7.7 million with 18.5 internationally recruited scientists (Johnson *et al.*, 2003).

Prorating budget expenditures by internationally recruited scientists over time shows that CIAT's resources allocated to bean improvement further declined from about US$7.0 million in 2002 to just under US$3 million in 2007. A substantial decrease in core funding and major restructuring processes in the CGIAR and in CIAT led to a reduction of major operations in the bean improvement programme. The most affected bean research operations were in CIAT headquarters in Colombia.

Recently, funding has increased and stabilized at about US$5.5 million. The entry of new donors who are interested in targeting genetic improvement as a gateway to consolidating food security gains reversed the downward trend and sparked a recovery in funding. Growth in the recent past is targeted at small-scale bean production in Africa and is mediated through the PABRA network.

The above estimates are given in nominal terms. In real prices, factoring in inflation, the decline in funding was steeper. The height of the programme's activities in the mid-1980s to the early 1990s was synonymous with breeding activities such as crossing and population development that subsequently led to many of the released varieties in the 1990s and early 2000s.

Bean improvement capacity in national programmes

Across the 10 countries, a total of 190 research staff work on bean improvement. They are equivalent to 120 FTE staff and 86 FTE scientists. Thirty-four FTE technicians were employed by bean improvement programmes in 2010.

The average-sized programme in Table 8.1 has 8–9 FTE scientists and ranges from 2–3 staff in Burundi to 21–22 scientists in Ethiopia. Relative to the economic importance of the crop and the size of the country, Burundi with no bean breeders appears to be woefully deficient in scientific capacity. The DR Congo also seems to be understaffed.

The majority of scientists are breeders and agronomists. Except for Burundi, all programmes have at least one FTE plant breeder. The programmes have also strived for a diversified allocation to disciplines that support breeding per se. Rwanda is the exception with a high concentration in breeding and agronomy (Table 8.1). Most programmes have also allocated resources to pathology and entomology, reflecting the importance of biotic constraints in bean production.

About 15% of the 86 FTE scientists have completed PhDs. The remainder is equally split

Table 8.1. Full-time equivalent staff by major specialization working on bean improvement in sub-Saharan Africa in 2009.

Country	Breeding	Pathology	Entomology	Agronomy	Seed production	Social science	Others	Total
Burundi	0	0.1	0.3	1.5	0	0.3	0.4	2.6
DR Congo	2.8	0	0.3	0.6	0	0.5	0.9	5.1
Ethiopia	4.8	1.9	0.8	9.7	1.7	1.7	0.8	21.4
Malawi	4	0.6	0.5	1.3	0	0	0	6.4
Mozambique	2.1	0	0.3	1.9	0	0.8	0	5.1
Rwanda	5.1	1	0	8.4	0	0	0	14.5
Tanzania	2.5	1.7	2.4	4.5	0	1.1	0	12.2
Uganda	3.3	2.3	0.5	1.7	0	0.9	0	8.7
Zambia	1	0.5	1	1	0	1	0.7	5.2
Zimbabwe	1.5	0	2	0.8	0.5	0.5	0	5.3
Total	27.1	8.1	8.1	31.4	2.2	6.8	2.8	86.5

between MScs and BScs. The bean programmes in Uganda and Rwanda are well endowed with PhD scientists; the programmes in Burundi, DR Congo, Zambia and Zimbabwe are staffed only with MSc and BSc scientists. These observations are best understood against the framework of the PABRA bean breeding strategy (see Appendix Fig. 8.A1). Countries with responsibility for more breeding programmes have been recipients of more training, mentoring and capacity-building inputs than those with fewer breeding activities. From this process, stronger NARS have emerged in Rwanda, Ethiopia and Uganda that have the ability to attract and sustain large projects and financial support from their governments and from major donors such as Alliance for a Green Revolution in Africa (AGRA), HarvestPlus, the Bill & Melinda Gates Foundation (BMGF), the McKnight Foundation and the Department for International Development, UK (DFID). These programmes have multiple or sole responsibilities, such as Rwanda for climbing beans, in the PABRA network (Appendix Fig. 8.A1).

For bean improvement, additional data were collected on age, length of service and gender of scientists. In general, bean improvement scientists in East and Southern Africa are considerably younger than scientists in other commodity improvement programmes that are primarily located in West Africa and that are described in Chapters 7 and 10. The average age of all bean researchers was 41 years with a modal age cohort of 31 to 50 years. Zimbabwe and Ethiopia had a large number of younger researchers, whereas DR Congo and Tanzania had scientists in the older category. Scientists in DR Congo and Tanzania were being retained after their retirement age. In the case of Tanzania, there were incidences of retired scientists being recalled to fill a technical gap (M. Mukuchu, Tanzania, 2012, personal communication). The average number of years at work were 17 in DR Congo and 16 in Tanzania.

In contrast, young people (30 or less) account for half of the work force among bean researchers in Zimbabwe, averaging 8 years on the job. A similar tendency was observed in Ethiopia, where about two-thirds of scientists fall in the 31–50 years age bracket. On average, scientists in Ethiopia have worked only 4 years in bean improvement. Stability is one of the hallmarks of successful agricultural research, especially in genetic improvement. An average of only 4 years would seem to jeopardize the conduct of even routine tasks.

The importance of beans in Ethiopia is a recent development, being an emerging crop in the 1990s. In Ethiopia, beans are now a popular export crop traded in the Ethiopian Commodity Exchange. Career choices in agribusiness and related research areas have been facilitated by political will, private–public partnerships and external donor investment. Young people with interests in agricultural-related careers perceive their career prospects as good.

About 30% of scientists in 2010 were women. Ethiopia was an outlier with only 1–2 FTE woman scientists from a total of 21–22 FTE scientists. Burundi, Rwanda and Uganda reported the highest incidence of women researchers in bean crop improvement. Most of these are in the 31–50 years age bracket. The current bean programme coordinator in Burundi is a woman and, until recently, the same positions in the bean programmes of South Tanzania and Uganda were also held by women. The impact of mentorship and role modelling in enhancing female representation in organizations is well documented (The Guardian, 2010).

CIAT has collected historical data on research resource allocation in bean crop improvement in SSA since 1980. Of the ten study countries in this section, only Malawi had a full-time bean breeder in 1980 (Johnson *et al.*, 2003). By 1990, 21 FTE bean breeders were employed in bean crop improvement (Table 8.2). All ten countries had at least one FTE breeder in their programme in 1990. Therefore, the 1980s was the decade of growth in bean genetic improvement. Twenty years later, in 2010, the number of bean breeders at 24 FTE scientists is about the same as it was in 1990. Maintenance of the same aggregate scientific capacity in bean breeding is actually quite an achievement in the face of declining funding in the late 1990s and early 2000s.

Although on aggregate the number of breeders has remained roughly the same, most countries experienced substantial changes in breeding capacity during the past 20 years. Burundi, DR Congo and Tanzania each lost two or more full-time bean breeders. In contrast, Malawi, Uganda and Rwanda have multiplied their breeding capacity by several-fold between 1990 and 2010 (Table 8.2). Overall, estimated breeding

Table 8.2. Number of bean breeders and estimated research intensities between 1990 and 2010 by country.

Year	Burundi	DR Congo	Ethiopia	Malawi	Tanzania	Rwanda	Uganda	Zimbabwe	Total/ weighted average
				Number of plant breeders					
1990	2	5	4	1	5	1	1	2	21
2010	0	2.8	4.8	4	2.5	5.1	3.3	1.5	24
	Estimated research intensities (number of breeders per million tonnes of production)								
1990	6.1	37.0	48.2	11.8	20.0	5.1	2.5	42.6	13.8
2010	0	24.3	13.2	26.0	2.9	15.6	7.3	69.2	9.6

intensity fell from 13.8 to 9.6 FTE breeders per million tonnes of production because the allocation of research resources did not keep pace with expanding production.

In several of the countries, higher breeding intensity corresponds to a specific effort to promote bean production as a new crop to exporters, and to local and regional traders, processors and seed companies. In Ethiopia, in particular, the emergence of white pea beans for export markets has led to increased supplies (Ferris and Kaganzi, 2008). In Malawi, the inclusion of the bean crop in the farm-input subsidy programme accounts for an increase in bean production levels prior to 2010. In Rwanda, climbing beans gained importance in the 1990s, offering significantly higher yields than bush beans (in part by escaping some of the moisture problems that cause disease) and providing a way of taking advantage of vertical space on very small farms (Sperling and Muyaneza, 1995; Muthoni *et al.*, 2007; Rubyogo *et al.*, 2010).

Varietal Output

In the 1998 Initiative, information was only collected on CIAT-related releases (Johnson *et al.*, 2003). Hence, the varietal release database had to be recollected for all varieties since CIAT's establishment. The earlier work did note that CIAT-related releases were increasing over time in Africa and, in comparing the 1990s to the 1980s, CIAT crosses were becoming more popular as a source of released varieties compared to direct releases from CIAT accessions from their bean genetic resources unit. Materials from germplasm exchange networks, such as the Eastern and Central Africa Bean Research Network (ECABREN) and the Southern Africa Bean Research Network (SABRN), were also used for the first time as a source of varietal output in the 1990s. Apart from one release in South Africa, however, no country in SSA had released a variety from a NARS cross with at least one CIAT parent.

In general, information on varietal release is reliable and informative of varietal production but it is not well defined and rigorous in some countries, such as DR Congo, where functioning varietal release boards do not exist. In contrast, Ethiopia, Rwanda, Uganda and Zambia have lengthy and bureaucratic release procedures and as a consequence also have more complete information on the identity of released cultivars.

The total number of released bean varieties across the nine study countries is impressive. Beginning in 1968 and ending in 2010, the nine national governments have released 250 cultivars. Varietal output for these countries in bean improvement is about 25% higher than varietal output for 17 countries in cowpea improvement as described in Chapter 6.

Parity is evident in Table 8.3 because all countries have released at least ten varieties. Few if any varieties were released before 1980, thereby bearing witness to the lack of investment in bean research during the colonial period.

Releases are trending strongly upward over time. About seven in eight releases occurred after 1990. The pattern of increasing release over time applies to almost all countries in Table 8.3. With access to the ECABREN network, even Burundi with its minimal breeding resources has been able to release varieties throughout the 2000s. Among

Table 8.3. Number of bean varieties released by decade and country.

Country	1970s	1980s	1990s	2000s	Undated	Total
Burundi	1	4	7	18	5	35
DR Congo	0	2	4	22	7	35
Ethiopia	2	1	7	22	0	32
Malawi	0	0	6	9	0	15
Mozambique	0	2	4	8	1	15
Rwanda	3	9	20	25	4	61
Tanzania	0	3	11	14	1	29
Uganda	1	0	11	6	0	18
Zambia	0	1	3	6	0	10
Total	7	22	73	130	18	250

producing countries with more than 200,000 hectares of growing area, Malawi and Uganda were the only ones that did not release more than 10 improved varieties in the first decade of the 2000s. In this same period, the programmes of these countries sent their four breeders for postgraduate breeding studies. This created a gap in the technical support available in-house to advance materials through the breeding process and to final release stages (C. Mukankusi, NARO Namulonge, 2012, personal communication). Several releases in both countries were expected in 2011.

Rwanda is the leading country in releases and has also released more varieties than any other country in each of the four decades reported in Table 8.3. Prior to the Genocide in 1994, Rwanda released 15 varieties in the early 1990s. After this tragedy, bean improvement in Rwanda was able to recover as indicated by the release of four varieties in 1997. The 'speedy' recovery in bean varietal output is a sharp contrast to stagnation in potato improvement (see Chapter 9, this volume).

All nine countries display variable release patterns with bursts of activity in a few years separated by long dry spells of consecutive years with no releases. Mozambique's release behaviour is typical of this erratic performance in varietal output. Only two varieties were released in the 1980s, four in the 1990s and eight in 2010.

Direct releases from germplasm accessions seemingly play a more important role in beans than in many other food crops. Over 40% of the releases for which information is available can trace their origins either to the release of local landraces or to internationally distributed germplasm accessions without further improvement in terms of breeding effort other than in-country selection (Table 8.4). Releases from local land races are more common than releases from internationally distributed germplasm, although substantial variation exists among the nine countries. Local landraces loom large in the releases from Burundi, Rwanda and Tanzania. The DR Congo has depended heavily on imported germplasm accessions as the finished material for varietal release.

The 31 imported germplasm accessions in Table 8.4 come from a handful of countries. Most numerous are accessions from the Beltsville Agricultural Research Center in Maryland, USA. Accessions from the Mexican national programme also figure prominently in internationally distributed germplasm that has eventually been released in SSA (S. Beebe, CIAT, Colombia, 2010, personal communication). The Colombian national programme has also contributed to varietal output in several countries. Within the region, Rwanda has been the most popular donor of landrace materials to recipient countries, such as Burundi. CIAT has brokered the distribution of many of these varieties via its germplasm network.

Purified landrace materials that are preferred and demanded regionally need to be officially released in order for them to be legally traded regionally in neighbouring countries. Seed of a commodity becomes a tradable commodity only if it has been officially registered and released (CIMMYT, 2008). Cross-border seed trade regulations therefore drive the purification and official release of local landraces in countries such as Uganda, Zambia and Rwanda.

Bred varieties contribute a 58% share of varietal releases (Table 8.4). CIAT is the main source of finished bred varieties that are

Table 8.4. Provenance of released varieties by country and genetic and improvement category.

	Pureline selections from germplasm accessions		Bred varieties				
Country	Local landraces	Imported	CIAT bred, NARS selected	Bred by other institutes, NARS selected	NARS cross, NARS selected	Missing information	Total
Burundi	8	7	6	2	0	12	35
DR Congo	1	10	8	2	0	14	35
Ethiopia	2	2	16	5	5	2	32
Malawi	0	0	12	0	0	3	15
Mozambique	3	0	10	1	0	1	15
Rwanda	20	9	16	3	10	3	61
Tanzania	7	2	8	8	0	4	29
Uganda	4	1	9	0	0	4	18
Zambia	2	0	4	4	0	0	10
Total	47	31	89	25	15	43	250

subsequently selected by bean improvement programmes in SSA. Some countries such as Malawi rely almost exclusively on CIAT-bred varieties as a source of material for varietal release.

Other institutional suppliers, mainly in developing countries, are a secondary source of materials for release and have been very important to countries like Tanzania. As discussed in the introduction, the participation of multiple alternative suppliers is one of the distinguishing features of bean crop improvement in SSA. Multiple smaller institutional providers have added internationality to CIAT's primary role as a source of genetic materials for the generation of bean varietal output in ESA. These include the Bean and Cowpea CRSP in the USA, the Institute of Horticultural Plant Breeding (IVT) in the Netherlands, the Escuela Agricola Panamerica (EAP) in Honduras, the Centro Agronomico Tropical de Investigacion y Ensenanza (CATIE) in Costa Rica, the NVRS/Wellsbourne Project in the UK and the Tokachi Agricultural Experimental Station in Japan.

The incidence of NARS bred and selected materials is restricted to Ethiopia and Rwanda (Table 8.4). Only 15 of 250 released varieties have their origins in a NARS cross. Although small in number, NARS bred materials are also increasing over time.

Figure 8.1 shows bean varietal releases across decades. Varieties that are bred with parents from the CIAT genebank or those that are a product of crossing materials with local lines are all identified as crosses. The trends for releasing CIAT-related crosses are compared against the performance of varieties that are a product of selection and purification of local landraces. The findings show a growth in both trends with varieties resulting from CIAT-related varieties increasing steadily from the 1980s. Landrace purification has also been gradually growing and seems to have stabilized from the 1990s onwards. The growing importance of bred varieties relative to the release of purified landraces and germplasm accessions is consistent with the increasing maturity and development of the breeding programmes in the region.

Varietal Adoption

As described earlier, data on area under beans in the nine study areas was obtained from expert panels and from triangulation of data obtained from recent adoption studies, from the central and national statistics bureaus, and from FAOSTAT. In Rwanda and Uganda, nationally representative surveys were carried out on the adoption and impact of improved bean varieties. Those results are reported by Larochelle *et al.* (Chapter 16, this volume). The adoption estimates from the expert panels and the surveys are compared by Walker (Chapter 20, this volume). Therefore, the national-level adoption estimates in Table 8.5 are taken from two sources: the national adoption and impact surveys for Rwanda

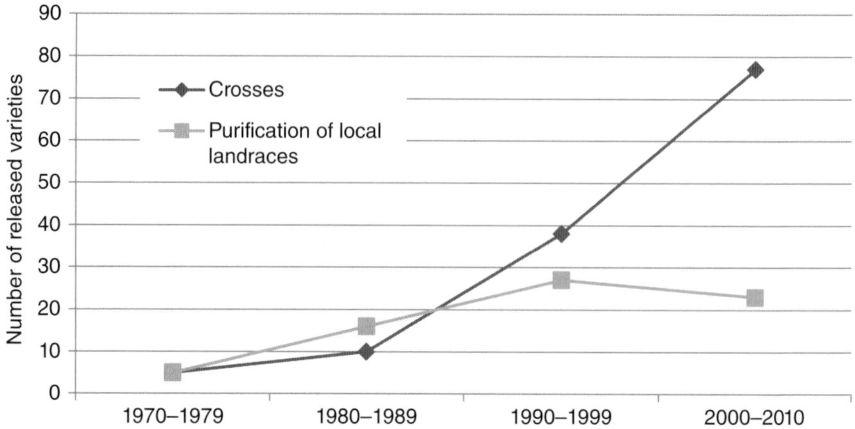

Fig. 8.1. Improved bean variety released in sub-Saharan Africa by type of improvement.

Table 8.5. Adoption of modern varieties of bean in sub-Saharan Africa, 2009.

Country	Area under bean cultivation (ha)	Area MVs (%)
Burundi	410,100[a]	8.1
DR Congo	224,584[a]	16.1
Ethiopia	207,494[b]	43.7
Malawi	260,287[c]	54.6
Mozambique	108,000[b]	13.5
Rwanda	285,000[c]	19.0
Tanzania	1,207,950[c]	45.8
Uganda	532,883[b]	31.0
Zambia	75,145[b]	9.5

[a]Recently completed nationwide adoption studies, specifically the CIALCA (Consortium for Improving Agriculture-based Livelihoods in Central Africa) Project impact assessment study for Burunda and DR Congo.
[b]National central bureaus of statistics such as the Ethiopia Central Statistics (ECSA), Trabalho de Inquito (TIA), Uganda Bureau of Statistics (UBOS) and Zambia Bureau of Statistics.
[c]FAOSTAT 2009/2010.

and Uganda and expert opinion for the other seven countries.

Estimates of adoption of improved varieties in Table 8.5 refer to the 129 bred varieties that originated from crosses described in Table 8.4. Adding in pureline selections, especially from local landraces, would lead to substantially higher adoption estimates in all countries except Malawi.

The area-weighted estimate of adoption of bred varieties and lines is 33% for the approximately 3.3 million hectares of growing area across the nine countries in Table 8.5. Based on their national estimates, the countries fall into three groups in Table 8.5. Burundi, DR Congo, Mozambique and Zambia lag behind in the uptake of improved varieties. Explanations for lagging adoption are most transparent in Mozambique where the majority of bred varieties were only released in 2010. If varietal output had been higher earlier, diffusion of modern varieties could have been greater.

The DR Congo epitomizes institutional uncertainty especially in Kivu, the prime bean-growing area. A state of perpetual civil disturbance is not conducive to the adoption of modern varieties. Somewhat paradoxically, the estimate for adoption of CIAT-related materials was the highest at 48% for the DR Congo in the 1998 Initiative (Johnson *et al.*, 2003). Either considerable disadoption must have taken place or the definition of CIAT-related materials must have changed between the two time periods.

Zambia to some extent may share the timing problem of Mozambique in that most of their releases have been recent. There might not be sufficient seed in the system for diffusion to reach a take-off point for the majority of cultivars released in the 2000s.

Given the importance of the crop and an adequate record on varietal output, it is hard to understand why adoption lags so far behind in Burundi, which suffers from several of the chronic problems of small NARS in SSA. The country has, however, done a good job of borrowing varietal technologies in other food crops, such as potatoes, where diffusion of improved varieties is substantially higher than in beans.

National adoption estimates from Uganda and Rwanda are quite close to the regional average. Uganda is slightly above the average; Rwanda's performance is the modal estimate with four countries above and four countries below its adoption level. Because of the stability of its bean-breeding programme over time and the productivity in releasing varieties for its own small-scale farmers and for those of the Great Lakes Region, an adoption estimate of 19% seems unduly low for Rwanda. The survey estimate of all released varieties and those that were considered improved was considerably higher at 47%. Much of this difference is attributed to the inclusion of pureline selections of local landraces, but some of it also stems from selected germplasm accessions that were distributed internationally. For example, Decelaya and Flor de Mayo from the Escuela Nacional de Agricultura (ENA) are released germplasm accessions of enduring popularity in the Rwandan highlands.

The 19% survey estimate in Rwanda does not convey the extent of progress in the diffusion of improved varieties. Strong linkages exist between the national programme, policy makers, civil society and CIAT. Confronted with decreasing areas under the crop and increasing demand on agricultural land, the government's priority has been to enhance bean productivity through the use of more productive bean technologies per unit of land. Climbing bean ideally satisfies this need for intensified production. The leading variety, RWV 2070, was recently released in 2007; it already approaches 5% of bean-growing area, which is an excellent diffusion performance for a new climbing cultivar (Table 8.6).

The survey findings in Rwanda suggest a spatially diversified portfolio of varieties in Table 8.6. Indeed, over 30 named varieties were reliably identified as improved, but the vast majority of these are sown on small areas amounting to less than 10,000 hectares. Rwandan farmers have been known to plant a range of bean types partly as a risk aversion strategy in the face of a range of biotic and abiotic stresses and partly for different consumption goals (Sperling and Loevinsohn, 1993).

In the late 1990s, CIAT-related improved varieties were reported on 15% of area in Uganda (Johnson *et al.*, 2003). In the period 2003 to 2005, impact studies in Uganda noted 31% of the total bean area was under improved varieties (Kalyebara *et al.*, 2007). In the same period, K132 (released in 1994) was the most widely grown improved variety due to its high yield potential and market demand, despite its susceptibility to bean root rot disease. The variety has maintained this position over time. Nabe 4 also figures prominently as an important variety in Table 8.6.

Malawi, Ethiopia and Tanzania are characterized by above average adoption in Table 8.5. From the perspective of the 1998 Initiative, Malawi has made the most progress in going from negligible adoption of improved varieties to a position in which they are sown on one in every two hectares (Johnson *et al.*, 2003). In Malawi, 10 of the 15 released varieties were perceived as being sown on 2% or more of the national area planted to beans. All of these successful varieties, according to expert opinion, were CIAT lines that were released in the 1990s and 2000s. Three varieties, Kalima, Malua and Napilira, released in the early 1990s accounted for most of the total volume of beans traded in the domestic markets in 2007 (Muthoni *et al.*, 2007).

Another very specific case of significant impact is Ethiopia, where the percentage of area under CIAT-related improved varieties rose from 8% in the late 1990s to 44% in 2010. In Ethiopia, the interaction between niche market varieties and market growth observed in the uptake of white pea beans for export markets is a significant and interesting event. Bean productivity increased from less than 500 kg per hectare in 2002 to about 1000 kg per hectare in 2009 (ESE, 2010). At the national level, value of bean exports rose from less than US$23 million in 2002 to more than US$60 million in 2009 (FAO, 2011). Improved cultivars Nasir and Awash-1 account for 12% and 10% of the total area under beans. Mexican 142, released in 1972 and once referred to as the most widely adopted variety in Ethiopia, is now in third place (Teshale *et al.*, 2006). Nasir and Awash 1 are market-class bean varieties, having the preferred colour, shape and size that dominate bean export and domestic markets.

CIAT related varieties were only estimated to cover 4% of Tanzania's bean-growing area in 1998 (Johnson *et al.*, 2003). A significantly higher estimate of 46% in 2009 is perhaps not that surprising because Tanzania has a history of using many institutional suppliers for their breeding materials (Table 8.4). Several non-CIAT

Table 8.6. Economically important improved bean varieties in SSA, 2009–2010, by national area.

Country	Improved variety	Area (%)	Country	Improved variety	Area (%)
Burundi	AND 10	4	Rwanda	Kitabi	0.09
Burundi	A410	2	Rwanda	Mvuyekumurenge	0.09
Burundi	VCB 81030	2	Rwanda	Karitasi	0.07
Burundi	Maharagisoja, VCB 81012, A321	<1	Rwanda	Kinyamanza	0.06
DR Congo	VCB 81012	6	Rwanda	Barahuruye	0.05
DR Congo	AND 10	4	Rwanda	Ruari	0.05
DR Congo	CODMLV056	2	Rwanda	Isar	0.04
DR Congo	VCB 81013	1	Rwanda	Cab2 (mata)	0.03
DR Congo	K132	1	Rwanda	Muhinzimworozi	0.03
DR Congo	Maharagisoja, Kalima	>1	Rwanda	Utugondo	0.03
Ethiopia	Nasir	12	Rwanda	Mwatsiye	0.02
Ethiopia	Awash-1	10	Rwanda	Rugandura	0.02
Ethiopia	Mexican-142	4	Rwanda	Rubona	0.01
Ethiopia	Dimitu	3	Rwanda	Munezero	0.01
Ethiopia	Dark Red Kidney	3	Rwanda	Padiri	0.01
Ethiopia	Deme	3	Rwanda	Nsuzumirurushako	0.01
Ethiopia	Dinkinesh	3	Tanzania	Lyamungu 85	8
Ethiopia	Awash Melka	2	Tanzania	Lyamungu 90	8
Ethiopia	Cranscope	1	Tanzania	Uyole 96	6
Ethiopia	Roba, Cherecher, Batu, Goberasha, Chore, Argane	<1	Tanzania	Jesca	5
Malawi	Kalima	12	Tanzania	Wanja	5
Malawi	Maluwa	10	Tanzania	Uyole 2004	4
Malawi	Napilira	9	Tanzania	Kabanima	4
Malawi	Mkhalira	7	Tanzania	Urafiki	2
Malawi	Kabalabala	6	Tanzania	Selian 97	2
Malawi	Kholophethe	4	Tanzania	Uyole 03	2
Malawi	Kambidzi	3	Tanzania	Calima 2009, Kablanketi II, Bilfa Uyole, Roba 1	<1
Malawi	Nagaga	2	Uganda	K132	16.43
Malawi	Sapatsika	2	Uganda	Nabe 4	9.69
Mozambique	Sugar 131	8	Uganda	Naads	2.15
Mozambique	Cal 143	3	Uganda	Nabe 6/Nailon/Obweru	0.94

Mozambique	Diacol calima	2	Uganda	Muzahuura/Muzahura/Muzahuura/Nabe 14/Nyiramuzahura	0.61
Mozambique	Ica pijao, Carioca	<1%	Uganda	Nabe 3	0.56
Rwanda	Rwv 2070	4.21	Uganda	Nyirakanada (red)(climbing)	0.22
Rwanda	Rwr221	3.87	Uganda	Roba 1/Taso	0.18
Rwanda	Shyushya	3.20	Uganda	K 121/Kabalila/K 131	0.10
Rwanda	Cab19	1.57	Uganda	Nyiramutuku/Nyiramutuku	0.05
Rwanda	Rozikoko	1.46	Uganda	Kajamarika/Kajamarike/ Kajjakarika/Kajjamarika	0.04
Rwanda	Minwari	1.08	Uganda	Nabe 5	0.03
Rwanda	Mukwararaye	0.84	Uganda	Nabe 12c/Sugar 31	0.03
Rwanda	Rwv 2409	0.46	Uganda	Masava/Masavu/Nabe 11	0.02
Rwanda	Kinigi	0.35	Zambia	Lyambai	3
Rwanda	Kenyerumpure/Kenyerunkuru	0.32	Zambia	Chambeshi	2
Rwanda	Tubura	0.32	Zambia	Lwangeni	2
Rwanda	Page	0.23	Zambia	Lukupa	1
Rwanda	Kiryumukungu	0.20	Zambia	Pan 148, Kabale, Kapisha, Carioca	<1
Rwanda	var11	0.15			
Rwanda	Maharagisoya	0.15			
Rwanda	Mvuyekumurenge	0.09			

related released varieties enjoy good acceptance in large geographic, bean-growing concentrations in Tanzania. The two leading varieties in Tanzania are early-maturing, bush types. Lyamungu 85 and 90 trace their origins to a long-term, bean-breeding programme at the Escuela Agricola Panamericana (EAP) in Honduras.

The incidence of varietal spill-overs in beans seems to be low where consumption preferences and production conditions are diverse bordering on niche specificity (Table 8.6). But there are exceptions. For example, AND 10, the leading variety in Burundi, is a CIAT-bred cultivar that was released in 1998. It was also the second most popular variety in the DR Congo, where the top-ranking variety VCB 813030 is also a CIAT-bred cultivar.

Because of the rapidly increasing output of released varieties documented in Table 8.3, slow varietal turnover and ageing varieties in farmers' fields should not be a major concern for bean improvement in the nine survey countries. However, several improved cultivars are more than 20 years old. The issue of slow varietal turnover seems to be most relevant to Tanzania, where the two most popular varieties were released more than 20 years ago.

Summary

CIAT began working on common beans in SSA in 1984 when a small breeding programme was established in the Great Lakes Region. It targeted small-farm households producing beans in low input conditions of declining soil fertility and focused on yield and genetic resistance to well-defined pests and diseases. Over time that programme has matured and blossomed into the umbrella PABRA that covers three genetic improvement networks. Over time the genetic improvement strategy has also widened its emphasis and placed more weight on consumer preferences and grain types as organizing constructs for breeding activities.

The gains in area adoption of improved varieties since 1998 suggest that the investment in bean genetic improvement in the Great Lakes Region continues to pay handsome dividends. Across the nine study countries, the uptake of bred cultivars was estimated at 30% of total bean-growing area. Enhanced adoption is most apparent in Ethiopia where bean exports have fuelled increased demand for modern market-class bean varieties. Rising adoption is also easy to spot in Rwanda in the increasing frequency of climbing cultivars in response to the growing demand for intensification.

Malawi and Tanzania have make impressive strides in the adoption of improved bean varieties. Experts assessed acceptance of these cultivars in farmers' fields at levels near or exceeding 50% of cultivated area. This estimate represents a sea change from the 1998 figures that placed diffusion of CIAT-related varieties at less than 5% of area in each country. Although smaller market and area-based project studies have confirmed the popularity of these varieties in concentrated geographic pockets, Malawi and Tanzania are two prime candidates for a national representative survey on the adoption of improved bean varieties during the next round of the DIIVA Project. The next version of the CIAT Bean Atlas should be a rich despository of data to facilitate sample design.

Uganda has registered a solid performance in adoption outcomes with an estimate slightly above the regional average. About half of the area of improved varieties is planted to K132, the leading variety in the region.

Burundi, DR Congo, Mozambique and Zambia are countries where local bean landraces prevail because adoption of improved varieties is less than 15% of area cultivated. Mozambique's lagging adoption is attributed to scanty output in the form of only a few released varieties since the civil war ended in 1992 and 2009. In contrast, varietal output has been more than adequate in the other three countries. Poor adoption performance is attributed to the non-transfer or rejection of released varieties. Lagging adoption would seem to be most problematic in Burundi, where beans are such an important dietary component and account for a sizeable share of cultivated area.

Mozambique aside, varietal output has been impressive in the 1990s and 2000s. Anchored by a strong and stable programme in Rwanda and facilitated by CIAT's regional breeding networks, the flow of varieties has been increasing over time since the mid-1980s. Even small NARS programmes, such as Burundi's, have actively participated in varietal release. Although participatory varietal selection was not discussed in this paper, its acceptance and use has contributed to varietal output in several countries such as Rwanda.

The lion's share of varietal releases have come from CIAT crosses and lines. CIAT's genetic resources unit also distributed germplasm accessions that were directly released as improved varieties following in-country selection. Other research institutions in developed countries have contributed pertinent materials that led to varietal releases. Programmes, institutes and universities that warrant mention are the Bean and Cowpea CRSP in the USA, the Institute of Horticultural Plant Breeding (IVT) in the Netherlands, the Escuela Agricola Panamerica (EAP) in Honduras, the Centro Agronomico Tropical de Investigacion y Ensenanza (CATIE) in Costa Rica, the NVRS/Wellsbourne Project in the UK and the Tokachi Agricultural Experimental Station in Japan.

The assessment of scientific capacity in the bean improvement programmes was positive. Most were adequately staffed, well-educated and experienced, with about 30% women scientists. Areas for improvement were also highlighted. Given the importance of beans to its economy and to the welfare of its poor, Burundi's programme needed more scientists in general and more breeders in particular. Younger scientists were at a premium in the programmes of DR Congo and Tanzania. Ethiopia had the most dynamic programme but gender balance and lack of experience by many of its younger scientists were areas of concern.

In closing, CIAT's contribution to methods development in establishing a benchmark on varietal output, adoption and impact in SSA warrants a few words. Arguably, research on estimating cultivar-specific adoption in beans is as difficult as any of the 20 crops in this study. Difficulties begin with the large discrepancies between the FAOSTAT data and the best available estimates from national surveys on the bean area planted and harvested for several key countries. At the field level, farmers often plant seed mixtures, thereby blurring the visibility of varieties even in very small plots. As with most crops in SSA, varietal names change from place to place; a variety named in one location may not be the same as a variety of the same name in another. Fortunately, within a location, names for an existing variety do not change over time.

With data from CIAT's Bean Atlas and with structured workshops that invited not only scientists but also experts knowledgeable about technology transfer in well-defined regions, CIAT scientists sought to overcome or shed light on the fuzziness of improved varietal adoption. In several settings, such as Rwanda, the net result of this effort was a long list of varieties that were adopted on substantially less than 1% of bean-growing area. The nationally representative surveys gave estimates with a distribution similar to expert opinion: a handful of varieties adopted on more than 1% area and many estimated with a very small extension. The varietal identity between the two sources was often different, but the distribution was the same in both Rwanda and Uganda where the 1500 household surveys were conducted.

Notes

[1] This paper is an abridged version of Muthoni and Andrado (2012).
[2] The description of breeding strategy draws heavily on Buruchara *et al.*, 2011.

References

Buruchara, R., Chirwa, R., Sperling, L., Mukankusi, C., Rubyogo, J.C. *et al.* (2011) Development and delivery of bean varieties in Africa: the Pan-Africa Bean Research Alliance (PABRA) model. *African Crop Science Journal* 19(4), 227–245.

CIMMYT (2008) *Technical Agreement on Harmonization of Seed Regulation in the SADC Region. Seed Variety Release, Seed Certification and Quality Assurance, Quarantine and Phytosanitary Measure for Seed*. SADC Secretariat, Gaborone, Botswana.

Ethiopian Seed Enterprise (ESE) (2010) *Annual Report*. ESE, Addis Ababa.
FAO (2011) *FAOSTAT 2011*. http://faostat.fao.org/?lang¼en. (Accessed 26 November 2011.)
Ferris, S. and Kaganzi, E. (2008) *Evaluating Marketing Opportunities for Haricot Beans in Ethiopia. Rural Agro-Enterprise Development Project Report*. International Livestock Research Institute (ILRI), Addis Ababa.
Johnson, N.L., Pachico, D. and Wortmann, C.S. (2003) The impact of CIAT's genetic improvement research on beans. In: Evenson, R.E. and Gollin, D. (eds) *Crop Variety Improvement and its Effect on Productivity: The Impact of International Agricultural Research*. CAB International, Wallingford, UK, pp. 257–274.
Kalyebara, R., Andima, D., Xavery, P., Mugabo, J., Njingulula, P. *et al.* (2007) *Impact of Improved Bean R&D Technologies in Sub-Saharan Africa – Evidence from 7 Countries.* Occasional Publication Series, Pan African Bean Research Alliance, CIAT Africa Region, Kampala.
Katungi, E., Farrow, A., Chianu, J., Sperling, L. and Beebe, S. (2009) *Common Bean in Eastern and Southern Africa: A Situation and Outlook Analysis*. CIAT, Kampala, Uganda. http://www.icrisat.org/what-we-do/impi/projects/tl2-publications/regional-situation-outlook-reports/rso-common-bean-esa.pdf (Accessed 5 May 2015.)
Muthoni, R., Barungi, M. and Rubyogo, J.C. (2007) *Bean Technology Development and Promotion in the Pan Africa Bean Research Alliance from 2003 to 2007*. M&E and Performance Measurement Manuscript. CIAT, Kampala.
Muthoni, R. and Andrade, R. (2012) *The Impact of Bean Genetic Improvement Research. Diffusion and Impact of Improved Varieties in Africa (DIIVA) – Objective 1 Report*. CIAT, Kampala, Uganda.
PABRA (2011) *Annual Report*. CIAT–Africa, Kampala.
Rubyogo, J.C., Sperling, L., Muthoni, R. and Buruchara, R. (2010) Bean seed delivery for small farmers in Sub-Saharan Africa: the power of partnerships. *Society & Natural Resources* 23(4), 285–302.
Sperling, L. and Loevinsohn, M. (1993) The dynamics of adoption: distribution and mortality of bean varieties among small farmers in Rwanda. *Agricultural Systems* 41, 441–453.
Sperling, L. and Muyaneza, S. (1995) Intensifying production among smallholder farmers: the impact of improved climbing beans in Rwanda. *African Crop Science Journal* 3, 117–125.
Teshale, A., Rubyogo, J.C., Sperling, L., Amsalu, B., Abate, T. *et al.* (2006) Creating partnerships for enhanced impact: bean variety delivery in Ethiopia. *Journal of Crop Science Society of Ethiopia* 12, 27–30.
The Guardian (2010) http://www.theguardian.com/lifeandstyle/2010/oct/31/role-models-for-young-women (Accessed 5 May 2015).
Walker, T.S. (2010) *Guidelines for Data Collection for Objective 1 of the DIIVA Project*. Fletcher, North Carolina.
Wortmann, C.S., Kirkby, R.A., Eledu, C.A. and Allen, D.J. (1998) *Atlas of Common Bean* (Phaseolus vulgaris *L.*) *Production in Africa.* CIAT, Cali, Colombia.

Appendix 8.1

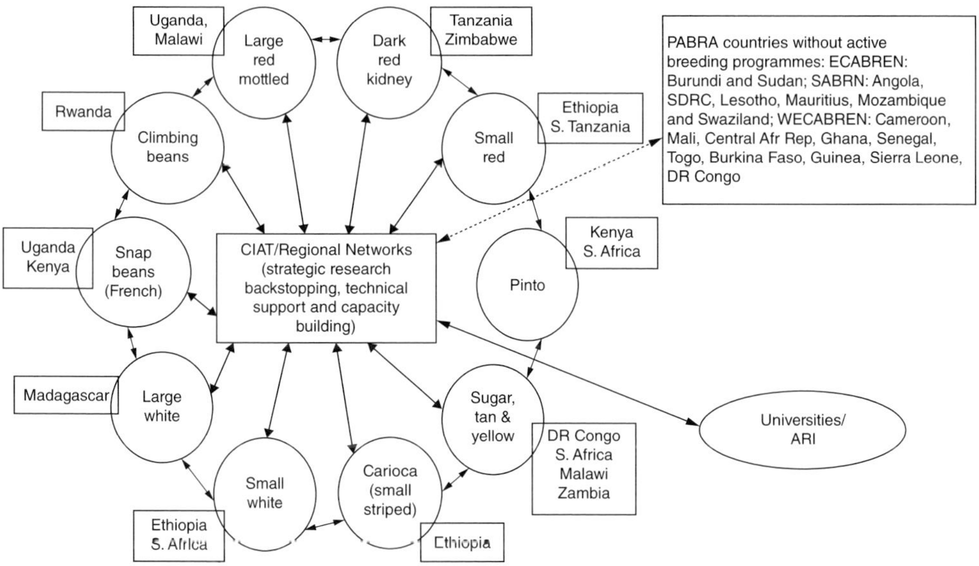

Fig. 8.A1. Breeding responsibilities and capacities under PABRA. The responsibilities involve: (i) CIAT headquarter breeding programme in Colombia and the regional breeding programmes of ECABREN and SABRN (central rectangle); (ii) national bean programmes responsible for different types of beans (peripheral rectangles attached to circles; and (iii) various universities and advances research institutes (ARIs). Arrows show the relationships among the different national and regional breeding efforts, which are interconnected. Countries without active breeding programmes (largest rectangle) do not have specific responsibilities but conduct adaptive testing of breeding lines or released varieties through the regional network. (Source: PABRA, 2011.)

9 The Effectiveness of Potato and Sweetpotato Improvement Programmes from the Perspectives of Varietal Output and Adoption in Sub-Saharan Africa

R. Labarta*
CIAT (formerly of CIP), Colombia

Introduction[1]

The International Potato Center (CIP) was one of the second wave of International Agricultural Research Centers established in the early 1970s. Its founding was based on the potential to improve human welfare via changes in potato productivity from applied research in developing countries. Although potato is viewed as a crop of the north, a tipping point was reached at about 10 million hectares and 150 million tonnes in the early 2000s when potato area and production in developing countries exceeded those in developed countries (Walker *et al.*, 2011). Sweetpotato was added to CIP's mandate in 1988. Until then, the International Institute of Tropical Agriculture (IITA) was responsible for the genetic improvement of sweetpotato in the Consultative Group on International Agricultural Research (CGIAR).

Globally, sub-Saharan Africa (SSA) is not a large producer of either crop but both potato and sweetpotato are characterized by a robust production growth that places them in the top quartile of the 20 DIIVA (Diffusion and Impact of Improved Varieties in Africa) crops in this aspect. During the past two decades, potato production in SSA has increased more than fourfold from about 3 to 13 million tonnes. Between 1993 and 2011, sweetpotato production has expanded from about 6 to 17 million tonnes. Growth in potato production is driven by strong market demand, which, in turn, is fuelled by urbanization. Growth in sweetpotato production is conditioned by rural population growth and the dynamics of the production environment for subsistence food crops. Increasing sweetpotato area is partially attributed to farmers' perceptions of declining growing-season rainfall and soil fertility. Sweetpotato is a good bet to produce food in 4–6 months in marginal locations with unassured rainfall and low soil fertility.

Aside from their production potential, both crops were characterized by antecedents that instilled hope that investing in crop improvement research could result in success and generate spillover benefits on a regional scale. For potato, the most relevant research that contributed to CIP's creation was John Niederhauser's finding that the Toluca Valley in Mexico was the origin of the late blight fungus. Late blight caused the Irish potato famine in the 1840s and still is the main biotic constraint to global potato production, especially in conditions where fungicides cannot be applied in a timely manner. Dr Niederhauser

* E-mail: r.labarta@cgiar.org

began his search for late-blight varietal resistance in the Rockefeller Foundation in Mexico in the late 1940s and was successful in his quest for broad-based resistance.[2] In the late 1960s, three late-blight resistant potato varieties were approved for release in Mexico. These were distributed quickly to several tropical and subtropical highland potato-growing regions including East Africa. These varieties provided CIP with a starting point for its potato improvement work in SSA. They and selections from their progenies were instrumental in generating an important success story in the widespread uptake of late-blight resistant clones in Rwanda, Uganda, DR Congo and Burundi (Rueda *et al.*, 1996).

For sweetpotato, the popularity of the clone Tanzania pointed to the potential for wide adaptability of improved varieties in the Great Lakes Region where sweetpotato is more intensively cultivated than in other countries in SSA. Tanzania was selected at the old agricultural research station for East Africa in Amani in the mid-20th century. Known by different names by farmers in the seven countries where it is grown,[3] Tanzania was the dominant variety in the Great Lakes when CIP began to invest resources in sweetpotato genetic improvement in the region early in the 1990s (Mwanga *et al.*, 2001). Tanzania combines virus resistance with drought tolerance in planting material in a white-fleshed, mealy background preferred by consumers in SSA. Unfortunately, white-fleshed materials have, at most, only negligible amounts of pro-vitamin A. In contrast, orange-fleshed sweetpotato is rich in beta-carotene. A daily intake of two small cups of orange-fleshed sweetpotato is sufficient to satisfy a child's recommended daily intake (RDA) and protect against Vitamin A deficiency, which is common in rural East Africa. With targeted donor support, CIP accepted the challenge of finding orange-fleshed clones in a high-yielding agronomic background similar to Tanzania. The breeding emphasis in sweetpotato on orange-fleshed materials in an adaptable agronomic background is analogous to the focus in potato genetic improvement on early maturing late-blight resistant varieties.

In other chapters in this volume, the issue of alternative suppliers is addressed. For sweetpotato, there are, for all intents and purposes, few if any alternative suppliers of genetic materials in addition to CIP. A few of the CIP-distributed elite clones have originated from the US genebank in Beltsville, Maryland, but no other institutional affiliations are detected in the pedigrees of released sweetpotato materials. For potato, the issue of alternative suppliers is relevant. Dutch seed from the private sector is usually an alternative to nationally bred and selected varieties or to CIP-bred materials. Imported Dutch varieties score high marks on wide adaptability and on preferred consumer traits, but they are expensive, susceptible to disease, difficult to grow without the intensive and timely use of inputs, and seed stocks degenerate rapidly when non-renewal is practised. Indeed, degenerated European tuber seed is one of the main sources of local potato varieties in SSA.

Country Coverage, Methods and Data Collection

In the 1998 Initiative, nine countries from SSA, Burundi, DR Congo, Ethiopia, Kenya, Madagascar, Rwanda, Sudan, Tanzania and Uganda, were included in the sample for potatoes (Walker *et al.*, 2003). In 2007, information on varietal output and adoption was updated for these same countries with the exception of Sudan (Thiele *et al.*, 2009). In 2010, five countries were selected based on the importance of potato production and on recent support and involvement by CIP in the country's potato improvement programme. The five countries, Ethiopia, Kenya, Malawi, Rwanda and Uganda, account for 65% of potato production in SSA.

Sweetpotato is represented by Burundi, Mozambique, Rwanda, Tanzania and Uganda. CIP has worked actively with the sweetpotato national programmes in these five countries that account for 54% of production in SSA. Nigeria is a major omission but the absence of a CIP presence at the time of the data collection made it difficult to keep this country in the final sample.

Both crops produce large quantities of food in the five countries selected for this study (5.3 million tonnes of potatoes and 6.4 million tonnes of sweetpotato). Based on estimates reported by FAOSTAT, the value of production of each crop was about US$1 billion in 2010.

For each of the identified ten priority country-by-commodity combinations, a list was prepared of key national organizations that are collaborating with CIP's crop genetic enhancement programme and that are involved in the dissemination of improved potato and sweetpotato varieties. The list included representation from public-sector research institutes, government extension agencies and non-governmental organizations (NGOs). (Private sector and university-related research is not common in potato and sweetpotato crop improvement in SSA.) A comprehensive questionnaire was designed for collecting data that would allow for the assembly of information on the three databases in the DIIVA Project. The questionnaire also included a detailed guideline for helping potato and sweetpotato experts fill out the questionnaires. CIP breeders, agronomists and social scientists provided input to the formulation of the questionnaire.

Once completed, the questionnaire was e-mailed to the pre-identified collaborators in different countries. In addition to the guidelines attached to the questionnaire, further instructions were provided on the type of information sought. Some collaborators raised specific queries about the questionnaire but, in general, the majority was able to complete most of the questionnaire. Many respondents, however, were unable to fully respond to the section on the estimates of the adoption of improved potato and sweetpotato varieties. It was very hard to estimate the area under each variety in different production regions.

Because reliable data on varietal adoption were not obtained with the e-mail questionnaire, small workshops were organized in each country to address any questions not answered. The estimation of adoption rates for improved cultivars was the focus of these meetings. A typical workshop was attended by six to ten participants and composed of researchers from potato and sweetpotato programmes and representatives from NGOs and the national extension service with experience in the dissemination of improved potato and sweetpotato cultivars. The workshops were facilitated by CIP's regional economist. With the input of all participants, adoption levels of improved varieties were estimated by major production areas previously defined by the same group of participants. Various CIP experts (especially potato and sweetpotato breeders) reviewed data relevant to their area of expertise and the preliminary report. They made contributions for better interpreting the results of the study.

Scientific Strength

Similarly to Evenson and Gollin (2003), CIP elicited information on the number of full-time equivalent (FTE) scientists by crop improvement programme, by major discipline and by degree-trained scientists. In 2010, data collection was expanded to measure the strengths of the national potato and sweetpotato programmes in the areas of (i) access to infrastructure and equipment for breeding work; and (ii) strategies of varietal dissemination pursued by each breeding programme.

In estimating research intensity, we relied on production data from FAOSTAT; however, for some estimates and countries, alternative sources that offer more reliable estimates were used (Labarta, 2012). For example, potato production in Malawi is markedly overestimated and in Ethiopia is severely underestimated in FAOSTAT compared to more reliable national sources.

Potato

The total scientific strength of the five potato improvement programmes approached 60 FTE scientists in 2010 (Table 9.1). Across countries, scientific staff strength was unevenly divided. Kenya accounted for about half of the FTE scientists. In both absolute and relative terms, Rwanda and Uganda with less than five FTE scientists were inadequately staffed in 2010. Investments in breeding at one or fewer FTE scientists were extremely low. Moreover, Uganda only had two technicians working in its potato research programme in 2010. Countries with small genetic improvement programmes either have scientists in very few disciplines (Malawi) or share FTE scientists across disciplines (Rwanda and Uganda).

Compared to their position in the late 1990s, the size of the programme in Kenya tripled and the programme in Ethiopia increased by 60%. Compared to a benchmark of the early 1990s for Rwanda and the late 1990s for Uganda, both programmes have maintained more or less their size but the strength of breeding has definitely decreased in Uganda. Moreover, research intensity has declined sharply in Rwanda from 14.7 researchers per million tonnes of production immediately prior to the genocide in 1994

Table 9.1. Number and intensity of FTE scientists by discipline and country for potato.

											Intensity indices	
Country	Breeding	Pathology	Molecular biology	Entomology\ Nematology	Agronomy\ Physiology	Seed production	Tissue culture	Postharvest	Social science	Total	FTE/million tonnes	FTE/US$ 100 million
Kenya	2	2	0	1	3	12	5	2	3	30	37.0	10.0
Malawi	1	0	0	0	1	0	1	0	0	3	7.4	4.4
Rwanda	0.5	0.8	0.1	0.1	1	1	1	0.1	0.3	4.6	6.7	5.7
Uganda	0.1	0.8	0	0.1	0.5	0.9	1	0	0.3	3.7	7.3	3.4
Ethiopia	5.76	1.28	0	0	4.48	2.56	1.92	0	0	16	26.9	26.6
Total	9.36	4.88	0.1	1.2	9.98	16.46	9.92	2.1	3.6	57.3	12.7	

to 2.7 researchers per million tonnes in 2010. Estimated research intensity also fell in Uganda from 9.1 researchers per million tonnes in 1999 to 4.7 in 2010. Largely because of the increase in staffing in Kenya, average research intensity across the four countries has not changed substantially between the 1990s and 2012. Staffing more than kept pace with increasing production in Kenya but lagged behind rising potato output in Ethiopia, Rwanda and Uganda.

Typical of root and tuber crops, the potato improvement programmes feature a diversified array of disciplines. Only about one FTE scientist in six is a plant breeder (Table 9.1). The Ethiopian Institute of Agricultural Research with a decentralized potato programme has the largest contingent of potato breeders. All programmes have invested in tissue culture, one of the oldest components of biotechnology. In contrast, only Rwanda has made a small investment in the newer component of molecular biology. The larger allocation of pathology compared to entomology suggests that diseases, caused by fungi, viruses and bacteria, are perceived as more important problems than insect pests and nematodes. Given the strong demand for potatoes and competitive alternative uses, it is not surprising that only one country, Kenya, has invested heavily in postharvest research. Kenya is also the only country to invest appreciably in social science in their potato improvement programme. In comparing allocation patterns between 1999 and 2010, Kenya has allocated most of its considerable additional scientific manpower to seed production and to a lesser extent to tissue culture, postharvest research and social science. Typical of crop improvement programmes, regardless of their commodity emphasis, all countries have made a commitment to agronomy. Once again, the decentralized research setting in Ethiopia results in a large contingent of agronomists.

The inclusion of technicians brings the number of total FTE staff in the five programmes to 92. Of these, only 6% have PhDs. Malawi, Rwanda and Uganda have less than 0.5 PhD-level scientists in each of their programmes.

As mentioned earlier, scientific staffing is a key component in assessing the capacity of crop improvement programmes to generate improved varieties demanded by farmers. Infrastructure is another. An infrastructural adequacy index was constructed from responses by different breeding programmes about the level of access they have to different labs, equipment and fields that they need to carry out their work. Possible responses were good access (1), just adequate access (2) and inadequate access (3). The average index presented in the last row of Table 9.2 represents a more desirable situation when the index approaches 1 and a less desirable situation when the index approaches 3.

In spite of being one of the smallest programmes on potatoes, the Rwandan programme is easily the programme that has better access to infrastructure for carrying out potato breeding. The government of Rwanda has engaged in supporting the research programmes of their priority crops (potato is one of the top five) with investments in new and high-quality equipment. With these investments, Rwanda is rebuilding the strong potato programme that existed before the 1994 Genocide. Unfortunately, as discussed, this investment in labs and equipment has not been accompanied by an increase in human resources.

Table 9.2. Adequacy of different access to infrastructure for potato breeding by country in 2010.

Infrastructure type	Kenya	Malawi	Rwanda	Uganda	Ethiopia
Experimental field	2	1	1	1	1
Greenhouse	3	3	1	3	2
Cytogenetic lab	3	3	2	3	3
In-vitro multiplication area	1	2	1	2	2
Seed multiplication plot	2	2	1	1	1
Molecular marker facility	3	3	1	3	3
Seed storage facility	2	3	1	2	1
Processing and quality lab	3	3	1	3	3
Hardware (database)	3	3	2	2	1
Vehicle	2	2	2	3	2
Average index	2.4	2.5	1.3	2.3	1.9

1, good access; 2, just adequate access; 3, inadequate access.

On average, the potato programme in Ethiopia can be considered to have just adequate access to equipment for breeding work. There has been a partial update of equipment in the potato programme in Ethiopia but there is still a need to invest in a molecular marker facility, a quality lab and in several other infrastructural areas. For several infrastructural categories, respondents in Kenya, Uganda and Malawi assigned a score of 3 in Table 9.2, indicating limited access to infrastructure in those areas.

Sweetpotato

The total staff strength in the five sweetpotato programmes approaches 33 (Table 9.3). In general, the two programmes in southern Africa are well staffed; the three programmes in the Great Lakes region are sparsely staffed. Relative to the country's size of production, Uganda is the programme that seems deficient in the number of scientists, with only four FTE scientists working in sweetpotato research. It is the leading producer of sweetpotato in East Africa yet its estimated research intensity in terms of number of FTE scientists per million tonnes of sweetpotato barely exceeds 1.4. On a positive note, Uganda's sweetpotato programme is well endowed with nine technicians but its low number of scientists dampens the average research intensity across the five country programmes. The weighted average research intensity is slightly more than 5 scientists per million tonnes of sweetpotato production, which is less than half of the estimate for potato improvement.

Every country has allocated at least one FTE scientist to sweetpotato breeding. Roughly the same situation applies to agronomy, which seems to be the other core area in the sweetpotato programme's disciplinary portfolio. The other disciplines trail breeding and agronomy by a wide margin.

Small programmes tend to cover most of the disciplines with few FTE scientists because these scientists have multiple appointments within the sweetpotato programme or across other crop programmes. Similarly to the potato programmes, the number of FTE scientists in molecular biology and social sciences are very limited. Only Uganda has a FTE scientist dedicated to molecular biology; only in Mozambique is research in social science undertaken to directly support investigations in sweetpotato improvement.

Several of the allocations in Table 9.3 seem puzzling. For example, Mozambique and Tanzania have allocated resources to pathology/virology; however, the sweetpotato virus complex is a substantially greater constraint to production in the Great Lakes region where the three sweetpotato programmes have not allocated resources to virology, according to the data in Table 9.3.

As in potato improvement, the sweetpotato programmes have a total of about six FTE PhD scientists; however, more than four of them work in the programme in Tanzania. In the other countries, the frequency of PhDs range from none in Burundi to one in Uganda.

Access to infrastructure is roughly the same in sweetpotato improvement as it was in potato improvement. Overall, access to infrastructure is only adequate for sweetpotato breeding research. All the programmes have good access to experimental and seed multiplication fields but, as expected, they have inadequate access to biotechnology facilities. Unlike in potato improvement (see Table 9.2), there were no outliers for sweetpotato improvement because the estimated country average infrastructural indices were tightly clustered across the five countries.

Crop genetic enhancement programmes use various strategies to broadly disseminate improved cultivars. The level of the use of different strategies for this purpose was assessed by analysing answers from potato and sweetpotato programmes related to their objectives in varietal dissemination. The answers included: not pursued at all (1), rarely pursued (2), sometimes pursued (3), commonly pursued (4) and very actively pursued (5). An index was built to assess how active breeding programmes have been in disseminating their improved material.

The varietal-transfer indices are presented in Table 9.4 for sweetpotato in which the diffusion of new clones, especially orange-fleshed materials, is presently of great interest. The Mozambican programme is by far the most active in pursuing strategies to broadly disseminate their new varieties. In the past 10 years, this programme, which is led by an international presence through SARRNET[4] (first with IITA and since 2006 with CIP), has aggressively disseminated their new planting material via all the

Table 9.3. Number and intensity for FTE scientists by discipline and country for sweetpotato.

Country	Breeding	Pathology	Molecular biology	Entomology\ Nematology	Agronomy\ Physiology	Seed production	Tissue culture	Postharvest	Social science	Total	Intensity indices	
											FTE/million tonnes	FTE/US$100 million
Mozambique	2	0.5	0	0.3	0.9	1.4	1.5	0.6	1.5	8.7	12.8	4.5
Uganda	1	0	0	0.1	1.3	0.5	0.6	0.5	0	4	1.4	2.1
Rwanda	1	0	0	0.1	1	0	0	0	0	2.1	2.5	2.8
Burundi	1	0	0	0	1	0	0	0	0	2	10.3	4.7
Tanzania	3.6	1.8	2.1	0	2.7	0.3	1.8	1.8	1.8	15.9	15.2	6.9
Total	8.6	2.3	2.1	0.5	6.9	2.2	3.9	2.9	3.3	32.7		

Table 9.4. Strategies pursued by sweetpotato breeding programmes to disseminate improved varieties

Strategies	Mozambique	Uganda	Rwanda	Burundi	Tanzania
Early involvement of farmers	4	3	2	3	5
Marketing of varieties	4	4	2	1	2
Use of mass communication	4	3	2	3	2
Promotion/public extension	5	3	3	3	4
Promotion/NGOs	5	5	4	4	4
Public–private partnerships	5	3	3	1	3
Average index	4.5	3.5	2.7	2.5	3.3

1, not pursued at all; 2, rarely pursued; 3, sometimes pursued; 4, commonly pursued; 5, actively pursued.

strategies described in Table 9.4. The Ugandan and Tanzanian sweetpotato programmes also have been active in the dissemination of new sweetpotato varieties. Tanzania's programme has as its highest priority the early involvement of farmers in the participatory varietal selection of new cultivars. Uganda has concentrated its efforts on marketing and promoting improved material.

For technology transfer, national programmes in the five selected countries use NGO dissemination programmes as the first option. Public–private partnerships to distribute new sweetpotato varieties are also becoming more common, but these still need to be supported in small programmes such as Burundi's, where market-related interactions are low.

Varietal Output

Potato

A total of 117 improved potato varieties have been released in the five selected countries (Table 9.5). The release data are consistent with increasing varietal output over time.

CIP's influence in potato breeding has increased and also changed over time. Before the mid-1980s, national potato breeders allocated most of their effort to screening finished varieties for adaptation to regional production conditions. Fifteen of these released clones came from Europe, mainly from the UK and the Netherlands (Table 9.5). During this early period, CIP distributed advanced materials that were previously released in developing countries and other elite clones from their germplasm bank. As discussed in the introduction, several national programmes in SSA also benefited from elite late-blight clones released in Mexico with the support of the Rockefeller Foundation (CIP, 1972).

Since the establishment of CIP's regional office in Nairobi and the launch of the PRAPACE[5] network in the late 1970s and early 1980s, CIP's role has expanded from that of an agent for germplasm distribution of elite clones to a purveyor of progenies and parents for selection and crossing by national programmes. Hence, sources of releases have shifted from advanced clones from developed and other developing countries to progeny selected by NARS from CIP crosses and from NARS crosses that use CIP parental materials. The change of emphasis documented in Table 9.5 is consistent with the increasing maturity of potato breeding programmes in the region.

The release data in Table 9.5 attest to a different release profile in each country over time. Until recently, Malawi has invested sparsely in potato improvement. It has only released varieties in 2 years since 1970. In 1980, Malawi approved ten varieties for release. These clones were previously released in Kenya, South Africa and Uganda or in Europe. One clone, Rosita, is still the dominant potato variety in Malawi and Mozambique. Rosita (or Sangema as it is known in Mexico) is a red-skinned clone released in Uganda from the aforementioned Mexican late-blight resistant varieties imported in the early 1970s. The other release year was 2011. Malawi released six varieties largely as a result of CIP establishing a presence in the country in 2006.

Kenya's record for varietal release is not as episodic as Malawi's, but it is erratic given fairly steady support for potato improvement. Kenya is characterized by an abrupt rise in potato varieties generated in the early 1970s, sporadic varietal

Table 9.5. Improved potato varieties released in Kenya, Malawi, Rwanda, Uganda and Ethiopia before 1990, in 1990–1999 and in 2000–2011.

	Kenya			Malawi			Rwanda			Uganda			Ethiopia		
	<1990	1990–1999	2000–2011	<1990	1990–1999	2000–2011	<1990	1990–1999	2000–2011	<1990	1990–1999	2000–2011	<1990	1990–1999	2000–2011
NARS released clones with no CIP participation	12	0	1	10	0	0	8	0	0	9	0	0	0	0	2
Developing country clones	1	0	1	0	0	0	5	0	0	9	0	0	0	0	0
Developed country clones	11	0	0	10	0	0	3	0	0	0	0	0	0	0	2
NARS released clone with CIP participation	1	3	6	0	0	6	7	6	1[a]	0	8	4	2	6	22
CIP cross	0	3	3	0	0	5	1	6	1	0	6	4	1	2	12
CIP progenitor	0	0	2	0	0	0	2	0	0	0	0	0	1	1	10
CIP distributed	1	0	1	0	0	1	4	0	0	0	2	0	0	3	0

[a]The potato programme in Rwanda is in the last stages of evaluating five promising clones from CIP genebank (CIP crosses) that have been already released in other African countries.

production in the 1980s and 1990s, and renewed varietal output in the last decade. The support of a seed distribution programme has been influential in the releases in 2010 (CIP, 2011a). Kenya has also moved upstream in variety breeding in the early 2000s by crossing new material using CIP progenitors.

Rwanda released 21 of their 22 potato varieties before 1994 when the Genocide destroyed the agricultural research efforts in the country. The first varieties came from Mexico and the late blight programme but soon after, with the establishment of the PRAPACE regional network, the Rwanda potato programme became an important user of material from the CIP potato genetic improvement programme. Since 1994, the Rwandan potato programme has only been able to release one variety, Victoria, which was previously released in Uganda with the same name. Hence, the dominant supply of improved potato varieties are from materials first introduced in the early 1980s and released on average 20–25 years ago. In 2004, the national programme received five advanced clones from CIP that are under evaluation. They were scheduled for release in 2012. Three of these varieties have already been released in Kenya.

Although seven selections from CIP crosses were released between 1998 and 2010, the pace of varietal release in Uganda has gradually slowed over time, which is consistent with the reduction of FTE scientists and a changing emphasis from the release of finished materials in the 1970s to the release of progeny selections in the 1990s and 2000s. Indeed, in the past 20 years, 100% of the potato varieties released in Uganda have been CIP crosses that were evaluated for many years in many locations in the country. The PRAPACE network and an in-country CIP breeding presence in Uganda in the late 1980s and early 1990s facilitated the release of several varieties, including elite materials from the Rwandan programme and Victoria, now the most widely adopted potato variety in East Africa.

Ethiopia has a time release profile that is the opposite of Rwanda's: few releases in the 1970s and 1980s juxtaposed to dynamic release activity since the mid-1990s. In the mid-1980s many European varieties, such as the Dutch clones Spunta and Alpha, were imported by NGOs but none was released officially, although they have remained in the seed system (Medhin *et al.*, 2001). Varieties released later in Ethiopia have largely originated from CIP's genetic improvement programme. Ethiopia is the only potato programme that markedly and consistently increased the number of potato varieties released in the 1990s and 2000s. Furthermore, the Ethiopian potato programme has moved upstream in potato breeding in the past 10 years: ten of their 24 released varieties have been crosses made in Ethiopia using CIP progenitors. In addition, the potato programme in Ethiopia has benefited from its decentralized set-up with regional centres that have participated in the release of new potato varieties. In addition to the Holetta Research Center, the potato programmes at Alemaya University and at the regional Adet Research Center have been instrumental in releasing varieties (Hirpa *et al.*, 2010).

Sweetpotato

Comparing the release data in Tables 9.5 and 9.6 highlights several differences in the output trajectories of potatoes and sweetpotatoes. Early varietal output is substantially less in sweetpotato than in potato. Only nine sweetpotato varieties were released prior to 1990. When CIP received the CGIAR's mandate for sweetpotato, there was not much international and national research to draw on that could benefit producing countries in the Great Lakes region and in southern Africa in the short-to-medium term. About 30% of sweetpotato varietal releases in Table 9.6 are landraces that were selected in the same country or neighbouring countries of the region. The 89 varieties include 18 sweetpotato varieties that were not officially released in Mozambique. The varieties selected in Mozambique were recommended for farmer use and multiplied extensively but not released officially. Fifteen selected and bred varieties were released in 2011. Mozambique has benefitted from SARRNET since the mid-1990s and is also the home to an international sweetpotato breeding presence since the early 2000s. Subtracting the released landraces and the Mozambican list of recent and official releases in southern Africa only gives a total of about 30 varieties bred and selected for the conditions of the Great Lakes region where high dry matter and virus resistance are traits of primary importance.

Table 9.6. Number of sweetpotato varieties released in Mozambique, Uganda, Rwanda, Burundi and Tanzania.

	Mozambique			Uganda			Rwanda			Burundi			Tanzania		
	<1990	1990–1999	2000–2011	<1990	1990–1999	2000 2011	<1990	1990–1999	2000–2011	<1990	1990–1999	2000–2011	<1990	1990–1999	2000–2011
NARS released clones with no CIP participation	0	0	2	0	6	2	4	1	3	1	1	0	0	0	10
Land races selected from own country	0	0	2	0	5	2	4	1	3	0	0	0	0	0	10
NARS cross	0	0	0	0	1	0	0	0	0	1	0	0	0	0	0
Developing country clones/landraces	0	0	0	0	0	0	0	0	0	0	1	0	0	0	0
NARS released clones with CIP participation	0	5	26	0	6	6	4	1	10	0	0	1	0	0	0
CIP cross	0	0	17	0	0	0	0	0	2	0	0	0	0	0	0
CIP progenitor	0	0	0	0	0	5	0	0	2	0	0	0	0	0	0
CIP distributed	0	5	9	0	6	1	4	1	6	0	0	1	0	0	0

The data in Table 9.6 also suggest that the inference of a bleak picture of very low sweetpotato varietal output is changing. The number of varieties released in the five countries has increased over time.

CIP participation in sweetpotato breeding in Africa before 2000 was restricted to the distribution of varieties already released in other countries, such as Peru, the USA and Taiwan, with mature breeding programmes. The varieties were further tested in each recipient country and released or recommended for further multiplication. With the growing realization that virus resistance was of paramount importance in the Great Lakes region, breeding and selection were increasingly carried out in that region and in southern Africa through the national programme in Mozambique. Of the 60 new varieties released in the 2000s, 19 were CIP crosses and seven varieties were crossed locally using CIP progenitors. Currently, CIP is leading a large breeding initiative to breed sweetpotato for Africa under the SASHA (Sweetpotato Action for Security and Health in Africa) project that is funded by the Bill & Melinda Gates Foundation.

The sweetpotato improvement programme in Uganda is one of the bright spots in sweetpotato breeding in East Africa. That programme has maintained a steady stream of output since the mid-1990s. Breeding outputs with limited intervention from CIP in Table 9.6 were led in Uganda by their National Agricultural Research Organization (NARO) (Mwanga *et al.*, 2001).

The Ugandan sweetpotato programme was the first among the five included in this study to carry out breeding with polycrosses. This programme released the variety Sowola in 1995. Sowola was selected from an open-pollinated field with 24 parents that included the preferred landraces in Uganda (Mwanga *et al.*, 2001). In 1999, Uganda continued with open-pollinated crosses in order to select and release improved varieties. That same year, the sweetpotato programme incorporated some varieties from IITA-Nigeria that were facilitated by CIP. Also in 1999, the first orange-fleshed sweetpotato cultivar was crossed in Uganda (NASPOT 5). This was the starting point for further breeding with this trait. In 2004, two orange-fleshed landraces were evaluated and released in Uganda and used as progenitors for materials selected for release in 2007. High beta-carotene content (associated with orange flesh) in new material is a priority in Uganda.

Rwanda was the first country to engage in sweetpotato breeding. Work began in the 1980s by collecting and evaluating a group of landraces that were later released. Varieties Rusenya and Mugande were released during this period and have been widely used as progenitors for later crosses. CIP started its involvement in sweetpotato breeding in Africa in Rwanda by introducing varieties released in the USA. After a long period in recovery from the 1994 Genocide, Rwanda has rebuilt its sweetpotato programme with international assistance. In the last decade, the Rwandan programme has released ten varieties that drew on CIP-related materials.

In spite of not being a country with high sweetpotato consumption, Mozambique has benefited from the efforts of SARRNET. It is the country breeding programme that CIP has most actively supported in the last decade in southern Africa. Early efforts by IITA in the 1980s produced a group of recommended varieties but they were not officially released. With the evaluation and recommendation for further multiplication of orange-fleshed varieties introduced by CIP in the last decade, the 'orange revolution' started in Mozambique with massive distribution of planting material of these varieties (Labarta, 2009). These varieties were not officially released. It was only in 2011 that with the support of the Alliance for the Green Revolution in Africa (AGRA) and in collaboration with CIP that the national sweetpotato programme in Mozambique was able to officially release 15 improved cultivars. These varieties were selected from open-pollinated fields where CIP progenitors and selected landraces were used.

The release of sweetpotato varieties in Burundi has been very limited so far in spite of the importance of the crop in this country. It is expected that Burundi can benefit from the sweetpotato breeding platform that the SASHA project has established in Uganda.

Tanzania is also another case with reduced efforts in sweetpotato genetic improvement. In spite of having a large programme in terms of number of scientists working in the crop, only ten varieties have been officially released. Furthermore, these ten varieties correspond to evaluation and selection of landraces that have been collected nationwide in Uganda. It is expected that they can be used in future crosses of new material. CIP has also recently distributed new sweetpotato varieties released in other countries that would help the breeding efforts in Tanzania.

Varietal Adoption

Potato

Two sources were used to assess varietal adoption: (i) expert opinion estimates for Kenya, Malawi and Uganda; and (ii) survey estimates for Ethiopia and Rwanda (Labarta *et al.*, 2012). The survey estimates for Ethiopia validated expert opinion; deviations between the two sources were very small. Expert opinion estimates for improved variety adoption, however, were substantially larger than the survey estimates for improved variety adoption in Rwanda. The size and possible explanations for this discrepancy are discussed in Chapter 20 of this volume. Nationally, the adoption estimates range from 1% in Malawi to 74% in Uganda (Table 9.7). Weighted average adoption across the five countries is about 35%.

With the exception of Malawi, adoption estimates are available for four countries for a paired comparison between the 1990s and 2010. Weighted average adoption declined from 49% in the 1990s to 37% in 2010. Adoption gains in Ethiopia and Kenya were not sufficient to offset what appears to be a massive disadoption of improved varieties in Rwanda. Prior to the 1994 Genocide, the adoption of improved varieties in Rwanda with its impressive potato seed programme was estimated at 100%. In 2010, the nationally representative survey estimate for adoption of improved varieties in Rwanda was

Table 9.7. Estimated adoption of improved potato varieties by country, region and source.

Country	Province/region	Total area (ha)	Adoption level (%)	Source
Ethiopia	Amhara	71,170	8	Survey
Ethiopia	Oromia	53,002	37.4	Survey
Ethiopia	SNNPR	37,537	29.2	Survey
Ethiopia	National	164,146	22.6	Survey
Rwanda	Volcanic Area	82,597	26.3	Survey
Rwanda	Bumveruka	26,281	51.8	Survey
Rwanda	Crescent Nile Congo	18,772	54.1	Survey
Rwanda	Other Areas	23,127	36.2	Survey
Rwanda	National	150,777	38.7	Survey
Kenya	Meru	19,467	71	Expert opinion
Kenya	Keiyo-Marakwet	8,012	75	Expert opinion
Kenya	Mt Elgon	4,354	22	Expert opinion
Kenya	Nakuru	19,473	50.1	Expert opinion
Kenya	Narok	10,209	4.2	Expert opinion
Kenya	Bomet	1,473	0	Expert opinion
Kenya	Nyandaru	19,141	26	Expert opinion
Kenya	Nyeri	22,598	20.5	Expert opinion
Kenya	Taita	1,473	56.4	Expert opinion
Kenya	Kiambu	19,467	15.6	Expert opinion
Kenya	National	152,998	29.1	Expert opinion
Uganda	Southwestern highlands	68,340	79	Expert opinion
Uganda	Southern highlands	8,160	60	Expert opinion
Uganda	Eastern highlands	10,200	48	Expert opinion
Uganda	Lake Albert Crescent	7,650	70	Expert opinion
Uganda	West Nile Zone	7,650	80	Expert opinion
Uganda	National	102,000	74	Expert opinion
Malawi	National	45,816	1	Expert opinion

only 36%. Uganda also has lost some ground in improved varietal adoption since 1999.

The 1994 Genocide in Rwanda was a severe blow to the survival and use of improved varieties in the Great Lakes region. Although farmers took improved planting materials to refugee camps in the DR Congo and although relief seed programmes were subsequently undertaken, it is apparent that recovery is still incomplete almost 20 years after the tragedy of civil war. Estimated adoption of improved varieties in the main volcanic-ash growing region is only 26% (Table 9.7). Local varieties such as Gashari, Makoroni, Cyunyu, Nyirakarayi and Gashara were ranked quite often among the top three varieties used by farmers in many communities in the focus-group community discussions (Labarta *et al.* 2012). But as potato is not a native crop in Rwanda, most of these local varieties may be old European clones that were introduced over 30 years ago. They have remained in the system and have acquired local names. After the destruction of the local capacity to produce seed and to release new varieties, many of the old European varieties that have remained in DR Congo, Burundi, Tanzania and other countries may have re-entered into Rwanda from the borders in order to deal with the shortage of seed potato in the post- genocide period.

The household survey results confirm the findings from the focus-group discussions on Rwanda. Two local varieties, Nyirakabondo and Nyabizi, account for 65% of the potato-growing area in the main vocanic-ash region.

In Ethiopia, the main potato-producing region, Amhara, also lags behind other smaller growing highland regions in the incidence of improved variety adoption. In Kenya, the variation in adoption levels across the ten potato-growing regions is marked. Many small growing regions with widely varying adoption levels suggests that finding materials with wider adaptability is more of a challenge in Kenya than Uganda where the regional adoption levels vary in the narrow range of 60–80% with the exception of the eastern highlands (Table 9.7).

CIP-related clones figure prominently in the list of adopted improved varieties described in Table 9.8. Spill-over varieties are common across the countries in the PRAPACE network. Victoria is the leading variety in Uganda with more than 50% of potato-growing area. It is the second most popular improved clone in Kenya. (Victoria was selected by Lyle Sikka, a CIP breeder posted in Uganda, in the 1990s.) Victoria has also spread to Burundi and the DR Congo.

Kinigi (selected by ISAR breeders from a CIP cross) and Cruza 148 (a Mexican cross selected by CIP researchers in a USAID project) are the most widely used varieties in Rwanda. Kinigi has about the same level of adoption in Uganda

Table 9.8. Adoption of specific potato improved varieties by country in 2010.

Country	Improved cultivar	Adoption (%)
Malawi	Six new clones released in 2011	1.00
Ethiopia	Jalene (2002)	7.51
Ethiopia	Gudene (2006)	4.90
Ethiopia	Menagesha (1993)	2.91
Ethiopia	Bule (2005)	2.60
Ethiopia	Holland (2009)	0.98
Ethiopia	Guassa (2002)	0.91
Ethiopia	Sisay (1987)	0.79
Ethiopia	Wechecha (1997)	0.49
Ethiopia	New clones	0.43
Ethiopia	Belete (2009)	0.31
Ethiopia	Tolcha (1993)	0.30
Ethiopia	Diagmeng (2002)	0.20
Ethiopia	Gera (2003)	0.18
Ethiopia	Gorobella (2002)	0.04
Ethiopia	Shenkola (2005)	0.02
Uganda	Victoria	53.60
Uganda	Kinigi	13.40
Uganda	Nakpot 1	0.80
Uganda	Kachpot 2	0.80
Uganda	Rutuku	0.70
Rwanda	Kinigi (1984)	14.18
Rwanda	Cruza 148 (1985)	11.17
Rwanda	Mabondo (1988)	4.58
Rwanda	Rutuku (1984)	2.59
Rwanda	Sengema (1980)	1.40
Rwanda	Gasore (1984)	0.67
Rwanda	Kirundo (1989)	0.66
Rwanda	Peloro (1984)	0.31
Rwanda	Gikungu (1992)	0.22
Rwanda	Victoria (2000)	0.10
Kenya	Tigoni	16.99
Kenya	Asante	6.39
Kenya	Purple Tigoni	3.46
Kenya	Tigoni red	2.22

as in Rwanda (Table 9.8). It was one of the first crosses introduced by CIP in Rwanda and has wide market acceptance for its processing characteristics and as a table potato. Although its late blight resistance has broken down, it continues to be a preferred variety in Rwanda. Cruza 148 was also introduced as a late blight resistant variety; however, farmers prefer this variety because of its latent resistance to bacterial wilt, which has become a major issue for farmers in Rwanda and neighboring countries. In spite of not having preferred taste and processing characteristics because of a blue streak in its flesh, the demand for Cruza is high especially among women potato growers in Rwanda, Burundi, and the Kivu region of the DR Congo.

Tigoni and its somatic mutations in a purple and red skin colour account for over 20% of potato-growing area in Kenya. Jalene and Gaudene are also CIP-related and are the two leading improved varieties in Ethiopia with survey adoption levels in the 5-10% range. These two varieties have benefited from technology-transfer efforts made by the public extension system in Ethiopia and from large projects that CIP has implemented with various partners in the region (CIP, 2011b). Local varieties remain important in Ethiopia and most of them refer to old degenerated varieties that were introduced from Europe between 20 and 40 years ago. The most important one is the variety Kie Abeba that accounts for about one-quarter of potato-growing area in Ethiopia.

Potato planting in Malawi is dominated by the old variety Rosita (introduced from Mexico) that is estimated to approach 60% of the total area under potato production. Likewise, a Dutch variety, Violeta, introduced in 1980, is common, accounting for 24% of total growing area. In this study, these two dominant cultivars are now considered to be degenerated local varieties even though they were officially released. These varieties have stayed in farmers' fields for many years and their non-replacement and lack of an effective clean seed programme have resulted in very low potato yields in the country (Demo *et al.*, 2009). Six varieties newly released in 2011 are in the very early stages of adoption.

Improved varietal turnover seems to have slowed in four of the five potato-growing countries in the DIIVA Project. The present batch of adopted improved varieties has been in the system on average for more than 20 years. The only country where slow turnover does not adversely affect programme performance is Ethiopia where the weighted average age of adopted improved varieties is 7.25 years. With the exception of Ethiopia and to a lesser extent Kenya, varietal age is trending upwards since the mid to late 1990s.

Sweetpotato

Sweetpotato estimates on varietal adoption came from expert opinion in Burundi, Mozambique, and Tanzania and from nationally representative surveys in Rwanda and Uganda. Expert estimates in Rwanda and Uganda were markedly higher than the survey estimates (Labarta *et al.*, 2012). Plausible explanations for these differences are discussed in Chapter 20 in this volume and focus on overly optimistic expectations from large technology transfer efforts and the invisibility of sweetpotato varieties in very small home-garden plots where piecemeal harvesting is common (Labarta *et al.*, 2012).

The weighted average for aggregate adoption of modern varieties across the five countries is 7% (Table 9.9). This estimate does not include landraces that were officially released. Inclusion of those materials gives a weighted average estimate in the range of 35–40% in most countries. In an extreme case, adoption in Tanzania increases from zero to 43% when officially released landraces are included.

Adoption of selected and bred materials in Rwanda, Uganda and Mozambique are moderate to negligible in most subregions of those countries in spite of large dissemination programmes, most recently by CIP and HarvestPlus (Arimond *et al.*, 2010). However, when considered from the perspective of number of households reached, adoption of improved varieties increases substantially. For example, in Uganda, 8% of farm households cultivated orange-fleshed sweetpotato compared with a modest 2.6% coverage in area. This supports the bio-fortification strategy to promote the consumption of small quantities of beta-carotene-rich varieties to overcome Vitamin A deficiency. Improved cultivars have achieved some penetration into Central Uganda and Zambezia, the central province in Mozambique. Survey estimates suggest negligible adoption in

Table 9.9. Estimated adoption of improved sweetpotato varieties by country, region and source.

Country	Province/region	Total area (ha)	Adoption level (%)	Source
Mozambique	Tete	37,816	9.6	Expert opinion
Mozambique	Zambezia	23,615	19.8	Expert opinion
Mozambique	Sofala	18,139	5.4	Expert opinion
Mozambique	Niassa	12,635	5.3	Expert opinion
Mozambique	South	20,302	9.8	Expert opinion
Mozambique	National	130,000	9.2	Expert opinion
Burundi	National	125,000	28.4	Expert opinion
Tanzania	National	480,000	0	Expert opinion
Rwanda	North	33,621	0	Survey
Rwanda	South	32,081	0.5	Survey
Rwanda	West	33,629	0	Survey
Rwanda	East	24,055	0	Survey
Rwanda	National	123,086	0.1	Survey
Uganda	Central	138,087	22.1	Survey
Uganda	Eastern	225,250	0.2	Survey
Uganda	Northern	85,303	2.7	Survey
Uganda	Western	171,360	2.1	Survey
Uganda	National	620,000	8.8	Survey

other regions of Uganda and in Rwanda for the country as a whole.

Varieties coming from CIP crosses are in the very early stages of their diffusion process and have only reached 1% of the total area under sweetpotato. NASPOT 1 is the only bred improved cultivar that lays claim to more than 5% of national area (Table 9.10). It was selected from an open-pollinated cross containing some progenitors distributed by CIP and preferred landraces. Known for its high dry matter content, its popularity is increasing in Central Uganda. However, its adoption level still pales in comparison to the extended coverage of several local varieties. For example, Bungunduza and Muwuulu are cultivated in over 40% of growing area in Eastern Uganda, the largest sweetpotato-producing region of the country (Labarta *et al.*, 2012).

In 2010, the majority of sweetpotato area in Mozambique was dominated by landraces that have adapted to the varying and difficult farming conditions of this impoverished coastal country in southern Africa. Five landraces (Admarc, Mudiliva, Muanagemela, Canasuma and Sector) are estimated to contribute to 55% of the total sweetpotato area of Mozambique. However, breeding work initiated by the national programme with support of CIP in 1996 is starting to produce desired impacts. Varieties introduced by CIP and evaluated by the national sweetpotato programme have reached 9% of the total area under sweetpotatoes according to expert opinion. The selection of introduced material has favoured the promotion of the vitamin-A-rich orange-fleshed varieties. The varieties that have achieved relatively larger success so far are Jonathan (from Peru), Persistente (from Mozambique), Resisto (from the USA) and LO-323 (from the USA). It is expected that the recently released 15 orange- fleshed varieties will reach a higher adoption rate considering that their breeding process has included progenitors with drought tolerance, which is a major constraint for most food crops in Mozambique.

In Burundi and Tanzania, adoption of sweetpotato varieties is largely dominated by landraces, with very little adoption of varieties that come from crossings or advanced clones from other countries. In Burundi, the variety Mugande that was officially released by the sweetpotato programme was estimated to have reached 20% of adoption. In Tanzania, adoption of all the released landraces has reached 43.9% of the total area under sweetpotato. Sinama (known as Tanzania in other countries) is the variety that has reached the highest adoption rate (14.5%) due to its high dry matter content, high virus resistance, and wide acceptance among farmers and markets. With similar characteristics, the variety Ukerewe is estimated to have reached 10% of the total area under sweetpotato in Tanzania.

Table 9.10. Adoption of specific sweetpotato improved varieties in 2010 by country with qualifying comments.

Country	Improved cultivar	Adoption (%)	Qualifying comment
Mozambique	Jonathan	4.00	Does not include Persistente, a released landrace from Mozambique, which is estimated at 3% of growing area. Hence, released landraces are estimated to account for only 3% of sweetpotato-growing area
Mozambique	Resisto	2.70	
Mozambique	LO-323	2.50	
Uganda	Naspot 1 (1999)	6.00	Released landraces account for an additional 9.1% of sweetpotato-growing area
Uganda	Naspot 9&10 (07)	1.20	
Uganda	Kakamega	1.10	
Uganda	Naspot 11 (2010)	0.30	
Uganda	Naspot 2 (1999)	0.10	
Rwanda	Caroline Lee	0.10	Released landraces from Rwanda are estimated to account for about 27.9% of area
Burundi	Mugande	20.00	
Burundi	Mnzovu Y'umugamba	7.20	
Burundi	Tanzania	1.20	
Tanzania	Modern Varieties	0.00	Released landraces from Tanzania are estimated to account for about 43% of area

Summary

The DIIVA Project offered a new opportunity to extend previous efforts to study the contribution of the CGIAR to varietal change in developing countries. It also offered CIP the opportunity to include sweetpotato in the group of crops studied and to draw some conclusions on the efforts for genetic improvement of this crop. This study included Kenya, Malawi, Rwanda, Uganda and Ethiopia for potatoes, and Mozambique, Uganda, Rwanda, Burundi and Tanzania for sweetpotatoes.

The results contain good and bad news on the performance of potato and sweetpotato improvement programmes from the perspectives of varietal output, adoption and change. In potato, varietal release and adoption is trending strongly upward in Ethiopia and is also heading in the right direction in Kenya. Increasingly, the five potato programmes are capable of selecting elite materials from their own crosses or from introduced CIP progenies. They no longer have to rely on adaptation trials of elite varieties from other countries for release. Sweetpotato farmers no longer have to depend exclusively on purified landrace materials that may be well adapted to their conditions but usually do not show much if any gains in desirable traits. They can now choose from more than ten recently bred varieties in the more developed breeding programmes in Uganda and Mozambique.

The bad news underscores the challenges confronting potato and sweetpotato breeding in SSA. Varietal change in potatoes is still reeling from the Rwandan Genocide in 1994. Rwanda was the regional hub for several important breeding activities in the PRAPACE network in the 1980s and early 1990s. The chain of events and consequences include the loss of life and the destruction of the Ruhringeri Station, insufficient scientific resources in the potato programme during recovery, no production of released varieties between 1994 and 2010, substantial and unexpected disadoption of varieties released in the 1980s, and the slowing of varietal turnover of modern varieties as farmers reverted back to old, degenerated, local cultivars. No new varieties replaced the first-generation PRAPACE cultivars in their fields. Gains in adoption in Ethiopia and Kenya were not sufficient to offset disadoption in Rwanda. As a result, the level of adoption in the region declined between the mid-1990s and 2010.

Reversing the incidence of disadoption and increasing the turnover of modern varieties are major challenges facing the potato programme in Rwanda and to a lesser extent in Uganda. A scarcity of PhD scientists in the programme and in the region makes that challenge more formidable. The imminent release of five new cultivars, the designation of potato as a priority commodity programme and recent investments in infrastructure enhance the odds of meeting the challenge.

For sweetpotato, the challenge is rooted in the realization that adoption of bred and selected white- and orange-fleshed materials has not yet reached appreciable levels in spite of aggressive and innovative efforts to transfer improved material on a large scale. More recent efforts, like the SASHA initiative, may markedly increase the uptake of modern sweetpotato clones, but this needs to be confirmed with time. Although the bio-fortification efforts that can significantly reduce Vitamin A deficiency have achieved some positive results in reaching resource-poor households, there is a need for reviewing and possibly revising the sweetpotato breeding strategy.

The results have also reaffirmed the role of the International Potato Center in contributing to varietal output, adoption, and change to the potato- and sweetpotato-growing regions in SSA. For example, in the last decade CIP was involved in the release of 42 of the 45 released potato varieties. The majority came from CIP crosses. CIP's influence on sweetpotato is more muted because of low scientific staffing in programmes such as Burundi and Rwanda and because of the tendency to release landrace materials in Tanzania. Success by the more mature breeding programmes where CIP is deeply involved, such as those in Uganda and Mozambique, should pave the way for a more active role in the region.

Documenting a new benchmark for potato and sweetpotato improvement programmes has highlighted the importance of regional networks and the timely posting of an international presence in a country programme. The PRAPACE potato programme was instrumental in the transfer and subsequent release of improved materials that spilled over to several countries in East and Southern Africa. SARRNET played the same role in sweetpotato improvement in Southern Africa. The work of both of these programmes has been periodically compromised by the shortage of funds.

Three examples in this chapter speak to the benefits from the occasional posting of an international presence in a national crop improvement programme. Varietal output in potato in Malawi and sweetpotato in Mozambique increased substantially as a result of positive interactions among national and international resident staff in the same country. The leading potato variety in East Africa today owes its origin to the posting of an internationally well-known potato breeder in Uganda's national programme in the late 1980s and early 1990s.

Lastly, this study shows the importance of monitoring the evolution of genetic improvement in the developing countries on a continual basis. Repeating the study every 5 years and including more countries would be desirable. The verification of adoption estimates from various sources with nationally representative surveys should also be encouraged. Verification sharpens the perception of reality and challenges conventional wisdom. Evidence for the unanticipated disadoption of improved potato varieties in Rwanda and much slower than expected early adoption of improved sweetpotato varieties in Uganda and Rwanda call for a revision of previous thinking and the subsequent taking of corrective action.

Notes

[1] This paper is a revised and abridged version of Labarta (2012).

[2] John Niederhauser won the World Food Prize for his applied research on potato pathology in 1990.

[3] Tanzania is called Sinama in Tanzania, Enaironi in Kenya, Tanzania in Uganda and Rwanda, Kenya in Malawi, ADMARC in Central Mozambique and Chingovwa in Zambia (Mwanga *et al.*, 2001).

[4] The Southern Africa Root Crops Network was established in 1994 by IITA and CIP and includes collaborators of the 12 Southern Africa Development Community (SADC) countries.

[5] French acronym for Regional Potato and Sweetpotato Improvement Network in Eastern and Central Africa that was first established in 1982 and includes Burundi, DR Congo, Eritrea, Ethiopia, Kenya, Madagascar, Rwanda, Sudan, Tanzania and Uganda.

References

Arimond, M., Ball, A., Béchoff, A., Bosch, D., Bouis, H. *et al.* (2010) *Reaching and Engaging End Users (REU) of Orange Fleshed Sweet Potato (OFSP) in East and Southern Africa.* HarvestPlus/IFPRI, Washington, DC.

CIP (1972) *The International Potato Center. Report to the Technical Advisory Committee of the International Group of Agricultural Research.* CIP, Lima.

CIP (2011a) *Tackling the Food Price Crisis in Eastern and Central Africa with the Humble Potato: Enhanced Productivity and Uptake through the '3G' Revolution.* USAID End of Project Report, 2011. CIP, Lima.

CIP (2011b) *Roadmap for Investment in the Seed Potato Value Chain in Eastern Africa.* CIP, Lima.

Demo, P., Pankomera, P., Connell, T. and Khumar, N. (2009) Potential of Potato farming in improving the livelihoods of small scale farmers in Malawi. *African Crop Science Conference Proceedings* 9, 761–765.

Evenson, R.E. and Gollin, D. (eds) (2003) *Crop Variety Improvement and its Effect on Productivity: The Impact of International Agricultural Research.* CAB International, Wallingford, UK.

Hirpa, A., Meuwissen, M., Tesfaye, A., Lommen, W., Lansink, A. *et al.* (2010) Analysis of seed potato systems in Ethiopia. *American Journal of Potato Research* 87, 537–552.

Labarta, R. (2009) *Reaching End Users (REU) Project.* End of Year Three Technical Report. International Potato Center, Quelimane, Mozambique.

Labarta, R. (2012) *The Genetic Improvement of Potato and Sweetpotato in Sub-Saharan Africa.* Diffusion of Improved Varieties in Africa (DIIVA) Project Report on Objective 1. International Potato Center, Nairobi.

Labarta, R., Wambugu, S., Yirga, C., Mugabo, J., Dieu Nsabimana, J., Mulwa, C., Sulz, S., Larochelle, C., Alwang, J., Andrade, R. and Yigezu, Y. (2012) *The Adoption of Improved Potato and Sweetpotato Varieties in Ethiopia, Rwanda and Uganda: Comparing Expert Opinion Estimates and the Results of Nationally Representative Surveys.* Diffusion and Impact of Improved Varieties in Africa (DIIVA) – Objective 2 Report. International Potato Center (CIP), Nairobi.

Medhin, G., Giorgis, W., Gebre, E., Bedane, K. and Kassa, B. (2001) *Country Profile on Potato Production and Utilization: Ethiopia.* Ethiopian Agricultural Research Organization (EARO), Holetta Research Center, Ethiopia.

Mwanga, R., Odongo, B., Ocitti p'Obwoya, C., Gibson, R., Smit, N. and Carey, E. (2001) Release of five sweetpotato cultivars in Uganda. *HortScience* 36(2), 385–386.

Rueda, J.L., Ewell, P.T., Walker, T.S., Soto, M., Bicamumpaka, M. and Berrios, D. (1996) Economic impact of high-yielding, late-blight-resistant varieties in the Eastern and Central African Highlands. In: Walker, T. and Crissman, C. (eds) *Case Studies of the Economic Impact of CIP-Related Technologies.* CIP, Lima, pp. 15–30.

Thiele, G., Hareau, G., Suarez, V., Chujoy, E., Bonierbale, M., and Maldonado, L. (2009) *Varietal Change in Potatoes in Developing countries and the Contribution of the International Potato Center 1972–2007.* Working Paper 2008-6. CIP, Lima.

Walker, T., Bi, Y., Li, J., Gaur, P. and Grande, E. (2003) Potato genetic improvement in developing countries and CIP's role in varietal change. In: Evenson, R.E. and Gollin, D. (eds) *Crop Variety Improvement and its Effect on Productivity: The Impact of International Agricultural Research.* CAB International, Wallingford, UK, pp. 315–336.

Walker, T., Thiele, G., Suarez, V. and Crissman, C. (2011) *Hindsight and Foresight about Potato Production and Consumption.* Social Sciences Working Paper 2011-5. International Potato Center (CIP), Lima.

10 Evaluating the Key Aspects of the Performance of Genetic Improvement in Priority Food Crops and Countries in Sub-Saharan Africa: The Case of Rice

A. Diagne,*[1] F.M. Kinkingninhoun-Medagbe,[2]
E. Amovin-Assagba,[2] T. Nakelse,[1] K. Sanni[1] and A. Toure[2]
[1](Formerly AfricaRice); [2]AfricaRice, Cotonou, Benin

Introduction[1]

Worldwide, more than 3.5 billion people depend on rice for more than 20% of their daily calorie intake (IRRI, AfricaRice and CIAT, 2010). Annual rice consumption can be very high, exceeding 100 kg per capita in many Asian countries and in some African countries (e.g. Madagascar and Liberia) as well. Rice consumption is growing faster than any other major commodity in Africa because it is a convenience food for the growing urban population.

Genetic improvement of rice in Africa is characterized by a rich if somewhat disjointed research history (Dalton and Guei, 2003). Two rice species (*Oryza sativa* from Asia and *O. glaberrima* domesticated in Africa), five rice growing environments (uplands, rainfed lowlands, irrigated lowlands, mangrove swamps and deep-water regions), several bilateral organizations (especially Institut de Recherches Agronomiques Tropicales (IRAT) and Centre de Coopération Internationale en Recherche Agronomique pour le Développement (CIRAD)), and three CG Centers (AfricaRice, the International Rice Research Institute (IRRI) and the International Institute of Tropical Agriculture (IITA)) figure prominently in that history, which has been shaped by investments in national programmes since the early 1950s in Nigeria and in the Rokupr Research Station in Sierra Leone for regional mangrove rice improvement since the mid-1930s.

Monitoring adoption of modern varieties also began on a continental scale in sub-Saharan Africa (SSA) earlier than for any other crop. The semi-dwarf, short-duration high-yielding varieties (HYVs) of rice from Asia had entered Africa as early as the late 1960s. Dalrymple (1978) estimated that the diffusion of modern rice varieties had reached 4% in 15 rice-growing countries by the late 1970s.

By the 1990s, diffusion of improved varieties was sufficient to support rate-of-return studies and impact assessment research with a specific focus in a handful of countries. For example, Adesina and Zinnah (1993) estimated that improved varieties from the Rokupr Research Station had reached 56% of rice-cultivated area in Sierra Leone in 1990. With somewhat different emphases, three other studies arrived at an estimated annual rate of return of between 18% and 34% to rice improvement research in Sierra Leone (Dalton and Guei, 2003).

* E-mail: aliouwork@gmail.com

 Crop Improvement, Adoption and Impact of Improved Varieties in Food Crops in Sub-Saharan Africa (eds T.S. Walker and J. Alwang)

Almost all the adoption and impact assessment research addressed improved cultivars in *Oryza sativa*, which was first introduced into West Africa more than five centuries ago by Portuguese explorers returning from India. Locally adapted varieties of Asian rice evolved from these initial and subsequent introductions mediated by centuries of farmer selection. Advantages of the African rice in terms of heartiness and local adaptation could not be transferred to Asian rice because of species incompatibility that was solved in the early 1990s when interspecific crosses became a viable option. Selections from these crosses are referred to by AfricaRice as the 'New Rice for Africa' or NERICA varieties. The first generations were targeted for the uplands where the expected productivity gains from hybridization of the two species should stimulate diffusion of improved varieties in a major rice growing environment that was lagging behind in adoption. NERICA varieties for the irrigated and lowland growing environments were subsequently released starting in 2005.

The NERICA varieties were too early in the breeding pipeline and in the adoption process to be considered in the 1998 Initiative (Dalton and Guei, 2003). Ten years later, AfricaRice was heavily involved in documenting the spread of the NERICA varieties throughout SSA. For the DIIVA Project, the emphasis on adoption and the timing of this effort could not have been better. Funded from its Japan Project, AfricaRice undertook national adoption surveys with national agricultural research and statistical agencies in 21 rice-growing countries. Complementary information was systematically collected on expert perception on varietal adoption, scientific capacity in national agricultural research systems (NARS) and on varietal releases so that performance in rice improvement could be assessed from the perspective of inputs (strength of NARS), outputs (varietal releases) and outcomes (adoption of modern varieties).

Survey Design and Data Collection

AfricaRice implemented the DIIVA Project in 22 countries, including 10 countries targeted by the project (Côte d'Ivoire, Ghana, Guinea, Madagascar, Mali, Nigeria, Senegal, Sierra Leone, Tanzania and Uganda) and 12 countries where AfricaRice was carrying out other projects (Benin, Burkina Faso, Cameroon, Central African Republic (CAR), Democratic Republic of Congo, The Gambia, Guinea Bissau, Kenya, Liberia, Mozambique, Rwanda and Togo). The seven countries in the 1998 Initiative, namely Côte d'Ivoire, Ghana, Guinea, Mali, Nigeria, Senegal and Sierra Leone, were priorities for targeting in the DIIVA Project. Madagascar was also a priority country because it and Nigeria are the largest rice producers in SSA. In the next three analytical sections, country coverage varies from a minimum of 11 on varietal output to a maximum of 16 on scientific capacity. Data reliability was the criterion guiding country inclusion or exclusion on each of the key aspects of performance in genetic improvement.

Two main activities were conducted:

1. Development of tools for data collection: Two types of questionnaire were developed for data collection. The first relates to rice varieties and the second to the country's scientific resources in rice genetic improvement.
2. Exchange and capacity building on data collection tools and methods: Two workshops were organized, a workshop for English-speaking countries was held in Addis Ababa (26–31 July 2010), and another for French-speaking countries was convened in Ouagadougou (16–21 August 2010). During these workshops, the project's country focal points were trained on the methodology and data collection tools.

After the workshops, the final versions of the two questionnaires were sent to the country focal points for the survey. Many e-mails were also sent to provide guidance on data collection and synthesis. Each country was requested to:

- Extract data related to rice varieties from the Rice Statistics Survey database collected by AfricaRice in 2009 (see AfricaRice, 2010) in order to fill in the first DIIVA questionnaire on rice varieties;
- Fill in missing data relating to scientific strength on both questionnaires by collecting complementary data from different researchers from the regional research centres of the country; and
- Organize a one-day national multidisciplinary workshop to validate all the results, with a particular focus on expert estimates on

rice variety adoption estimates. The aim of the workshop was to gather together experts working in rice research and extension who were familiar with rice varieties and the areas under each variety in the country.

AfricaRice received the required information from some of the countries through repeated communications by e-mail and telephone and during meetings of the Africa Rice Breeding Task Force held in Cotonou, Benin. However, many countries did not provide all of the requested information and for these countries some of the missing data for the first questionnaire, the rice variety questionnaire, have been completed using information available in the expert opinion data set from the scientist questionnaire of AfricaRice's 2009 Rice Statistics Survey, mainly for adoption estimates.

Scientific Strength

The treatment by AfricaRice of scientific capacity is more detailed than most other crop-related chapters in this volume. AfricaRice first elicited information on total number of scientists in rice improvement and then converted that information to full-time equivalent (FTE) scientists. We follow the same sequence in the discussion below. All data refer to public-sector scientists in national agricultural research programmes. Private-sector participation in rice research is negligible in SSA and university researchers do not play a prominent role in rice research in the vast majority of countries of interest for this assessment.

Total number of NARS scientists in rice improvement

Data from 16 countries are considered in this section. In 2010, these countries accounted for about 75% of rice production in SSA. In total, 289 research scientists are involved in rice improvement (Fig. 10.1). Most of the country programmes are moderately large with 16 to 24 scientists working in them. On the basis of casual knowledge and interactions over time, several, like Kenya and Sierra Leone, were larger than expected in 2010.

Rice improvement displays a well-diversified portfolio of disciplines (Table 10.1). Only about one quarter of rice scientists (24%) are geneticists or plant breeders, 13% are research support staff (programme coordinator, head of research division, biometrician, research assistant, research technician with no specific discipline, etc.), 11% are agronomists, 9% are social scientists,

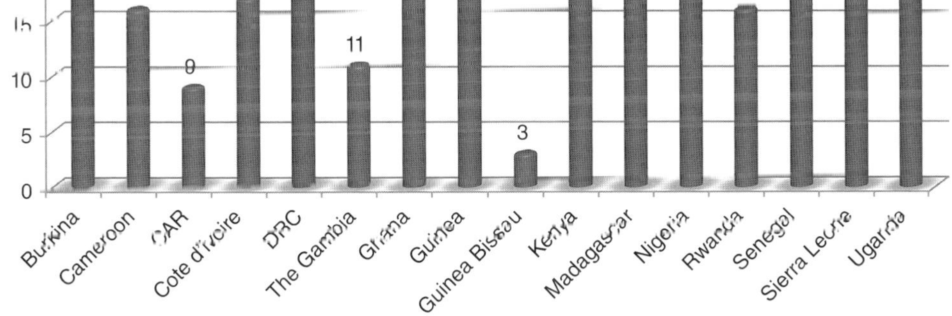

Fig. 10.1. Number of scientists by country. CAR, Central African Republic; DRC, DR Congo. Source: DIIVA expert opinion survey and AfricaRice (2010).

Table 10.1. Distribution of rice scientists by main disciplines, all 16 countries.

Discipline	Rice scientists (%)	Discipline	Rice scientists (%)
Genetics/plant breeding	24	Physiology	6
Research support staff	13	Weed science	6
Agronomy	11	Water management	3
Social science	9	Postharvest	2
Entomology	8	Biotechnology	1
Phytopathology	8	Climatology/GIS	1
Soil science	8	Food science	1

8% are in soil sciences, 8% in entomology and 8% in pathology. The remaining disciplines are sparsely populated with the least represented areas in biotechnology, climatology/geographic information systems (GIS) and food science each with 1% of scientists.

This diversified disciplinary portfolio confirms conventional wisdom and also generates a few surprises. Rice is the most diversified cereal improvement programme in the DIIVA Project with the lowest relative importance of plant breeding narrowly defined. Given the rice plant's sensitivity to abiotic stress, soil and water management are well represented in most programmes. Pathology and entomology respond to well-known biotic stresses from diseases and insect pests. Weed science figures prominently in the scientific allocation to agronomy. Small investments in food science and postharvest research reflect a strong demand for rice and the absence of a need to explore alternative uses.

The main surprise in Table 10.1 centres on the 9% share in social science research, which is far heavier in rice than for any other crop in this volume. With the exception of the Central African Republic, all NARS have allocated at least one social scientist to rice improvement research. This seemingly high allocation to social science in rice research itself warrants a small, focused investigation to determine if rice improvement was too inclusive in its measurement or if the other crop programmes were too exclusive in their assessment of what constituted participation by social scientists in research support to crop improvement.

More importantly, plant breeding seems to be poorly represented in several major-producing countries in West Africa. For example, Nigeria has less than two FTE plant breeders. Guinea with 11 is the only country with more than five FTE plant breeders. Given the potential importance of rice in West Africa, plant breeding seems to be thin on the ground relative to their allied research support disciplines. The low allocation in molecular biology was also unanticipated both because investments in this area are very small and because they are not concentrated in the major-producing countries of Madagascar and Nigeria.

The majority (63%) of the 289 scientists were employed more than 4 years in rice research. Sixteen per cent are beginners in this field with less than 1 year of service, whereas 13% have spent between 1 and 2 years (13–24 months). On average, rice scientists have an average of 12 years of experience in rice research. These results show that most scientists involved in rice improvement have substantial experience.

In general, programmes in West Africa are characterized by older, more experienced scientists than those in East and Southern Africa. In the countries targeted in the DIIVA Project, the mean length of experience exceeded 14 years, whereas a comparable estimate for the other countries was 10 years. Rice is proportionally a more important crop in West Africa than in other regions of SSA, and an older cadre of scientists in West and Central Africa is a recurring theme in other chapters in this volume.

FTE scientists

Most scientists involved in rice improvement were also engaged in other activities. Only 28% devoted 75–100% of their time to rice improvement and 9% allocated 50–75% of their time. The remaining five scientists in eight spent less than 50% of their time on rice research.

The 289 rice scientists translated into 123.3 FTE scientists. Twelve of the 16 programmes were characterized by conversion percentages from total to FTE scientists that oscillated in the narrow band of 40–60%. In Burkina Faso, the DR Congo, Ghana and Guinea Bissau, part-time scientists were the norm as the conversion interval fell to 15–30%. Across the disciplines, part-timers were more likely to be found in entomology and agronomy. Full-time scientific staff were more common in plant breeding and research support than in other areas of investigation.

The allocation of FTE scientists by discipline is presented in Table 10.2. Subtracting out those involved in research support, the number of FTE scientists declines to 103.2. Except for agronomy, research support, entomology and, to a lesser extent, plant breeding, the FTE disciplinary composition in Table 10.2 is very similar to that described in Table 10.1. The distribution across countries varied from a total of just 0.9 FTE scientists in Guinea Bissau to 12.9 FTE scientists in Sierra Leone, with an average of 7.7 FTE among the 16 SSA countries included in the analysis. Only six of the 16 programmes approached or exceeded a threshold of ten FTE scientists. As mentioned earlier, the low number of FTE plant breeders in several principal-producing countries is a cause for concern (Table 10.2).

With regard to educational level, the analysis shows that more than half (57%) of the rice research scientists have an MSc degree, more than a quarter (28%) have a PhD and 15% have a BSc. These proportions are roughly the same as those estimated by Dalton and Guei (2003) in 1998. The proportion of rice scientists with a PhD degree is higher in West and Central Africa than in East and Southern Africa. This level of education is equivalent to about 35 PhD scientists for SSA as a whole.

The Democratic Republic of Congo (DR Congo) had no involvement of PhD holders in rice research. In contrast, Nigeria had the highest level of involvement by scientists with PhDs (8 FTE), followed by Ghana (6 FTE), and Madagascar (5 FTE). At the MSc level, Kenya ranked highest (10 FTE scientists). Burkina Faso, The Gambia, Ghana, Nigeria and Guinea Bissau have no BSc-level scientists in rice improvement research, whereas Côte d'Ivoire, Guinea and Uganda seem adequately endowed at this level with 3.5 FTE each. The absence of BSc-level scientists in the former countries is potentially another area of weakness in scientific capacity, because a chronological transition exploiting learning by doing and mentoring cannot be established if young scientists are not available for hands-on training.

More than half of the rice scientists are over the age of 50 years, 24% are 41–50 years old, 20% are 31–40 years old and 4% are aged 30 or below. Nine per cent of the rice scientists are over the age of 60 years and will be retiring in a few years' time. Only about 15% of the rice scientists are women. This percentage does not seem to vary by region.

Research Intensity

The weighted average estimate for research intensity for rice improvement is 9.0 FTE scientists per million tonnes of production (Table 10.2). If scientists engaged in research support are not included, this estimate falls to 7.6. This level of research intensity places rice in roughly the same position as maize in West and Central Africa. Investment in rice improvement is substantially more intense than comparable allocations to sorghum and pearl millet but considerably less intense than the scientific attention paid to maize and wheat in East and Southern Africa.

The largest producers are characterized by values between 2 and 3 FTE scientists per million tonnes of production (Table 10.2). These estimates are considerably less than those calculated for the largest rice producers in South Asia (Chapter 13, this volume). Almost all of the national programmes in South Asia have, however, a substantially larger crop of FTE scientists than Nigeria and Madagascar. Their low research intensities stem from very large production volumes characteristic of rice's dominance as the staple food crop in South Asia.

Some of the small-producing countries, such as Kenya, CAR and Rwanda, exhibit very high research intensities that are more than 100 in Table 10.2. These three countries produce less than 50,000 tonnes of rice per annum. Their intensive level of investment with 5.0 to 12.5 FTE scientists in rice improvement would not seem to be economically defensible even using favourable assumptions in a cost–benefit analysis. In contrast, Guinea Bissau is the rare case of a small producer with a low research intensity

Table 10.2. Full-time equivalent (FTE) of scientists by discipline in rice research programmes in sub-Saharan Africa, 2010.

Country	Genetics/ plant breeding	Biotech-nology	Phyto-path-ology	Ento-mology	Physi-ology	Weed science	Climat-ology / GIS	Water manage-ment (irrigation)	Soil science	Food science – grain quality	Posthar-vest technol-ogy	Social science	Agron-omy	Research support	Total	Research intensity
Burkina Faso	0.38	0.00	0.38	1.00	0.00	0.50	0.00	0.38	0.13	0.00	0.00	1.75	0.25	0.00	4.75	20.4
Cameroon	1.50	0.00	0.00	0.13	0.00	0.38	0.00	0.00	0.38	0.00	0.00	1.50	3.38	0.00	7.14	40.8
CAR	0.00	0.38	0.00	0.00	0.00	0.00	0.00	0.00	0.00	0.00	0.00	0.63	0.00	4.13	5.12	131.3
Côte d'Ivoire	0.75	0.00	1.25	0.00	0.88	0.38	0.38	0.00	0.75	0.00	0.00	0.00	1.13	4.38	9.87	15.2
DR Congo	0.50	0.00	0.13	0.00	0.25	0.13	0.00	0.00	0.13	0.13	0.00	1.50	0.38	0.00	3.17	10.0
The Gambia	1.75	0.13	0.25	1.00	0.00	0.00	0.00	0.00	0.00	0.00	0.00	0.00	2.38	0.88	6.37	63.8
Ghana	1.13	0.00	0.50	0.13	1.13	0.75	0.00	0.13	0.50	0.13	0.00	0.38	0.00	0.00	4.75	9.7
Guinea	8.75	0.00	0.38	0.00	0.00	0.88	0.00	0.00	0.63	0.00	0.00	1.25	0.38	0.00	12.25	7.6
Guinea Bissau	0.63	0.00	0.00	0.00	0.00	0.00	0.00	0.00	0.25	0.00	0.00	0.00	0.00	0.00	0.88	5.0
Kenya	3.75	0.00	0.88	0.75	1.25	0.38	0.00	0.88	0.00	0.00	0.00	0.75	0.38	3.50	12.50	156.2
Madagascar	3.88	0.00	0.88	1.00	1.50	0.00	0.88	0.00	1.13	0.00	0.00	0.00	0.38	0.63	10.25	2.2
Nigeria	1.88	0.00	1.88	1.13	0.00	1.88	0.00	0.88	1.00	0.13	0.00	0.13	0.00	1.00	9.87	3.1
Rwanda	1.13	0.00	0.38	0.00	0.00	0.88	0.00	0.00	1.00	0.00	0.88	0.75	0.00	2.25	7.43	110.5
Senegal	1.00	0.00	0.63	0.13	0.00	0.63	0.00	0.13	1.25	0.00	0.38	1.13	0.00	2.25	7.54	12.5
Sierra Leone	4.25	0.00	0.63	0.88	0.75	0.38	0.00	0.00	0.88	0.50	0.88	1.25	2.50	0.00	12.87	14.2
Uganda	2.19	0.00	1.88	0.13	1.75	0.00	0.00	0.00	0.00	0.00	0.38	0.25	0.88	0.25	8.53	39.1
Total	33.91	0.50	9.93	6.25	7.50	7.23	1.25	2.37	9.00	1.04	2.55	11.94	12.00	20.12	126.10	9.0

Source: DIIVA expert opinion survey and AfricaRice (2010). Research intensity measured as FTE scientists per million tonnes of production.

because slightly less than one FTE scientist is allocated to rice research in this country where rice is the primary staple.

Comparing 1998 to 2010

Seven rice-growing countries in West Africa were covered in the 1998 Initiative (Dalton and Guei, 2003). Paired comparisons between 1998 and 2010 can be conducted for six of those countries where data are available for both periods. Between 1998 and 2010, scientific capacity in five of the six countries increased substantially (Table 10.3). Ghana was the exception where the decline was less than 1.0 FTE scientist. On aggregate, the increase was 83% of the base level in 1998.

In spite of a decline in research intensity in four of the six countries, the aggregate research intensity increased from 5.6 in 1998 to 7.6 in 2010. This moderate increment is attributed to scientific capacity rising faster than production. Underlying this increase is the fact that rice production in Nigeria was roughly the same at 3.2 million tonnes in 1998 as in 2010 according to FAOSTAT.

Varietal Output

Impediments to varietal release in SSA are not unique to rice; however, concern was expressed at the start of the DIIVA Project that documenting varietal output would be cumbersome if not impossible in many rice-growing countries because of absent registries and ineffective practices conditioning varietal release. AfricaRice invested in a study (Sanni *et al.*, 2011) to describe the variation in release regulations and their adherence to them in more than 20 rice-growing countries in SSA. In this section, we first examine the evidence on the process of varietal release and then we assess data, mainly from release registries, on the national availability of improved varieties to farmers.

Current status of varietal release and registration of new rice varieties

The release of new varieties, which is governed by the seed laws of each country, is one of the most important goals of any breeding programme. The seed laws that regulate the varietal release process vary from country to country in terms of requirements and institutional responsibilities (Table 10.4). Under a regulated system of varietal release, the basic regulatory components that are usually mandatory are: demonstration of value for cultivation and use (VCU), distinctness, uniformity and stability (DUS) testing, recommendation by the varietal release committee, and official registration.

Variety registration is the last activity of a breeding programme. It means that the national authorities officially recognize that the candidate variety is distinct, uniform and stable, and performs better than the standard check (usually a local variety) in some characteristics (e.g. productivity, resistance to diseases and product quality). If these criteria are met, the variety is officially released for use and is listed with its accompanying characteristics in the national register or catalogue of released varieties. It allows for the official production and marketing of seeds of the released variety.

The analysis shows that although a majority of the countries have a formal varietal release

Table 10.3. Differences in FTE scientists and research intensities between 1998 and 2010 by country.

	FTE scientists			Research intensities		
Country	1998	2010	Difference	1998	2010	Difference
Côte d'Ivoire	3.1	9.87	6.77	5.2	15.2	10.0
Ghana	5.5	4.75	–0.75	28.4	9.7	–18.7
Guinea	9.4	12.25	2.85	9.0	7.6	–1.4
Nigeria	5.3	9.87	4.57	1.6	3.1	1.5
Senegal	1.7	7.54	5.84	13.8	12.5	–1.3
Sierra Leone	6.3	12.87	6.57	19.2	14.2	–5.0
Sum or weighted average	31.3	57.15	4.31	5.6	7.6	2.0

Table 10.4. Current rice varietal release and registration practice in sub-Saharan African countries.

Country	Formal release system in place	Official release required before growing commercially	Varietal register in place	Authority that maintains the list of registered varieties	List in common with any other list	Varietal registration required for seed sale	Varietal registration required for certified seed sale	Varietal descriptor part of registration process
Benin	No	Yes	Yes	MOA	ECOWAS	Yes	Yes	Yes
Burkina Faso	Yes	Yes	Yes	MOA	ECOWAS	Yes	Yes	Yes
Cameroon	No	No		None	No	No	No	Yes
CAR	No	No	No	None	No	No	No	Yes
Côte d'Ivoire	Yes	Yes	Yes	MOA	ECOWAS	Yes	Yes	Yes
DR Congo	Yes	Yes	No	NSS	No	Yes	Yes	Yes
Ethiopia	Yes	Yes	Yes	MOA	No	Yes	Yes	Yes
The Gambia	No	Yes	Yes	NSC	ECOWAS	No	No	Yes
Ghana	Yes	No		MOA	No	No	Yes	Yes
Guinea	No	Yes	Yes	Each institute keeps their own register	ECOWAS	Yes	Yes	Yes
Kenya	Yes	Yes	Yes	KEPHIS	No	Yes	Yes	Yes
Mali	Yes	Yes	Yes	Seed Laboratory	No	Yes	Yes	Yes
Mozambique	Yes	Yes	Yes	MOA	No	Yes	No	Yes
Niger	No	Yes	Yes	DGA and INRAN	ECOWAS	Yes	Yes	Yes
Nigeria	Yes	Yes	Yes	NACGRAB	ECOWAS	No	No	Yes
Rwanda	No	No	Yes	ISAR	No	No	Yes	Yes
Senegal	Yes	Yes	Yes	MOA	ECOWAS			
Tanzania	Yes	Yes	Yes	MOA	No	Yes	Yes	Yes
Togo	No	No	No	ITRA	No	No	No	No
Uganda	Yes	Yes	Yes	NSC	EAC	Yes	Yes	Yes

Source: AfricaRice rice variety survey, Sanni *et al.* (2011). DGA, Directorate General for Agriculture; EAC, East African Community; ECOWAS, Economic Community of West African States; INRAN, Niger National Institute of Agricultural Research; ISAR, Rwandan Agricultural Research Institute; ITRA, Institut Togolaise de Recherche Agronomique; KEPHIS, Kenya Plant Health Inspectorate Services; MOA, Ministry of Agriculture; NACGRAB, National Centre for Genetic Resources & Biotechnology; NSC, National Seed Council; NSS, National Seed Service.

system, a large minority do not (Table 10.4). Official release is required in most countries before a variety can be grown commercially. The varietal release systems vary greatly among countries. Cameroon, CAR, Rwanda and Togo do not have a formal varietal release system in place; a variety can be produced and commercialized without any formal release. Ghana has a formal varietal release system but a variety can be commercialized without a formal release. Although The Gambia does not have a formal varietal release system, the official release of a variety is required before it can be grown commercially.

In countries with a formal varietal release system, a national variety release committee (NVRC) is established to perform the task of reviewing the description and performance of varieties nominated for release by both public and private breeders. The membership of the NVRCs mainly comprises representatives from national agricultural research systems (NARS), seed companies and universities. In most countries the national seed authority (NSA) is responsible for convening and chairing the NVRC meetings. On average, the NVRC meets once a year except in Uganda where it meets twice a year to evaluate and approve variety releases. There are no incentives given for NVRC membership, which is on a voluntary basis. Due to financial constraints, the NVRCs may not meet for several years in some countries.

The effectiveness of the NVRCs varies from country to country. Based on the degree of functioning of the varietal release systems, countries can be grouped into three categories. The first group with a functional varietal release system includes Burkina Faso, Kenya, Mali, Mozambique, Nigeria, Senegal, Tanzania and Uganda. The second group, comprising countries with a varietal release system that is either nonfunctional or ineffective, includes Benin, DR Congo, Ghana and Guinea. The third group contains countries that do not have a varietal release system: Cameroon, Chad, CAR, The Gambia, Rwanda and Togo.

With the exception of Cameroon, CAR and Togo, the surveyed countries have varietal registers of all the released varieties and their descriptors. These registers are maintained by either the Ministry of Agriculture or another body appointed by the state, except in Guinea where each national agricultural research institute keeps a register of its own released varieties. The lists of varieties in the varietal registers of most countries in West Africa have now been harmonized with the ECOWAS (Economic Community of West African States) crop catalogue.

The variety descriptor forms part of the registration process in most countries. Even countries without a crop register require the varieties to be documented somewhere. Although the varietal register exists in most countries, there is some variation in the consistency and accuracy of these registers. In some countries, the register is not regularly updated as new varieties are released.

To meet the minimum requirements for varietal release, rice breeding programmes routinely assemble breeding nurseries and test variety performance in national and regional variety multi-location trials with the objective of generating important agronomic data to identify the best rice varieties for release. Among the countries surveyed, Togo is the only one that does not require the descriptor of new varieties before they are released. The testing may take 1–3 years before the data are sufficient to be submitted to the NVRC.

The traits that are used for DUS are those that are not affected by the environment. The DUS tests are mostly conducted by national seed authorities (NSAs). The required test duration varies from one to three seasons, depending on the country. Most SSA countries have neither DUS nor VCU published guidelines. The lack of published guidelines creates a bottleneck for breeders and seed companies because the important traits that should be presented for variety release are not clear.

VCU is required for rice varietal release in all the surveyed countries, whereas DUS is only required in five countries. This could be because most African countries produce inbred rice and only a few large companies commercialize rice seeds. Only a few countries in SSA (Kenya, Uganda and Ethiopia) have a number of well-established seed companies in place. The VCU data are recorded on important agronomic traits such as grain yield, disease resistance and plant height. The number of traits for VCU varies from one country to another.

The recording of agronomic traits is time-consuming and only important ones should be collected. VCU test requirements range from two to three seasons of multi-location trials in most countries and the number of locations varies depending on the mega-environments for which the variety is being recommended. In most countries, the NSA is responsible for assembling and conducting national performance trials (NPTs) from which VCU data are obtained. Once the VCU data have been recorded, they are submitted to the NVRC for consideration. However, NPTs do not guarantee that the variety will be released once the trials are completed. In countries like Kenya and Tanzania, NPTs are conducted by the NSA for a given fee. The fees paid for the NPT may dissuade some breeders, especially those from the public sector, from submitting their rice varieties for release in those countries, thus delaying the release of a new variety. To complement the VCU data from NPTs, independent and on-farm trials are required. Some countries (e.g. Benin, Mali, Mozambique and Uganda) accept VCU data from other countries with similar agroecological zones to add to in-country data. Data collected from participatory varietal selection (PVS) trials are acceptable as credible data for varietal release in 11 countries, most of which are countries without a formal varietal release system. Amongst 11 countries that have adopted PVS as a source of data for varietal release, Mozambique, Nigeria and Senegal are the only ones with a formal varietal release system. The institutionalization of PVS has greatly helped to accelerate the rate of varietal release in Senegal in recent years.

The following constraints have been identified that hinder the smooth release of varieties, thereby creating bottlenecks in the system:

- Most NVRCs lack good coordination and do not hold regular meetings for the release of recommended new varieties thus delaying their release. In most countries, this is due to the funds not being released by the government for the exercise (the NVRCs are mostly based in the public sector).
- There is a lack of clear guidelines (unified protocols and monitoring system) for parallel external trials. Researchers may therefore use different strategies, making it difficult for the committee to compare the results.
- The varietal release process is costly because the same variety has to be tested each time it is to be released in another country, even if the agroecological characteristics of the country are similar to those of countries where the variety has been released earlier.
- The variety release procedure is cumbersome and duplicative and it delays the introduction of new rice varieties. Retesting in a similar growth environment in another country delays the time required for a new variety to get to farmers and seed companies. Even within the same country, the release process delays the registration of new varieties because of the number of seasons required to collect VCU and DUS data.

The historical record of varietal output

The historical record of release is presented for 11 of the most important rice-producing countries (Table 10.5). Each of these has a record of varietal release or confirmed varietal availability for adoption that dates from the 1970s or earlier. Seven of these countries belong to those studied in the 1998 Initiative (Dalton and Guei, 2003). Burkina Faso, Cameroon, Madagascar and Tanzania all have longer-term credible information on varietal output, although their release policies vary sharply from country to country.

In total, 454 rice varieties were released in these 11 countries between 1932 and 2009. One of the first varieties released in 1954 was BG 79 named as FARO 1 in Nigeria. Like many varieties released prior to 1980, BG 79 was a rice line selected in Asia, in this case Sri Lanka. Several of these earlier releases focused on the floating deep-water rice-growing environment and they mainly came from Asia. Of the 11 countries, Nigeria has the most consistent record of varietal release over time (Table 10.5). FARO 57 was released in 2005.

Although all countries in Table 10.5 have released more than ten varieties, Guinea with 124 varieties has by far the highest incidence of release over time. This output of released materials seems astounding, but it can be explained. Multiple institutions release varieties in Guinea, which has long benefited from bilateral assistance from North Korea in rice genetic improvement. And, as we have seen in the previous section, Guinea has more rice breeders than any other rice-producing country.

The incidence of varietal releases peaked in the 1980s and 1990s when two CGIAR institutions, IITA and WARDA (the West Africa Rice

Table 10.5. Number of releases by time period from 1954 to 2009 by country.

Country	Time periods: Prior to 1970	1970s	1980s	1990s	2000s	Undated	Total
Burkina Faso	1	5	9	13	6	0	34
Cameroon	0	8	8	4	9	4	33
Côte d'Ivoire	2	8	12	24	2	0	48
Ghana	0	3	7	1	4	0	15
Guinea	2	2	38	77	5	0	124
Madagascar	4	2	7	1	6	0	20
Mali	4	3	7	7	9	0	30
Nigeria	12	13	19	8	12	0	64
Senegal	0	4	6	12	23	0	45
Sierra Leone	1	5	17	2	0	4	29
Tanzania	1	0	3	1	7	0	12
Total	27	53	133	150	83	8	454

Development Association), were actively pursuing rice improvement and germplasm exchange in West and Central Africa. With the exception of Senegal where the aforementioned commitment to participatory varietal selection has resulted in increased output, country-specific output has either fallen or stagnated in the 2000s. Part of this drop in productivity is attributed to civil strife in the late 1990s and early 2000s in Côte d'Ivoire and Sierra Leone. Widespread civil disturbance adversely affected both national and international rice-breeding activities in West Africa. Part of this aggregate decline is also derived from Guinea's failure to maintain its phenomenal release rate of the 1980s and 1990s. Net of Guinea, total varieties released for the other ten countries also reached a high in the 1980s, but slightly more improved materials were produced in the 2000s than in the 1990s.

Several smaller rice-growing countries and the DR Congo and Uganda, which did not have many antecedents in varietal release prior to 1980, were prolific in releasing varieties since 2000. Notable among these countries were Benin with 11, DR Congo (14), The Gambia (8), Rwanda (24), Togo (8) and Uganda (10). These releases are important, but they do not compensate for the slowing down or stagnancy of the release rate in 10 of the 11 larger-producing countries described above.

In the 1998 Initiative, the decline in releases in the 2000s was not envisaged. From 2000 to 2004, a total of 122 rice varieties were targeted for release in Nigeria, Guinea, Côte d'Ivoire, Sierra Leone, Mali, Ghana and Senegal (Dalton and Guei, 2003). This total more than doubled the number released in any previous 5-year period since 1980. With the benefit of hindsight, only 13 varieties were released across the seven countries from 2000 to 2004. It is likely that some of the 109 unreleased, targeted-for-release cultivars were approved for cultivation after 2004. But this large discrepancy between forecast and actual releases suggests that predicting releases is a risky business that can create the illusion of and lead to over-optimism about impending varietal change.

Other aspects of varietal output

Information on the recommendation domain for the cultivar is available for 416 of the 454 released varieties. Across the 11 countries, rainfed lowland and rainfed upland are the dominant rice-growing environments each with a 40% share of total rice cultivated area. About one-third of the upland area has access to groundwater or supplementary irrigation. Lowland irrigated rice claims about 10% of area and the other 10% is divided between mangrove swamp and floating deep-water cultivation.

The relative importance of releases by targeted rice-growing environment is not congruent with these area allocations. The major deviation suggests an emphasis on irrigated lowland in variety release. About one-third of the 416 varieties were targeted to this rice-growing environment. This emphasis is expected because of greater varietal availability in Asia for irrigated rice cultivars. Irrigated regions also have higher production potential. Small irrigated areas, such as Chokwe in the South of Mozambique, are often characterized by heavy government subsidies that enhance farmers' lobbying power to more forcefully articulate their demands relative to farmers in larger but more marginal rice-growing environments.

Of the 416 released varieties, 175 were targeted for the upland rice-growing environment. This congruence between release and area shares is encouraging because the upland rice-growing environment was identified in the 1998 study as the major rice-growing environment lagging behind in adoption (Dalton and Guei, 2003). Relative to its share in area cultivated, fewer varieties (80 in total) have been released in the rainfed lowlands than for any other rice-growing environment. About 7% of released varieties target the mangrove and deep-water rice-growing environments but no varieties recommended for these rice-growing environments were released recently between 2000 and 2009.

About 45% of the 454 releases are related to materials from the three CG Centers that work on or have worked on rice. Most of these are elite finished varieties that were bred at the CG-Center research stations of AfricaRice, IITA and IRRI. CG-related germplasm and elite lines have figured more prominently in national programme releases over time. Their share has risen from 26% in the 1970s, to 35% in the 1980s, to 54% in the 1990s and to 77% in the 2000s.

The evidence for a transition in breeding signifying more applied research and less adaptive testing over time is still scanty in rice research. By the late 1990s, most countries did not maintain ex-situ germplasm banks in rice with sufficient parental lines to permit crossing and subsequent evaluation on an annual basis (Dalton and Guei, 2003). The emphasis on adaptive testing of elite introduced lines still prevails today in most of the 11 countries. The exceptions are Senegal and Madagascar. In particular, the recent flurry of release activity in Senegal is based on crossing parental lines from AfricaRice and subsequent in-country selection. Several recent releases in Madagascar were selected from progenies made from crosses with IRRI parental materials.

The recent absence of releases from in-country crosses and subsequent selection is more puzzling in Nigeria than in any of the other ten countries. In the 1970s and 1980s, several varieties were released via conventional breeding from both Nigerian and IARC-related parents in Nigeria. The dominance of introduced elite lines in recent varietal-release outcomes is unexpected in a large national programme with a steady record of varietal releases in a commodity whose output substitutes for imports.

Varietal Adoption

Largely because of nationally representative surveys funded by AfricaRice's Japan project, more is known about estimated adoption of improved varieties in rice in SSA than for any other crop in this volume. Estimates are available for 19 of the 36 rice-growing countries that cultivated more than 1000 hectares in 2009 in SSA. According to FAOSTAT, these 19 countries accounted for 90% of rice harvested area of 8.7 million hectares in 2009. Liberia, Mozambique and Chad, with areas between 100,000 and 300,000 hectares, were the largest omissions from the 19 included countries. The sources of the adoption estimates for modern varieties (MVs) are both from expert opinion estimates, for the first 12 countries in Table 10.6, and data from national representative surveys, for the last 7 countries. In the largest country, Madagascar, the adoption estimates were based on expert opinion. One of the adoption surveys was funded by the DIIVA Project in Nigeria in 2010 (Diagne *et al.*, 2013b). The other national surveys were carried out prior to the initiation of the DIIVA Project and are taken from the database known as AfricaRice's Rice Statistics Survey in 2009, which covered 19 countries (including Nigeria). The results from the Ghana survey are complemented by those from a recent International Food Policy Research Institute (IFPRI) survey on improved rice varieties in 2012 (Ragasa *et al.*, 2013). The adoption estimates of the seven last countries in Table 10.6 (Benin, Cote d'Ivoire, DR Congo, Mali, Sierra Leone, Tanzania and The Gambia) were obtained by summing the individual variety adoption

Table 10.6. Adoption of improved varieties of rice in sub-Saharan Africa, 2009.

Country	RGV/ National	Area	Share area MVs (%)	Country	RGV/ National	Area	Share area MVs (%)
Burkina Faso	Irrigated	21,216	100	Nigeria	Irrigated	32,974	73.6
	Upland s.	9,224	100		Upland w/si	59,655	82.8
	Lowland	61,803	75		Upland s.	464,589	40.3
	National	92,243	83.25		Upland w/gw	201,590	51.4
Cameroon	Irrigated	20,076	98.5		Lowland	185,926	48.8
	Upland s.	54,222	77		Mangrove	149,612	38.1
	Lowland	25,355	82.8		Other	107,001	87.19
	National	99,653	82.81		National	1,201,347	50.4
CAR	Irrigated	635	100	Rwanda	Irrigated	12,775	69
	Upland s.	15,000	70		National	12,775	69
	Lowland	334	95	Senegal	Irrigated	46,000	100
	National	15,969	71.72		Upland s.	1,500	77.5
Ghana	Irrigated	189,000	95		Upland w/gw	2,000	80
	Upland s.	71,000	60		Mangrove	1,500	70
	Lowland	92,000	70		Lowland	50,000	80
	National	352,000	81.41		National	101,000	88.92
Guinea	Upland w/gw	240,000	20	Togo	Irrigated	10,471	100
	Upland s.	232,000	10		Upland s.	3,782	40
	Dry plain	33,000	20		Lowland	22,239	70
	Mangrove	152,000	10		National	36,492	75.5
	Lowland	55,800	20	Uganda	Irrigated	5,000	80
	National	712,800	14.61		Upland s.	50,000	90
Kenya	Irrigated	78	82.86		Upland w/gw	25,000	75
	Upland s.	875	89.5		Lowland	40,000	80
	Lowland	113	97.5		National	120,000	83.13
	National	1,066	89.86	Benin	National	38,700	83
Madagascar	Irrigated	100,000	60	Côte d'Ivoire	National	569,000	33
	Upland s.	200,000	50	DR Congo	National	482,400	28
	Upland w/gw	5,000	90	Mali	National	646,100	25
	Lowland	1,095,000	30	Sierra Leone	National	434,200	16
	National	1,400,000	35.21	Tanzania	National	627,600	11
				The Gambia	National	73,000	17

Upland: s. = strict; w/gw = with groundwater; w/si = with supplemental irrigation; RGV = rice-growing environment.

estimates in Appendix 2 in Diagne *et.al.* (2013a). But that summation is likely to underestimate the share of modern varieties because they did not include MVs that were classified as unknown or unidentified varieties, which occupied 14%, 7%, 1%, 53%, 58%, 10% and 23% of the rice areas for these countries, respectively.

National and agroecological estimates

For 12 of the 19 countries, adoption estimates are available by rice-growing environment. For the other seven countries, adoption estimates are available nationally.

The national and agroecological adoption estimates for improved rice varieties are presented in Table 10.6. The area-weighted mean national adoption level of modern varietal adoption is 37%. Nine of the 19 countries had more than half of their rice area planted to modern varieties in 2009 but eight of these higher-adopting countries cultivated less than 125,000 hectares. Nigeria was the only larger-producing country with modern variety adoption exceeding 50% of cultivated area. In the larger-producing countries of Tanzania, Sierra Leone and Guinea, less than one hectare in four was sown to a modern variety.

The finding that adoption of modern varieties is higher in smaller than in larger-producing countries could be attributed to the selection of higher adoption outcomes among smaller countries in the conduction of nationally representative surveys. Of the 17 other small countries, several may not have been sampled because it was known that well-adapted modern varieties were not available to farmers. Alternatively, negligible adoption, as in the case of Mozambique, where modern varieties are confined to a heavily subsidized small irrigation district in the south, was a factor in lack of performance in carrying out the nationally representative survey. Nevertheless, this potential selection bias does not unduly influence the aggregate 35% adoption estimate. Moreover, a few low-adopting, small rice-producing countries such as the Central African Republic and The Gambia are represented in Table 10.6.

Across the 12 countries with detailed subregional information in Table 10.6, the mean area-weighted level of modern variety adoption was 46%. In terms of their importance to rice-growing area, the rice-growing environments are ordered as follows: (i) lowland rainfed (39%); (ii) upland (27%); (iii) upland with groundwater (13%); (iv) irrigated (11%); (v) mangrove (7%); and (vi) other (3%). In terms of their uptake of modern varieties, the described rice-growing environments are ranked: (i) irrigated (85%); (ii) upland with or without groundwater potential (42%); (iii) rainfed lowland (40%); and (iv) mangrove (24%).

Eleven of the 12 countries contain production tracts characteristic of the irrigated subregion. In all these countries, the level of modern variety adoption exceeds 60% of cultivated area in this higher-production-potential rice-growing environment. Several countries are approaching or have arrived at full adoption.

Farmers in most of these countries also cultivate rice in the rainfed lowlands and in the uplands. A few countries have achieved moderately high levels of adoption of modern varieties across the irrigated, lowland rainfed, and upland production tracts. But most countries are characterized by one or two lagging agroecological subregions in adoption. And, within the lowlands and uplands, there is one dominant lagging country in modern variety adoption. For the lowlands, that country is Madagascar, by far the most extensive subnational rice-growing environment in Table 10.6. Experts assigned a 30% estimate for the uptake of modern varieties in this rice-growing environment in this key subregion of the largest-producing country in SSA. For the uplands, that country is Guinea where the survey estimate for adoption of MVs was in the range of 15–20% in an area of about 475,000 hectares.

Modern varieties cultivated in 2009–2010

A quick reading of the variety-specific estimates in Table 10.7 reveals that all surveyed countries contain one or more aggregated categories where the variety in question could be credibly called improved but its identity among improved varieties could not be verified. In the extreme case of Guinea, only three aggregate improved groupings could be tabulated. Because of the difficult and persuasive problem of accurate identity

Table 10.7. Economically important improved rice varieties in sub-Saharan Africa, 2009–2010, by national area.

Country	Improved variety	Share of area (%)	Country	Improved variety	Share of area (%)
Burkina Faso	FKR 19/TOX 728-1	27.15	Burkina Faso	FKR 29/1215-1-5	0.59
Burkina Faso	TCS 10	23.47	Burkina Faso	FKR 16/4456	0.5
Burkina Faso	FKR 62N	3.25	Burkina Faso	FKR 18/SC 27	0.37
Burkina Faso	FKR 8/IR 8	2.77	Burkina Faso	FKR 3/IRAT 10	0.12
Burkina Faso	FKR 28/ITA 123	2.16	Burkina Faso	FKR 17/FKR 17	0.12
Burkina Faso	FKR 60N	1.94	Burkina Faso	FKR 33/1195-5-2	0.12
Burkina Faso	FKR 56 N	1.76	Burkina Faso	FKR 45 N	0.08
Burkina Faso	FKR 41/WAB 56-125	1.73	Burkina Faso	FKR 44/IR 13240-108-2-2-3	0.06
Burkina Faso	FKR 14/4418	1.41	Burkina Faso	FKR 49 N	0.04
Burkina Faso	FKR 43/CNA 6675	1.11	Burkina Faso	FKR 9/FKR 9	0.03
Cameroon	IR 46	6.75	Cameroon	SEBOTA 1141	1.17
Cameroon	TOX 3145-34-3-2	5.81	Cameroon	IR 20	0.84
Cameroon	B1 285	3.4	Cameroon	M2	0.84
Cameroon	CICA 8	3.32	Cameroon	SEBOTA 36	0.79
Cameroon	NERICA 1	3.02	Cameroon	I 5	0.75
Cameroon	NERICA 2	2.74	Cameroon	NERICA 60	0.68
Cameroon	ITA 300	2.63	Cameroon	WAB 189 HB	0.58
Cameroon	Tainain 5	2.43	Cameroon	ITA 306	0.39
Cameroon	MBANGA KERRI	2.42	Cameroon	PRIMAVERA	0.36
Cameroon	SEBOTA 33	2.41	Cameroon	WAB 35	0.34
Cameroon	VARIETY 14	2.21	Cameroon	NERICA 56	0.31
Cameroon	CHINO S	1.66	Cameroon	Taï chou	0.31
Cameroon	B 22	1.47	Cameroon	ITA 222	0.29
Cameroon	BKN 7033	1.27	Cameroon	FARO 49	0.26
Cameroon	IRAT112(FY 150)	1.24	Cameroon	ITA 312	0.19
Cameroon	YARK	1.21			
Côte d'Ivoire	Chinois/TS2	11.14	Côte d'Ivoire	IR5	0.21
Côte d'Ivoire	Bouake 189	5.98	Côte d'Ivoire	Adrao	0.19
Côte d'Ivoire	Gambiaca	2.71	Côte d'Ivoire	SORO	0.17
Côte d'Ivoire	WITA 9	1.53	Côte d'Ivoire	ANADER	0.17
Côte d'Ivoire	Akadi	1.45	Côte d'Ivoire	DJABATE	0.12
Côte d'Ivoire	NERICA (others)	1.37	Côte d'Ivoire	WAB 638-1/AKADI/DR 2/DIALLO	0.12

Continued

Table 10.7. Continued.

Country	Improved variety	Share of area (%)	Country	Improved variety	Share of area (%)
Côte d'Ivoire	Sanogo	1.32	Côte d'Ivoire	WAB	0.12
Côte d'Ivoire	NERICA 2	1.03	Côte d'Ivoire	MOSSI	0.09
Côte d'Ivoire	WITA 4	0.78	Côte d'Ivoire	RIZ NOUVEAU	0.08
Côte d'Ivoire	WITA (others)	0.58	Côte d'Ivoire	IDSA 6	0.08
Côte d'Ivoire	Soberdjo	0.54	Côte d'Ivoire	IR 3	0.06
Côte d'Ivoire	IR4	0.52	Côte d'Ivoire	WITA 12	0.05
Côte d'Ivoire	Gnakrouba	0.5	Côte d'Ivoire	MASSANDJE	0.04
Côte d'Ivoire	IR 32-237/Kaoulaka	0.5	Côte d'Ivoire	THAILLANDAIS	0.04
Côte d'Ivoire	Guidibo	0.45	Côte d'Ivoire	HARICOT	0.02
Côte d'Ivoire	3 Mois	0.27	Côte d'Ivoire	WITA 6	0.02
Côte d'Ivoire	2 Mois	0.27	Côte d'Ivoire	ITA	0.02
Côte d'Ivoire	NERICA 1/Bon Fani	0.26			
DR Congo	SIPI 4	8.78	DR Congo	JASMINE	0.68
DR Congo	IRAT13(RY 7)	6.05	DR Congo	IR 5	0.53
DR Congo	R66	2.26	DR Congo	PNR-1	0.51
DR Congo	HUBEI-6	1.62	DR Congo	IRAT112(RY 150)	0.49
DR Congo	R5	1.61	DR Congo	NERICA 7	0.33
DR Congo	IR-8	1.6	DR Congo	PEKIN 725	0.21
DR Congo	IRAT-233	1.44	DR Congo	NERICA 6	0.14
DR Congo	NERICA-4	1	DR Congo	LIENGE	0.05
DR Congo	ANDARO-NERICA	0.87			
Ghana	JASMINE 85/Gbewaa/Lapaz	27.05	Ghana	IR20	0.62
Ghana	Togo Marshall	10.76	Ghana	NERICA 1	0.59
Ghana	Jet 3	4.41	Ghana	WITA 7	0.52
Ghana	Agric (cannot be named)	4.3	Ghana	FARO 15	0.39
Ghana	Digang (also called Abirikukuo or Aberikukugu)	2.73	Ghana	Agric-Perfume	0.31
Ghana	Aromatic Short	1.95	Ghana	Ashiaman Perfume (eit her JASMINE 85 or Togo Marshall)	0.19
Ghana	GR 18 (Afife)	1.67	Ghana	Bodia	0.16
Ghana	TOX 3107	0.84	Ghana	NERICA 14	0.1
Ghana	GR 21	0.65	Ghana	NERICA 9	0.06
Ghana	Sikamu/TOX 3108	0.62			

Madagascar	X265	8.21	Madagascar	FOFIFA 161	1.35
Madagascar	2067	5.85	Madagascar	NERICA 4	1.35
Madagascar	FOFIFA 160	3.6	Madagascar	SEBOTA 70	1.35
Madagascar	3737	2.25	Madagascar	X360	1.12
Madagascar	NDR80	2.25	Madagascar	X398	1.12
Madagascar	X243	1.8	Madagascar	1632	0.9
Madagascar	B22	1.35	Madagascar	2787	0.9
Madagascar	FOFIFA 159	1.35	Madagascar	ON333	0.45
Mali	Gambiaka	9.73	Mali	Jama Jigi (Leizong 52)	0.21
Mali	DM16	3.39	Mali	Wat 310 (Sambala malo)	0.17
Mali	BG90-2	2.16	Mali	BR4	0.08
Mali	Kogoni 91-1 (Gambiaka suroni)	1.84	Mali	Nerica 4 (Dususuma malo)	0.05
Mali	AD 9216	1.03	Mali	WAB189.B.B.B.8.HB (Kumabani)	0.02
Mali	Khao Dawk Mali 105	0.92	Mali	IRAT63	0.013
Mali	RPKN 2 (Téliman)	0.9	Mali	WAB181-18 (Kikasoka)	0.003
Mali	NERICA [illegible]	0.58	Mali	Sahélika (ECIA)	0.001
Mali	IR32307-107-3-2-2 (Wasa)	0.45			
Nigeria	FARO 44 (SIPI 4)	10.67	Nigeria	FARO 50	0.18
Nigeria	FARO 15	6.88	Nigeria	FARO 54 (WAB189)	0.17
Nigeria	FARO 46 (ITA150)	6.14	Nigeria	FARO 57 (TOX400)	0.15
Nigeria	EX CHINA	4.85	Nigeria	CAROLINA	0.14
Nigeria	OTHER IMPROVED	4.33	Nigeria	FARO 35 (ITA212)	0.14
Nigeria	FARO 52 (WITA4)	3.84	Nigeria	FARO 45 (ITA257)	0.14
Nigeria	GARUWAYE	3	Nigeria	WILLY RICE	0.11
Nigeria	FARO 55 (NERICA1)	2.91	Nigeria	FARO 13	0.1
Nigeria	MAI ADA	1.81	Nigeria	UTSIYA	0.05
Nigeria	OTHER NERICA	1.44	Nigeria	JIRI JIRI	0.04
Nigeria	FARO 37 (ITA306)	1.38	Nigeria	FARO 49	0.04
Nigeria	ROK	0.72	Nigeria	SUA KOKO 8	0.03
Nigeria	FARO 29 (BG90-2)	0.49	Nigeria	FARO 21	0.02
Nigeria	FARO 23 (IR5)	0.35	Nigeria	FARO 27	0.01
Nigeria	FARO 56 (/NERICA2)	0.3	Nigeria	FARO 14	0.01
Senegal	Sahel 108	22.06	Senegal	Ousmane	0.33
Senegal	Sahel 202	7.77	Senegal	Wankaro	0.32
Senegal	Sahel 201	4.94	Senegal	NERICA 1	0.3

Continued

Table 10.7. Continued.

Country	Improved variety	Share of area (%)	Country	Improved variety	Share of area (%)
Senegal	Chinois	0.74	Senegal	Opa	0.24
Senegal	IR 1529	0.48	Senegal	Niada	0.23
Senegal	Diamarang	0.41	Senegal	Mansa Mno	0.22
Senegal	Bg-90-2	0.39	Senegal	Other farmer responses believed to be MVs	50.18
Senegal	Satin	0.39			
Sierra Leone	IMPROVED ADRAO NERICA	3.69	Sierra Leone	CP 4	0.24
Sierra Leone	ROK 3	2.86	Sierra Leone	ROK 22	0.23
Sierra Leone	ROK 5	2.13	Sierra Leone	NERICA 3	0.22
Sierra Leone	ROK 10	1.74	Sierra Leone	ROK 2	0.15
Sierra Leone	IR 8	1.41	Sierra Leone	CHANGAI	0.1
Sierra Leone	IMPROVED NARS	0.71	Sierra Leone	ROK 4	0.08
Sierra Leone	CHINESE 501	0.7	Sierra Leone	BANGLADESH	0.01
Sierra Leone	IVS	0.64	Sierra Leone	ROK 1	0.004
Sierra Leone	IR 841	0.62	Sierra Leone	NERICA 6	0.001
Sierra Leone	GP 9	0.62			
Tanzania	IR 64	2.7	Tanzania	TXD 85	0.4
Tanzania	TXD 306/SARO 5	2.52	Tanzania	SHIKALI	0.38
Tanzania	CHINA	1.29	Tanzania	IR 36	0.27
Tanzania	NERICA	1.24	Tanzania	CHAMOTO	0.05
Tanzania	TXD 220	0.95	Tanzania	TXD 88	0.03
Tanzania	IR 54	0.83	Tanzania	IR 56	0.02
Tanzania	MBAWA 2	0.43			
The Gambia	NERICA 3	11.11	The Gambia	ROK 5	0.39
The Gambia	PEKING	2.02	The Gambia	FKR 19	0.25
The Gambia	NERICA 4	1.9	The Gambia	JASMINE 85	0.16
The Gambia	NERICA 1	0.85	The Gambia	TAIWAN	0.04
The Gambia	NERICA 6	0.42			
Uganda	NERICA 4/NARIC 3/ SUPERICA 2	32.07	Uganda	NERICA 10	0.95
Uganda	NERICA 1	6.42	Uganda	NERICA 2	0.78
Uganda	WAP	1.64	Uganda	NERICA LOWLAND	0.08

recognition, estimates of specific improved varieties are likely to be understated in Table 10.7. However, even if identity attribution could be resolved, its resolution would not change the general finding in Table 10.7 that few improved rice varieties are dominant in most countries. Adoption of many improved varieties at a low rate of uptake seems to be the norm prevailing in most countries. Nigeria epitomizes a diversified portfolio of adopted varieties where 13 released FARO varieties have positive adoption outcomes but they each contribute less than 1% to rice cultivated area (Table 10.7). Collectively, these varieties account for only slightly over 2% of rice area.

In spite of the absence of dominant varieties that have large spill-over effects across two or more countries, the leading varieties in Table 10.7 warrant a brief description in the larger-producing countries. FARO 44, known in West Africa as SIPI 4, is an early-maturing, semi-dwarf variety released in Nigeria in 1996 for the irrigated lowland production tracts. SIPI 4 is also the dominant cultivar in the DR Congo where it is planted on about 9% of cultivated area. FARO 15 is a selection from a cross made by NCRI using FARO 1 (BG 79) and IR-8 as parents. This medium-duration variety was released in 1974 and is suited to irrigated lowland production. FARO 46 was crossed and selected by IITA. It is a short-statured, early-maturing variety that is easy to thresh and appropriate for upland production. Ex China as its name implies was introduced from Asia and made available for upland cultivation in 1988. NERICA varieties are also finding acceptance in Nigeria. Taken as a group, they occupied about 8% of cultivated area in the 2009 survey of more than 10,500 rice-growing households.

JASMINE 85, an aromatic rice variety prized for its desirable market traits, was officially released in Ghana in the wake of its popularity among rice producers and consumers. Jasmine 85 originated in Thailand where it was bred by IRRI in the mid-1960s. It has been released in the USA where it has been the subject of property-rights lawsuits.

X265, called Mailaka, is an IRRI bred and NARS (FOFIFA) selected variety that is targeted for the huge rainfed lowland growing region of Madagascar. It was released in 1986. The second leading variety, 2067, is considerably older, released in 1970. It was bred by IRAT and selected by FOFIFA for the irrigated lowlands.

Three leading rice varieties in Senegal were released in 1994 for irrigated cultivation. Sahel 108 (IR 13240) is an introduction from IRRI, Sahel 201 (BW 293) was introduced from Sri Lanka, and Sahel 202 (ITA 306) was bred and selected at IITA.

In Côte d'Ivoire, Asian rice varieties that are not officially released and originate mainly in Taiwan are leading the group of MVs. Bouake 189 is a popular lowland rice variety, bred in Côte d'Ivoire, and released in Mali in 1981. Gambiaka comes from Madagascar and still occupies more area than any other released rice cultivar in Mali where it is produced in the deep-water rice-growing environment.

Two varieties break an adopted-area threshold of 5% in Cameroon. IR 46 was introduced from IRRI and released in 1973. Tox 3145-34 3-2 was released even earlier in 1971. It is an elite line from IITA targeted for lowland rice-growing environments.

Tanzania lags behind in adoption of improved varieties. Two of the three leading improved cultivars that only claim a small share of area were introduced from Asia. SARO 5 is a promising semi-aromatic variety that was bred in Tanzania at ARI. It was released in 2003.

Compared to progress in the late 1990s, Sierra Leone has lost considerable ground in the adoption of improved varieties. Nonetheless, ROK 5 and ROK 10 are still important varieties in the mangrove rice-growing environment for which they were recommended in 1978. These varieties were bred and selected in Sierra Leone. ROK 5 has also been released in Guinea and Senegal. The survey data also suggest that NERICA varieties are beginning to penetrate into the rice-growing regions of Sierra Leone. However, Pa Kiamp, which is not listed as a modern variety in Table 10.7, is now diffusing more rapidly than any other variety in Sierra Leone. The origins of Pa Kiamp are uncertain, but it is believed to be from Guinea and is attributed to a farmer selection (Richards, 2012). It seems especially well adapted to in-land valleys (Spencer, 2010).

Among rice-growing countries cultivating more than 100,000 hectares, the NERICA varieties are most widely adopted in Uganda where they account for about 40% of area. NERICA 4 is the dominant variety in the uplands where it has been referred to as a new crop.

Burkina Faso is the only country in Table 10.7 where the leading two MVs contribute to more than 50% of area. Tox 728-1 is an IITA selected line that was released in Burkina Faso in 1986; TSC 10 is a more recent introduction from Taiwan. It was released in 2002 for irrigated cultivation.

Progress in MV adoption since 1998

Comparing the national adoption results in the 1998 Initiative to those in Table 10.7 for the same countries suggests that not much progress has been made in MV uptake in the recent past. In 1998, the area-weighted MV adoption level for rice was estimated at 47% for the seven study countries. In 2010, the comparable adoption estimate for the same countries was 37%. Gains in Senegal and Côte d'Ivoire did not offset losses in the adoption level in the other five countries. Reduced adoption is easy to explain in Sierra Leone, which was wracked by civil war in the recent past.

Decreased adoption between the two periods is harder to explain for Ghana, Guinea, Mali and Nigeria. The most likely explanation is that the source of the adoption estimates changed between the two periods. In the late 1990s, all of the adoption estimates were elicited from expert panels. In 2010, the adoption estimates came from nationally representative surveys. It is likely that adoption levels were overestimated in 1998 with expert opinion because disadoption of MVs has not been noted in the literature. Adoption levels were probably severely overestimated in Sierra Leone, which had been in the grip of a civil war beginning in 1991. (Civil war is not conducive to crop production and is even less amenable to arriving at ground truth on adoption in an abnormal time of crisis.) Moreover, two nationally representative adoption surveys in Ghana – one carried out by AfricaRice and one implemented by IFPRI – arrive at an aggregate adoption level for MVs that is less than the estimate derived from the information presented in Dalton and Guei (2003). Therefore, there are good reasons to speculate that the mean level of MV was less than 37% in 1998 across the seven countries in the 1998 Initiative.

Varietal age

Many of the important economic varieties were released in the 1970s and 1980s suggesting that varietal turnover has not been rapid in rice in many countries. Indeed, in the 1998 Initiative, rice was characterized as the crop having the slowest rate of improved varietal replacement among those in the study. Area-weighted mean varietal age exceeded 20 years.

The results in 2009 indicate that the pace of turnover of new varieties is still slow in rice vis-à-vis other crops. Area-weighted mean varietal age still averages about 21 years across the 12 large-producing countries that were the subject of adoption analysis in this section. Varietal age ranged from a low of 4 years in Ghana where JASMINE-85 was released officially in 2009 after rice farmers had access to it for many years to 28 years in Cameroon where first-generation, semi-dwarf varieties are still very much in evidence. The larger-producing nations of Madagascar and Nigeria are still characterized by slow rates of improved varietal turnover. Estimates of weighted average age centre on 22 years in both countries.

Summary and Conclusions

Scientific capacity in rice improvement scores well on a number of aspects. In 16 countries that accounted for 75% of production in SSA in 2010, a total of 289 scientists were carrying out research in rice improvement. Most of the 16 programmes were staffed with at least ten scientists. The programmes featured a well-diversified portfolio of disciplines and featured strength in soil and water management, social science, and weed science in several countries. On average, the scientists had 12 years of experience in rice research. More than half had an MSc degree and over one-quarter held PhDs. The 289 total scientists translated into 123 FTE scientists. Thirty-five of these FTE scientists held PhD degrees. The weighted average estimate for research intensity for rice improvement was 9.0 FTE scientists per million tonnes of production. This level of research intensity places rice in roughly the same position as maize in West and Central Africa.

Scientific capacity also seems to be rising over time. Between 1998 and 2010, scientific capacity in five of the six countries, where pairwise comparisons could be carried out, increased substantially. On aggregate, the increase was 83% of the base level in 1998.

Several areas for strengthening scientific capacity were also highlighted. Only about one-quarter of rice scientists are geneticists or plant breeders. Plant breeding seems to be poorly represented in several major-producing countries in West Africa. For example, Nigeria has less than two FTE plant breeders. Guinea was the only country with more than five FTE plant breeders. The low allocation in molecular biology was also unanticipated both because investments in this area are very small and because they are not concentrated in the major-producing countries of Madagascar and Nigeria. The advanced age of many rice scientists in West African programmes was also a cause for concern, as was the low numbers of younger BSc scientists in several programmes to serve as potential replacements when their mentors retire. To ensure the sustainability of rice improvement research, it will be necessary for these countries to recruit young scientists, as well as to increase the proportion of women scientists who presently represent only 15% of total scientific capacity.

Enhanced scientific capacity has not translated into increased varietal releases in several of the larger rice-producing countries. Focusing on 11 countries that cultivated more than 100,000 hectares of rice and that had a release record in the 1960s and 1970s shows that varietal output in the form of releases peaked in the 1980s and 1990s. Many varieties that were targeted for release shortly after the 1998 Initiative in 2000–2004 were not released during that period. Part of this drop in productivity is attributed to civil strife in the late 1990s and early 2000s in Côte d'Ivoire and Sierra Leone. Widespread civil disturbance adversely affected both national and international rice breeding activities in West Africa. Part of this aggregate decline is also derived from Guinea's failure to maintain its phenomenal release rate of the 1980s and 1990s.

In contrast to these larger producers, several small-producing countries have been prolific in releasing modern varieties since 2000. And not all large producers are characterized by a stagnating or declining record of varietal output. Senegal with a dynamic breeding programme featuring an emphasis on participatory varietal selection has officially released more than 20 improved cultivars for public consumption since 2000.

Historically, relative to its area allocation of about 10%, a disproportionately large share (about one-third) of releases was targeted at the irrigated lowlands. On a more encouraging note, the difficult and complex upland rice-growing environment has not been neglected from the perspective of varietal output. Of the 416 released varieties 175 were aimed at the upland rice-growing environment, which accounts for about 40% of rice area. Relative to its share in area cultivated, fewer varieties (80 in total) have been released in the rainfed lowlands than for any other rice-growing environment. About 7% of released varieties target the mangrove and deep-water rice growing environments but no varieties recommended for these rice-growing environments were released between 2000 and 2009.

About 45% of the releases are related to materials from the three CG Centers that work on or have worked on rice. Most of these are elite finished varieties that were bred at the CG-Center research stations of AfricaRice, IITA and IRRI. CG-related germplasm and elite lines have figured more prominently in national programme releases over time. Their share has risen from 26% in the 1970s, to 35% in the 1980s, to 54% in the 1990s and to 77% in the 2000s.

There is not much evidence in the release database to support the hypothesis that national programmes increasingly release varieties selected from crosses that they have made or segregating materials that they have received. Most releases still come from elite lines bred and selected outside the country of release. On a more positive note, the release of purified landraces that was common in the 1960s and 1970s is increasingly rare.

The assembly of the release database was a daunting challenge in rice improvement in SSA, so much so that AfricaRice invested in a process-based study in 25 rice-growing countries to describe release procedures in an effort to subsequently make recommendations for their improvement. Release laws, practices, and registries and their application varied markedly from country to country. Despite this variation, a common conclusion was appropriate for most countries: present varietal release systems restrict the flow of improved varieties to rice-farming communities. Less restrictive procedures are called for. Clear guidelines for these varietal release systems are also needed in all countries to ease the work of breeders and seed companies. The national release committee may also be involved in all stages of the varietal selection process in their country (participatory varietal selection for instance) in order to facilitate varietal release at end of the process.

Estimates of adoption of improved rice varieties were available for 19 of the 36 rice-growing countries that cultivated more than 1000 hectares in 2009 in SSA. These 19 countries accounted for 90% of rice-harvested area of 8.7 million hectares in 2009. Nationally representative survey data were the source of adoption estimates in 18 of the 19 countries.

The area-weighted mean national adoption level of modern varietal adoption is 37%. Nine of the 19 countries had more than half of their rice area planted to modern varieties in 2009 but eight of these higher-adopting countries cultivated less than 125,000 hectares. Nigeria was the only large-producing country with modern variety adoption exceeding 50% of cultivated area. In Tanzania, Sierra Leone and Guinea, less than one hectare in four was sown to a modern variety.

Across a subset of 12 countries with detailed subnational information, the mean area-weighted level of modern variety adoption was 46%. In terms of their importance to rice-growing area, the rice-growing environments were ordered as follows: (i) lowland rainfed (39%); (ii) upland (27%); (iii) upland with groundwater (13%); (iv) irrigated (11%); (v) mangrove (7%); and (vi) other (3%). In terms of their uptake of modern varieties, the main rice-growing environments were ranked: (i) irrigated (85%); (ii) upland with or without groundwater potential (42%); (iii) rainfed lowland (40%); and (iv) mangrove (24%). For many countries, the irrigated lowlands are now nearing or at full adoption.

Most countries are characterized by one or two lagging agroecological subregions in adoption. And, within the lowlands and uplands, there is one dominant country lagging in modern variety adoption. For the lowlands, that country is Madagascar, by far the most extensive subnational rice-growing environment in SSA with more than 1 million hectares. Experts assigned a 30% estimate for the uptake of modern varieties in this key subregion. For the uplands, that country is Guinea where the survey estimate for adoption of MVs was in the range of 15–20% on an area of about 475,000 hectares.

All surveyed countries contained one or more aggregated categories where the variety in question could be credibly called improved but its identity among improved varieties could not be verified. Because of the difficult and pervasive problem of accurate identity recognition, estimates of specific improved varieties are likely to be understated in this study. However, few if any improved rice varieties are dominant in most countries. Adoption of many improved varieties at a low rate of uptake seems to be the norm prevailing in many countries.

Comparing the national adoption results in the 1998 Initiative to those for the same countries suggests that not much progress has been made in MV uptake in the recent past. In 1998, the area-weighted MV adoption level for rice was estimated at 47% for the seven study countries. In 2010, the comparable adoption estimate for the same countries was 37%. Gains in Senegal and Côte d'Ivoire did not offset losses in the adoption level in the other five countries. However, this comparison is not valid because the expert opinion panels were the source of adoption estimates for 1998; in 2010, the source changed to nationally representative surveys. It is likely that surveys would have resulted in a lower estimate of adoption in 1998 in all countries, especially Sierra Leone, which was in a civil war in 1991–2001.

The recent adoption evidence suggests that newer improved cultivars are not rapidly replacing earlier first- and second-generation modern varieties. Estimates of weighted average varietal age exceed 20 years indicating that representative improved varieties found in farmers' fields today were released more than 2 decades ago. The pace of varietal turnover was the same in 2010 as it was in 1998. Even with the impressive dissemination of the NERICA varieties and other recent improved cultivars, varietal age is not declining in most of the larger-producing countries. More investment and efforts are warranted to exploit the potential of rice genetic improvement to accelerate the pace of varietal change in a crop that is unconstrained by demand in SSA.

Note

[1] This paper is a revised and abridged version of Diagne *et al.*, 2013a.

References

Adesina, A. and Zinnah, M.M. (1993) Technology characteristics, farmer's perceptions, and adoption decisions: A Tobit model application in Sierra Leone. *Agricultural Economics* 9, 297–311.

AfricaRice (2010) *Rice Data Systems for Sub-Saharan Africa: Contribution to the Japan-AfricaRice Emergency Rice Project*. AfricaRice, Cotonou, Benin.

Dalrymple, D. (1978) *The Development and Spread of the High-Yielding Varieties of Wheat and Rice among Less-Developed Nations.* 6th edn. Foreign Agricultural Economic Report No. 95. USDA/USAID, Washington, DC.

Dalton, T.J. and Guei, R.G. (2003) Ecological diversity and rice varietal improvement in West Africa. In: Evenson, R.E. and Gollin, D. (eds) *Crop Varietal Improvement and Its Effect on Productivity: The Impact of International Agricultural Research*. CAB International, Wallingford, UK, pp. 109–134.

Diagne, A., Kinkingninhoun-Medagbe, F.M., Amovin-Assagba, E., Sanni, K. and Toure, A. (2013a) *To Attain a Wider Understanding of Key Aspects of the Performance of Food-Crop Genetic Improvement in Priority Country-by-Commodity Combinations in Sub-Saharan Africa – The Case of Rice Crop.* Diffusion and Impact of Improved Varieties in Africa (DIIVA), Objective 1 Report. AfricaRice Center, Cotonou, Benin.

Diagne, A., Kinkingninhoun-Medagbe, F.M., Ojehomon, V.T., Abedayo, O.O., Amovin-Assagba, E. and Nakelse, T. (2013b) *Assessing the Diffusion and Adoption of Improved Rice Varieties in Nigeria*. Diffusion and Impact of Improved Varieties in Africa (DIIVA), Objective 2 Report. AfricaRice Center, Cotonou, Benin.

International Rice Research Institute (IRRI), AfricaRice and International Center for Tropical Agriculture (CIAT) (2010) *Global Rice Science Partnership (GRiSP)*. CGIAR Thematic area 3: sustainable crop productivity increase for global food security. A proposal for a CGIAR research program on rice-based production systems. IRRI, AfricaRice and CIAT. IRRI, Los Banos, Philippines.

Ragasa, C., Dankyi, A., Acheampong, P., Nimo Wiredu, A., Tripp, R. *et al.* (2013) *Patterns of Adoption of Improved Rice Technologies in Ghana*. DRAFT REPORT IFPRI: Washington, DC.

Richards, P. (2012) Comment to SaloneFAB members, 2/1/12. https://groups.google.com/forum/#!msg/salonefab/fGo_sYsZSE0/d1TJVhheP_wJ (Accessed 7 May 2015).

Sanni, K., Toure, A. and Diagne, A. (2011) *Variety Release and Registration Practices for Rice in African Countries.* Study Report. AfricaRice, Cotonou, Benin.

Spencer, D.S.C. (2010) *Farmer Assessment of Rice Varieties in Northern Sierra Leone.* Enterprise Development Service Ltd, Freetown, Sierra Leone.

11 Assessing the Effectiveness of Maize and Wheat Improvement from the Perspectives of Varietal Output and Adoption in East and Southern Africa

H. De Groote,[1]* Z. Gitonga,[1] S. Mugo[1] and T.S. Walker[2]
[1]*International Maize and Wheat Improvement Center (CIMMYT), Nairobi, Kenya;*
[2]*Independent Researcher, Fletcher, North Carolina, USA*

Introduction[1]

Throughout much of East and Southern Africa (ESA), maize is the staple food crop. Excluding South Africa, domestic wheat production in ESA only looms large in Ethiopia where wheat has been cultivated since ancient times. Ethiopia is a secondary centre of origin for bread wheat and is the centre of diversity for durum wheat, which is used to make pasta.

In the 20th century, maize and wheat improvement have followed markedly different paths in their quest for varietal change. Building on Norman Borlaug's research at the Rockefeller Foundation, the Centro International de Mejoramiento de Maiz y Trigo (CIMMYT), since its establishment in 1968, adopted a centralized breeding approach to wheat improvement (Lynam, 2010). That approach, featuring crossing and pedigree breeding at its headquarters in Mexico and subsequent distribution of elite lines for testing, has worked exceptionally well in the production of spring bread wheat varieties widely adapted to the environmental conditions in ESA. By the late 1990s, improved varieties accounted for 90% of area planted to bread wheat (Heisey and Lantican, 2000). The vast majority of these high-yielding varieties (HYVs) were semi-dwarf in stature and CIMMYT-related in origin. They replaced tall local varieties and even tall improved varieties.

The same centralized approached that CIMMYT followed for wheat did not work that well in maize. Few elite lines and populations were directly released and their acceptance by farmers was not widespread. Improved varieties in the region were synonymous with high-yielding, late-maturing hybrids, such as SR 52, from the Rhodesian national programme, which invested in maize breeding as early as the 1930s (Eicher, 1995). There was demand for CIMMYT germplasm, especially for tropical maize landraces from the Andes. One of these landraces collected by the Rockefeller Foundation in 1953 is a parental line in the leading maize hybrids H614D in Kenya and BH-660 in Ethiopia.

In the mid-1980s, CIMMYT embarked on a more decentralized strategy and invested in a regional breeding programme in Zimbabwe, followed by the establishment of breeding programmes in Kenya and Ethiopia in 1997. All three programmes have increased their research staff

* E-mail: h.degroote@cgiar.org

substantially in the last decade. That strategy initially emphasized improved open-pollinated varieties (OPVs) and composites. Greater specific adaptation and drought resistance are still two of the main planks in that breeding platform.

More was known about the varietal output and adoption of maize and wheat cultivars in ESA than for any other staple food crop in any region in sub-Saharan Africa (SSA) at the start of the Diffusion and Impact of Improved Varieties in Africa (DIIVA) Project in 2010. Beginning in the early 1990s, CIMMYT economists had periodically established benchmarks on varietal release and adoption in their research on impact assessment (Heisey and Lantican, 2000; Hassan *et al.*, 2001). Their work on the performance of maize and wheat crop improvement in ESA is updated in this chapter.

Country Coverage, Methods and Data Collection

Data in this study on maize and wheat were collected during 2010–2011. The work was subcontracted to ten collaborators from the national agricultural research system (NARS). Workshops were organized to provide training in the methodology used to collect information on the key parameters of the investigation.

In four countries (Ethiopia, Kenya, Zambia and Zimbabwe), data on both maize and wheat were collected. In five countries (Angola, Malawi, Tanzania, Mozambique and Uganda), only maize was covered. Hence, four crop-by-country combinations for wheat and nine for maize results in a total of 13 crop-by-country observations.

Excluding South Africa, the nine maize countries accounted for 97% of maize area and 93% of production in ESA in 2009. Comparable estimates for wheat coverage are 92% of area and 93% of production. The distribution of area and production in maize in ESA was the most equitable among the 20 study crops in the DIIVA Project. None of the nine maize-growing countries contributed more than a 20% share to production in 2009. Maize area ranged from about 900,000 hectares in Uganda and Zambia to around 3 million hectares in Tanzania. In contrast, 90% of the wheat grown in the four study countries was cultivated in Ethiopia.

Coverage was even more extensive in the 1998 Initiative than in the DIIVA Project. In maize, the 1998 study also included Lesotho, South Africa and Swaziland (Hassan *et al.*, 2001). For wheat, Heisey and Lantican (2000) covered seven countries: in addition to the four countries covered in this study, they also included South Africa, Sudan and Tanzania. In this study, wheat and maize in South Africa were not covered because no contract was established with the collaborator and because the emphasis was on small-scale, rainfed production typical of sub-Saharan Africa. Nevertheless, South Africa offers a regional perspective on crop improvement in a more developed economic and institutional setting. Lesotho and Swaziland in maize and Tanzania in wheat were too small nationally to warrant coverage. Wheat in the Sudan was not included for reasons related to security concerns.

The same country coverage in the DIIVA Project and the 1998 Initiative provides the opportunity for a time series comparison for nine countries in maize and for four countries in wheat. The size of this opportunity depends on the extent that quality data were collected on each of the three main aspects recorded in 2010 database.

Scientific Strength

The consultants for each study visited key informants in the public sector (national agricultural research programmes and universities) and the private sector to obtain a list of staff working in maize and wheat research in each country.[2] Three categories were distinguished: (i) scientists, defined as holders of a Bachelor of Science (BSc) degree or above; (ii) technicians, defined as holders of a post-high school diploma; and (iii) others, holders of at least an agricultural related certificate.

Maize

The total of full-time equivalent (FTE) scientists in the maize programmes of the nine countries summed to 237. They were supported by 151 technicians and 183 other staff (Fig. 11.1).

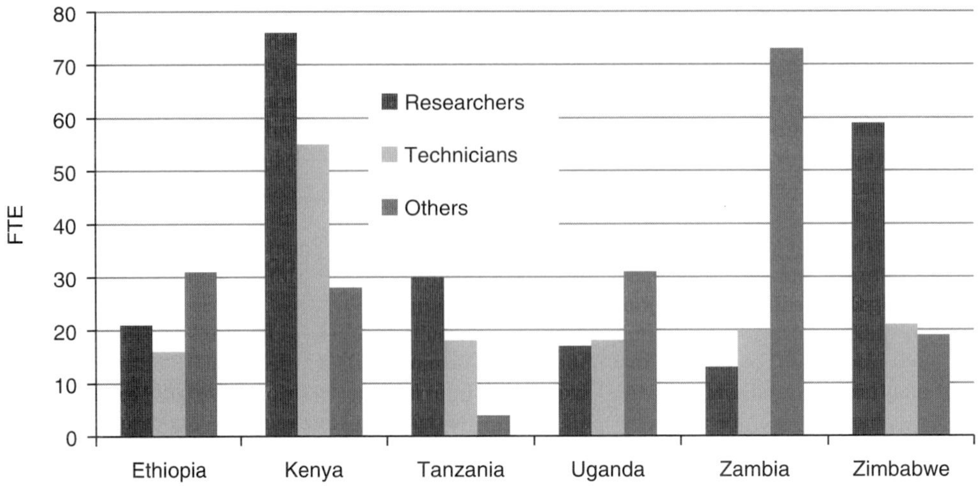

Fig. 11.1. Maize research staff in selected countries of East and Southern Africa by category and country.

Large differences among countries were found: two countries, Kenya and Zimbabwe, had more than 50 scientists; two, Ethiopia and Tanzania, had between 20 and 50 scientists (Ethiopia and Tanzania); and the remainder employed between 10 and 20 FTE scientists. Angola and Mozambique, each with 12 scientists, ranked last in scientific capacity (Table 11.1). Technicians were more evenly distributed across countries than researchers. In contrast, more than 70 of the other staff represented in Fig. 11.1 were located in Zambia where most were employed by two of the larger private-sector seed companies.

The largest group of scientists, by discipline, was plant breeders (41%), followed by agronomists (26%) and seed specialists (14%) (Fig. 11.2). There were relatively few social scientists (8%, with one lonely farming systems specialist) or scientists from other disciplines (11%, including entomologists, pathologists and post-harvest specialists).

Kenya, Tanzania and Zimbabwe reflected this typical distribution with a heavy representation in seed production (Table 11.1). Ethiopian scientists, on the other hand, were almost all plant breeders, most likely in response to a decentralized research infrastructure stemming from diverse crop agroecologies. Angola and Malawi had more agronomists, probably reflecting the early stage of maize research in these countries. Uganda had a very even distribution of scientists across disciplines.

To account for the size of the maize sector in the different countries, we normalized the number of scientists per country over that country's maize production in million tonnes. According to this metric, the weighted average research intensity was 11.75 FTE scientists per million metric tonnes. The estimated research intensity of 4.3 was the lowest in Malawi where the loss of senior breeders is a cause of grave concern (Smale *et al.*, 2011). Zimbabwe with about 45 scientists per million tonnes of maize production in 2010 (which was a more normal production year than 2009) overshadowed all other countries in research intensity. This high research intensity reflects not only the advanced stage of the maize seed industry, which exports seed of many maize varieties to the rest of the continent, but also the strong secular decline in maize production during the last decade.

Most maize improvement programmes experienced increases in FTE scientists between 1997–1998 and 2009–2010 (Table 11.2). Only Angola and Ethiopia lost scientific staff; Tanzania, Uganda and Zimbabwe made large gains that exceeded ten FTE researchers between the two periods. Across the nine countries, the net gain in scientists totalled 97.5, which is equivalent to about two-thirds of the base level of 145.7 in 1997–1998 (Hassan *et al.*, 2001).

Table 11.1. FTE scientists by discipline working in maize improvement in the East and Southern Africa in 2009.

Country	Breeding	Pathology	Entomology	Agronomy	Seed production	Postharvest	Social science	Total	FTE scientists per million t of production
Angola	1	1	1	6	2	0	1	12	12.4
Ethiopia[a]	17.3	0.65	0.8	1.3	0.2	0.2	1	21.45	5.5
Kenya [a]	23	1	2	14	10	5	7	62	25.4
Malawi	4	2	0	8	1	1	0	16	4.3
Mozambique	8	1	0	2	1	0	0	12	8.4
Tanzania	11.5	0.5	1	8.55	5.55	1.05	1.6	30.75	9.3
Uganda	3	3	2	2	3	2	2	17	13.4
Zambia	10	0	0	3	0	0	0	13	6.9
Zimbabwe[a]	20	1	0	19	12	1	0	53	44.5
Total	97 8	10.15	6.8	63.85	34.75	10.25	12.6	237.2	

[a]Excludes international scientists working in CIMMYT's Global Maize Program.

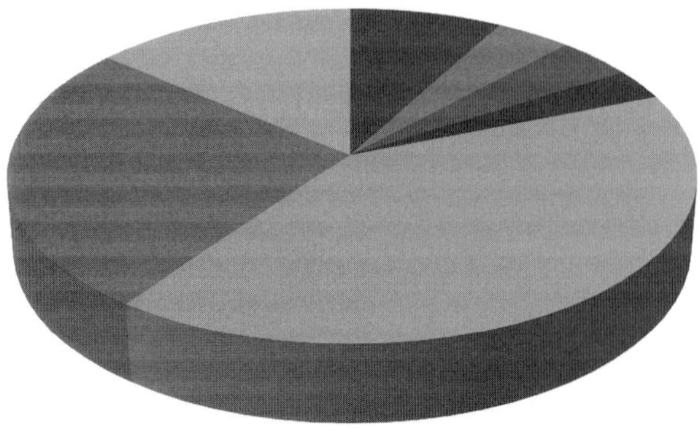

Fig. 11.2. Distribution of maize scientists in East and Southern Africa by discipline.

Table 11.2. Comparing FTE scientists and estimated research intensities in maize improvement in East and Southern Africa between 1997–1998 and 2010 by country.

Country	FTE scientists in 2010	FTE scientists in 1997–1998	Difference in FTE scientists	Estimated research intensity in 2010	Estimated research intensity in 1997–1998	Difference in estimated research intensities
Angola	12.0	14.0	–2.0	12.4	27.7	–15.4
Ethiopia	21.5	24.0	–2.6	5.5	10.2	–4.7
Kenya	62.0	54.0	8.0	25.4	21.9	3.5
Malawi	16.0	9.0	7.0	4.3	5.1	-0.8
Mozambique	12.0	5.0	7.0	8.3	4.4	3.9
Tanzania	30.8	14.0	16.8	9.3	5.2	4.0
Uganda	17.0	4.0	13.0	13.4	4.3	9.0
Zambia	13.0	6.0	7.0	6.9	9.4	–2.5
Zimbabwe	53.0	15.7	37.3	44.5	11.1	33.4

The aggregate increase in scientific capacity did not result in a significant rise in maize research intensity, which declined substantially in Angola, Ethiopia and Zambia between the two periods. This could be partially explained by staff who went for training; reduced research funding could have also played a contributing role. However, large relative gains in research intensity were registered in Mozambique, Tanzania and Uganda where increases in scientific staffing were proportionally larger than the upward trend in production between the two periods. As discussed earlier, rising research intensity in Zimbabwe stems from the economically counterintuitive trends of declining production accompanied by documented increases in researcher staffing mainly in the private sector, where some of the research is guided towards seed export. Overall, the estimated research intensity improved very modestly from 10.5 scientists per million tonnes of production in 1997–1998 to 11.75 scientists per million tonnes of production in 2009–2010.

In 1997–1998, only about 17% of FTE scientists were employed in the private sector (Hassan *et al.*, 2001). The majority of these were in Zimbabwe. Moving forward to 2009/10, 110 or 46% of the FTE scientists working on maize improvement in ESA were in the private sector. Five countries, Kenya, Tanzania, Uganda, Zambia and Zimbabwe, had invested in more than ten FTE scientists in maize improvement. The resource allocation in the private sector was more heavily tilted toward agronomy and seed production than the disciplinary emphasis in the public sector.

Wheat

The wheat programmes of the four countries in this study employed 64 FTE scientists, supported by 60 technicians and 98 other staff. On aggregate, wheat scientists seem better supported than maize scientists because the ratios of technicians to scientists and other staff to scientists are higher in wheat than in maize.

Wheat improvement programmes are heavily concentrated in plant breeding. Slightly more than half the scientists were breeders, followed by agronomists (22%), pathologists (11%) and seed production specialists (9%) (Table 11.3). The higher number of pathologists in wheat as compared to maize is probably due to the importance of rust diseases in wheat production. There were few social scientists (5%) or scientists from other disciplines including entomology (3%) seed production (6%) and postharvest research (<1%).

Although the number of researchers in wheat appears modest compared to maize, the estimated researcher intensities in Table 11.3 are very high in Kenya, Zambia and Zimbabwe because of the small size of production. The weighted average research intensity was 18.75 FTE scientists per million metric tonnes.

As in maize, wheat crop improvement was also characterized by an increase in scientific staffing between 1998 and 2009 (Table 11.4). Proportionally, the size of the aggregate gain of 23 scientists between the two periods is also similar to maize. Weighted average research intensity rose by two FTE scientists from 16.75 in 1998.

Varietal Output

A release list of maize and wheat varieties was obtained through interviews with key informants using the same standardized structured questionnaire in the different countries on both crops, supplemented with data from national registries of released crop varieties. The questionnaire also included questions on the institutions and procedures for varietal testing and registration.

The key actors of each country's seed industry were interviewed, in particular those involved in breeding, variety registration and release. Institutionally, informants were employed by seed companies, national agricultural research institutes, ministries of agriculture, regulatory authorities and universities. The data collected were triangulated with literature reviews, information from personal contacts and searches on the Internet. The Internet searches revealed that many seed companies advertise their varieties available for production and that several regulatory agencies publish lists of released varieties.

Maize

On the basis of the key informant surveys and secondary information, 802 improved maize varieties were identified that were released from 1958 to 2010. With about 200 varieties each, Kenya and Zambia are responsible for about half of this impressive varietal output. Zimbabwe with 100 releases and Tanzania with 90 have also contributed substantially to varietal output in ESA. With only about 30 releases each, Angola and Mozambique lag behind the other study countries.

More than two-thirds of these varieties (69%) were released in the last decade. Between 2000 and 2010, slightly more than 50 maize varieties were released by the nine countries as a group per annum. Kenya also led in the average number of new improved maize varieties released per year in the last decade (16), followed by Zambia (12), Zimbabwe (6) and Tanzania (5). Ethiopia and Uganda released, on average, three new varieties per year, and the others less than that.

Most of the released cultivars (83%) were hybrids. In countries with well developed private seed sectors, including Kenya, Zambia, Zimbabwe and Malawi, at least three-quarters of the released varieties were hybrids. In countries where the maize industry is in its infancy, such as Angola, most of the varieties were OPVs. Ethiopia is somewhere in between: although its seed sector is well developed, most of it is in the hands of the public sector, which explains why the proportion of hybrids in released maize varieties is only 50%. Even in Mozambique, the majority of releases were hybrids.

Table 11.3. FTE scientists by discipline working in wheat improvement in East and Southern Africa in 2009.

Wheat	Breeding	Pathology	Entomology	Agronomy	Seed production	Postharvest	Social science	Total	FTE scientists per million t of production
Ethiopia	11.2	4.45	0.15	0.75	0.3	0.05	1.15	18.05	5.9
Kenya	11	2	2	6	5	0	2	28	216.7
Zambia	8	1	0	2	1	0	0	12	69.7
Zimbabwe	4	0	0	2	0	0	0	6	150.0
Total	38.2	7.45	2.15	10.75	6.3	0.05	3.15	64.05	

The private sector is responsible for 53% of the releases. NARS partnering with CG Centers account for 20% of releases and NARS by themselves for 13%. The other 14% of releases comprise NARS and the private sector (8%); and the CG Centers and the private sector (6%). Some caution is in order, however, because several private companies started out as public companies or parastatals, such as Tanseed in Tanzania; Zamseed in Zambia; and Kenya Seed Company (KSC). KSC started out as a private company, but currently more than half of its shares are owned by the government through parastatals. Moreover, KSC has been allocated the property rights to major varieties developed in the past by the Kenya public research institutes including the Kenya Agricultural Research Institute (KARI) and its predecessor, the East African Agriculture and Forestry Research Organization . Therefore, it is not always easy to determine if a variety should be classified as private or public. In our analysis, we classified varieties as public if they were developed by a public research institute, either the NARS (including national research institutes, ministries of agriculture, universities and other public institutions) or research institutes from the CG Centers, including CIMMYT and IITA.

The contribution of the private sector markedly increased over time (Fig. 11.3). Liberalization of agricultural input markets in many ESA countries took place in late 1980s and early 1990s. Before then, most maize varieties released came from the public sector (44 varieties). In the past two decades, however, the private sector clearly dominated, with 139 releases in the 1990s and 263 in the 2000s. In the public sector, most new varieties were developed in partnerships of

Table 11.4. Comparing FTE scientists and estimated research intensities in wheat improvement in East and Southern Africa between 1997–1998 and 2010 by country.

Country	FTE scientists in 2010	FTE scientists in 1997–1998	Difference in FTE scientists	Estimated research intensity in 2010	Estimated research intensity in 1997–1998	Difference in estimated research intensities
Ethiopia	18.0	17	1.0	5.9	9.1	–3.2
Kenya	28.0	17	11.0	216.7	55.7	161.0
Zambia	12.0	4	8.0	69.7	70.2	–0.5
Zimbabwe	6.0	3	3.0	150.0	13.6	136.4

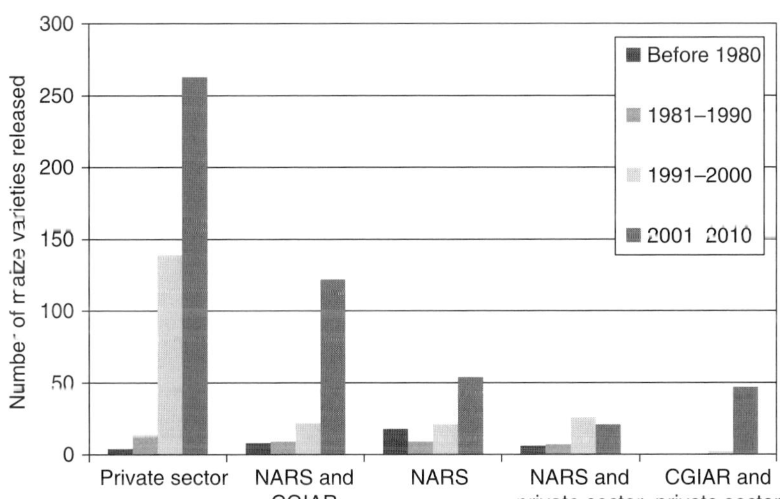

Fig. 11.3. Number of improved maize varieties released by decade and institutional source.

NARS with CIMMYT (122 varieties in the past decade). There also has been an increase in private–public partnership (PPP) releases, more with the NARS in the 1990s (26 varieties) but increasingly with CIMMYT after 2000 (47 varieties). Larger national seed companies are increasingly concentrating their variety releases and dissemination on hybrids. Multinational seed companies such as Monsanto and Pioneer, as well as regional companies such as Pannar and Seed Co only release hybrids. The largest Kenyan company, KSC, has recently released hybrids for the lowlands and the drylands, after which it stopped producing its OPVs, which were targeted for those areas.

Institutional sources have varied widely by country. Kenya is the only country that has released at least ten varieties from the five institutional sources discussed above. Zambia and Zimbabwe have relied heavily on the private sector for varietal release. In contrast, NARS and CGIAR partnerships have loomed large in the generation of varietal output in Malawi (Fig. 11.4).

Wheat

Varietal output in wheat reflects a more nuanced, country-specific story than the maize narrative, which centred on a sharp expansion of released varieties in almost all study countries in the recent past. Across the four study countries, varietal output relative to the size of production is a hefty 244 improved cultivars released since the early 1960s. These 244 varieties represent the full historical record that was initially compiled for the 1997 wheat impact study that was updated in 2002 and updated yet again in the DIIVA Project.

Of these varieties 104 were released in Ethiopia where political repression, terror and famine influenced the level and pace of varietal release from the late 1970s to the early 1990s. This time period corresponded to the reign of the Mengistu Regime. During the Regime, varietal output dropped sharply in the 1980s as the state of agricultural research in general and crop improvement in particular deteriorated (see Fig. 11.5). No improved wheat varieties were released between 1988 and 1992. An alternative hypothesis for the absence of released varieties in the late 1980s was the appearance of new rust races that made much of the Ethiopian germplasm base vulnerable (Heisey and Lantican, 2000). Wheat varietal releases rebounded in the mid- to late 1990s. Political stability further contributed to the release of 43 varieties since 2001. Like their predecessors, most of these releases are for spring bread wheat. They are semi-dwarf in stature. The majority show CIMMYT parentage

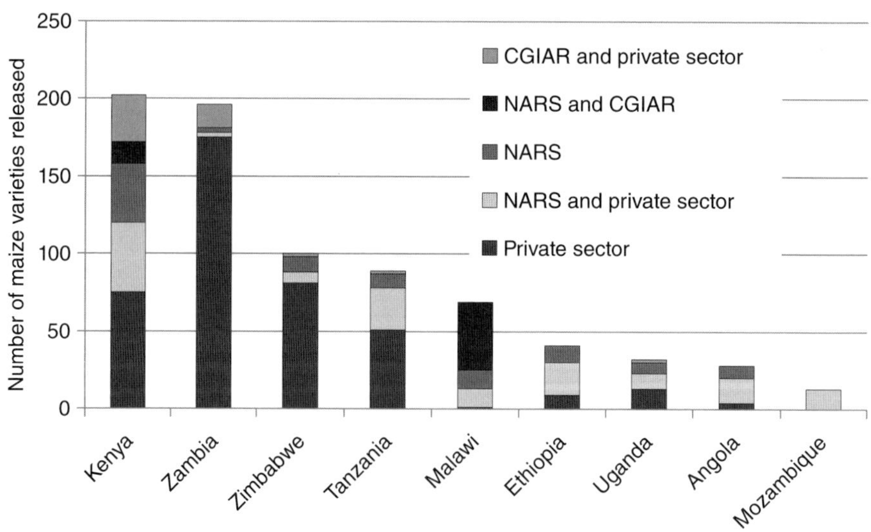

Fig. 11.4. Number of maize varieties released (1958–2010), by country and origin.

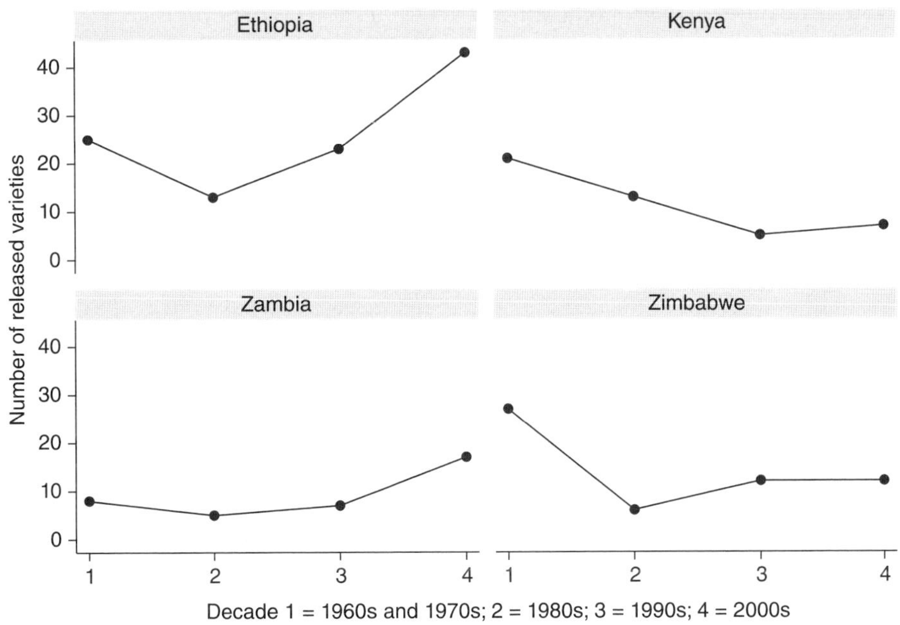

Fig. 11.5. Number of wheat varieties released in selected countries of East and Southern Africa by country and time period.

or a CIMMYT-related pedigree. Twenty of the 43 releases refer to spring durum wheat. Eleven of these can be traced to the collaborative work of a CG Center, about evenly divided between CIMMYT and the International Center for Agricultural Research in the Dry Areas (ICARDA).

Zambia is the other country that shows an upward trend in wheat varietal output since the 1980s. Zambia's wheat is destined for bread flour. All of its 17 releases since 2000 are targeted for cultivation in the winter season under irrigation. In the 1980s and 1990s, a few heat-tolerant varieties addressed the recommendation domain of rainfed summer season production. Most of the 17 releases were generated by private sector companies.

Both Kenya and Zimbawbe are characterized by a stagnating or even declining trend in wheat production since the 1970s and 1980s (Fig. 11.4). In the past, Kenya has relied heavily on CIMMYT materials for its bread wheat releases. From 1973 to 1993, 28 of 33 releases featured a CIMMYT parent or ancestor in its pedigree. Recently, the record of varietal release has been patchy in Kenya. No wheat varieties were released between 1994 and 1998. Only seven varieties have been released since 2001; no releases occurred between 2002 and 2006.

After a steep decline in varietal output in the 1980s, Zimbabwe's varietal production seems to have stabilized. Twelve of its 15 wheat releases since 2000 came from the private sector where Seed Co is a major player.

Varietal Adoption

CIMMYT economists in undertaking impact assessment of maize modern varieties (MVs) established two adoption baselines in the 1990s in ESA. Adoption levels of improved OPVs and hybrids were estimated nationally in 1990–1992 and 1996–1998 from two sources: expert opinion of key informants – usually national programme scientists – and seed sales from public- and private-sector companies. During the project implementation workshop, concern was voiced that the rapid expansion of private-sector participation in maize hybrid production made seed-sales inquiries a daunting undertaking as many companies refused to release cultivar-specific information. Therefore, expert opinion was viewed as the primary source for adoption estimates for maize in ESA, as it was for the majority of the crop-by-country observations in the rest of the DIIVA Project.

Similar to the earlier work in the 1990s, a structured questionnaire was administered in recent years to key informants and experts in all study countries, thereby eliciting a list of improved maize or wheat varieties, as well as an estimate of the proportion of farmers growing them and the proportion of maize or wheat area allocated to specific improved varieties. Some experts consulted the literature of their particular country–crop combination and provided estimates from different studies and reports. Unlike the research that established the 1990s baselines, this elicitation was assigned to consultants who were not closely supervised and who did not have a stake in the outcomes of the work. Moreover, the surge in releases in the 2000s substantially complicated the task of key informants who often expressed ignorance about the adoption prospects of numerous varieties available for planting. Expert opinion returned viable information only for maize in Uganda and maize and wheat in Ethiopia.

Summing up, the additionality of the project to the adoption estimates for modern varieties of maize and wheat in ESA was not as large as expected. Nationally representative surveys were undertaken on the diffusion of modern varieties of maize and wheat in Ethiopia. A seed-related inquiry was carried out in Tanzania. And seemingly reliable estimates on maize adoption were elicited by both CIMMYT and IITA expert panels in Uganda.

Fortunately, since John Gerhart's diffusion surveys on the uptake of maize hybrids in western Kenya in 1973 and 1974, adoption of maize MVs has received more attention in ESA than any other food crop in any other region of sub-Saharan Africa (Gerhart, 1974). Where possible, the DIIVA-related data were complemented with estimates from the literature, a recent summary of that literature (Smale *et al.*, 2011), and with recent household survey and adoption studies (which were not covered in that review). Recent surveys took place in the major maize-growing areas of Kenya (Aflatoxin control project survey in 2010–2011), Zambia (Harvestplus baseline survey in 2012) (De Groote *et al.*, 2011) and Ethiopia (DIIVA objective 2 survey, 2011) (Jaleta *et al.*, 2013), and in selected districts of Uganda (QPMD project survey in 2009) (De Groote *et al.*, 2009), of Tanzania and Mozambique (SIMLESA project survey in 2010) and of several countries in Southern Africa participating in the Drought Tolerance Maize in Africa (DTMA) initiative.

In maize, we begin this section by assessing the earlier adoption baselines prior to their update. This simple time-series analysis is followed by a description of the state of adoption in each country in 2009/10. Cultivar-specific estimates figure prominently in this brief overview. In wheat, the focus is on improved spring bread and durum varieties in Ethiopia. Kenya, Zambia and Zimbawbe had already achieved full adoption of modern varieties in 1998 (Heisey and Lantican, 2000).

Maize

Evaluating adoption estimates over time

In comparing the 2009 estimates to the 1990s benchmarks, Ethiopia, Tanzania and Malawi display an unamibiguous upward trend in adoption (Fig. 11.6). Adoption in Kenya has plateaued at a level substantially below full adoption in the recent period (Olwande and Smale, 2012). Some disadoption of hybrids appears to be taking place in Zimbabwe and not much seems to be happening in Angola.

The sub-graphs of Mozambique, Uganda and Zambia present some anomalies that require comment if not clarification. A 65% estimate of adoption of improved OPVs in Mozambique in the early 1990s is almost assuredly a gross overstatement of the extent of the penetration of these materials – such as the OPV Matuba released in 1984 – into the central and nothern regions of the country. Mozambique was in the concluding phase of a 14-year civil war in the early 1990s.

According to the benchmark data, adoption of improved varieties in Uganda dropped from 40% in 1991 to 9% in 1997. This difference is most likely a result of using seed sales as a source of information in 1997 when national programme scientists perceived that improved OPVs accounted for 50% of harvested area (Hassan

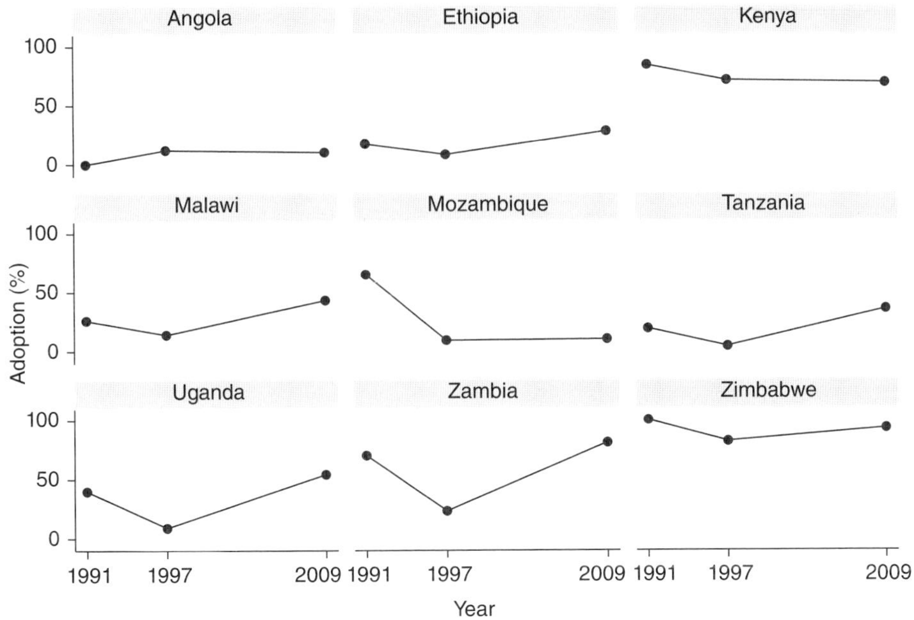

Fig. 11.6. Comparing adoption of modern maize varieties by country in 1991, 1997 and 2009.

et al., 2001). The discrepancy between the two sources for Uganda was easily the largest among the countries in the 1998 study. In general, the disparity between estimates widened as the importance of OPVs increased and hybrids decreased. Uganda was characterized by the largest share of improved OPVs among the study countries in 1997.

The v-shape in the Zambia sub-graph (Fig. 11.6) is policy driven. Specifically, it is derived from rapid adoption of improved OPVs and hybrids released in the 1980s fuelled by government programmes entailing heavy expenditures in the form of subsidies that could not be sustained (Howard and Mungoma, 1997). Structural adjustment and market liberalization initially ushered in an era of contracting maize area and production with disadoption of hybrids and improved varieties in the mid- to late 1990s. Subsequently, a more supportive policy environment paved the way for an cycle of expansion with intensive private-sector participation and renewed adoption of recently released hybrids (Howard and Mungoma, 1997).

On aggregate, as the sub-graphs in Fig. 11.6 show, the relationship between area-weighted adoption across the nine countries and time is nonlinear. Adoption at 39% was higher in the early 1990s than in the mid-1990s when it dipped to 29%. (The area-weighted estimate is further based on the reasonable assumption that adoption in Mozambique in 1991 was in reality, at most, only 10%.) The reduction in this aggregate estimate between 1991 and 1997 is primarily attributed to a decline in adoption in Kenya, Tanzania and Zambia and, to a lesser extent, Malawi between the two periods.

In 2009, area-weighted adoption had reached 44%. Country-specific estimates are presented in Table 11.5. From the perspective of about 30% in 1997, attaining a level approaching 45% in 2009 seems like a noteworthy achievement. From the viewpoint of a level close to 40% adoption of modern varieties in 1991, arriving at 45% in 2009 seems like painfully slow progress.

Country and variety profiles in adoption in 2009

ANGOLA. The estimated adoption rate of 10% in Table 11.5 comes from Table 3 in Langyintuo *et al.* (2008), which is also cited in Smale *et al.* (2011). Most of the released varieties were developed by the national programme. The consultant's

Table 11.5. Adoption of improved maize varieties in the East and Central area of sub-Saharan Africa, 2009.

Country	National/ agroecology	Area (ha)	Area MVs (%)
Angola[a]	National	1,554,100	10
Ethiopia[b]	Highland	434,958	28.5
Ethiopia	Mid-altitude	1,096,234	28.5
Ethiopia	Rift Valley	233,392	18.5
Ethiopia	National	1,768,120	27.9
Kenya[c]	National	1,884,370	69
Malawi[a]	National	1,609,000	43
Mozambique[a]	National	1,612,000	10
Tanzania[d]	National	2,961,330	35
Uganda[e]	National	887,000	54
Zambia[f]	National	911,942	84
Zimbabwe[e]	National	1,508,000	93

[a]Source: Langyintuo *et al.* (2008); [b]DIIVA household survey (Jaleta *et al.*, 2013); [c]Aflatoxin Control Project household survey (Swanckaert *et al.*, 2013); [d]Majebelle (2013); [e]DIIVA expert opinion survey; [f]De Groote *et al.* (2011).

report lists CM-1 as the most important variety, followed by Branco Redondo, SAM-3 and ZM521. Matuba and Dente de Cavalo were perceived to be cultivated on smaller areas. ZM521 was developed in collaboration with CIMMYT. Branco Redondo, SAM-3 and Dente de Cavalo are old landrace materials/improved OPVs released in the mid-to-late 1960s.

In a recent DTMA survey of about 1050 households allocated across five important maize-growing countries in Southern Africa, the diffusion of improved varieties was by far the lowest in Angola (Kassie *et al.*, 2012). Only two districts were surveyed in each country. The districts selected in Angola were in the lowlands just above sea level. Only 5% of household heads were aware of the difference between improved OPVs and hybrids. Adoption of improved cultivars was estimated at 3%.

ETHIOPIA. A large, stratified (on production potential) random sample of 2455 farm households from 39 districts in five regional states of Ethiopia (Tigray, Amhara, Oromia, Benishangul-Gumuz and SNNPR) shows that modern maize varieties are steadily finding a home in the fields of Ethiopian farmers (Jaleta *et al.*, 2013). National adoption increased from an estimated 8% in 1997 to 28% in 2009. Adoption is higher in the highlands and the mid-altitude region that are characterized by higher production potential than in the more marginal Rift Valley region (Table 11.5).

The leading two improved cultivars are hybrids BH-660 and BH-540, each with a 7–8% share of maize area (Table 11.6). Of the 28% adoption share of modern varieties, 26% is attributed to hybrids and only 2% to improved OPVs. Ethiopia seems to have made the transition to a country where hybrids dominate among improved maize cultivars. This finding also appears to be the culmination of a trend detected in Hassan *et al.* (2001) who found that hybrids contributed about two-thirds to improved MV adoption, compared to about one third for improved OPVs in the late 1990s. Earlier, in 1991, OPVs were estimated to be substantially more important than hybrids. In the 2009 survey, hybrids were also well represented in the environments of lower production potential, but they were most frequently found in the mid-altitude districts of medium-production potential.

One of the striking aspects of the list of varieties for Ethiopia in Table 11.6 is the realization that more than half of the 42 improved varieties released since 1973 have recorded positive adoption outcomes in 2009. This result suggests that maize crop improvement scientists are releasing superior cultivars that satisfy farmers' demands for specific varietal characteristics. Yield is one of the most important characteristics to farmers and differences in productivity, particularly between released hybrids and local varieties, were transparent in the large-scale, DIIVA-Project related survey (Jaleta *et al.*, 2013) and in the impact assessment that is based on those data (Zheng *et al.*, 2013). The greater productivity of hybrids translated into appreciable reductions in the extent and severity of poverty for both maize producers and consumers in Ethiopia (Chapter 15, this volume).

One concern is the advancing age of the leading hybrids that were released in the early to mid-1990s. Overall, the improved cultivars in Ethiopia are not old, but they would definitely be classified as mature because mean area-weighted varietal age is 16–17 years in the higher production potential districts and drops off to 27 years in the districts of more marginal production potential (Jaleta *et al.*, 2013). Apparently, breeding for the former agroecologies has been considerably more productive than genetic improvement in the latter zones.

Table 11.6. Economically important improved maize varieties and hybrids (H) in selected countries in ESA, 2009.

Country	Improved variety	Area (%)	Country	Improved variety	Area (%)
Ethiopia	BH-660 (H)	7.86	Ethiopia	Morka (H)	0.23
Ethiopia	BH-540 (H)	7.08	Ethiopia	ZAMA (H)	0.20
Ethiopia	PIONEER (H)	2.78	Ethiopia	Fetene	0.20
Ethiopia	Shone (H)	2.02	Ethiopia	Gibe-1	0.13
Ethiopia	Tabor (H)	1.85	Ethiopia	BHQP-542 (H)	0.10
Ethiopia	BH-140 (H)	1.57	Ethiopia	Melekasa-4	0.09
Ethiopia	Jabi (H)	0.92	Ethiopia	BH-670 (H)	0.06
Ethiopia	BH-543 (H)	0.81	Ethiopia	BHQPY-545 (H)	0.06
Ethiopia	Awasa 511	0.66	Ethiopia	AMH-800 (H)	0.03
Ethiopia	Agar (H)	0.36	Ethiopia	Welel (H)	0.01
Ethiopia	Katumani	0.33	Ethiopia	Gutto	0.01
Ethiopia	Melekasa-1	0.25	Ethiopia	Abo-bako	0.01
Ethiopia	Melekasa-2	0.25	Ethiopia	Melekasa-7	0.01
Kenya	H614 (H)	22.60	Kenya	H9401 (H)	0.21
Kenya	SC DUMA 43 (H)	7.17	Kenya	H627 (H)	0.19
Kenya	H624 (H)	4.72	Kenya	Pan 5195 (H)	0.19
Kenya	Katumani	3.84	Kenya	H615 (H)	0.18
Kenya	H6210 (H)	3.13	Kenya	Pan 691 (H)	0.17
Kenya	PHB 4 (H)	3.04	Kenya	DK513 (H)	0.17
Kenya	PHB 3253 (H)	3.03	Kenya	CG 5252 (H)	0.12
Kenya	H513 (H)	2.74	Kenya	H516 (H)	0.12
Kenya	H6213 (H)	2.69	Kenya	H611 (H)	0.12
Kenya	PHB 1 (H)	2.22	Kenya	H612 (H)	0.12
Kenya	DK 3081 (H)	2.09	Kenya	WH 501 (H)	0.12
Kenya	WH 505 (H)	2.04	Kenya	WH 504 (H)	0.12
Kenya	H628 (H)	1.83	Kenya	DLC	0.10
Kenya	H625 (H)	1.75	Kenya	H622 (H)	0.10
Kenya	PAN&M-97 (H)	1.59	Kenya	H526 (H)	0.10
Kenya	DH04 (H)	1.49	Kenya	WH 105 (H)	0.08
Kenya	H629 (H)	1.07	Kenya	WH 503 (H)	0.08
Kenya	DH02 (H)	0.80	Kenya	DK8053 (H)	0.06
Kenya	H6213 (H)	0.65	Kenya	CG 4141 (H)	0.04
Kenya	H626 (H)	0.62	Kenya	IR Maize (Ua Kayongo)[a]	0.04
Kenya	DH01 (H)	0.62	Kenya	WS 502 (H)	0.04
Kenya	H511 (H)	0.54	Kenya	H616 (H)	0.04
Kenya	PH2 (H)	0.48	Kenya	WS 402 (H)	0.03
Kenya	H613 (H)	0.42	Kenya	H623 (H)	0.02
Kenya	Coast Composite (H)	0.39	Kenya	Pan 5355 (H)	0.02
Kenya	DK 8071 (H)	0.39	Kenya	WH 500 (H)	0.01
Kenya	SC DUMA 41 (H)	0.39	Kenya	PH 1033 (H)	0.01
Kenya	WH 403 (H)	0.36	Kenya	WH 404 (H)	0.01
Kenya	Pan 612 (H)	0.33	Kenya	H520 (H)	0.01
Kenya	H513 (H)	0.29	Kenya	PAN63 (H)	0.01
Kenya	Pan 67 (H)	0.24	Kenya	KH500-21A (H)	0.01
Kenya	H512 (H)	0.23			
Tanzania	Situka	6.765	Tanzania	SC 513 (H)	0.332
Tanzania	TMV-1	3.915	Tanzania	UH 6303 (H)	0.321
Tanzania	SC 627 (H)	3.686	Tanzania	TMV-2	0.268
Tanzania	Staha	3.410	Tanzania	Lishe K1	0.248
Tanzania	PAN 4M-19 (H)	2.712	Tanzania	PHB 3257 (H)	0.106
Tanzania	Kilima	1.769	Tanzania	Katumani	0.099
Tanzania	PAN 63 (H)	1.605	Tanzania	Longe 6H (H)	0.083

Continued

Table 11.6. Continued.

Country	Improved variety	Area (%)	Country	Improved variety	Area (%)
Tanzania	Longe 4	1.487	Tanzania	DK 8055 (H)	0.073
Tanzania	H614-D	1.484	Tanzania	PHB 3255 (H)	0.053
Tanzania	Situka 1	1.363	Tanzania	PHB 3258 (H)	0.053
Tanzania	PAN 691 (H)	0.885	Tanzania	PHB 3254 (H)	0.052
Tanzania	SC 13(H)	0.775	Tanzania	Kito	0.033
Tanzania	DK 8031 (H)	0.664	Tanzania	PHB 3256 (H)	0.027
Tanzania	PHB 3253 (H)	0.657	Tanzania	PHB 3261 (H)	0.027
Tanzania	UH 615 (H)	0.553	Tanzania	PHB 3262 (H)	0.027
Tanzania	UH 6303 (H)	0.520	Tanzania	DK 8054 (H)	0.022
Tanzania	DK 8053 (H)	0.503	Tanzania	PHB 3259 (H)	0.001
Tanzania	SC 403 (H)	0.410	Tanzania	PHB 3260 (H)	0.001
Tanzania	PAN 67 (H)	0.387			
Uganda	Longe 5 (H)	19.92	Uganda	Zimbabwe (H)	1.94
Uganda	Longe 4 (H)	15.55	Uganda	Longe 2H (H)	1.46
Uganda	Kawanda composite	8.26	Uganda	622 (H)	1.46
Uganda	Longe 1 (H)	7.77	Uganda	DK (H)	1.46
Uganda	H614-D	5.35	Uganda	Longe 3 (H)	0.97
Uganda	Longe 6 (H)	4.37	Uganda	H625 (H)	0.49

[a]IR, Imazaphyr resistant (seed is coated with the herbicide to make it resistant to *Striga*).

KENYA. Kenya is blessed with multiple, good-quality surveys conducted on maize varietal adoption and with rural household panel data from the Tegemo Institute (Olwande and Smale, 2012). The estimates in Tables 11.5 and 11.6 are based on a recent 1342-household survey that focused on aflatoxin control in maize. H614-D, released in 1986, from KSC was found to be the leading variety with a 26% adoption followed by SC DUMA43 from Seed Co (19%) and Katumani composite B (KCB) (11%) from KARI. The seed production of the latter has declined as KSC has withdrawn all its OPVs from the market and replaced them with hybrids specific for their target areas, the coast and the drylands. The improved maize varieties from local companies including KSC and Western Seed Company competed favourably with seed varieties from regional companies including Seed Co and multinationals including Monsanto.

The results from this survey were compared to two earlier benchmark surveys in 1992 and 2001 in Swanckaert *et al.* (2013). Increasing land scarcity was evident in their comparative analysis. Mean farm size declined from 3.70 hectares in 1992 to 1.76 hectares in 2001 to 1.13 hectares in 2010. Between 2001 and 2010, adoption only increased significantly in the dry transitional zone that accounts for about 5% of production. Adoption was stagnant in the highlands and the moist mid-altitude zones, the principal maize agroecologies. In spite of surging varietal releases and availability in the 2000s, varietal age continued its upward trend as the new varieties seem not to be replacing older hybrids, especially H614-D. Area-weighted mean varietal age was estimated at 17 years in 1992, 22 years in 2001 and 24 years in 2010. Several dimensions of the perceived problem of slow hybrid turnover in Kenya are probed in depth in Olwande and Smale (2012) who used the panel household data of the Tegemo Institute of Egerton University.

MALAWI. The Malawi seed system is dominated by hybrids from private seed companies except MH18, a CIMMYT OPV released by National Seed Company of Malawi (NSCM). Most of the improved maize seed varieties come from three main private seed companies namely Seed Co, Pannar Seed and Monsanto. Cultivar change in Malawi and interaction with supportive policy is well documented in several studies synthesized in Smale *et al.* (2011).

MOZAMBIQUE. Adoption of improved maize OPVs and hybrids is too low to justify a national survey. Two sources provide useful and contrasting information on the uptake of improved varieties.

The Trabajos de Inqueritos Agropecuarias (TIAs) are nationally representative rural household surveys that are carried out by the statistical wing of the Ministry of Agriculture. The sample size is about 5000 households. More than half the districts in the country are sampled. Respondents are asked if they purchased improved and treated maize seed in packets from a dealer or store during the past agricultural year. Responses are only broadly indicative of adoption, because they estimate the number of adopting farmers instead of area under improved varieties, therefore underestimating adoption if larger farmers are more likely to adopt, and because improved OPVs do not have to be renewed every year, leading to underestimating the area in improved MVs. Because these two biases are in opposing directions, it is difficult to judge if TIA estimates understate or overstate adoption of improved varieties. The survey-weighted national average estimates for 2005–2008 and 2012 suggested that 5.6, 9.3, 10.0, 9.9 and 8.7% of maize farmers bought improved seed each year (Benedito Cungara, Mozambique, 2013, personal communication). Purchases were higher in the mid-altitude central provinces than in the lowland north or south. Manica Province bordering Zimbabwe and Tete Province bordering Malawi both recorded levels of improved variety or hybrid purchases in the 20–30% range in 2 of the 5 years. In Tete, Malawi is the primary source of seed and fertilizer.

A recent four-district survey financed by the Australian Government under the auspices of the SIMLESA Project is the other source of information on adoption of improved varieties in Mozambique (Woldemariam *et al.*, 2012). The 510-household survey was conducted in four central Mozambique districts known for their relatively more intensive use of agricultural inputs in rainfed agriculture. During the survey, improved seed was considered to be seed that was purchased in a sealed packet and treated, which is similar to the TIA definition. During the analysis, improved OPV varieties were those recycled not more than three times and hybrids were those not recycled. The estimated adoption level was 18% for hybrids and 10% for improved OPVs; this summed to a national figure of 28%. The leading OPV was short-duration Matuba, which is now severely outcrossed and requires purification.

Summing up, there is no substantive evidence that the national adoption estimate of hybrids and improved OPVs is significantly greater than 10% in Mozambique. The TIA national estimates on maize are not consistent with an upward trend in production. The survey year when production attained its maximum of about 1.4 million tonnes was 2006. An extensive programme by SG 2000 in the mid-1990s to the early 2000s to boost adoption of improved maize technologies via high-input demonstrations was unsuccessful in leveraging diffusion.

TANZANIA. Estimates in the literature of modern variety adoption for maize in Tanzania displayed more variation than for any other country in ESA in 2009. They ranged from 18% of area coverage in Langyintuo *et al.* (2008), more than 46% in Lyimo *et al.* (2014), to 78% in the SIMLESA four-district survey in 2010. Given this high level of uncertainty, the DIIVA Project invested in a consultancy on seed production in Tanzania. Adoption of improved OPVs and hybrids has definitely increased because improved-cultivar seed production has risen markedly in recent years from 10,500 tonnes in 2005/06 to 26,500 tonnes in 2010/11 (Majebelle, 2013). Adjusting for the possibility of renewing seed once in 3 years in OPVs, the recent seed production data are consistent with an adoption level of modern varieties of 35% (Table 11.5). OPVs with a 55% share of improved cultivar adoption maintained a small edge over hybrids in 2010/11. The leading improved cultivars are the OPVs Situka, TMV-1 and Staha (Table 11.6). The leading hybrid is SC627 from Seed Co. With 38 improved cultivars listed in Table 11.6, Tanzania is beginning to show the varietal richness and complexity that one finds in Kenya and Zambia where the seed industry is more developed. These estimates are based on seed produced and not on seed sold; therefore, they would overstate adoption if seed produced is not eventually used.

UGANDA. The estimates in Tables 11.5 and 11.6 for Uganda are taken from expert opinion panels. Most varieties are from the national breeding programme in Namulonge (called Longe varieties), the top one of which is the OPV Longe 5 or Obatampa (a quality protein maize (QPM) variety from Ghana), followed by Longe 4. The second most important source is Kenya, in particular varieties from Mt Elgon Seed Company, the Ugandan branch of KSC. Their popular varieties are all late-maturing hybrids, the most popular of which is H614. Most popular varieties were released in 1999–2000. Except from an unspecified DeCalb variety, no varieties from regional or international seed companies seem to be popular in Uganda.

ZAMBIA. In the late 1990s, Zambia is arguably the most important success story of hybrid change in maize in ESA. Based on a HarvestPlus survey of 1128 households covering five provinces in 35 districts in 2011, about 84% of maize-growing area is planted to modern varieties (De Groote *et al.*, 2011). Recently available hybrids from both local (MRI, ZAMSEED) and regional seed companies (Pannar Seed, Seed Co, Monsanto) performed well, with no single company dominating the market. PAN53 had the highest adoption rates (10.4%), followed closely by MRI 624 (9.6%) and SC627 (8%). More than 20 hybrids were adopted on at least 1% of area. A total of 105 improved cultivars were planted by farmers. This richness in varietal change is a testimony to the competition in the Zambia maize seed market. A young estimated varietal age of only 10 years is indicative of moderately high varietal turnover and is another positive feature in this compelling success story.

ZIMBABWE. Most of improved maize seed (43%) comes from Seed Co followed by varieties from Pannar seed (24%) and Pioneer Seed Company with 16% of the maize seed market share (Langyintuo and Setimela, 2007). The 93% estimate of adoption in Table 11.5 is taken from Langyintuo *et al.* (2008).

Wheat

Participants in the DIIVA Project attempted to elicit expert opinion on MV wheat adoption in Kenya, Zambia and Zimbabwe but they were not successful in generating reliable data. A DIIVA-supported nationally representative survey on wheat varietal adoption was carried out by EIAR and CIMMYT in Ethiopia in 2010/11 (Yirga *et al.*, 2013). In the survey, 2098 households were interviewed from eight wheat-growing agroecologies. The primary sampling unit was the kebele or peasant association. From 15 to 18 households were sampled per kebele. The vast majority (or 1839) sample households cultivated wheat in 2009/10. About 82% of the interviewed households cultivated bread wheat. Only 15% of the sample grew durum wheat. About 60% of the durum-producing households also cultivated bread wheat.

Although wheat is a self-pollinated crop, breeders believe that seed needs to be renewed at least once in 5 years for the variety to maintain its yield potential. With this criterion defining adoption, improved varieties account for 62% of the area planted to bread wheat. Relaxing this criterion, 78% of the bread wheat area was planted to improved varieties. Irrespective of seed renewal, only about 0.5% of durum area was sown to improved varieties.

Assuming that about 25% of wheat-growing area is in durum wheat (C. Yirba, 2012, Kenya, personal communication) and that 78% is the estimate used for comparative purposes for bread wheat, the national estimate for adoption of improved wheat varieties is 59% for 2009/2010.

Comparing these estimates to those for 1997 is a complex undertaking. Heisey and Lantican (2000) provide estimates from expert opinion that improved varieties were planted to 80% of bread wheat area in 1997. Expert responses to the CIMMYT questionnaire resulted in a 20% adoption estimate for improved durum varieties. The distribution of area to the two types of wheat was 60% bread to 40% durum, according to the experts. Therefore, the national estimate for varietal adoption in 1997 was 56%, which is not that much different from the 59% survey estimate described above.

There are good reasons to suspect that a direct comparison of these two national estimates understates the extent of progress in varietal change. First, adoption of improved durum varieties was most likely considerably lower than 20% in 1997. In 2009, the expert perception was 13% for the area presence of improved durum cultivars; however, the 28 released durum varieties were infrequently mentioned in the national survey as being cultivated (Yirba *et al.*, 2012). Foka, the leading durum variety, was only found on one-tenth of one per cent of wheat-growing area (Table 11.7). Secondly, experts

Table 11.7. Economically important improved wheat varieties in Ethiopia in 2009–2010 by national area.[a]

Type	Improved variety	Area (%)
Bread wheat	Kubsa	28.54
Bread wheat	Tussie	9.30
Bread wheat	Galema	7.90
Bread wheat	Digelu	2.42
Bread wheat	Hawwi	0.89
Bread wheat	Sofumer	0.51
Bread wheat	Millenium	0.38
Bread wheat	Other Bread Improved	27.90
Durum wheat	Foka	0.10
Durum wheat	Ude	0.04
Durum wheat	Yerer	0.02
Durum wheat	Kilinto	0.01
Durum wheat	Other Durum Improved	0.20

[a]Source: DIIVA household survey, wheat areas, Ethiopia.

were even more optimistic about the prospects for improved bread wheat in 2010 than in 1997. In 2010, their perception centred on an improved adoption estimate of 88.5% or an 11% difference with the survey results. Adjusting the 1997 estimate downward to reflect the likely reality that (i) adoption of released durum varieties is as limited then as it is now and (ii) that expert opinion probably overstated adoption of spring bread released cultivars in 1997 as it did in 2010 leads to a revised national adoption estimate in 1997 of 43%. Part of the gain in adoption between 43% in 1997 and 59% in 2009 is attributed to the trend in the conversion of area from durum to bread wheat over time.

The ongoing replacement of spring bread landraces, such as Israel, with improved semi-dwarf varieties also has contributed to rising adoption over time. Comparing rates of varietal change also indicates that more recently released wheat varieties are replacing older improved varieties. Most of the agronomically important varieties, such as Kubsa and Galema, listed in Table 11.7, were released in the mid-1990s or early 2000s. The survey results yield an area-weighted average varietal age of 13 years for improved spring wheat varieties in 2009/10. Comparing this result to an estimate of 16.25 years for improved bread wheat cultivars in Heisey and Lantican (2000) points to increasing varietal turnover over time. With the threat of resistance breaking down to leaf rust, we have every reason to expect rapid varietal turnover in wheat cultivars.

One of the disappointments in the analysis of the survey data was the inability to provide a cultivar-specific identity to a large group of varieties labelled in a residual other category in Table 11.7. This lack of transparency in improved varietal identification in large-scale, representative surveys is not unique to wheat in Ethiopia. It is a recurring theme that is extensively discussed in Chapter 20 on validation of expert opinion with survey results.

Conclusions

Although investing small amounts of resources in consultancies did not advance the state of the art in estimating adoption of modern maize and wheat varieties in ESA, sufficient information was generated in other areas of the DIIVA Project and other projects to tell a coherent story about the recent diffusion of those cultivars. The storyline is supported by an increasingly rich primary and secondary literature. Of the 14.7 million hectares in the nine maize study countries in 2009/2010, 6.5 million hectares (equivalent to 44%) were sown to modern cultivars. Measured from two benchmarks in the 1990s, solid gains in adoption were made by Ethiopia (now 28%), Malawi (43%) and Tanzania (35%). Zambia also rebounded (to 84%) from a decline in adoption in the mid-1990s.

Evaluation of progress in adoption depends upon which benchmark is used. The impact assessment research of the CIMMYT economists shows that estimated adoption of improved maize cultivars was higher in 1990–1992 than in 1996–1998. The average of the two periods gives a benchmark of about 34%, which implies an increase of about 10 percentage points between the 1990s and 2010.

More robust growth in adoption of maize modern varieties was not observed because Kenya has not been able to reach full adoption and Zimbabwe is starting to fall short of full adoption of maize hybrids. More importantly, adoption of maize modern varieties has still not taken off in the other countries. With adoption levels at a low equilibrium of 10%, Angola and Mozambique each contributed only slightly more than 1% to the aggregate level of 44% MV adoption in ESA.

In maize, over 80% of adoption of improved cultivars is attributed to hybrids. And the share of hybrids in improved cultivars is growing over time. In countries where the private sector dominates the seed sector, hybrids have become dominant, but also in countries such as Ethiopia, where improved OPVs and composites were perceived to account for a larger share of modern variety area than hybrids in the 1990s. Several factors contribute to this. First, hybrids have a clear yield advantage and over the years farmers have come to appreciate their superior performance over OPVs. Second, hybrids are more interesting for the private sector because good practice requires that farmers purchase new seed every season, as compared to OPVs, which can be recycled three times. Moreover, seed companies can keep strict control over the intellectual property rights of their hybrids, unlike OPVs. Finally, while more expensive than OPV seed, typically about twice the price (Erentstein *et al.*, 2011), this cost is low relative to the extra yield. Research policy should also be informed by the trend of large seed companies to drop OPVs altogether from their portfolio: even the publicly held KSC has followed its multinational (Monsanto and Pioneer) and regional (Seed Co and Pannar) colleagues in this regard.

The most transparent element of the story centres on the explosion of maize varieties released in the recent past. Since 2000, the nine maize-growing countries as a group have released the equivalent of 50 varieties annually. More maize varieties have been released in the 2000s than in the previous three decades combined. Consistent with this rising output, the number of scientists working in genetic improvement has also risen in seven of the nine study countries.

Fuelled by economic liberalization and structural adjustment in the late 1980s and early 1990s, the private sector now enjoys an overwhelming lead in varietal releases in Kenya, Zambia and Zimbabwe. But with the exception of Zambia with its Farmers Input Support Programme (FISP), what has amounted to a sea change in varietal availability has not translated into a marked increase in varietal adoption. Indeed, in Kenya increasing varietal age over time suggests that varietal turnover is decreasing because the dominant hybrid is still H-614D, which was released in 1984. Bottlenecks have been identified that affect the timely commercialization of newly released hybrids (Langyintuo *et al.*, 2008), in particular the lack of access to credit for new seed companies, the slow transfer of genetic materials between public and private sectors, the implementation of the harmonized regional seed laws and regulations.

Historically, the maize varietal diffusion story is well known and has often been told. The diverse and evolving maize seed sector in Southern Africa, from emerging in Angola and Mozambique, strongly centralized in Ethiopia, to fully liberalized and highly competitive in Zambia, is ideally suited to study the diversity of seed systems and its impact on adoption of improved maize varieties.

In all countries, the initial development of improved maize varieties was politically controlled. The resulting inefficiencies and budget burden led to liberalization efforts in the last two decades. Most countries started with public seed development and distribution by one national seed company and moved to sector liberalization, starting in the early 1990s. All countries in this study have now liberalized their seed sectors but are at different stages in seed-sector development. Lately, regional seed harmonization efforts have also been added (Waithaka *et al.*, 2011).

Liberalization resulted in many seed companies entering the market. In Ethiopia, the liberalization started only very recently and its effect is not yet felt. Overall, the number of seed companies and available varieties has increased tremendously and the private sector now takes a much larger share in variety development and distribution than the public sector. Increasingly, however, these companies are not interested in OPVs.

The effect of the maize seed industry's development on maize productivity, however, has been limited. Only in Ethiopia, Malawi and Zambia have yields increased over the past decade and surpassed 2 t/ha. In the other countries, productivity levels have stagnated around 1 t/ha (Southern Africa) or 1.5 t/ha (East Africa). Maize production per capita, on the other hand, has stayed remarkably constant over the last 20 years (around 80 kg in Kenya and Tanzania, and around 40 kg in Uganda and Ethiopia). Only in Malawi and Zambia has production per capita increased to over 200 kg/person during the last decade. The three countries with increased maize yields share a strong government involvement in maize research and dissemination.

Liberalization dramatically increased the number of options in terms of seed companies and varieties. This, in turn, increased the adoption rate in countries with an intermediate seed sector, including Tanzania and Uganda, whereas in countries with a mature seed sector, such as Kenya and Zambia, adoption rates have basically plateaued. Maize varieties by themselves have only a limited effect on yield; they have to be accompanied by improved crop management practices including use of fertilizers, pesticides and weed control to realize their potential. Liberalization in the fertilizer sector has, however, not led to increased adoption of fertilizer. Mozambique epitomizes the worst-case scenario in the region: the incidence of inorganic fertilizer use in maize is still less than 5%.

The disappointing lack of yield increase and impact of the liberalization of agricultural input markets might have played a role in the reversal of liberalization policy as observed in the increasing importance of input subsidy programmes. These programmes had dramatic effects in Malawi and Zambia, where yields surpassed 2 t/ha and a per capita maize production of 200 kg. The only other country in SSA with yields higher than 2 t/ha is Ethiopia, where the increase in productivity cannot be attributed to the liberalization either but rather to the heavy investment in agricultural research as well as in dissemination, including a centralized agricultural input order and distribution system.

The story about the adoption of improved varieties of wheat focuses on the diffusion of semi-dwarf, spring bread wheat varieties throughout the region beginning in 1960s. Both CIMMYT and Ethiopia, where the bulk of wheat is produced in SSA, figure prominently in the history of varietal change in wheat. On the basis of data generated from the DIIVA Project's large-scale survey, Ethiopia has made steady progress in replacing tall local varieties with higher yielding semi-dwarf varieties in the production of spring bread wheat. The estimated rate of adoption now approaches 80%. However, the leading improved varieties are the same as those documented in CIMMYT's 2002 impact study. Hopefully, the new initiative on wheat stem rust will be an incentive for renewed varietal change.

In contrast to the robust adoption of improved varieties in spring bread wheat, varietal adoption has been negligible for the 20 plus released varieties of spring durum wheat. The survey results suggest an adoption level of only 0.5%. The survey estimate is substantially lower than the adoption level perceived by expert opinion in 1998 and in 2009. Varietal change in durum wheat remains a daunting challenge.

Notes

[1] This paper is a revised and abridged version of De Groote *et al.* (2012). It builds on earlier work by Hassan *et al.* (2001) on maize crop improvement in East and Southern Africa and by Heisey and Lantican (2000) on wheat breeding in Eastern and Southern Africa.

[2] Private-sector firms interviewed included six in Kenya, seven in Tanzania, three in Uganda, five in Zambia and six in Zimbabwe. These all had invested in maize breeding in addition to private-sector participation in maize improvement in Malawi. The university refers to Alemaya University in Ethiopia.

References

De Groote, H., Gunaratna, N.S., Kebebe, E., Lyimo, S.D., Ouma, J.O. *et al.* (2009) *Indicators of QPMD Project Performance: A Survey of Potential Impact of QPM on Nutrition, Health and Wellbeing of Farm Households and the Impact of the QPMD Project on QPM Awareness and Adoption in Target Areas in Ethiopia, Kenya, Tanzania and Uganda*. Project report prepared for the Canadian International Development Agency (CIDA), 2009. CIMMYT, Nairobi.

De Groote, H., Gitonga, Z., Smale, M., Asare-Marfo, D., Kasuta, E. *et al.* (2011) *Adoption and Diversification of Modern Maize Varieties in Zambia in 2011: A Baseline Report Prepared for HarvestPlus*. CIMMYT, Nairobi; HarvestPlus, Washington, DC.

De Groote, H., Gitonga, Z. and Mugo, S. (2012) *Diffusion and Impact of Improved Maize and wheat varieties in East and Southern Africa*. Diffusion and Impact of Improved Varieties in Africa (DIIVA), Objective 1 Report. CIMMYT, Nairobi.

Eicher, C.K. (1995) Zimbabwe's maize-based Green Revolution: preconditions for replication. *World Development* 23(5), 805–818.

Erentstein, O., Kassie, G.T. and Mwangi, W. (2011) *Challenges and Opportunities for Maize Seed Sector Development in Eastern Africa*. Paper presented at the conference 'Increasing agricultural productivity and enhancing food security in Africa: new challenges and opportunities', 1–3 2011, Addis Ababa.

Gerhart, J.D. (1974). *The Diffusion of Hybrid Maize in Western Kenya*. PhD thesis, Princeton University, Princeton, New Jersey.

Hassan, R.M., Mulugetta, M. and Mwangi, W. (2001) *Maize Breeding Research in Eastern and Southern Africa: Current Status and Impacts of Past Investments Made by the Public and Private Sectors 1966–97*. CIMMYT, Mexico (DF).

Heisey, P.W. and Lantican, M.A. (2000) *International Wheat Breeding Research in Eastern and Southern Africa 1966–97*. CIMMYT Economics Working Paper 00-01. CIMMYT, Mexico (DF).

Howard, J.A. and Mungoma, C. (1997) Zambia's stop-and-go revolution. In: Byerlee, D. and Eicher, C.K. (eds) *Africa's Emerging Maize Revolution*. Lynne Rienner Publishers, Boulder, Colorado, pp. 45–61.

Jaleta, M., Yirga, C., Kassie, M., De Groote, H. and Shiferaw, B. (2013) *Knowledge, Adoption and Use Intensity of Improved Maize Technologies in Ethiopia*. CIMMYT's Report on Objective 2. Part I. CIMMYT, Nairobi.

Kassie, G.T., Erenstein, O., Mwangi, W., La Rovere, R., Setimela, P. and Langyintuo, A. (2012) *Characterization of Maize Production in Southern Africa: Synthesis of CIMMYT/DTMA Household Level Farming System Surveys in Angola, Malawi, Mozambique, Zambia and Zimbabwe*. Socio-Economics Program Working Paper 4. CIMMYT, Mexico (DF).

Langyintuo, A.S. and Setimela, P. (2007). *Assessment of the Effectiveness of Maize Seed Assistance to Vulnerable Farm Households in Zimbabwe*. International Maize and Wheat Improvement Center. CIMMYT, Mexico (DF).

Langyintuo, A.S., Mwangi, W., Diallo, A.O., MacRobert, J., Dixon, J. and Banziger, M. (2008) *An Analysis of the Bottlenecks Affecting the Production and Deployment of Maize Seed in Eastern and Southern Africa*. CIMMYT, Harare.

Lyimo, S., Mduruma, Z. and De Groote, H. (2014) The use of improved maize varieties in Tanzania. *African Journal of Agricultural Research* 9, 643–657.

Lynam, J. (2010) *Evolving a Plant Breeding and Seed System in Sub-Saharan Africa in an Era of Donor Dependence*. Prepared for the Global Partnership Initiative for Plant Breeding Capacity (GIPB), FAO, Rome.

Olwande, J. and Smale, M. (2012) *Is Old Better? Maize Hybrid Change on Farm Households in Kenya*. Selected paper prepared for presentation at the International Association of Agricultural Economists (IAAE) Triennial Conference, Foz do Iguaçu, Brazil, 18–24 August 2012. http://purl.umn.edu/126669 (Accessed 8 May 2015)

Majebelle, B. (2013) *Consultancy Report for the DIIVA Project on Tracking Varietal Change in Maize in Tanzania*. Mimeo. Report to Bioversity International on the DIIVA Project.

Smale, M., Byerlee, D. and Jayne, T. (2011) *Maize Revolutions in Sub-Saharan Africa*. Policy Research Working Paper 5659. The World Bank, Washington, DC.

Swanckaert, J., De Groote, H. and D'Haese, M. (2013) *Mapping the Effect of Market Liberalisation Policies on the Maize Seed Systems in Kenya Based on Micro-Evidence from 1992 to 2010*. Paper presented at the 138th EAAE seminar, 'Pro-poor innovation in food supply chains', 11–13 September 2013, Ghent.

Waithaka, M., Mzuma, J., Kyotalimye, M. and Nyachae, O. (2011) *Impacts of an Improved Seed Policy Environment in Eastern and Central Africa*. ASARECA, Kampala.

Woldemariam, H., Cachomba, I., Munyaradzi, M. and Kassie, M. (2012) *Characterization of Maize-Legume Farming Systems and Farm Households in Mozambique: Analysis of Technology Choice, Resource Use, Gender, Risk Management, Food Security and Poverty Profiles*. SIMLESA Project, CIMMYT and ACIAR, Maputo.

Yirba, C., Jaleta, M., De Groote, H. and Gitonga, Z. (2012) *Diffusion and Impact of Improved Maize and Wheat Varieties in East and Southern Africa*. CIMMYT's Report on Objective 2. Part II – Wheat in Ethiopia. CIMMYT, Nairobi.

Yirga, C., Jaleta, M., Shiferaw, B., De Groote, H., Kassie, M. *et al.* (2013) *Analysis of Adoption and Diffusion of Improved Wheat Technologies in Ethiopia*. Research Report 101 Ethiopian Institute of Agricultural Research and the International Maize and Wheat Improvement Center, Addis Ababa.

Zheng, D., Alwang, J., Norton, G.W., Shiferaw, B. and Jaleta, M. (2013) *Ex-Post Impacts of Improved Maize Varieties on Poverty in Rural Ethiopia*. Diffusion and impact of improved varieties in Africa (DIIVA) – Objective 3 Report. VPI, Blacksburg, VA, and CIMMYT, Nairobi.

12 Varietal Output and Adoption in Barley, Chickpea, Faba Bean, Field Pea and Lentil in Ethiopia, Eritrea and Sudan

Y.A. Yigezu,[1]* C. Yirga[2] and A. Aw-Hassan[1]
[1]International Center for Agricultural Research in the Dry Areas (ICARDA), Amman, Jordan; [2]Ethiopian Institute of Agricultural Research (EIAR), Addis Ababa, Ethiopia

Introduction[1]

The International Center for Agricultural Research in Dry Areas (ICARDA) was established in 1977 to undertake agricultural research relevant to the needs of people living in North Africa and West Asia. It has a global responsibility for the improvement of barley, lentil and faba bean in the CGIAR (Consultative Group on International Agricultural Research). Since its founding, it also has a regional responsibility for the improvement of chickpea. Even though ICARDA does not have a global or regional mandate for field peas, its former field pea programme has done some work in genetic improvement of this crop from which Ethiopia has benefited.

These five crops are also important in the Horn of Africa especially in Ethiopia where barley, chickpea, faba bean, field pea and lentil are prized both for their grain and their straw for animal feed. Ethiopia is the dominant producer in sub-Saharan Africa (SSA) where cultivated area in 2011 ranges from close to 1.0 million hectares for barley to only about 90,000 hectares for lentil. Total cultivated area for chickpea, faba bean and field pea approached 1.0 million hectares in SSA in 2011.

The five crops in this paper rank in the top half of the 20 commodities in the DIIVA Project in terms of production growth in SSA over the past 20 years. Most of this increase is due to area expansion that reflects strong market demand for the crop. From the perspective of small-producing households, the development and acceptance of improved varieties in these crops has excellent potential to alleviate poverty in the Horn of Africa where they are mostly cultivated in the longer 'meher' rainy season in the Highlands of Ethiopia and the uplands of Eritrea. They are hardy crops that tolerate cold temperatures and drought. Of the five, chickpea is perhaps characterized by the harshest production environment. As in India, it is planted in the post-rainy season and is cultivated on residual moisture in deeper black clay soils.

Survey Design and Data Collection

Initially, ICARDA's participation in the DIIVA Project centred on one cereal, barley, and three pulses, chickpea, faba bean and lentil, in one country, Ethiopia. Because ICARDA has supported genetic improvement in these crops in other countries in East Africa, Eritrea and Sudan were added to the study. Given the importance of field peas in Ethiopia, ICARDA

* E-mail: y.yigezu@cgiar.org

also recommended that field pea be included during the project initiation workshop.

Five crops and three countries give 15 possible crop-by-country observations. Of these, six crop-by-country combinations are not relevant because of negligible area and production. The nine relevant observations pertain to the five crops in Ethiopia, barley and chickpea in Eritrea, and faba bean and chickpea in the Sudan.

Only one observation, barley in Ethiopia is available for a reliable before-and-after analysis. Ethiopia was the only country in SSA in the barley chapter in Evenson and Gollin (2003) (Aw-Hassan *et al.*, 2003a). Chickpea, faba pea and field pea were not covered in the 1998 Initiative. They are regarded as new crops for the DIIVA Project (Walker, Chapter 4, this volume). Even though lentil in Ethiopia and Sudan is included in both the 1998 and the current initiatives, estimates on certain aspects in the 1998 Initiative were incomplete (Aw-Hassan *et al.*, 2003b). Moreover, Sudan is traditionally not a lentil-growing country. In the 1980s in collab oration with ICARDA, a programme to promote lentil production started in the Sudan. Promotional policies to encourage lentil production in northern Sudan, especially in the Rubatab area, were launched. Consequently, cultivated area increased to about 9240 ha in 1992/1993; however, after removal of support, lentil crop area decreased sharply. Most farmers have abandoned lentil cultivation.

Assembly, elicitation and collection of project-related data on scientific staffing, varietal release and varietal adoption in Ethiopia involved a high degree of interaction between ICARDA and the Ethiopian Institute of Agricultural Research (EIAR). Multiple visits were made to Ethiopia, several workshops were held, and several surveys of scientists and farmers were carried out. Given that barley, faba beans and potato share the same agroecologies, a nationally representative adoption inquiry on improved varieties was conducted jointly with the International Potato Center (CIP) in the Ethiopian Highlands. In contrast, the degree of interaction with national scientists in Eritrea and the Sudan was low. Information in those countries was obtained from responses to questionnaires via e-mail.

Estimates of scientific staffing were collected by the national crop coordinators from their respective human resources departments. In Ethiopia, national estimates were compiled from two federal research centres and six EIAR regional research centres. The number of scientists involved in and the level of funding for research on the five crops at universities and by private companies are negligible (at most 5% of total national research investment for each crop). Therefore, only public-sector investments were included in the 2010 database on the scientific strength of crop improvement programmes.

Data on scientific strength in pulse improvement obtained from the Sudan and Eritrea were not disaggregated by crop type. Likewise, in the case of Ethiopia, data were available for two groups: lentil and chickpea were in one regional research subprogramme, and faba bean and field pea were in another. These grouped estimates were disaggregated using shares allocated to each crop estimated by national counterparts.

For Ethiopia, elicitation of adoption estimates was carried out in a systematic manner during a workshop held in Addis Ababa in mid-2011. Breeders, research managers and economists from EIAR headquarters met with economists from ICARDA. Major agroecological zones were identified for improved varieties as a group for the four ICARDA mandate crops into one of three (high, medium and low) levels of uptake (see Appendix Table 12.A1). Generating estimates of the area under local and improved varieties of each crop by agroecological zone was the third step in the process. These figures were further disaggregated by variety. After variety-specific estimates were obtained, EIAR's crop-research coordinators were asked to disaggregate the adoption estimates into administrative zones, which are easier to understand and more immediate.

At every stage of the estimation and disaggregation process, the adoption estimates were refined and seemed to better reflect reality. During the initial stages, there were sharper disagreements among the different experts on the estimates, whereas, at the later stages, when the estimates started to become smaller (especially when the zonal level adoption estimates were generated), most experts arrived at a consensus, suggesting that estimates are more realistic when conducted for smaller geographic areas.

Scientific Strength of the Barley, Chickpea, Faba Bean, Field Pea and Lentil Improvement Programmes

Crop improvement is defined broadly and includes not only plant breeding but also all other disciplines that contribute to the programme, such as pathology, physiology, entomology, social science and postharvest technology (Table 12.1). A quick look at the total number of full-time equivalent (FTE) scientists suggests that the crop improvement programmes in Ethiopia and the Sudan are adequately staffed. Each programme has at least 6.85 FTE scientists and, relative to its economic importance, the chickpea improvement programme in the Sudan is one of the largest in the DIIVA Project. Better-staffed programmes in Ethiopia and Sudan were expected in large agricultural countries in SSA with strong national programmes at EIAR and ARC (Agriculture Research Corporation, Sudan).

The same cannot be said for the level of investment in agricultural research in Eritrea, which reinforces the stereotype of a very small NARS setting that is at best able to maintain some germplasm and engage in borrowing from neighbouring countries. Staff is only deployed in breeding. The pulse programme does not conduct any research on chickpea or faba bean.

Ethiopian crop improvement programmes are characterized by a heavy concentration on plant breeding that approaches or exceeds half of FTE positions. The emphasis on breeding is especially marked in the improvement of field pea and faba bean in Table 12.1. A decentralized regional research setting may explain the apparent concentration on breeding relative to other disciplines. In contrast, Sudan has invested heavily in pathology and agronomy in a more diversified disciplinary setting where the number of pathologists and agronomists equals the number of plant breeders. In Sudan, the main production region of chickpea is in the River Nile State where farmers realized that traditionally grown local varieties are highly susceptible to wilt disease, which reduces yield considerably. Hence, searching for resistance to *Fusarium* wilt and *Ascochyta* blight is an important component of the chickpea improvement programme, which, in turn, partially explains the strong demand for pathologists.

Scientific staff in both Ethiopia and the Sudan are well educated relative to their peers in other crop improvement programmes in SSA. In the pulse programmes, PhD and MSc staff outnumber BSc staff in both countries. In contrast, the three members of staff in Eritrea listed in Table 12.1 are at the BSc level.

Likewise, estimated research intensities compare favourably with estimates for other crops discussed in earlier chapters of this volume. For example, in Ethiopia, the number of FTE scientists per million tonnes of production ranges from a low of about 11 in field pea and faba bean to a high of 43 in lentil.

Between 1998 and 2010, scientific staffing in the barley programme in Ethiopia has expanded from 4.8 to more than 20 FTE scientists (Aw-Hassan *et al.*, 2003a). This significant expansion did not, however, usher in a change in the disciplinary composition of scientific staff. In 1998, barley improvement was concentrated in breeding as it is today.

Table 12.1. Full-time equivalent (FTE) staff by major specialization in 2010.

Crop	Country	Breeding	Pathology	Entomology	Agronomy	Seed production	Postharvest	Social science	Total
Barley	Eritrea	1.00	0	0	0	0	0	0	1.00
Barley	Ethiopia	9.10	3.00	2.60	3.60	0	0	2.80	21.10
Chickpea	Ethiopia	3.90	1.08	0.84	0.60	0.36	0	1.62	8.40
Chickpea	Sudan	6.00	4.80	0.60	5.40	0	0.60	1.20	18.60
Faba bean	Ethiopia	4.85	0.25	0.13	0.68	0	0.13	0.83	6.85
Faba bean	Sudan	2.80	2.24	0.28	2.52	0	0.28	0.56	8.68
Field pea	Ethiopia	4.85	0.25	0.13	0.68	0	0.13	0.83	6.85
Lentil	Eritrea	2.00	0	0	0	0	0	0	2.00
Lentil	Ethiopia	2.52	0.64	0.52	0.36	0.24	0	1.00	5.28

Recent data on the number of scientists (FTE) have been collected for ICARDA in three time periods: 1999, 2005 and 2009. The figures show that there has been a marginal increase (10%) in the manpower deployed in barley and faba bean and a marginal reduction (7%) in the manpower deployed in lentil research over the last decade. On the other hand, staffing in the chickpea programme remained more or less the same.

Comparison of even earlier data for barley (Aw-Hassan *et al.*, 2003a) suggests that the size of the ICARDA barley programme has been stable at about 4.5 to 5.5 FTE scientists since 1990. In 1980, the programme was somewhat smaller at 3.5 FTE scientists. Lentil improvement at ICARDA follows a somewhat different trajectory. Small staff investments in pathology, genetic resources and biotechnology drove an increase equivalent to 0.5–0.75 FTE scientists in the late 1990s that has been maintained over time. Project changes related to agronomy and postharvest projects have resulted in fluctuations and a small decline in staffing since 1999.

Varietal Output from the Barley, Chickpea, Faba Bean, Field Pea and Lentil Improvement Programmes

During the past 30 years, there has generally been an increasing trend in the total number of varieties released. About half of the total of 140 varietal releases occurred in the most recent 5-year periods (Table 12.2). Since 2001 releases have also showed fewer fluctuations over time as the incidence of zero-release events has declined. Enhanced stability in varietal output applies mainly to Ethiopia in barley, chickpea and faba bean.

Eritrea is characterized by few releases. Indeed, no chickpea varieties have been released officially, although it is believed that improved cultivars from Ethiopia and the Sudan have been smuggled across the border and are being planted by farmers.

Releases in the Sudan display a different temporal behaviour. They peaked in the late 1980s and the 1990s for both faba bean and chickpea. Three lentil varieties were also released in the Sudan in the 1990s. Given the apparently well staffed, pulse improvement programmes described in Table 12.1 for Sudan, the decline in recent activity in varietal release is puzzling.

The 38 barley releases in Table 12.2 mostly came from pureline selections of local landraces in Ethiopia and populations of local landraces in Eritrea. Seven varieties were selected from ICARDA-bred elite materials. The first of these was released only a decade ago in 2003. Three varieties were introductions from developed countries. Six were generated from national programme crosses. For one released variety, information on the role of national agricultural research system (NARS) is missing.

Although the barley programme in Ethiopia is a mature crop improvement programme, the path to maturity was non-linear and cannot be readily inferred from the varietal release information over time. Pureline selections from local

Table 12.2. Number of varieties released by crop and country over 5-year periods from 1980 to 2010.

Crop	Country	1980 or earlier	1981–1985	1986–1990	1991–1995	1996–2000	2001–2005	2006–2010	Total
Barley	Ethiopia	4	1	1	2	2	16	12	38
Barley	Eritrea	na	na	na	0	0	3	0	3
Lentil	Ethiopia	1	3	0	2	1	3	1	14
Faba bean	Ethiopia	4	0	0	3	1	7	6	21
Faba bean	Sudan	0	0	3	3	0	0	1	7
Chickpea	Ethiopia	3	1	0	2	2	5	6	19
Chickpea	Sudan	0	0	4	4	6	0	1	15
Chickpea	Eritrea	na	na	0	0	0	0	0	0
Field pea	Ethiopia	2	2	0	6	7	3	6	26
Total		14	7	8	22	19	37	33	140

na, prior to 1991 Eritrea was part of Ethiopia.

landraces were released as late as 2007, and intensive direct crossing and selection was carried out as early as the mid- to late 1970s.

In general, the pulse improvement programmes show a greater reliance on the selection of elite materials and finished varieties and less emphasis on the release of pureline selections of local landraces. Released Desi and Kabuli chickpea in Ethiopia are predominantly selected from ICRISAT and ICARDA elite lines. In Sudan, all released chickpea varieties are either ICARDA-bred materials or are derived from ICARDA crosses. Direct introduction of elite germplasm from ICARDA figures prominently in lentil releases in Ethiopia. In contrast, genetic improvement in faba bean shows a wide range of activities in Ethiopia from the release of a few purified landraces to the direct utilization of ICARDA germplasm to the crossing and selections from ICARDA germplasm progenies. The Sudan has carried out breeding activities on faba bean since 1960. Most released varieties are the result of selection from its own crosses. Lastly, finished varieties from several international sources, including India and Australia, loom large in the production of released field pea cultivars in Ethiopia.

The varietal output data in Ethiopia are rich in information on the notable traits of released varieties. The majority of barley releases are tolerant to the major leaf diseases of scald and net blotch. Some show tolerance to shootfly and early vigour that translate into healthy early-season, stand establishment. Others are prized for their good malting quality and potential for making local beer. Still others show specific regional and wide seasonal adaptation and can be cultivated in both the meher and belg seasons in Ethiopia. Tolerance to drought and resistance to lodging are other desirable traits that are mentioned in several of the release descriptions. A few releases are characterized by good biomass production, tall plant height and soft straw that confer advantages in feeding livestock. In Eritrea, tolerance to drought is a very important trait in the selection of varieties from landrace populations.

For the grain legumes, disease resistance, early maturity and large seed size figure prominently, in addition to high-yield potential, as desirable traits in East Africa. Tolerance to waterlogging is frequently mentioned in the faba bean release descriptions. Resistances to rust and wilt are valuable traits in lentil where root rot can be a severe constraint to production. There is also a strong demand for *Ascochyta* blight and *Fusarium* wilt resistances in chickpea. Market-related criteria, such as seed size and colour, are prized in field pea. In general, tolerance to and escape from abiotic stresses, resistance to major diseases and desirable market characteristics are foremost considerations in these released pulse varieties. Like the barley-released cultivars, most of these modern grain-legume varieties can stake a claim to scoring well on two to three of these desirable traits.

Adoption of Improved Barley, Chickpea, Faba Bean, Field Pea and Lentil Varieties

The uptake of improved barley varieties is substantially higher than the adoption of released varieties of ICARDA-mandated grain legumes in Ethiopia (Table 12.3). The adoption estimate of 39% is calculated from a survey of 1278 barley-cultivating households in the three main growing regions in Ethiopia (Yigezu *et al.*, 2012b). The estimate from the elicitation of expert opinion was 29% for the adoption of improved varieties in the same regions and administrative zones where the survey was conducted and 23% for the country as a whole when marginal geographic regions in the Rift valley and the short-duration rainy belg season are considered. Extrapolating the difference between the survey and expert estimates gives an estimate of about 26% for a mixed-source national estimate. Aggregating up secondary data at the lowest administrative level, the 'kebele', generates a national estimate of 16%; however, the uptake of the leading household-survey cultivar, M-21, a recently released variety rapidly gaining favour with farmers in the Orimoya region, is barely visible in the secondary data (Yigezu *et al.*, 2012b). When more marginal producing regions and seasons are considered, that estimate could decline to 26% based on expert opinion in all geographic areas in the two growing seasons. Both estimates are also substantially higher than the 11% national estimate in the late 1990s (Aw-Hassan, *et al.*, 2003a). Therefore, barley improvement has made steady progress in modern varietal adoption in the recent past.

Table 12.3. Adoption of improved cultivars in the ICARDA commodity mandate in East Africa in 2010 by crop, country and contribution of the IARCS.

			Percentage of area under the crop planted to:		Materials containing ICARDA germplasm or directly related to CG Center activities	
Crop	Country	Average area 2008–2010 (ha)	Local varieties	Improved varieties	Area (ha)	MV area (%)
Barley[a]	Ethiopia	969,000	61	39	132,000	40
Barley[b]	Eritrea	42,000	85	15	6,300	100
Chickpea[a]	Ethiopia	222,000	81	19	29,000	100
Chickpea[b]	Eritrea	6,000	98	2	120	100
Chickpea[b]	Sudan	21,000	0	100	21,000	100
Faba bean[a]	Ethiopia	537,000	89	11	3,650	17
Faba bean[b]	Sudan	77,000	13	87	0	0
Lentil[a]	Ethiopia	92,000	85	15	9,200	100
Field pea[b]	Ethiopia	225,000	98	2	0	0

[a]Nationally representative survey; [b]expert opinion.

With about 15–19% of area under improved varieties, the diffusion of modern chickpea and lentil cultivars is superior to the uptake of released materials for faba bean and field pea in Ethiopia. Adoption differences in chickpea and lentil on the one hand and faba bean and field pea on the other do not seem to be related to research output in the form of released varieties. All four crops are characterized by at least ten varieties in their opportunity set for adoption. Indeed, more varieties have been released in faba bean and field pea than in chickpea and lentil.

The improved varieties of the four pulse crops have one thing in common. Their cultivation is restricted to small geographic pockets in larger growing regions. This clustered spatial pattern is usually the result of well defined, project-related technology transfer programmes.

As a cereal with multiple uses, the prospects for improved varieties are brighter in barley than in grain legumes. First, barley has a substantially higher multiplication ratio than pulse crops where one planted seed seldom if ever generates more than 100 seeds of output. More output from the same weight of propagation materials potentially accelerates the diffusion of improved varieties. Secondly, local barley varieties are not suitable for brewing; therefore, as local and national brewing expands, the demand for improved malting barleys increases.

Eritrea lags behind Ethiopia in the adoption of barley and chickpea. Sudanese scientists reported very high levels of adoption of improved chickpea and faba bean varieties. Relatively small areas under cultivation with irrigation may partially account for these perceived levels that approach or equal full adoption.

The Highlands between 1700 and 3000 masl (metres above sea level) is the dominant agroecology for the cultivation of the ICARDA-mandated crops in Ethiopia. Their production relies on seasonal rainfall because they are rarely irrigated. Rainfed barley and chickpea in Eritrea is produced at a lower altitude in upland conditions. Faba bean and chickpea are irrigated in the Sudanese lowlands.

Within the highlands of Ethiopia, spatial variation in adoption of improved varieties is marked. For example, adoption of improved barley varieties in the three main highland regions varies from about 13% in Amhara to 53% in Orimiya (Yigezu *et al.*, 2012b). In several 'weredas' (districts) in Orimiya, adoption exceeds 75%, especially in areas that specialize in the production of malting barley.

In terms of production potential, barley can be divided into three recommendation domains in Ethiopia: high potential highlands, low potential highlands and the low moisture region. The high potential highlands contribute an area share of 75% and an adoption share of 85% which is greater than comparable shares for the two lower potential production domains but less than expected as improved varieties have also penetrated into areas of the lower potential regions.

Improved barley adoption varies by altitude. Barley is mainly cultivated between 2300 and 3200 masl. The household survey data show that this range of elevation accounts for 88% of growing area and 96% of adopted area. Altitudes below 2300 masl in Ethiopia are characterized by lower levels of improved variety adoption similar to the uplands in Eritrea.

About half of improved variety area is planted to two leading barley varieties in Ethiopia Miscal-21 (M-21) and Holker. M-21 is an improved malting variety that was released in 2006 (Table 12.4). It was bred and selected in Mexico in ICARDA's barley improvement programme hosted by CIMMYT. According to focus-group interviews in sampled communities, M-21 is prized for its high yield and desirable qualities for food and feed. In 2010, M-21 was increasing in area in 34 of the 35 communities where it was grown in the sample survey. It was claiming some area from traditional varieties but it was also replacing Beka, an early first-generation malting variety. Beka was introduced by EIAR from France in 1973.

From the perspective of the household survey of 1280 barley producers, experts markedly underestimated the economic importance of this recently released variety. Experts gave an estimate of 2–3% of area planted to M-21; the survey estimate was 9–11%. This difference partially explains why experts believed that the aggregate improved food varieties contributed more to improved varietal area than the total improved malting varieties. The split in expert opinion between these two types was about 14% for food and 9% for malting varieties. Comparable estimates on improved variety adoption from the household survey were 18% for food and 21% for malting varieties.

The second leading barley variety is also a malting variety. Holker is valued for its high yield, attractive price and its desirable food quality. Although released in 1979, it still seems to be spreading in the Ethiopian Highlands, albeit at a slow pace. Holker was bred at EIAR. Research in Kenya contributed to its pedigree.

None of the improved food varieties could stake a claim to more than 5% of area in the household survey but the two most popular, HB 42 and HB 1307, merit a brief description. HB 42 was the leading food variety. Its economic importance was equivalent to slightly less than 10%, or about 32,000 of the 330,000 ha of total area sown to improved barley varieties. Aside from its high yield and attractive price, it is valued for its large-sized grain.

Of all the released barley food varieties, HB 1307 has the most promising adoption prospects. Respondents in 12 of 13 communities where it was grown believed that it was expanding in area. Heavy yield, desirable grain colour and tolerance to lodging were cited as its strong points. It was released in 2006, and experts appeared to be aware of its potential. They placed it slightly ahead of HB 42 in adopted area.

Although released more than two decades apart, HB 42 and HB 1307 were both selected from EIAR crosses. Of the five top-ranked improved varieties in adoption, three were bred nationally and two were bred internationally. Collectively, they account for about 70% of adopted area. The remaining 30% is contributed by 14 other improved varieties that were usually identified by name in the household survey. These varieties tended to be pureline landrace selections and were more regional and location specific than the top five, which seemed to be characterized by wider adaptation.

Perceived adoption of two of the improved pulse varieties in Ethiopia also exceeded 5% in Table 12.4. Arerti is the most widely diffused modern variety of chickpea. It is an early-maturing, *Ascochyta*-resistant elite line from ICARDA. Alamaya is the leading improved lentil variety. It too is an elite ICARDA line, in this case incorporating rust resistance in a heavy-yielding background.

In Sudan, the leading varieties are Basabeer (BB7) in faba bean and Burgeig in chickpea. The share of their adopted area is believed to exceed 30% in each crop. Although breeding food legumes in Sudan started in 1960, information on the pedigrees of these leading varieties was not reported.

No improved faba bean or field pea variety was adopted on more than 5% of cultivated area in Ethiopia. Only 3 of 19 released faba bean varieties and 3 of 26 released field pea varieties were perceived as being cultivated in farmers' fields in 2010. For faba bean, West Shewa was the administrative zone with the highest level of adoption at only 7% of cultivated area. For field pea, the adoption level in the zone with the fastest uptake was even lower at 3% in Gurage.

Table 12.4. Economically important improved varieties of commodities in the ICARDA mandate in East Africa in 2010 by crop, country and ecology.

Crop	Country	Ecology	Variety name	Release year	Area coverage (% of ecology)	Total area of ecology ('000 ha)	Age
Faba bean	Ethiopia	Highland under rainfed conditions	CS-20DK	1977	3.34	15,650	33
			Degaga	2002	1.41	3,183	8
			Wolki	2007	2.04	55	3
Ethiopia total faba bean					11.29	18,888	
Faba bean	Sudan	Irrigated lowlands	Basabeer (BB7)	1993	35.00	27,000	17
			Hudeiba 93	1993	30.00	23,000	17
			Selaim-(SML)	1987	22.00	17,000	23
Lentil	Ethiopia	Highland under rainfed conditions	Alamaya	1997	14.43	6,977	13
			Teshale	2004	0.00	954	6
			Alem Tena	2004	0.00	156	6
			Adaa	1995	0.79	1,142	15
			Others		0.36	330	15
Ethiopia total lentil					15.58	9,559	
Lentil	Eritrea	Rainfed highlands	Bir Selam	2008	0	0	2
Barley	Ethiopia	Highlands under rainfed conditions	HB 42	1984	4.30	31,696	26
			HB1307	2006	2.50	16,284	4
			Aruso	2005	2.21	16,187	15
			Ardu1260-B	1986	1.67	7,367	24
			3336-20	1995	1.77	7,367	12
			Others	1975–2010	5.71	60,194	17.5
			Total food barley		18.15	139,095	
			Miscal-21 (M-21)	2006	10.6	87,625	4
			Holker	1979	7.95	81,712	31
			Beka	1973	2.59	19,289	37
			Total malting barley		21.14	188,626	
Ethiopia total barley					39.29	327,721	
Barley	Eritrea	Upland	Tekonda	2004	7.00	2,923	6
			Rahwa	2004	6.00	2,505	6
			Shishay	2004	2.00	835	6
			Total		15	6,263	
Chickpea	Ethiopia	Highlands under rainfed conditions	Arerti	2000	10.19	14,852	10
			Shasho	2000	5.63	3,482	10
			Habru	2004	0.88	4,683	6
			Natoli	2007	0.03	698	3
			Others	1974–2010	2.13	5,317	18
Ethiopia total chickpea					18.8	29,023	
Chickpea	Eritrea	Upland	Names not known	NK	2.00	118	NK
Chickpea	Sudan	Irrigated lowlands	Burgeig (ICCV 91-302)	1998	40.00	8,400	12
			Atmor (ICCV 89-509)	1996	15.00	3,150	14
			Shendi (ILC 1335)	1987	10.00	2,100	23
			Hawata (ICCV 92-318)	1998	10.00	2,100	12
			Salawa (FLIP 89-82C)	1996	10.00	2,100	14
			Matama-1 (FLIP 91-77C)	1998	10.00	2,100	12
			Wad Hamid (ICCV-2)	1996	5.00	1,050	14
Field pea	Ethiopia	Highlands under rainfed conditions	Adi (G22763-2C x 305PS210813-2)	1995	1.10	2,473	15
			Mohanderfer	1979	0.37	827	31
			Tegegnech	1994	0.06	124	16
Ethiopia total field pea					1.44	3,424	

NK, not known.

In general, data in the last column of Table 12.4 suggest that slow rates of varietal turnover are most problematic for the faba bean improvement programme in the Sudan where the varieties in farmers' fields were released 17–23 years ago. An ageing variety could also be a cause for concern for faba bean in Ethiopia as the most popular improved cultivar was released in 1977. However, the issue of slow varietal turnover is overwhelmed by the importance of increasing varietal adoption from a very low level. Prior to the release and adoption of M-21, slow varietal turnover was an increasingly pressing issue for the genetic improvement of malting barleys in Ethiopia. The rapid early acceptance of M-21 has quickened the pace of varietal turnover and has mitigated the seriousness of that issue, as age should be trending downwards over time with the substitution of M-21 for Beka.

Summary

Several empirical results in this paper provide grounds for optimism about the prospects for improved varietal change in the ICARDA-mandated crops in East Africa especially in Ethiopia. The uptake of improved varieties of barley has increased from about 10% of cultivated area in the late 1990s to about 39% today. Improved malting barleys account for most of the gains in adoption; however, released food barleys are also more readily finding a home in farmers' fields. The release of several ICARDA-related varieties in the 2000s has added to the dynamism of varietal production in EIAR's barley improvement programme. The rapid early adoption of M-21 is an emerging success story.

Better-staffed crop improvement programmes in Ethiopia in barley and in grain legumes and in the Sudan in grain legumes represent another encouraging development. Not only are the programmes strong in numbers, but also a high proportion of scientists have MScs and PhDs.

In Ethiopia, varietal output in all five of the ICARDA-mandated crops has been sufficient to drive improved varietal change. Except for lentil, each crop improvement programme has released at least ten varieties since 2001.

The results have also shed light on multiple areas for improvement in generating improved varieties characterized by better adoption outcomes. For Ethiopia, poor past adoption performance in the four pulses is a cause for concern. The diffusion of improved varieties in chickpea and lentil is not substantially different from 10%; the adoption of modern faba bean and field pea cultivars is less than 5%. In contrast, the results in Chapter 8 on bean, called haricot bean in Ethiopia, show estimated adoption approaching or exceeding 40%. This difference begs the question: why is adoption of improved bean varieties, such as Awash and Nasir, substantially higher than farmer acceptance of released varieties of the other four economically important grain legumes in Ethiopia? In the same vein, why is modern variety adoption higher in chickpea and lentil than in faba bean and field pea when varietal output has been greater in the latter crops than in the former? Informative responses to these two questions could help redirect crop improvement and/or technology transfer strategies.

Barley in Ethiopia faces the common problem of low acceptance of improved varieties in a large-producing region. Nationally representative survey results are consistent with a difference in estimated adoption of about 40% in the two major-producing regions in the Ethiopian Highlands. Amhara lags far behind Oromiya in improved cultivar adoption. Arriving at a better understanding of the causes of these spatial differences in adoption could also be informative for fine-tuning plant breeding strategies and tactics.

The absence of released varieties since 2000 seems to be the most problematic issue for pulse improvement in the Sudan. As a result, first-generation improved varieties are not being replaced by more productive younger varieties in farmers' fields. Slow varietal turnover of improved varieties translates into stagnating outcomes for the faba bean crop improvement programme. Given fairly high staff numbers and education levels, a paltry performance in varietal release is puzzling.

Eritrea needs to make more of a commitment to agricultural research. Its staffing pattern seems too low to make sustained progress in varietal development, adoption and impact. Without more commitment, it will have to depend on varieties released in neighbouring countries.

The results also have two methods-related implications. First, expert opinion elicitation of cultivar-specific adoption estimates is best performed at a low level of spatial aggregation. It is easier to arrive at a consensus at a lower administrative level. Staying at a more aggregate level, such as an agroecology, would have led to higher and most likely more unrealistic estimates of improved cultivar adoption. Secondly, extension agents and non-governmental organization (NGO) transfer specialists should also be included in the pool of experts. In this study, experts may have been defined too narrowly to include only scientists.

Note

[1] This paper is a revised and abridged version of ICARDA's Objective 1 Report submitted to the DIIVA Project in June 2012 (Yigezu *et al.*, 2012a).

References

Aw-Hassan, A. and Shideed, K., with Ceccarelli, S., Erskine, W., Grando, S. and Tutwiler, R. (2003a) The impact of international and national investment in barley germplasm improvement in the developing countries. In: Evenson, R.E. and Gollin, D. (eds) *Crop Variety Improvement and Its Effect on Productivity: The Impact of International Agricultural Research.* CAB International, Wallingford, UK, pp. 241–256.

Aw-Hassan, A., Shideed, K., with Sarker, A., Tutwiler, R. and Erskine, W. (2003b) Economic impact of international and national lentil improvement research in developing countries. In: Evenson, R.E. and Gollin, D. (eds) *Crop Variety Improvement and Its Effect on Productivity: The Impact of International Agricultural Research.* CAB International, Wallingford, UK, pp. 275–292.

Yigezu, Y.A., Yirga, C., Aw-Hassan, A. and March, A. (2012a) *Varietal Output and Adoption in Barley, Lentils, Chickpeas, Faba Beans, and Field Peas in Ethiopia, Eretria, and Sudan.* Diffusion and impact of improved varieties in Africa (DIIVA) – Objective 1 Report. ICARDA, Aleppo, Syria.

Yigezu, Y.A., Yirga, C., Aw-Hassan, A. and Labarta, R. (2012b) *Barley in Ethiopia: Verifying the Adoption Estimates from a Panel of Experts with Secondary Data and Household and Community Focus Group Surveys.* Diffusion and Impact of Improved Varieties in Africa (DIIVA) – Objective 2 Report. ICARDA, Aleppo, Syria.

Appendix 12.1

Table 12.A1. Priority areas for survey on adoption/diffusion studies: areas of high potential and high likelihood of diffusion.

	Crops							
	Lentils and chickpeas		Faba bean and field pea[a]			Barley		
Location	HP	LM	LT	RS	BS	HP	LM	Belg[b]
Shewa								
North Shewa (Oromiya)	***				**	***		*
North Shewa (Amhara)			*					
Enewari								
West Shewa								
East Shewa	***	*						
SW Shewa					***			
Gurage zone				***				
Arsi	*			***		***		*
Gonder								
North Gonder	*		*	**		***		
South Gonder						**		
Bale	*					**		*
North Ethiopia								
Tigray		*					**	
North-east Ethiopia								
Afar region								
Wollo								
South Wollo	*							*
North Wollo							**	
East Wollo								
Hadiya zone								
Rift valley								
Gojam	*							

[a]The distribution of field peas is exactly the same as that of faba beans except that field peas are not grown in the black soil (vertisol) regions.[b]Belg is the local name of the shorter rainy season in Ethiopia which usually is received between March and April. ***,**,* represent high, medium and low levels of diffusion of the new crop varieties, respectively. Ranking (using the number of asterisks) is done only within each crop (column). Hence, three asterisks for faba beans show high adoption relative to other zones producing faba beans but it does not imply equality in the level of adoption of lentils in zones with three asterisks in the lentils column. HP, high potential; LM, low moisture; LT, low temperature; RS, red soil; BS, black soil.

13 Scientific Strength in Rice Improvement Programmes, Varietal Outputs and Adoption of Improved Varieties in South Asia

S. Pandey,* Ma. L. Velasco and T. Yamano
International Rice Research Institute (IRRI), Manila, Philippines

Introduction

Rice is the staple crop of South Asia. The Green Revolution resulting from the spread of improved rice varieties and associated technologies, such as irrigation and fertilizers, led to a rapid growth in rice production over the past six decades. This has resulted in improvements in food security for growing populations and in poverty reduction throughout the region (Hazell, 2010).

A key ingredient for the success of the Green Revolution has been the development of improved rice cultivars. International and national rice breeding programmes have developed a large number of improved varieties (often known as modern varieties or MVs) during the past six decades. A productive breeding programme that generates a continuous flow of improved varieties is needed to overcome existing and evolving constraints to growth in rice productivity. A deceleration in growth in yield during the late 1990s and early 2000s (Dawe *et al.*, 2010) has raised questions about the effectiveness of rice-breeding programmes. Two major indicators of the strength of rice improvement programmes are the scientific capacity of the staff involved and the annual flow of varietal outputs. Therefore, it is important to investigate how the scientific strength and the varietal flows of rice are changing over time. These indicators provide information on the potential impact of crop improvement programmes.

The actual impact of crop improvement programmes, however, depends on the extent of adoption of improved varieties. Estimates of cultivar-specific adoption levels are needed to judge how well the recent outputs of breeding programmes are contributing to productivity growth or to overcoming other evolving constraints in rice production.

A project complementary to DIIVA, but focused on rainfed areas of south Asia, was implemented during 2010 to 2013 to assess the effectiveness of crop improvement programmes of important food crops. The project 'Tracking Improved Varieties in South Asia' (or TRIVSA) covered rice for the humid/subhumid region and sorghum, pigeon pea, pearl millet, groundnut and chickpea for the semi-arid areas. The main objectives of the project were to develop recent estimates of scientific capacity of crop improvement

* Sushil Pandey, formerly a senior economist at IRRI, is currently an independent researcher based in Manila, Philippines.
E-mail: sushilpandey056@gmail.org

programmes in national systems, assess the varietal outputs and estimate cultivar-specific adoption of improved varieties. The project was implemented in collaboration with the national programmes of these countries and generated a comparative benchmark for the results from sub-Saharan Africa for predominantly rainfed food crops.

This chapter focuses on rice in India, Bangladesh, Bhutan, Nepal and Sri Lanka (Fig. 13.1). In India, the project covered the eastern states of Chhattisgarh, Odisha and West Bengal. General background information on rice production systems is briefly described in the next section. The study approach is outlined subsequently in the third section. Sections four to six provide empirical results on scientific strength of publicly funded rice improvement programmes for 2010, varietal outputs until 2010 and cultivar-specific adoption levels for the cropping season 2010–2011. These results

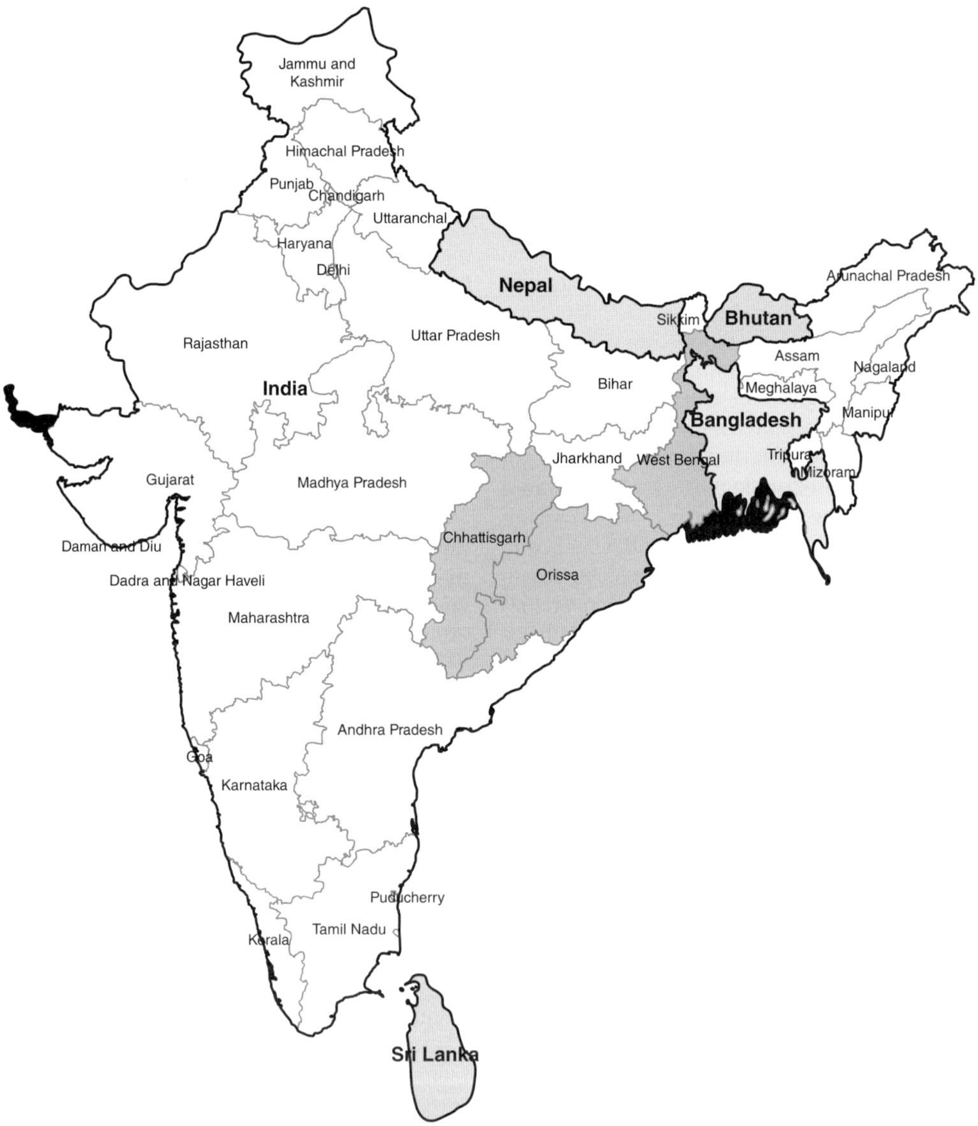

Fig. 13.1. South Asian countries covered in the project.

sections are followed by a broader discussion of their implications and of the methods used. This concluding section also contains a summary of substantive findings.

Characteristics of Rice Production Systems of South Asia

Rice production of South Asian countries covered in this project varies from Bhutan with only 23,000 hectares of rice to India with 43 million hectares (Table 13.1). These five countries account for over 55 million hectares of rice area and over 190 million tonnes of production.[1] The total rice area in the three eastern Indian states included in the project is approximately 13 million hectares, which accounts for 30% of the total rice area in India.[2] Average yield is highest in Bangladesh at 4.1 t/ha. In India, the national average yield is 3.3 t/ha, but the average yields in Odisha and Chhattisgarh are about 1 t/ha lower.

Rice production has increased over time in all states/countries over the past four decades, with the average growth rate being in the range 1.64% per year to 2.67% per year (Table 13.2). In all cases, yield growth has been the main contributor to production growth.

Rice production in West Bengal and Bangladesh achieved relatively higher growth rates compared to other states/countries. This impressive increase in production and yield was the result of a combined effect of expansion of irrigation and the adoption of improved varieties, particularly early maturing ones that facilitated the cultivation of a second crop in the dry season (or boro crop) using irrigation. As a result, the average rice yield in West Bengal increased to 3.86 t/ha. Similar factors also contributed to higher rice productivity in Bangladesh.

Chhattisgarh is a relatively new state in eastern India carved out of the south-eastern districts of Madhya Pradesh in 2000. Aggregate rice production in the state, based on district-level data prior to 2000, showed that rice production almost doubled between 1970 and 2010 due mainly to yield growth. Being a drought-prone state, yield fluctuations in the state are, however, high (Bhandari *et al.*, 2007). This characteristic is also shared by Odisha where yield is severely affected not only by drought but also by submergence in low-lying areas and salinity in the coastal belts.

Table 13.1. Area, production and yield of rough rice (triennium average, 2008–2010).

	Harvested area ('000 ha)	Production ('000 t)	Yield (t/ha)
India	43,370	142,139	3.28
West Bengal	5,503	21,212	3.86
Odisha	4,349	10,279	2.36
Chhattisgarh	3,703	6,898	1.86
Bangladesh	11,553	46,955	4.06
Nepal	1,529	4,276	2.80
Sri Lanka	957	3,611	3.83
Bhutan	23	73	3.20

Data sources: RNR Statistics, Ministry of Agriculture for Bhutan; Indiastat.com for Indian states; USDA for others.

Table 13.2. Growth rates of rice area, production and yield (% per year), 1970–2010.

	Area	Production	Yield
India	0.39	2.33	1.94
West Bengal	0.26	2.67	2.40
Odisha	0.01	1.64	1.64
Chhattisgarh	0.57	1.91	1.34
Bangladesh	0.33	2.77	2.43
Nepal	0.70	1.90	1.19
Sri Lanka	0.55	2.23	1.68
Bhutan	–0.56	1.86	2.41

Early-maturing improved varieties were readily accepted by Sri Lankan farmers boosting the rice productivity growth from as early as the 1970s. Adoption of improved rice varieties in Sri Lanka is almost 100%, with traditional varieties being grown in some small pockets only. Despite such high adoption rates, productivity levels in less favourable rainfed areas are still low and unstable.

Nepal experienced a slow but steady growth in productivity. The overall productivity level is still low relative to other countries. This is mainly attributed to the dominance of a rainfed environment that accounts for over 70% of rice area. Most of the productivity growth is driven by the southern plains (known as the Terai), the main granary of Nepal. There is a substantial inflow of improved rice varieties into the Terai from the adjoining states of India.

In the case of the mountainous country Bhutan, rice production is spread over diverse climatic conditions from high-altitude northern mountains to low-altitude southern plains. Productivity varies across these climatic zones accordingly. This is the only country where there has been some contraction in rice area, but statistical information on rice area and production vary widely depending on the source of data. Bhutan has benefited from substantial inflow of rice varieties from Nepal and India.

The Study Approach: Methods and Data

To achieve the three objectives laid out earlier, data for the most recent years from multiple sources were assembled. Collaborative arrangements with eight national agencies involved in rice research were established in the five study countries (Table 13.3).

Assessing scientific strength in rice improvement

Information on several aspects of scientific strength of national agricultural research systems (NARS) in rice improvement in 2010 was collected through a survey of rice scientists. The survey questionnaires were distributed to pre-identified researchers affiliated with public research organizations and agricultural universities involved in rice improvement programmes. Information was collected on the number of scientists involved in crop improvement programmes, their full time equivalents (FTEs), the level of education of rice scientists, thematic areas of research and the target region/ecology. Only those scientists who were engaged directly in breeding, pre-breeding and other breeding-related activities were included in the survey.[3] In spite of this narrow definition, the included disciplines extend to pathologists, entomologists, physiologists, agronomists, and other areas that directly support and participate in breeding-related activities. Scientists were defined as those with at least a Bachelor of Science degree.

Assembling data on varietal output

Information on varietal releases is assembled from national data sources. The database typically consists of the release name, year of release, method used for developing the variety (e.g. hybridization, pureline selection), parentage, institution that developed the variety, target ecosystem/season, average grain yield and duration. In some cases, other characteristics such as tolerance/resistance to pest/diseases, grain type and grain quality are also included.

Estimating cultivar-specific adoption

Various methods are available for generating cultivar-specific adoption data. These have been mostly based on household-level surveys that may or may not generate nationally representative reliable estimates of cultivar-specific adoption. Ad-hoc surveys that are project-specific mostly generate some detailed data on adoption levels of various cultivars but the coverage of

Table 13.3. NARS (national agricultural research system) partners.

	Collaborating institute	Key collaborator
India		
West Bengal	NZFDO (NGO)	Dr Bhanudeb Bagchi
Odisha	Orissa University of Agriculture and Technology	Dr Debdutt Behura
Chhattisgarh	Indira Gandhi Agricultural University	Dr Ajay Kumar Koshta
Bangladesh	Bangladesh Rice Research Institute	Dr Md. Abu Bakr Siddique
Nepal	Nepal Agricultural Research Council	Mr Sudeep Gautam
	Institute of Agriculture and Animal Science	Mr Hari Panta
Sri Lanka	Department of Agriculture	Ms Buddhini Ranjika Walisinghe
Bhutan	Ministry of Agriculture	Mr Mahesh Ghimiray

such surveys is mostly limited. Large-scale surveys with broader coverage are mostly expensive and take considerable time to implement and process the data. In addition, such surveys are likely to miss improved cultivars in their early stages of adoption. Another alternative is to derive cultivar-specific adoption levels from seed sale data. However, these provide reliable estimates only when farmers do not keep their own seed and purchase the seed every year. In the case of rice, farmers in Asia mostly use their own rice seed, thus the seed sale data can only provide a lower bound estimate of adoption levels. Information can be gleaned similarly from other large-scale surveys such as cost-of-cultivation surveys or crop-cut surveys that are an integral part of agricultural statistics in several countries. But cultivar-specific adoption data are mostly not collected or even if collected such information is generally lost in the process of aggregation. In addition, the sampling design for such surveys is aimed at minimizing the error associated with the primary variable of interest (cost of cultivation or yield) and does not necessarily generate precise estimates of cultivar-specific adoption.

Given these limitations of the more usual approaches, an alternative may be to use an expert panel to elicit the cultivar-specific adoption data. Following the broader approach used in DIIVA, this study also relies on the use of expert panels to generate cultivar-specific adoption data for improved varieties of rice. Household surveys on limited scales were also conducted with the aim of validating the results from the expert panels.

A panel of experts knowledgeable about the adoption of rice cultivars in the domain (ecosystem, season or administrative zone) was formed. Typically, a panel consisted of 10–15 experts including breeders, extension workers, seed traders, local agricultural officers and others knowledgeable about rice production systems in the locality. Experts were requested to provide their initial estimates of percentage area under all improved varieties and landraces in the domain. Subsequently, they were asked to allocate the estimated area under improved varieties into cultivar-specific proportional areas for the top ten improved varieties. Improved varieties that did not fall in the top-ten category were grouped as 'other improved varieties' and given the residual share in the total percentage area under improved varieties. Each expert initially provided their own estimates.

The elicitation process based on an expert panel broadly follows a condensed version of the 13-step process outlined in Chapter 4 of this volume (Walker, 2010). The steps are:

1. Initial estimates from individual experts (without consultation with others) for the country/state/district, based on their prior knowledge and without any reference to supplementary notes;
2. Revised individual estimates after a list of varietal releases is provided;
3. Initial group estimate by ecosystem/season for the given geographical domain, with the available experts grouped into 3–6 groups that provide a good mix of expertise in each group of 3–5 members;
4. Revised group estimates for each ecosystem/season when there are more than one group for each ecosystem/season; and
5. Overall consensus group estimate with the involvement of all experts as a single group.

These steps, with some local adaptations, were applied to generate cultivar-specific adoption levels of improved rice cultivars. Adoption was measured in terms of area under the improved cultivar. An improved cultivar was defined as the rice cultivar listed in the official release list of the country or if it is known to be an improved cultivar that is sourced from another country.

The expert panel approach was used in all three states of eastern India and the other four countries included in the project. Adoption of rice cultivars is generally specific to agroecology (upland, medium land, lowland, rainfed/irrigated, coastal/inland etc.) and seasons (pre-monsoon, monsoon or dry). Hence, estimates based on expert panels were elicited for each of the major agroecologies or seasons and these estimates were aggregated to generate state/national level adoption levels using area weights. In some cases, elicitation was done at the district level (e.g. West Bengal and Nepal) and the information was aggregated to larger geographical units using area weights. Details of implementation of these expert elicitations are described in Velasco *et al.* (2013). The overall coverage is summarized in Table 13.4.

Validation of cultivar-specific adoption estimates generated via the expert-panel method was achieved by comparing the results from household-level adoption surveys supported by TRIVSA or household-survey data available from other sources. The TRIVSA-supported surveys included information on household demographics, rice varieties grown in 2010–2011 cropping year and their respective areas, varietal adoption in the past, sources of varietal information and major traits of cultivars grown. In addition, information on varietal abandonment and replacement was collected.

Table 13.4. Spatial coverage of expert panel elicitation.

	Elicitation at the country/ state level	Elicitation at the district/ ecosystem level	Number of districts covered
India			
West Bengal	No	Yes	17
Odisha[a]	Yes	Yes	29
Chhattisgarh	Yes	Yes	8
Bangladesh	Yes	Yes	6
Nepal	No	Yes	29
Sri Lanka	Yes	No	0
Bhutan	Yes	No	0

[a]Odisha had the most complete coverage including elicitation at the state, ecosystems and district levels.

Sampling for the adoption surveys was designed to capture spatial variability as much as possible. This was achieved for a fixed budget by an extensive survey approach in which more villages/districts were included by limiting the number of households in each village. A multistage sampling with stratification in selected districts was used. Sampling design ensured the inclusion of two to six blocks per district, one to six villages per block and two to ten households per village. At least one district in each ecosystem was included to capture the ecosystems-level diversity. The total sample size consisted of 7286 households across five countries (Table 13.5).

Prior to household-level surveys, focus group discussions were conducted at the village (or community level) to broadly describe the recent changes in rice varietal composition and establish broader information on rice production systems. Open-ended questions formed the basis for eliciting qualitative information from focus groups typically consisting of four to eight key

Table 13.5. Coverage of household and community surveys.

	Districts (#)	Blocks (#)	Villages (#)	Farmers (#)
Household surveys				
India				
West Bengal	17	34	126	1,262
Odisha	29	159	307	3,139
Chhattisgarh	8	19	120	902
Bangladesh	18	53	61	522
Nepal	29	174	265	1,160
Sri Lanka	na	na	na	na
Bhutan	8	40	154	301
Total	109	479	1,033	7,286
Community surveys				
India				
West Bengal	17	34	126	
Odisha	29	158	302	
Chhattisgarh	8	19	78	
Bangladesh	18	53	53	
Nepal	29	68	116	
Sri Lanka	na	na	na	
Bhutan	na	na	na	
Total	101	332	675	

Note: Not applicable (na) because no surveys supported by TRIVSA were conducted in Sri Lanka, which had cultivar-specific adoption data from other sources. Community surveys were not conducted in Bhutan.

informants. These surveys were conducted in about 70% of the villages that were included in the household survey. Key informants were also asked to provide information on area coverage of major improved varieties grown in the village in 2010 and shifts in cultivars during the previous 10 years (from 2000 to 2010).

Scientific Capacity in Rice Breeding

The results indicated that among the three states of eastern India and the four countries, Bangladesh has the highest number of FTE scientists and Bhutan the lowest (Table 13.6). These absolute values are not strictly comparable owing to differences in the total rice production among states/countries included. Normalized FTE scientists per million tonnes of production indicates very small values (<2 FTE scientists/million tonnes) for West Bengal, Chhattisgarh, Bangladesh and Nepal. The very high figure of normalized value in Bhutan is the result of a very small rice production base even though the absolute value of FTE scientists is the lowest. This is a typical result when the production domain is small. The value for Sri Lanka is high partly because of the small rice output but the country has also invested relatively more in rice improvement as indicated by the absolute value of FTE scientists. In the case of Odisha, the presence of a large breeding programme at the Central Rice Research Institute (located in Odisha) with its broader mandate for rainfed environments of India contributed to a higher normalized value of FTE scientists. In terms of the normalized FTE scientists, West Bengal has the lowest scientific strength.

Overall, the research intensities (using normalized FTE scientists as proxy) are quite small, with less than one FTE scientist invested in crop improvement programmes per million tonnes of rice output in three large production areas (West Bengal, Chhattisgarh and Bangladesh) (see Table 13.6). The research intensity for Nepal is less than two FTE scientists. Comparable data for rice research in Africa indicates 7 FTE scientists/million tonnes (Chapter 10, this volume). Thus, these regions in Asia where rice is a major staple have invested at a substantially lower level than in Africa. If 5 FTE scientists/million tonnes is considered to be a threshold level of investment for an effective crop improvement programme, almost all cases included fall below this level, indicating a case of under-investment. Other analyses that have considered both actual costs and potential benefits of breeding programmes have also concluded that in general there is an under-investment in rice research (Pandey and Pal, 2007; Gauchan and Pandey, 2011).

In terms of the educational attainment, there is a dominance of PhD graduates in the Indian states, whereas MSc graduates account for over 50% of FTE scientists in the other four countries (Table 13.7). The share of PhD graduates in FTE scientists is quite small in Bhutan and Sri Lanka at only 14%. Scientists in India, in general, have a higher level of education than in other countries.

In terms of disciplinary specialization of scientists working on rice improvement programmes, plant breeding and genetics accounts for more

Table 13.6. Full time equivalent scientists in rice genetic improvement research by Indian state or country.

	West Bengal	Odisha	Chhattisgarh	Bangladesh	Nepal	Sri Lanka	Bhutan
Scientists (total)	14	44	8	48	10	24	5
FTE scientists	12	40	6	41	7	18	2
FTE scientists (%)	88	91	69	86	74	77	42
FTE scientists/ million tonnes[a]	0.6	3.9	0.7	0.9	1.8	5.0	29.2
FTE scientists/ million hectares[a]	2.5	9.4	1.5	3.5	5.0	16.5	89.6

[a]2008–2010 triennium average rice production and area was used to obtain normalized FTE scientists. The empirical information on TRIVSA presented here and in rest of the chapter is based on individual country reports prepared by Salam *et al.* (2013), Ghimiray *et al.* (2013), Koshta and Chaudhary (2013), Gautam *et al.* (2013), Behura *et al.* (2013), Walisinghe *et al.* (2013) and Bagchi *et al.* (2013).

than 50% in most cases (Table 13.8). This is not surprising as the survey included only those scientists who are directly involved in crop improvement programmes. Other broader disciplines (agronomy, entomology, plant physiology and plant pathology) also account for a substantial share in FTE scientists. This may indicate the presence of a broad-based and integrative crop improvement programme. It was not possible to ascertain if this was really the case using the data available. What is somewhat surprising though is that biotechnology (and molecular biology) accounts for a very small share and is almost non-existent in several cases. It seems that the rice improvement programmes have not developed adequate in-house capacity to utilize these modern tools.

Breeding for high yield and abiotic stress tolerance are the main thematic foci of crop improvement programmes (Table 13.9). Although

Table 13.7. Educational attainment of rice scientists (% share of total FTE scientists).

	West Bengal	Odisha	Chhattisgarh	Bangladesh	Nepal	Sri Lanka	Bhutan
Bachelor of Science (BSc)[a]						10	
Master of Science (MSc)		16		66	54	76	86
Doctor of Philosophy (PhD)	100	84	100	34	46	14	14

[a]BSc graduates are not reported except in Sri Lanka presumably because they are considered as research support staff rather than research scientists.

Table 13.8. Disciplines of rice scientists (% share of total FTE scientists).

Discipline	West Bengal	Odisha	Chhattisgarh	Bangladesh	Nepal	Sri Lanka	Bhutan
Plant breeding and genetics	60	44	85	51	94	63	57
Plant pathology		10	15	11		5	14
Plant physiology	8	13		6			
Entomology		12		10		9	
Agronomy		3		5	6	12	29
Biotechnology		7		13			
Others[a]	32	12		4		10	

[a]Others include agricultural biology, agricultural botany and agricultural statistics.

Table 13.9. Rice research themes (% share of total FTE scientists).

Research themes	West Bengal	Odisha	Chhattisgarh	Bangladesh	Nepal	Sri Lanka	Bhutan
Breeding							
Abiotic stress	17	14	–	28	60	9	7
Higher yield	30	15	28	23	34	21	42
Grain quality	15	8	11	5	–	10	–
Pest/disease	–	15	26	19	5	13	–
Other breeding	17	11	15	12	1	10	–
Subtotal	79	63	80	87	100	63	49
Breeding related							
Germplasm conversation	2	7	16	1	–	5	16
Crop management/ seed production	19	30	4	12	–	32	35
Subtotal	21	37	20	13	–	37	51

there are variations across locations, grain quality improvement is an important thematic area. Consumer demand for improved grain quality generally tends to increase with an increase in income, and breeding programmes are probably responding to such demands. Pest management is another area where breeding programmes have traditionally paid considerable attention. Broader non-specific breeding activities, such as genomics, nutrient efficiency and microbial diversity, account for around 10–17% of FTE scientists.

It is also notable that crop improvement programmes allocated between 4% and 35% of FTE scientists on general crop management work. To the extent that these are not an integral part of the breeding programme, this may indicate some 'diversion' of the programme resources into other general agronomic work. There may be some opportunity to reduce such diversions through better rationalization of programmes.

FTE scientist allocation by agroecological targets for crop improvement programmes indicates an implicit prioritization. The agroecological classifications are not uniform across states/countries. Summary results are presented in Table 13.10. It is interesting to note that the share of FTE scientists targeted to irrigated ecosystems is quite substantial across eastern Indian states despite the domain being mainly rainfed. For example, over 80% of FTE scientists are allocated to crop improvement for irrigated ecosystems in West Bengal.

Although dry season irrigated rice has become quite important with the expansion of boro production in West Bengal, allocation of such a large proportion of FTE scientists indicates potential over-investment in this ecosystem because the share of boro rice in total production is only around 60%. The share of irrigated rice production in total rice output in the remaining two states is also lower than the share of FTE scientist allocation to this ecosystem. Hence, a re-examination of resource allocation patterns across ecosystems in these predominantly rainfed areas of eastern India seems desirable.

For the four countries, the relative shares of FTE scientists allocated to various agroecological zones are approximately proportional to their respective area shares.[4] In this sense, relative prioritization may be closer to optimal. However, a closer analysis may provide guidance for some efficiency gains through re-allocation.

Varietal Output

A summary of rice varietal output is provided in Table 13.11. The rate of release of new varieties for countries other than India is 1–1.6 varieties per year. The total number of releases for Bangladesh, Nepal and Sri Lanka are in the range 60–70, with the number of releases for Bhutan being 24.

In the case of India, the patterns are somewhat different. For India as a whole, the average annual rate of release is 20 varieties per year.[5] The release rates for Odisha and West Bengal are about three varieties per year, and for Chhattisgarh, it is low at one variety per year. Except for Chattisgarh, the varietal release rate for eastern Indian states is broadly on a par with the overall country-level release rate for India. For comparative purposes across states/countries, the total releases are normalized by current rice area. As expected, the normalized values indicate an inverse relationship between varietal output and the total rice area. The output intensity indicates that Bangladesh has the lowest intensity, at six varieties per million hectares. At the other end of the spectrum, Bhutan has a ridiculously high value of 1044 per million hectares – the result of a very small area (23,000 ha) of rice in that country. Nepal has moderate output intensity at 41 varieties per million hectares.

In the case of India, the overall output intensity is 23 varieties per million hectares. The output intensity for Odisha is higher than the overall India average due mainly to the inclusion of varieties developed by the Central Rice Research Institute (CRRI) with its mandate for improving rice varieties for rainfed areas of all of India. The nature of changes in the breeding programme can be judged by the pattern of temporal shifts in varietal output (Table 13.12). Varietal output generally increased over time but there is some slowing down in recent decades, although the pattern is not consistent across countries. This slowing down is probably due to the lagged effect of the resource squeeze in agricultural research experienced during the 1990s. It may take several years to counteract this lagged effect, although research investment has increased in recent years.

In terms of target ecosystems, the primary target of rice varieties released in India is the irrigated environment (Table 13.13). Overall,

Table 13.10. Share of FTE scientists and rice area of each agroecology.

	Agroecology	Share of FTE scientists (%)	Share of rice area (%)
India			
West Bengal	Irrigated	81	43
	Rainfed lowland	8	37
	Rainfed upland	3	11
	Rainfed deepwater	8	9
	Total (%)	100	
Odisha	Irrigated	40	23
	Rainfed lowland	20	39
	Rainfed upland	12	15
	Rainfed deepwater	10	23
	Across ecosystem	18	
	Total (%)	100	
Chhattisgarh	Irrigated	35	24
	Rainfed lowland	27	49
	Rainfed upland	11	24
	Across ecosystem	27	
	Total (%)	100	
Bangladesh	Aman	39	49
	Boro	39	41
	Aus	16	10
	Boro, Aus, Aman	6	
	Total (%)	100	
Nepal	Terai	70	70
	Mid-hills	23	26
	High-hills	7	4
	Total (%)	100	
Sri Lanka	Dry zone	16	60
	Dry and intermediate zone	56	
	Intermediate zone	3	22
	Wet zone	25	18
	Unfavourable land	0.2	
	Total (%)	100	
Bhutan	High altitude	39	20
	Mid altitude	39	45
	Low altitude	22	35
	Total (%)	100	

Sources: West Bengal – CACP, Government of India (various issues); report on cost of cultivation of principal crops in India. Odisha – IRRI (1992); Singh (2000). Chhhattisgarh – Agriculture Statistics, Yearly Published from Commissioner of Land Records and Settlements, Government of Chhattisgarh, Raipur (Chhattisgarh). Bangladesh – AED (2010). Nepal – MoAC (2010). Sri Lanka – Department of Census and Statistics, Sri Lanka. Bhutan – country report.

the irrigated ecosystem is the target for two-thirds of the varieties released so far in India. The temporal changes indicate some increases in the varietal output share of rainfed environments up until 2000 but the share declined slightly during 2001–2010. Within the rainfed environment, the varietal output share of deep water ecosystems has declined. Overall, this ecosystem has become less important in terms of area. This pattern is broadly similar in the eastern Indian states, except for West Bengal.[6]

In Bangladesh, the varietal release pattern indicates a shift from varieties with broad adaptation to all seasons to specific seasons (Table 13.14). The share of varieties adapted to all seasons regardless of season has decreased considerably over time. This is an indication of increasing maturity of the breeding programme.

Table 13.11. Summary of varietal releases in different countries/states.

	Period of release	Varieties released (#)	Annual release rate	Varieties/million ha[a]
India[b]	1961–2010	992	19.8	22.9
West Bengal	1969–2007	120	3.1	21.8
Odisha	1968–2010	144	3.3	33.1
Chhattisgarh	1996–2010	15	1.0	4.1
Bangladesh	1966–2010	72	1.6	6.2
Nepal	1966–2010	62	1.4	40.5
Sri Lanka	1958–2010	69	1.3	72.1
Bhutan	1988–2010	24	1.0	1043.5

[a]2008–2010 triennium average rice area was used to obtain the number of varietal release per unit area. [b]Central releases for the whole of India include 57 varieties released from 1962 to 1984, which were denotified in later years. No information on denotification of varieties released after 1984 was recorded in the database.

Table 13.12. Number of varieties released by decade in different countries/states.

	1961–1970	1971–1980	1981–1990	1991–2000	2001–2010	All years
India[a]	45	159	224	271	280	979
West Bengal[b]	1	14	63	25	17	120
Odisha	6	22	42	36	38	144
Chhattisgarh	na	na	na	3	12	15
Bangladesh	6	11	13	20	22	72
Nepal	6	11	17	10	18	62
Sri Lanka	6	14	17	17	14	68
Bhutan	0	0	6	6	12	24
Total for five countries	63	195	277	324	346	1205

[a]Central releases for whole India include 57 varieties released from 1962 to 1984, which were denotified in later years. No information on denotification of varieties released after 1984 was recorded in the database. This list excludes 13 varieties for which the information on the year of release was not included in the database; hence, the total number of varieties indicated in this table is 979 (i.e, 992 – 13 = 979). [b]The database for West Bengal includes varieties released during the period 1969–2007. na, indicates not applicable for the case of Chhattisgarh because the state was formed in 2000.

Table 13.13. Percentage frequency of varieties released in India by agroecology and decade, India.[a]

	1961–1970	1971–1980	1981–1990	1991–2000	2001–2010	All years
Irrigated	71	67	56	58	67	62
Rainfed lowland	18	18	23	20	16	19
Rainfed upland	4	15	13	16	15	14
Rainfed deepwater	7	0	8	7	3	5

[a]Central releases for the whole of India include 57 varieties released from 1962 to 1984, which were denotified in later years. No information on denotification of varieties released after 1984 was recorded in the database. This list excludes 13 varieties for which the information on the year of release was not included in the database; hence, the total number of varieties indicated in this table is 979 (i.e. 992 – 13 = 979).

In Nepal and Sri Lanka, no clear pattern of shifts in varietal release across ecosystem is discernible. Varieties are targeted mainly to the production environment of Terai, the southern plain belt, in Nepal. In Sri Lanka, varieties released for general cultivation still account for more than 50% of the total releases.

In the case of Bhutan, a clear shift in emphasis from the mid- to high-altitude zone is apparent, although adoption data indicate that the share of

Table 13.14. Percentage of varieties released by agroecology and decade.

	1961–1970	1971–1980	1981–1990	1991–2000	2001–2010	All years
Bangladesh						
Aman	50	36	23	55	55	46
Boro	33		31	30	27	25
Aus			15	15	14	11
Boro, Aus, Aman		64	31		5	18
Total varieties	6	11	13	20	22	72
Nepal						
Terai	33	100	47	70	50	60
Mid hills	67		47	10	39	32
High hills		0	6	20	11	8
Total varieties	6	11	17	10	18	62
Sri Lanka						
Dry zone	17				7	3
Dry and intermediate			12	6		4
Intermediate zone			6			1
Wet zone		21	35	29	21	25
Unfavourable land	16	15	23	18	7	16
General cultivation	67	64	24	47	64	51
Total varieties	6	14	17	17	14	68
Bhutan						
High altitude			17	17	42	29
Mid-altitude			50	83	17	42
Low-altitude			33	0	42	29
Total varieties			6	6	12	24

MVs in this zone is higher than in other parts of Bhutan. Hence, the rationale for this shift is not clear given that the high altitude zone accounts for only about 10% of the total rice area in the country.

The maturity of a crop improvement programme can be judged by analysing the sources of varieties released. In the early stages, national crop improvement programmes are likely to directly release improved varieties developed by international research centres and advanced research organizations in developed countries. There may also be a direct flow of varieties from neighbouring countries with more mature breeding programmes. As the crop improvement programmes mature, the proportion of improved varieties developed locally can be expected to increase.

Direct introduction of varieties developed at IRRI accounted for 22% of the varieties released in Bangaldesh, 32% in Nepal and 17% in Bhutan (Table 13.15). Most of these direct introductions were made prior to 1990 (Table 13.16). After 1990, the proportion of direct introductions decreased substantially, thereby indicating the increased capacity of national programmes to carry out crossing and selection for local conditions.

The domestic source of origin includes all varieties developed locally though crossing. Both parental lines may be local or at least one parental line may have originated from IRRI (or from other countries). All improved varieties in Sri Lanka are of domestic origin. Substantial cross-flow of varieties is evidenced by the fact that 18% of the releases in Nepal are of Indian origin and 17% of the releases in Bhutan are of Nepalese origin. The direct introduction of varieties developed outside of South Asia is also important for Nepal and Bhutan. Such cross-flows are, however, limited in the case of Bangladesh and non-existent in Sri Lanka.

Overall, the contribution of IRRI to the varietal output is substantial. Full pedigree analysis was not possible with the data available, but direct release of varieties developed at IRRI and local crosses made with parental lines developed at IRRI can provide an initial indication of the germplasm content originating from IRRI. The share of IRRI germplasm is substantial with the lowest being 26% for Sri Lanka and highest at 58% for Bhutan (Table 13.17). These are lower-bound estimates because only the varieties with an IRRI

Table 13.15. Original source of released varieties by country (% share in total releases).

	India[a]	Bangladesh	Nepal	Sri Lanka	Bhutan
Domestic	95	74	27	100	33
Direct introduction of IRRI lines	5	22	32		17
Foreign origin					
South Asia					
India		1	18		8
Bangladesh			2		4
Nepal					17
Sri Lanka			5		4
Bhutan					
Other Asian regions		3	16		17
Total number of varieties	992	72	62	69	24

[a]Central releases for the whole of India include 57 varieties released from 1962 to 1984, which were denotified in later years. No information on denotification of varieties released after 1984 was recorded in the database.

Table 13.16. Original source of released varieties by country before and after 1990 (% share in total releases).

	India[a]	Bangladesh	Nepal	Sri Lanka	Bhutan
1990 and before					
Domestic	93	60	12	100	
Direct introduction of IRRI lines	7	30	41		33
Foreign origin					
South Asia		3	29		33
Other Asian regions		7	18		33
Total varieties	428	30	34	38	6
After 1990					
Domestic	97	83	46	100	44
Direct introduction of IRRI lines	3	17	21		11
Foreign origin					
South Asia			18		33
Other Asian regions			14		11
Total varieties	551	42	28	31	18

[a]Central releases for the whole of India include 57 varieties released from 1962 to 1984, which were denotified in later years. No information on denotification of varieties released after 1984 was recorded in the database. This list excludes 13 varieties for which the information on the year of release was not included in the database; hence, the total number of varieties indicated in this table is 979 (i.e. 992 – 13 = 979).

parent, not IRRI ancestral lines, are considered in this analysis.

Adoption of Modern Rice Varieties

Estimates of cultivar-specific adoption for 2010

A summary of the results from expert elicitation aggregated to the country/state level is presented in Table 13.18. The percentage area under improved varieties for the country/state and the top five improved varieties together with their percentage share in the area under improved varieties are indicated in the table.

The overall adoption level of rice in MVs in 2010 ranges between 53% for Bhutan and 100% for Sri Lanka, with adoption levels generally exceeding 80%. This result is in contrast to the adoption estimates in Hossain *et al.* (2003) in which adoption in 1999 was estimated at 60% for Bangladesh, 73% for India, 36% for Nepal

Table 13.17. IRRI materials in locally released varieties by country/state (% share in total releases).

	Releases (#) (A)	Direct release of IRRI lines (B)	One parent is IRRI variety (C)	Both parents are IRRI varieties (D)	With IRRI material (B+C+D)
India[a]	992	5	26	4	34
West Bengal	120	5	23	2	30
Odisha	144	1	31	3	34
Chhattisgarh	15	0	33	0	33
Bangladesh	72	22	22	10	54
Nepal	62	32	19	0	52
Sri Lanka	69	0	26	0	26
Bhutan	24	25	29	4	58

[a]Central releases for the whole of India include 57 varieties released from 1962 to 1984, which were denotified in later years. No information on denotification of varieties released after 1984 was recorded in the database.

Table 13.18. Experts' estimates of percentage area under improved varieties.

	Percentage area under improved varieties	Variety	Share (%) in total area under improved variety	Year of release
West Bengal	92	Swarna	43	1979
		Satabdi	11	2000
		Cottondora Sannalu	6	1995
		Khitish	6	1982
		IR 36	6	1982
Odisha	89	Swarna	37	1979
		Pooja	11	1999
		Vijetha	9	1995
		Lalat	9	1988
		Pratikshya	4	2005
Chhattisgarh	85	Cottondora Sannalu	25	2000
		Swarna	17	1979
		Mahamaya	10	1996
		IR 64	8	1988
		Hybrids	3	
Bangladesh	80	BRRI dhan 28	19	1994
		BRRI dhan 29	14	1994
		BR 11	14	1980
		Swarna	6	1979
		Hybrids	5	
Nepal	84	Radha 4	15	1994
		Sona Mahsuri	13	1982
		Hardinath 1	10	2004
		Hybrids	7	
		Bindeshwori	6	1981
Sri Lanka	100	Bg 352	18	1992
		Bg 300	17	1987
		Bg 358	14	1999
		Bg 94-1	8	1975
		At 362	6	2002
Bhutan	53	BR 153	25	1989
		IR 64	17	1988
		Khangma Maap	15	1999
		Yusi Ray Maap 1	6	2002
		Bajo Maap 1	3	1999

and 91% for Sri Lanka. Except for Sri Lanka, which was approaching full adoption in 1999, MV uptake has increased substantially between 1999 and 2010 in all countries. The annual average increase in adoption level is in the range of 1–3 percentage points.[7]

The adoption estimates by cultivar indicate the dominant position of some varieties. The rice variety Swarna alone accounts for 30% and 43% of the area under improved varieties in Odisha and West Bengal, respectively, indicating that this variety is a widely adopted 'mega' variety. For other locations, such mega varieties do not exist, with more diversified areas claimed by the top five varieties.

The importance of agroecological factors is clearly discernible in the results for Bhutan (Table 13.19). There is no overlap in varieties among the three altitude zones, indicating that these agroecologies are quite distinct in terms of varietal adaptation. This clearly highlights the importance of conducting expert panel elicitation at distinct agroecological levels rather than at the aggregate national level.

The typical result at the disaggregated level is illustrated by the case of Bhutan where estimates are derived from three expert panels, one for each agroecology. Each of the panels had members familiar with that agroecology and, after the elicitation was done by each group, they held a joint plenary discussion to obtain the final consensus estimate. These joint discussions were very insightful because each group provided a detailed justification of their estimates and there were instances when they made some adjustments on the basis of the group feedback.

Table 13.19. Experts' estimates of percentage share in total area under improved varieties by altitude zones in Bhutan.

	High altitude	Mid-altitude	Low altitude
Khangma Maap	50		
Yusi Rey Maap	23		
Yusi Rey Kaap	13		
No 11	3		
Jakar Rey Naab	3		
IR 64		50	
Wengkhar Rey Kaap		20	
Bajo Maap		10	
BR 153			70
Bhur Rey Kaap 2			10
Bhur Kamja			6
Other MVs	8	20	14

Source: Ghimiray *et al.* (2013).

Validation of expert panel-based estimates

How do the estimates based on expert panels compare with the results from household surveys? A comparative analysis is presented in Table 13.20. The results indicate that aggregate MV adoption estimates derived from expert panels are generally lower than those obtained from household surveys, but are quite close overall except for Bhutan. At the individual cultivar level, the error is measured by the mean absolute error (MAE), which is the absolute value of differences between the two estimates for each cultivar averaged across all cultivars. The estimated MAE ranges between 1 and 5 percentage points. Overall, the difference may be considered to be at an acceptable level, given the cost and time efficiency of expert panel-based estimates.

The examination of detailed data can be useful in understanding the factors leading to the observed discrepancies. Again using the example of Bhutan for illustrative purposes (Table 13.21), the closest estimates are for the high-altitude zone, followed by the low-altitude zone, with the discrepancy being largest for the mid-altitude zone. Upon further investigation of the possible causes for this large discrepancy for the mid-altitude zone, it was found that some key knowledgeable people were not able to participate in the panel despite being invited. Prior information indicated that the household survey results are likely to be closer to reality than the expert panel for this zone. This clearly highlights the importance of ensuring that the panel is well constituted.

Some additional features of results from expert panel elicitation are illustrated by the case of Odisha (Table 13.22). Data generation for this state were the most detailed and complete. In Odisha, expert elicitation was done at the state level as a whole, initially, and this was repeated for each of the 29 districts. Similarly, household surveys and community surveys were also implemented in each of the 29 districts. In addition,

Table 13.20. Comparison between estimates from expert elicitation and household survey.

	Expert elicitation[a] (A)	Household survey (B)	Difference (percentage points) (A–B)	MAE (percentage points) (C)
India				
West Bengal	92.4	92.4	0.0	4.56
Odisha	89.3	87.0	2.3	1.33
Chhattisgarh	85.5	93.8	–8.3	2.24
Bangladesh	79.5	89.5	–10.0	2.54
Nepal	83.7	86.7	–3.0	3.58
Sri Lanka[b]	99.6	100.0	–0.4	1.10
Bhutan	53.3	42.0	11.3	3.64

[a]Adoption estimates from expert panel elicitation were derived by aggregating district-level estimates except for Bangladesh, Sri Lanka and Bhutan where estimates elicited for each ecosystem/season were aggregated. [b]No household surveys were conducted in Sri Lanka under TRIVSA because cultivar-specific adoption data were available from agricultural statistics generated annually by the Department of Agriculture. These adoption data were used to estimate the differences with expert estimates for Sri Lanka. MAE, mean absolute error.

Table 13.21. Estimates of percentage area under all improved varieties by altitude zone in Bhutan.

Altitude zone	Expert estimate	Household survey
High altitude	80	84
Mid altitude	40	19
Low altitude	55	46
All	53	42

Source: Ghimiray *et al.* (2013).

elicitation for the state was also done by agroecology (irrigated inland, irrigated coastal, and rainfed).

Overall, there is a good correspondence between the estimate of improved varieties obtained through expert elicitation (EE) at the state level and the aggregate estimate obtained from the household survey (HH). The difference between the estimates is only 2 percentage points. Given the low cost and time efficiency of EE relative to HH, this level of discrepancy may be considered to be tolerable.

At the individual variety level, the correspondence again is very good, at least for the top five varieties. The ranking of the top five varieties is matching between EE and HH. When using the state-level EE, varieties with estimated area of less than 1% jointly account for 25% of the total area. Not all of these varieties were individually identified by the expert panel, with the 'other MVs' category representing the residual category of minor MVs. In the case of the household survey, the area under the residual category is smaller (18%) but consists of more than 100 individually identified MVs, each with very small area (less than 1%).[8] Many of these varieties grouped as 'other MVs' are recent releases that are in early stages of adoption, whereas others are older varieties that have not spread much and have mainly remained confined to some specific areas.

In comparing the varietal list in the EE and HH surveys, it was generally found that the expert panel picked up some of the recent releases that are in early stages of adoption. Plant breeders and extension experts are likely to be aware of the initial spread of such varieties that may be missed in surveys with broad spatial coverage. On the other hand, experts often missed older releases that are being replaced by newer varieties or that did not spread much from the start.

Comparison between state-level and district-level EE aggregated to the state level

Expert estimates generated at lower levels of aggregation (such as districts or ecosystems) can be expected to reflect the cultivar-specific adoption

Table 13.22. Estimates of percentage area under improved varieties in Odisha derived from different methods.

Variety	Year of release	Percentage area		
		State-level EE[a]	HH[b]	Aggregated district-wise EE[a]
All improved varieties		85	87	89
Cultivar-specific adoption (% share in total area under improved variety)				
Swarna	1979	30	31	37
Pooja	1999	12	14	11
Vijetha	1995	9	10	9
Lalat	1988	6	8	9
Pratikshya	2005	6	4	4
Savithri	1982	4	1	2
Khandagiri	1992	3	4	5
Naveen	2005	2	1	2
Samba Mahsuri	1986	2	0.5	0.3
Moti	1988	1	1	0.3
Cottondora Sannalu	2000		4	4
Gayatri	1988		2	2
Sarala	2000		2	1
Other MVs		25	18	13

[a]EE, expert elicitation method; [b]HH, household survey. Source: Behura *et al.* (2013).

levels better than those generated at the aggregate level (country or state) because of the familiarity of local experts with specific information relating to adoption in their localities. This hypothesis can be examined by comparing the state-level EE and district-level EE aggregated to state level. Again using the case of Odisha for illustration, the results in Table 13.22 indicate a slight overestimation of aggregate MV adoption. The adoption level of Swarna is over-estimated by 6–7 percentage points in the district-level EE. The ranking of the top five varieties still remains the same. However, a notable observation is that the district-level EE picked up several varieties (Cottondora Sannalu, Gayatri and Sarala) that were not identified in the state-level EE, and the estimated proportions of area under these varieties are closer to those generated by the HH surveys. In addition, the share of the residual category 'other MVs' was lower in the district-level elicitation than in the state-level elicitation, indicating that more cultivars are identified at the district level. This indicates that the disaggregated district-level EE could provide a better alternative than the state-level EE for capturing varietal diversity.

Comparison between household- and ecosystems-level elicitation

The adoption of rice cultivars tends to be ecosystem specific. Irrigated areas tend to have different adoption patterns than rainfed areas. Within the rainfed areas, varietal adoption patterns differ across lowland, midland, upland and deepwater areas as adoption is critically dependent on the rice field hydrology (Hossain *et al.*, 2003; Singh *et al.*, 2003; Pandey *et al.*, 2012). Thus expert elicitation at the ecosystems level could produce more accurate results. Again using Odisha as an illustration, the household survey and expert elicitation produced similar results for irrigated environments (coastal and inland) but the differences were more noticeable in the rainfed environment where expert elicitation produced a lower estimate of adoption and had several varieties that were specific to the environment (Table 13.23). For the rainfed ecosystem, several varieties that were captured in the household survey were not included in expert elicitation. Examples are Khandagiri, Pratikshya, Cottondora Sannalu and Annapoorna. This indicates that, in the case of diverse and heterogeneous environments

Table 13.23. Estimates of percentage area under improved varieties by ecosystem in Odisha.

Irrigated inland	EE[a]	HH[b]	Irrigated coastal	EE[a]	HH[b]	Rainfed	EE[a]	HH[b]
All improved varieties	91	91	All improved varieties	90	88	All improved varieties	77	84
Cultivar-specific adoption (% share in total area under improved varieties)			Cultivar-specific adoption (% share in total area under improved varieties)			Cultivar-specific adoption (% share in total area under improved varieties)		
Swarna	38	23	Swarna	23	27	Swarna	32	33
Vijetha	11	15	Lalat	12	4	Pooja	13	11
Pooja	10	21	Pooja	12	19	Khandagiri	6	5
Lalat	7	8	Savithri	9		Lalat	6	9
Cottondora Sannalu	6	7	Pratikshya	5	4	Pratikshya	5	5
Pratikshya	4	2	Naveen	4		Vijetha	5	9
Naveen	3		T 141	4		Savithri	4	
Khandagiri	2	4	Khandagiri	4	3	Naveen	3	2
Moti	2	2	Gayatri	3	7	Gayatri	3	
PKV HMT	2		CR 1014	3		T 141	3	
Jagabandhu	1		Samba Mahsuri	3	2	Moti	1	
Padmini	1		Vijetha	2	2	Sarala	1	
Ramachandi	1		Sarala	2	6	Surendra	1	
Gayatri		4	Moti	1		Cottondora Sannalu		3
Savithri		2	Chakaakhi		3	Annapoorna		2
Other MVs	12	12	Cottondora Sannalu		2	Other MVs	17	21
			Tiki Mahsuri		2			
			Parijat		2			
			Other MVs	13	17			
MAE = 3.43			MAE = 3.21			MAE = 1.67		

[a]EE, expert elicitation method; [b]HH, household survey. MAE, mean absolute error. Source: Behura *et al.* (2013).

(such as rainfed), expert elicitation could miss several varieties that are captured in the household survey. Despite this observation, the estimated MAE was lower for rainfed than for irrigated areas of Odisha indicating that the predictive accuracy of the expert panel method for rainfed areas was not lower in this instance.

Variations in expert estimates

Do experts vary widely in their estimates of cultivar-specific adoption? A greater confidence can be placed in EE if estimates from individual experts are closer together. On the other hand, a wide variability in individual estimates is an indication of poor reliability.[9]

Again using the case of Odisha as an illustration, the MAE for individual expert estimates relative to the household survey data for each variety ranges from 1 to 6 percentage points (Table 13.24).

Table 13.24. Comparison of estimates of percentage share in total area under improved varieties of each variety across experts in Odisha.

	EE[a]	HH	MAE (%)	MAPE (%)
Swarna	31.9	30.6	6	21
Pooja	9.9	13.6	5	37
Vijetha	9.1	10.1	3	32
Lalat	11.7	8.2	5	61
Pratikshya	4.9	4.1	2	60
Khandagiri	7.3	4.0	4	108
Cottondora Sannalu	6.3	3.9	3	86
Gayatri	3.6	2.4	2	64
Savithri	5.1	1.5	4	240
Naveen	3.6	1.5	2	145
Moti	2.1	1.0	1	108
CR 1030	2.0	0.5	1	271
Parijat	3.3	0.5	3	529
Samba Mahsuri	2.0	0.5	2	326
Swarna Sub1	1.0	0.3	1	292
T 141	6.2	0.1	6	10,111
Kalinga III	1.3	0.0	1	10,662
Other MVs	12.0	17.3		

[a]Expert estimates (EE) are averages of estimates provided by each expert for each variety in Step 2 of the elicitation process conducted at the state level. Estimates provided by each expert for each variety were compared with the estimate derived from household survey (HH) for the same variety. Source: Behura *et al.* (2013).

These estimates of MAE are within a tolerable range overall. The absolute values of errors can be converted to relative terms by using the mean absolute percentage error (or MAPE) which is obtained by expressing MAE as a percentage of the adoption estimate derived from HH. MAPE provides error estimates relative to the base values for each variety and makes comparisons across varieties more meaningful. The MAPE values indicate that errors are inversely correlated with the adoption levels, with relative errors being small (although absolute errors are more) for those varieties that are widely adopted. Thus expert panel-based estimates are likely to be less precise for cultivars that are grown in smaller areas only.

It is interesting to highlight the case of Sri Lanka in which results of expert estimates were almost matching with the survey results currently available from the Department of Agriculture.[10] After completing the country-level EE in which nine experts participated, the team discussed with these experts the EE estimates and varietal adoption data available from the Department of Agriculture. It was surprising to find that the adoption estimates from these two sources were almost identical. Apparently, the experts had internalized this statistical information and basically used this as the basis for their subjective elicitation. Thus, the estimates obtained from experts were not independent of the statistical estimates already available. Obviously, there was no point in conducting EE in Sri Lanka but the team only became aware of this during the EE process.[11]

Varietal age and replacement of varieties

Varietal age is an important parameter for characterizing the adoption process. The age of a variety was defined as the number of years elapsed between the year of the official release of the variety and 2010. The average varietal age of a set of improved varieties is calculated as the weighted average, with the weights being the area share of the variety in the total MV area. The average varietal age can be expected to decrease over time as older varieties are replaced by newer varieties.

The average varietal age was calculated from both sources of adoption data (i.e. EE and HH) and the estimates are presented in Table 13.25. Overall, both sources result in fairly close estimates of the average varietal age. This is expected because the average varietal age is determined mainly by those varieties that are adopted widely and, as indicated above, errors in estimating those dominant varieties are relatively small. For minor varieties, the errors may be large but their area shares are likely to be too small to affect the average varietal age to any significant extent.

Table 13.25. Average varietal age (years).[a]

	Based on EE[b]	Based on HH[c]
India		
West Bengal	25	24
Odisha	21	20
Chhattisgarh	17	18
Bangladesh	22	20
Nepal	20	24
Sri Lanka	18	18
Bhutan	14	16

[a]Modern varieties without the year of release information or those that were not in the release list were excluded from this calculation. [b]EE, expert elicitation method; [c]HH, household survey.

The average age is found to be more than 10 years in all cases, with the average age being more than 20 years in several cases. This indicates that the popular improved varieties are mostly old and released before 2000. Newer varieties released after 2000 are not being adopted to any meaningful extent in most cases.

This result is supported by the information collected during community surveys on shifts in varietal composition. Older improved varieties are being replaced by newer improved varieties (Fig. 13.2) resulting in a reduction in the overall average age of varieties adopted (Table 13.26).[12] Such a replacement by newer cohorts is desirable because it indicates the diffusion of varieties coming out of the breeding programme in recent years. However, newer varieties that are replacing older varieties are also at least 10 years old. Thus it is a case of 'very old' varieties being replaced by 'somewhat old' varieties, with the newer releases (those released after 2000) still not being widely adopted.

Area under unknown varieties

There were several varieties identified as improved but not included in the release list. This was the case for both expert elicitation and

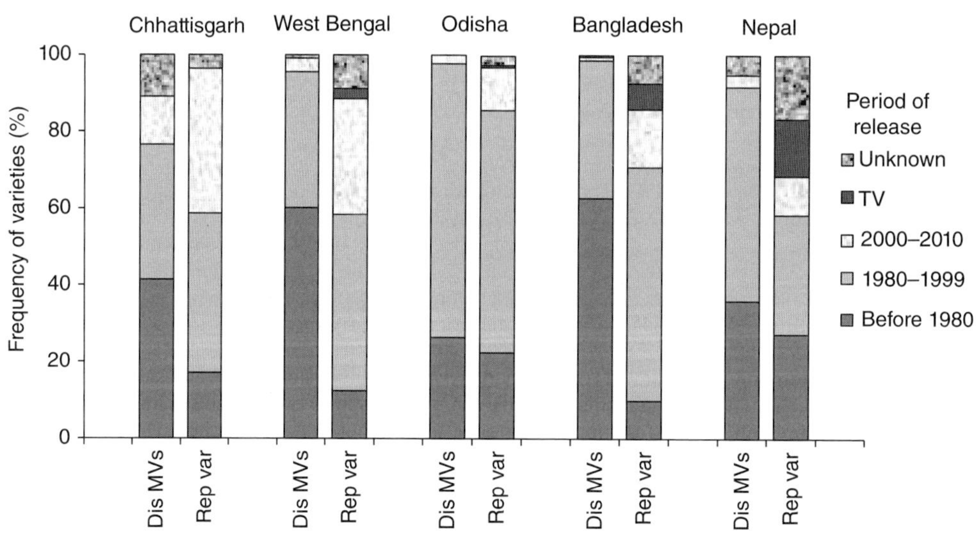

Fig. 13.2. Varietal replacement pattern during 2000–2010. TV, traditional varieties; Dis MVs, disadopted modern varieties; Rep var, replacement varieties.

household surveys (Table 13.27). Experts had access to the release list so they could check it to ensure that no mistakes were made in using the correct names of varieties. The presence of such named varieties that are not in the release list indicates that these are unreleased improved varieties that may have come from across the border or they may be simply 'escapes' from breeding programmes. The area share of such varieties in Nepal is high at almost 10%, with as many as 15 such varieties. Upon checking, it was found that these are mostly the varieties brought in through open borders with neighbouring states (Bihar and Uttar Pradesh) of India.

Table 13.26. Average varietal age (years) of disadopted improved varieties and replacement varieties.

	Disadopted improved varieties	Replacement varieties
India		
West Bengal	29	18
Odisha	27	19
Chhattisgarh	27	17
Bangladesh	34	19
Nepal	26	24
Bhutan[a]	19	

[a]Focus group discussions at the village level were not conducted in Bhutan because data collected from the HH survey conducted under TRIVSA were sufficient to provide information on varietal adoption, given the small size of the country. However, data on varieties grown in place of disadopted MVs were not recorded in the HH survey.

These varieties had not gone through the formal varietal release process in Nepal and, hence, were not on the release list. For Bangladesh, Bhutan and Sri Lanka, there were no such cases. For West Bengal, Odisha and Chhattisgarh, the numbers are very small. These are more likely to be escapes or truthfully labelled varieties promoted by private seed producers using their own brand names.

As expected, the number of these unknown improved varieties found in household surveys is much larger. In the case of household surveys, misclassification (farmers wrongly identifying a landrace as an improved variety) or the use of local names for identification of an improved variety may be important reasons in addition to those mentioned above in the context of expert elicitation. The available information did not permit identification of the dominant factors but past surveys indicate that farmers often misclassify a variety into improved or traditional categories, especially when varieties are identified by local names and are being grown for many years. Without proper genetic analysis, it is impossible to scientifically identify the correct category when the local names do not provide any clues.

The presence of a large number of unknown improved varieties in household surveys is an important source of discrepancy between cultivar-specific adoption levels derived from expert estimates and household surveys. Experts mostly put those into the residual 'other MVs' category. So, the expert elicitation method is likely to be less accurate in cases where 'other MVs' account for a large proportion of area.

Table 13.27. Number of unknown MVs (modern varieties) and percentage area.[a]

	Expert elicitation method		Household survey	
	Share in MV area (%)	Number of unknown MVs	Share in MV area (%)	Number of unknown MVs
India				
West Bengal	3.16	7	5.36	31
Odisha	0.01	1	3.54	45
Chhattisgarh	1.10	3	2.66	10
Bangladesh	0.00	0	5.01	44
Nepal	9.59	15	7.24	21
Bhutan	0.00	0	6.25	9

[a]Unknown MVs are named cultivars declared as MV by experts or by farmers but not found in the varietal release list. Sri Lanka had no unknown varieties so the country is not listed.

Summary and Discussion

This chapter included information on the scientific capacity of rice improvement programmes, varietal outputs resulting from such improvement programmes and adoption of improved varieties of rice in south Asia. This section provides a summary of the main findings and discussions on the validity and potential usefulness of the expert elicitation method for estimating cultivar-specific adoption.

Information on scientific capacity (in terms of FTE scientists) of rice improvement programmes of publicly funded research organizations in 2010 was readily collected through a survey of scientists involved. It was not possible to make a similar assessment of private sector research owing to the difficulties in accessing the information. The results indicated that the research intensities, defined as FTE scientists per million tonnes of output, are quite low even though there has been a substantial increase in investments in agricultural research in recent years. The research intensities in several cases were less than 1 FTE scientist per million tonnes of output; none exceeded 5 FTE scientists per million tonnes (except for the small country of Bhutan). Results also point towards under-investment in rainfed areas and in biotechnology work. There was also some evidence of diversion of crop improvement programme resources to general crop management work. Previous studies that reported on scientific capacity were either not focused on crop improvement programmes or used different indicators (number of scientists rather than FTE scientists). Hence, it was not possible to make a temporal comparison to analyse how the programmes have changed over time.

Regarding varietal output, cumulative output has increased over time but the rate of increase has declined in recent years. This possibly could be due to the resource squeeze faced by agriculture during the late 1990s and early 2000s. Annual release rate varies substantially across countries and Indian states reported here, with the median release rate being 1.4 cultivars per year. The national programmes are clearly becoming more mature as newer varieties are mostly developed within the national programmes, whereas direct introduction through international transfer of germplasm was the main mechanism earlier. Despite these indicators of stronger national programmes, the main issues of concern remain as low investments in general, possible suboptimal allocation of resources across various priority areas and limited capacity in biotechnology work.

Estimates of cultivar-specific adoption for 2010 were derived using both the expert elicitation method and household surveys. The similarity of results from these two methods partly validates the expert elicitation method used. The results indicate that adoption of improved varieties in the rainfed domain has increased substantially since 1998, with the adoption levels in most cases in excess of 80%. This translates to the average annual increase in adoption level in the range of 1–3% during 1998–2010. Despite this increase in adoption, a major concern is that the average yield levels in the rainfed domain have remained low and are unstable.

A new phenomenon that was not apparent in earlier major adoption studies is the spread of what are known popularly as 'mega' varieties. One such variety, Swarna, has spread widely throughout the rainfed areas in south Asia, with the area under this variety in some states of eastern India being in excess of 30%. Obviously, this is a widely adopted variety that, although developed in Andhra Pradesh and released in 1979, ultimately found its way into rainfed areas of eastern India.

The adoption and varietal replacement patterns indicate that the average varietal age is more than 13 years in all cases and more than 20 years in several cases, with the average ages of replacing and replaced varieties also being high at 17 years and above. Thus older varieties, such as Swarna, have remained dominant and newer varieties do not seem to be sufficiently widely adapted to replace these older varieties. Where some replacement has taken place, it is generally the case of an old variety replacing even older varieties, with the adoption of varieties released after 2000 being relatively low. This situation poses important challenges to rice breeding, seed multiplication and extension programmes in south Asia.

The results of the study demonstrate the potential usefulness of the expert panel approach for obtaining cultivar-specific adoption levels for rice. Expert panel estimates compared well with the estimates derived using household surveys. Proportional areas under total modern varieties were very similar with either method.

The expert panel provided reliable estimates of the area under the dominant varieties and similar estimates of area under the top five to seven varieties. The time required to organize an expert panel, especially when done at the aggregate country/state or ecosystem/season level was very short, with the actual expert panel discussions mostly lasting only about half a day. This is a very low-cost and time-efficient method of generating updated information on varietal adoption relative to the implementation of household surveys.

There are a number of factors that determined the success of the expert panel-based method. The first and foremost is the composition of the expert panel itself. 'Experts' in this context are people knowledgeable about farm-level realities and farmers' practices, not necessarily scientists or organization heads who may be too time-constrained to keep themselves updated about farmers' varietal choices. A panel of 8–12 experts who represent various groups such as extension workers, seed producers/traders, local leaders and representatives of farmers' groups would be important for bringing in a diversity of perspectives and knowledge bases needed to properly assess the changes in varietal adoption patterns. Our experience in implementing the expert panel approach in various locations clearly demonstrated this. The expert panel approach did not work well when panel members had very incomplete or fragmentary knowledge about spatial adoption of improved varieties. The case in point is the example of Bhutan in which experts from the mid-altitude zone were not able to participate in the panel discussion. This resulted in questionable estimates of adoption for this zone.

Skill at facilitation is another crucially important factor in generating high-quality estimates from expert panels. The facilitator should have a good understanding of the rice production system to guide the discussions effectively. At the same time, the facilitator needs to have good facilitation skills to ensure that all experts are fully engaged and the views of the more vocal members do not unduly bias the results. The use of the expert panel approach is not common in the agricultural sector of Asia; most NARS are not familiar with this approach. So the facilitator's role in carefully explaining the process and guiding the discussions is even more important.

Past adoption studies have clearly indicated that field hydrology is a critical determinant of adoption of different rice varieties in rainfed environments. Farmers identify different land types such as upland, mid-land or lowland based on the toposequence that determines the field hydrology. Farmers attempt to match rice varieties of different durations with field types and normally grow short-duration varieties in upper fields, whereas long-duration varieties are grown in lower fields. In irrigated areas, such hydrological factors are not important and varietal mix tends to be less heterogeneous spatially. This type of adoption pattern means that information on cultivar-specific adoption levels for rice are better generated by organizing expert-panel estimates at the ecosystem/season level. This situation is unique to rice because non-rice food crops are not grown under flooded field conditions.

Elicitation at the disaggregated level for different ecosystems/seasons is desirable, but local expertise/information is mostly organized at the administrative unit, such as districts. This suggests the need for a second-level stratification based on districts once the first-stage stratification on the basis of ecosystems is done. This two-stage stratification may be desirable, especially for states or countries with a substantial rice area. In fact, this strategy was used in Nepal and all three states of eastern India with the expert panel organized at the district level. In Bangladesh, this approach was used only partially and, in Bhutan and Sri Lanka, elicitation was done at the ecosystem/season level only.

There were some notable advantages to conducting expert panels at the district level. One notable advantage was the credibility of the adoption estimate derived. This is exemplified by the case of Nepal where district authorities who participated in the panel came up with adoption estimates that were substantially different from those reported in the national statistics. The elicited information was more credible to them because it was based on their current knowledge, whereas the basis for the adoption estimates reported in national statistics was unclear.

The downside of elicitation at the district level is, however, the difficulty in organizing the logistics, higher cost, and longer time needed to accomplish the task for all districts. In some cases (such as in West Bengal), local authorities were too busy to participate in expert panels, whereas in others administrative approval from higher authorities was set as a pre-condition for participation. The task was seen as an extra work load without additional rewards.

The difficulties mentioned above, however, can be expected to diminish over time as the expert panel approach becomes institutionalized. Once incorporated in the institutional work plan, it becomes an institutional activity and can be expected to be implemented in the same manner as other regular activities at the district level. Investments in capacity building of national organizations, however, would be desirable to help institutionalize this method.

A key issue on the value of elicited information is the extent to which the estimates provided by experts are independent of the information contained in government statistics. There is no value addition if experts base their estimates on published data rather than on their independent judgement about the current adoption levels. In fact, the elicited adoption level will merely mirror the government statistics with all its limitations in such situations. This is in fact what happened in the case of Sri Lanka where expert estimates matched government statistics very closely. There is obviously no need to conduct expert elicitation when government statistics reflect the reality well. Expert estimates may still be useful, however, for filling in information gaps in areas where data deficiencies are known to exist. Overall, it is important to ensure that experts do not merely repeat what is already available. Validation using farm-level surveys is an important step in this regard.

In addition to deriving adoption estimates, expert panels could also play an important role in assembling information on varietal flows across boundaries (districts, states and countries). Even qualitative information of this kind can be very important in designing focused surveys to measure adoption of such varieties.

A major challenge in all adoption studies is the correct identification of cultivars. For rice in Asia, farmers are mostly able to identify the varieties they grow as landraces or introduced improved varieties. For varieties identified as improved, a cross-check with the release list can be made to ensure that a correct classification is used. The problem is that often one ends up with a substantial number of varieties that farmers identify as being improved but that are not in the release list. Such varieties simply could have been misidentified or confusion could have resulted from farmers' practice of using local names, which often vary across communities. In addition, such varieties could be 'escapes' from breeding programmes or could have moved across the borders. These varieties also could have been sold using new brand names given by the private sector seed companies. It would be difficult to identify such varieties correctly without proper genetic analysis. Thus, it is difficult to estimate precisely cultivar-specific adoption of all cultivars even with farm surveys. Similar identification problems also affect the expert-elicitation method but to a lesser extent. A cheap and efficient method of identification on the basis of genetic fingerprinting of plant samples will be essential to overcome this constraint.

Notes

[1] Unless otherwise stated, all estimates for rice production and yield are expressed in terms of rough (or unmilled) rice throughout this paper.

[2] For India, the project focused on eastern India where rice is grown mainly under rainfed conditions as a monsoon crop. The three major states West Bengal, Odisha and Chhattisgarh were included to represent eastern India.

[3] For this reason, the FTE scientist estimates thus derived are not strictly comparable with other information sources such as the ASTI database where scientists involved in all aspects of rice research are included.

[4] Information on production shares by agroecology is not available; hence the use of area share here.

[5] In India, varieties may be released by the Central Varietal Release Committee (CVRC) or by the State Varietal Release Committee (SVRC). Varieties that are found to be adaptable to several states or ecosystems are released by CVRC, whereas SVRC releases varieties that are adapted to the particular state only. The same varieties that are included in the central list may also be included in the state release list, thus resulting in duplicate entries in some cases.

[6] It is to be noted, however, that there is considerable adoption of varieties targeted to irrigated environments in rainfed areas. This spill-over effect is not captured in the table above.
[7] The country coverage and the method of estimation in Hossain *et al.* (2003) are quite different relative to TRIVSA. The estimates are, therefore, not strictly comparable. Specifically, adoption estimates in Hossain *et al.* were derived from area estimates provided by NARS breeders for the four to six most popular improved varieties. The geographical coverage was only partial. For example, information for India was based on estimates for Punjab, eastern Madhya Pradesh and Tamil Nadu only.
[8] A larger proportion under 'Other MV' in EE is partly the result of the way EE was conducted. Experts were asked to list the top ten MVs only; hence, they lumped all other MVs into this residual category. If they were asked to list top 15 or even top 20 MVs, the residual category would have shrunk but the errors will probably increase when experts are asked to estimate areas under more varieties each with a very small proportion of area.
[9] There may be several reasons for wide variability – one being that the experts in the panel have very different knowledge base, with some being better informed (and hence more accurate) than others. Wide variability in early steps of elicitation might indicate the need to reconsider the composition of the panel.
[10] No household surveys were conducted in Sri Lanka under TRIVSA because detailed data on varietal adoption were available from agricultural statistics generated annually by the Department of Agriculture.
[11] The Sri Lankan case is unique because farmers are apparently required to indicate their intention to grow specific rice varieties by name and respective planned area while availing fertilizer subsidy from the local agricultural office. These data are processed rapidly, aggregated nationally and published in a statistical bulletin identifying area by variety. If farmers generally carry out their intentions in actual planting time, these ex-ante estimates provide a good estimate of ex-post adoption in Sri Lanka. In addition to adoption estimates based on this data source, the Department of Agriculture carries out limited sample surveys annually to monitor the returns to rice production, and these data also contains information on rice varieties and area coverage.
[12] Cultivar-specific area data were not available from focus-group discussions so the frequency of the specific MV cited was used as weights.

References

AED (2010) Report presented in Annual Review workshop, Agricultural Economics Division, Bangladesh Rice Research Institute, Gazipur, Bangladesh.

Bagchi, B., Velasco, M.L. and Pandey, S. (2013) *Tracking of Improved Rice Varieties in West Bengal*. Country report for rice component of the project Tracking of Improved Varieties in South Asia. IRRI monograph. International Rice Research Institute, Los Baños, Philippines.

Behura, D., Velasco, M.L. and Pandey, S. (2013) *Tracking of Improved Rice Varieties in Bhutan*. Country report for rice component of the project Tracking of Improved Varieties in South Asia. IRRI monograph. International Rice Research Institute, Los Baños, Philippines.

Bhandari, H., Pandey, S. and Hardy, B. (eds) (2007) *Economic Costs of Drought and Rice Farmers' Drought Coping Mechanisms in Eastern India*. International Rice Research Institute, Los Baños, Philippines.

Dawe, D., Pandey, S. and Nelson, A. (2010) *Emerging Trends and Spatial Patterns of Rice Production.* In: Pandey, S., Byerlee, D., Dawe, D., Dobermann, A., Mohanty, S. *et al.* (eds) *Rice in the Global Economy: Strategic Research and Policy Issues for Food Security*. International Rice Research Institute, Los Baños, Philippines, pp. 15–35.

Gauchan, D. and Pandey, S. (2011) Is investment in rice research in Nepal adequate and balanced across production environments? Some empirical evidences. *Quarterly Journal of International Agriculture* 50(4), 305–324.

Gautam, S., Panta, H.K., Velasco, M.L., Ghimire, Y.N., Gauchan, D. and Pandey, S. (2013) *Tracking of Improved Rice Varieties in Nepal*. Country report for rice component of the project Tracking of Improved Varieties in South Asia. IRRI monograph. International Rice Research Institute, Los Baños, Philippines.

Ghimiray, M., Pandey, S. and Velasco, M.L. (2013) *Tracking of Improved Rice Varieties in Bhutan*. Country report for rice component of the project Tracking of Improved Varieties in South Asia. IRRI monograph. International Rice Research Institute, Los Baños, Philippines.

Hazell, P. (2010) *Asia's Green Revolution: Past Achievements and Future Challenges*. In: Pandey, S., Byerlee, D., Dawe, D., Dobermann, A., Mohanty, S. *et al.* (eds) *Rice in the Global Economy: Strategic Research and Policy Issues for Food Security*. International Rice Research Institute, Los Baños, Philippines, pp. 61–92.

Hossain, M., Gollin, D., Cabanilla, V., Cabrera, E., Johnson, N. *et al.* (2003) International research and genetic improvement in rice: evidence from Asia and Latin America. In: Evenson, R.E. and Gollin, D. (eds) *Crop Variety Improvement and Its Effect on Productivity: The Impact of International Agricultural Research*. CAB International, Wallingford, UK, pp. 71–108.

IRRI (1992) *Annual Report*. International Rice Research Institute, Los Baños, Laguna, Philippines.

Koshta, A.K. and Chaudhary, V.K. (2013) *Tracking of Improved Rice Varieties in Chhattisgarh*. Country report for rice component of the project Tracking of Improved Varieties in South Asia. IRRI monograph. International Rice Research Institute, Los Baños, Philippines.

Ministry of Agriculture and Co-Operatives (MoAC) (2010) *Statistical Information on Nepalese Agriculture*. Ministry of Agriculture and Co-Operatives, Government of Nepal, Singh Durbar, Kathmandu, Nepal.

Pandey, S. and Pal, S. (2007) Are less-favored environments over-invested? The case of rice research in India. *Food Policy* 32, 606–623.

Pandey, S., Gauchan, D., Malabayabas, M., Bool-Emerick, M. and Hardy, B. (2012) *Patterns of Adoption of Improved Rice Varieties and Farm-Level Impacts in Stress-Prone Rainfed Areas in South Asia*. International Rice Research Institute, Los Baños, Philippines.

Salam, M.A., Siddique, M.A.B. and Islam, M.A. (2013) *Tracking of Improved Rice Varieties in Bangladesh*. Country report for rice component of the project Tracking of Improved Varieties in South Asia. IRRI monograph. International Rice Research Institute, Los Baños, Philippines.

Singh, B.N. (2000) Rainfed lowland rice improvement: current status and future strategy. *Journal of Rice Research* 1(1), 104–112.

Singh, H.N., Pandey, S., Velasco, L.E. and Kshirsagar, K.G. (2003) *Patterns of Adoption of Modern Rice Varieties and Constraints to Their Diffusion in Eastern India*. Poster presented in International Rice Congress 2003. 16–20 September, Beijing.

Velasco, M.L., Tsusaka, T. and Yamano, T. (2013) *Project Synthesis Report: Rice (Tracking of Improved Varieties in South Asia)*. IRRI monograph. International Rice Research Institute, Los Baños, Philippines.

Walisinghe, B.R., Rajapaksa, R.D.D.P., Pandey, S., Dissanayake, D.M.N., Velasco, M.L. and Perera, T.H.C.S. (2013) *Tracking of Improved Rice Varieties in Sri Lanka*. Country report for rice component of the project Tracking of Improved Varieties in South Asia. IRRI monograph. International Rice Research Institute, Los Baños, Philippines.

Walker, T.S. (2010) *Guidelines for Data Collection for Objective 1 of the DIIVA Project*. Fletcher, North Carolina.

14 Analysing Scientific Strength and Varietal Generation, Adoption and Turnover in Peninsular India: The Case of Sorghum, Pearl Millet, Chickpea, Pigeonpea and Groundnut

D. Kumara Charyulu,* M.C.S. Bantilan, A. Raja Laxmi and D. Shyam Moses
International Crops Research Institute for the Semi-Arid Tropics, Patancheru, India

Introduction

The importance of crop genetic improvement research is demonstrated by the Green Revolution, which led to a rapid increase in food production in Asia. Those productivity gains contributed to a reduction in poverty directly through increased farm-household income and indirectly through a long-term decline in the prices of food grains, which account for a large share of poor consumers' expenditure. The success of crop genetic improvement research that led to the development of improved varieties of food crops is well documented (Evenson and Gollin, 2003; Bantilan *et al.*, 2013).

Despite the rapid progress made in the past, poverty is still concentrated in South Asia with around 571 million or one-third of the world's poor, estimated at about 1.29 billion in 2011 (World Bank, 2012). Substantial scope exists for further reducing poverty through crop genetic improvement by increasing or stabilizing the yield of major food crops, particularly the dryland crops in South Asia. Modern varietal change by itself may not lift large numbers of people out of poverty, but greater dynamism in this area can go a long way toward moving poor people closer to that threshold. Moreover, modern varietal change can set the stage for the adoption of improved crop management practices, thereby making it possible for farmers to reduce the cost of production substantially.

Modern varietal change is addressed in this chapter for the five dryland crops in the mandate of the International Crops Research Institute for the Semi-Arid Tropics (ICRISAT): sorghum, pearl millet, chickpea, pigeonpea and groundnut. These results from peninsular India are complementary to those presented for sorghum, pearl millet, groundnut and pigeonpea in Chapter 7 and for chickpea in Chapter 12 for sub-Saharan Africa. Indeed, this work, like that described for rice in Chapter 13, was undertaken to establish a benchmark for evaluating the performance of genetic improvement in sub-Saharan Africa. However, our principal objective is to assess the effectiveness of crop improvement in India beginning in the mid 1960s when the first short-statured, high-yielding, early-maturing,

* E-mail: d.kumaracharyulu@cgiar.org

photoperiod-insensitive sorghum and pearl millet hybrids were released for cultivation. Like the other earlier chapters in this volume, this assessment is carried out from the perspectives of inputs (scientific capacity of national programmes), outputs (released varieties and hybrids) and outcomes (aggregate and cultivar-specific adoption and the velocity of varietal turnover). In conducting this evaluation, we also update the findings for India in the 1998 Initiative for sorghum (Deb and Bantilan, 2003), pearl millet (Bantilan and Deb, 2003) and groundnut (Bantilan *et al.*, 2003).

One of the unique areas and strengths of this paper is the reporting and analysis of variety-specific levels of adoption in 2010 for each of the five crops in their major-producing states. These estimates were generated via structured expert elicitation. Their validation from the perspectives of community focus groups and household surveys is described later in the chapter after the main analytical section on the evaluation of scientific capacity, varietal output, varietal adoption and the velocity of varietal turnover. Substantive and methodological implications are discussed in a concluding section where the main results are summarized. Before results are presented and discussed, we briefly describe state coverage, institutional linkages and methods of data collection followed by crop-specific background information that provides context for the assessment of the key aspects of genetic improvement during the past 50 years.

Crops coverage, institutional linkages and methods of data collection

Coverage is at the all-India level for the databases on scientific capacity and varietal release. For the adoption database, five to six of the largest-producing states were selected for each crop based on 2007–2009 cropped area (Table 14.1). These states accounted for about 90% of cultivated area in each crop in India during 2007 to 2009.

Institutional linkages

ICRISAT has implemented this research on the performance of genetic improvement for its mandated crops in close collaboration with the Indian Council of Agricultural Research (ICAR), New Delhi, and crop-specific AICRPs (All-India Coordinated Research Projects). Stakeholders from the State Agricultural Universities (SAUs) were involved in the elicitation process. Representatives of ICRISAT's Hybrid Parents Research Consortium (HPRC) and scientists from other major private companies also contributed. Crop-specific research collaborations among major stakeholders are summarized in Table 14.2.

Methods of data collection

Information on cultivar releases was compiled from the Central Varietal Release Committee (CVRC) and State Varietal Release Committee (SVRC) and from compiled annual reports published by Seed Division, Government of India. Similarly, information was also validated with the crop-specific Directorates or respective AICRP publications and databases.

The ICRISAT research team officially took part in the crop-specific AICRP Annual Meetings and explained the research and collected feedback from each centre. All the scientists (around 150 per crop) who work on crop improvement in India attend these planning meetings that are organized annually crop

Table 14.1. States covered by crop.

Sorghum	Pearl millet	Chickpea	Pigeonpea	Groundnut
Maharashtra (54)	Rajasthan (56)	Madhya Pradesh (34)	Maharashtra (31)	Gujarat (30)
Karnataka (18)	Maharashtra (12)	Maharashtra (16)	Karnataka (18)	Andhra Pradesh (29)
Rajasthan (8)	Gujarat (9)	Rajasthan (16)	Andhra Pradesh (13)	Karnataka (14)
Madhya Pradesh (6)	Uttar Pradesh (9)	Uttar Pradesh (7)	Uttar Pradesh (10)	Tamil Nadu (8)
Andhra Pradesh (4)	Haryana (7)	Karnataka (9)	Madhya Pradesh (9)	Maharashtra (6)
–	–	Andhra Pradesh (8)	Gujarat (8)	Rajasthan (5)

Note: Percentage shares of area cultivated in 2007–2009 are given in parentheses.

Table 14.2. Institutional partnerships by crop.

Crop	NARS, including ICAR collaborations with:			Others
Sorghum	DSR, Hyderabad	AICSIP	SAUs	HPRC
Pearl millet	–	AICPMIP	SAUs	HPRC
Pigeonpea	IIPR, Kanpur	AICRP on Pigeonpea	SAUs	HPRC
Chickpea	IIPR, Kanpur	AICRP on Chickpea	SAUs	–
Groundnut	DGR, Junagadh	AICRP on Groundnut	SAUs	–

AICSIP: All-India Coordinated Sorghum Improvement Project; AICPMIP: All-India Coordinated Pearl Millet Improvement Project; DGR: Directorate of Groundnut Research, Junagadh, Gujarat; DSR: Directorate of Sorghum Research, Rajendranagar, Hyderabad; HPRC: Hybrid Parents Research Consortium, ICRISAT; ICAR: Indian Council of Agricultural Research; IIPR: Indian Institute of Pulse Research, Kanpur; NARS: National agricultural research system; SAUs: State Agricultural Universities.

by crop. Participation in these meetings was a cost-effective means to elicit information on the adoption of improved varieties for each state listed in Table 14.1.

ICRISAT conducted the expert elicitations on cultivar adoption in two rounds. Experts in the first round were canvassed from scientists of the respective AICRP centre located in that state. In general, each expert elicitation was attended by a minimum of four to five scientists based at that centre. The elicitation group was represented by scientists with diverse backgrounds (breeding, plant protection, agronomy, extension, seed science, etc.). On the basis of knowledge and skills in the group, estimates were elicited at either the regional or state level. After obtaining these preliminary adoption estimates from each state during the first round, a second round of elicitations was carried out with state/national-level experts in separate crop-wise workshops.

Additional secondary sources of information were also gathered from the State Department of Agriculture, State Seed Development Corporation (SSDC) and State Seed Certification Agency (SSCA) for the same period. A 'varietal identification protocol' was also developed for increasing the accuracy in the identification of improved cultivars at the farm-level. The protocol was extensively used in the conduct of the adoption validation surveys that are described later in this chapter.

The Five Dryland Crops

The five dryland crops are made up of two cereals, sorghum and pearl millet, two pulses, chickpea and pigeonpea, and one oilseed, groundnut. In India, they have several things in common. With the exception of chickpea, the dryland crops are planted at the onset of the south-west monsoon in the rainy or kharif season. (Rabi or post-rainy season sorghum and irrigated summer groundnut and pearl millet are other important seasonal cropping systems in regionally compact areas of peninsular India.) They share a low historical level of productivity that ranged from about 400 kg per hectare for pearl millet to 800 kg per hectare for pigeonpea at the start of the Green Revolution in the mid-1960s. With the exception of groundnut, where China has eclipsed India as the largest global producer, more area is sown to these crops in India than in any other country. Major diseases and insect pests influence productivity in these dryland crops. Most of the diseases can be managed with resistant cultivars in all five crops. In contrast to rice and wheat, none of these crops has received sustained direct policy protection since independence. Lastly, although irrigation has steadily expanded in peninsular India over the past 50 years, these crops have not benefited substantially from this expansion. Their rainfed character has not changed. Additionally, empirical evidence suggests that the quality of their production environment has declined with regional shifts in production over time. The trend towards a lower quality production environment may not apply equally to all five crops but it is a recurring theme in this chapter.

Sorghum

In the past, sorghum was even more important in India than it is now. Shortly after independence in the early 1950s, sorghum ranked as the second most extensively grown cereal in the country after rice. Nowadays, more area is sown to wheat, pearl millet and maize than to sorghum.

Sorghum is grown in both rainy (2.6 million hectares) and post-rainy (3.5 million hectares) seasons. An estimated 2 million ha is also sown to forage sorghum cultivated in the summer season. Over half of rainy-season sorghum is cultivated as an intercrop with pulses and oilseeds. In contrast, 90% of the post-rainy sorghum is produced as a sole crop on black soil on residual moisture in fields that are fallowed during the monsoon from June to October.

Sorghum is produced for a variety of uses but it is mainly consumed as food, feed, fodder and forage. The end uses have evolved over time. Food and fodder have decreased in importance. Feed and forage have increased in importance. With increasing urbanization, the demand for sorghum as a food grain has sharply declined. The widespread replacement of bullocks with tractors has also reduced the demand for sorghum residue, stalks and leaves, as stover.

The rising demand for sorghum for animal feed and forage has not compensated for the declining demand for sorghum as a food grain and as stover. As a result, sorghum area has declined since it peaked at 18 million hectares in the late 1960s. Its seasonal composition in relative importance has also changed over time. In the 1960s, rainy-season sorghum accounted for about two-thirds of cultivated area. Today, the share of rainy-season sorghum in total harvested area has shrunk to about 40%. The post-rainy season is the dominant source of area and production.

Production reached its maximum in the late 1980s and early 1990s when it approached 13 million tonnes. Since their generation and release in the early to mid-1960s with help from the Rockefeller Foundation and the CSH (Coordinated Sorghum Hybrids) public sector, and later the private sector, improved cultivars have fuelled positive productivity gains in the rainy season in India. For example, in the dominant producing state of Maharashtra, yield growth was 1.87% between 1970 and 2009. Thus, productivity gains from rainy-season production partially offset the declining trend in area. However, productivity growth was overwhelmed by the strong decline in area that has accelerated since the early 1990s. Productivity growth in the post-rainy season has been negligible during the past 50 years because technological change, for all intents and purposes, has not taken place, i.e. adoption of improved cultivars and intensification of management practices is limited.

Pearl millet

In India, pearl millet is the third most important cereal after rice and wheat. It is predominantly grown as a grain crop but is also valued for its stover and fodder. Pearl millet production in India was characterized by subsistence cultivation during the 1970s with a small marketable surplus. But in recent years, its uses are expanding from food to animal feed, potable alcohol, processed food, etc.

In spite of systematic pearl millet research in India since the 1960s, area under cultivation witnessed a continuous reduction from 12.23 to 9.61 million hectares between 1966 and 2010. The reduction was attributed to frequent outbreaks of downy mildew disease, changing food consumption habits, lower remuneration in pearl millet cultivation compared to other commercial crops and weak demand for grain, resulting in farmers moved away from pearl millet cultivation to other commercial crops. Despite the decline in acreage, production has more than doubled from 4.5 to 10.36 million tonnes in the same period. This was made possible through the adoption of short-duration hybrids and their response to fertilizer. Sustained growth of production is a typical Green Revolution success story in the atypical circumstances of rainfed agriculture in the arid and dry semi-arid tropics (Pray and Nagarajan, 2009).

Aridity in pearl millet production is also increasing as the crop has shifted to dryer environments. In districts where pearl millet was produced in the mid-1960s, mean annual average rainfall was about 900 mm (Walker, 2009). In districts where the crop was cultivated in 2008, mean annual average rainfall was only 600 mm. This shift to aridity was especially noted in Rajasthan where pearl millet is traditionally cultivated. It has lost ground to other crops in the wetter eastern and central part of the state and has maintained its share of area in arid Western Rajasthan.

Chickpea

India is the largest chickpea producer as well as consumer in the world. India mainly produces small-seeded desi chickpea although bold-seeded kabuli chickpea, mainly grown in the Middle East, is gradually gaining in popularity. The demand for chickpea is strong and it is characterized by an array of end uses.

Chickpea was one of the main casualties of the expansion of wheat area during the Green Revolution in the 1960s and 1970. Wheat and chickpea compete for land in the post-rainy season in North India. Since the mid-1960s, chickpea area in North and North-eastern India has declined steadily from 4.5 million hectares to around 0.5 million hectares. Conversely, chickpea has increased by over 3.0 million hectares in the central and southern states.

In 2010 and 2011, the area under chickpea was estimated to be around 9.18 million hectares and harvested produce about 8.22 million tonnes with estimated yield approaching 900 kg per hectare. More than 70% of chickpea is grown in the post-rainy season as a rainfed crop; the remaining area is cultivated under irrigated conditions. During the last five decades (1960–2010), chickpea area has registered a slightly negative annual growth rate of –0.4% (acreage declined from 9.28 to 9.18 million hectares), whereas production has increased from 6.25 to 8.22 million tonnes with an average annual growth rate of 0.42%. Despite the decline in acreage, production has increased and this increase is attributed to the introduction of high-yielding and disease-resistant varieties.

Pigeonpea

Pigeonpea is a very plastic crop from the perspective of the length of its growing season. It is characterized by four common durations: early or extra early of about 110–120 days, medium duration of about 180 days, long duration of 240 to 270 days, and it also grows as a perennial. Long duration pigeonpea in North India, especially in Uttar Pradesh, was common in the 1950s and 1960s but, with the advent of the Green Revolution in rice and wheat, long-duration pigeonpea was replaced by more profitable sequential cropping systems. Nowadays, medium-duration pigeonpea is the dominant maturity group; it is usually produced as an intercrop with cotton and other cash crops in Central and South India. Pigeonpea is also cultivated on field bunds or as a backyard crop where it does not receive much if any purchased inputs.

Aside from its plasticity, pigeonpea is technologically interesting because it is one of the first grain legumes to benefit from marked productivity gains from in-breeding and subsequent hybridization. After many years of research, commercial hybrids from ICRISAT parental materials have been released and are now available in India.

In contrast to its diverse and novel traits in production, pigeonpea, unlike chickpea, does not have diverse end uses; it is consumed in India almost entirely as dhal.

During 2010–2011, pigeonpea was cultivated on about 4.42 million hectares with 2.89 million tonnes of production, representing 16% of the national pulse acreage and 15% of production. National average pigeonpea yield is hovering in the range of 650–800 kg per hectare; this has remained more or less stagnant from 1960 to 2010 despite extensive research efforts. This sluggish growth in productivity can be attributed to slow uptake of improved cultivars and production technologies and to the shift in crop area from more favourable to marginal environments.

Groundnut

In terms of consumption, groundnut is the fifth most important oilseed in India after oil palm, soybean, rapeseed and mustard. Groundnut is produced in arid and semi-arid regions characterized by low and erratic rainfall, poor irrigation, frequent droughts and sandy soils. It is largely grown in India in the kharif season under rainfed conditions. Only about 20% of the total groundnut area in India is irrigated, mainly in the summer season. Groundnut is cultivated on 5.85 million hectares, which is about one-fifth of the total area under oilseeds. Groundnut seed production contributes around 25% of total oilseed production, which was 8.26 million tonnes during 2010–2011. Between 1981 and 2010, groundnut production registered a positive but meager annual growth rate of 0.1%. However, the rate of growth was higher (2.2%) during

1981–1995, but decelerated by –0.3% afterwards when the level of protection against Malaysian palm oil was diminished and imports increased. With declining profitability, the sown area began trending downwards in the early 1990s. Declining area has been accompanied by markedly increasing variability in production and yield over time. Since the early 1990s, national average yield has fluctuated between 700 and 1450 kg per hectare.

Key Aspects of the Performance of Food-Crop Genetic Improvement

Scientific strength in dryland crop improvement programmes, modern varietal output, and perceived adoption of improved varieties and hybrids are the main themes described in this section.

Scientific capacity in dryland crop improvement

Scientific capacity for improvement of the five dryland crops focuses on the public sector. The private sector is very active in breeding pearl millet and to a lesser extent in sorghum hybrids. Information is presented on the number of full-time equivalent (FTE) scientists in companies developing pearl millet hybrids but comparable data were not available for sorghum where the public sector is still the dominant institutional player in agricultural research. Private-sector participation in grain legume research is limited in India.

Multiple institutes contribute to public-sector research on dryland crops in India. The human resources data presented here refer to those institutes listed in the second and third column of Table 14.2. The descriptive analysis in this subsection is conducted at the all-India level.

Comparing FTE scientists across crops by discipline

Parity across the five dryland crops in research investment and in varietal output is one of the principal findings of this chapter. Four of the five crops are characterized by a level of total capacity in the very narrow range of 84–86 FTE scientists (Table 14.3). Among the crops in this interval – sorghum, chickpea, pigeonpea and groundnut – the total number of scientists does differ and ranges from 103 in sorghum to 134 in chickpea. But lower FTE scientist conversion rates in the two pulse crops result in the same level of FTE scientists as that found in sorghum and groundnut. The mean conversion rate across the five crops was 72%. The 28% difference is devoted to other purposes such as working on other crops, teaching, guiding students, conducting training programmes and extension.

Pearl millet is the outlier in Table 14.3 with a total FTE complement of slightly over 50 scientists. Historically, sorghum has been a significantly stronger crop improvement programme than pearl millet; therefore, a relatively low estimate for pearl millet improvement was expected. Moreover, private-sector investment in pearl millet research is equivalent to 28 FTE scientists. Therefore, the total investment for pearl millet approaches the amount of scientific input in public-sector sorghum improvement.

Another indication of parity is the degree to which the five crop improvement programmes are concentrated in four core disciplines: plant breeding, agronomy, pathology and entomology. Collectively, these disciplines account for about 83% of scientific resources, ranging from 77% in sorghum to 91% in pearl millet. The other 16 disciplinary categories in Table 14.3 are only sparsely represented in these public-sector crop improvement programmes. With only a 4% share, physiology leads this group of minor disciplines that support dryland crop improvement.

Sorghum exhibits the most diversification in its disciplinary portfolio, featuring an investment in social science, postharvest research, biochemistry, genetic resources and genetics that exceeds that of the other programmes. Implicitly, this higher level of diversification partially responds to demand constraints that have led to falling production, a trend unique to sorghum among the five dryland crops.

Although similar in their disciplinary composition, it is easy to identify crop-wise differences attributed largely to biotic constraints in investments in pathology and entomology. Insect pests figure prominently as yield reducers in the dryland crops except in pearl millet, which is associated with less investment in entomology than the other four crops. In pigeonpea, pod borer consistently causes more economic damage than any single insect pest in these dryland crops. In sorghum, infestations of shoot flies, stem borers

Table 14.3. Full-time equivalent (FTE) scientists by discipline by crop for 2010.

Discipline	Sorghum	Pearl millet	Chickpea	Pigeonpea	Groundnut
Agricultural engineering	0	0	0.6	0.6	0
Agronomy	10.7	9.55	13.8	13.8	12
Biotechnology	2.5	0	1.5	1.6	1.6
Biochemistry	2.5	0.75	0.3	0.3	0
Computer application	0.8	0	0.3	0.3	0
Ecobotany	0.8	0.85	0	0	0
Entomology	13.9	2.55	10.8	14.1	12
Genetic resources	1.6	0	0.3	0.8	0
Genetics/cyotgenetics	2.5	0	0	0	1.6
Microbiology	0	0	4.6	3.8	1.6
Nematology	0	0	0.6	1.5	0.8
Pathology	10.7	8.55	17.2	12.4	11.2
Physiology	3.3	1.45	2.8	2.1	6.4
Plant breeding	30.3	25.7	29.2	32	32
Postharvest technology	0.8	0	0	0	0
Seed technology	0.8	0	0	0	1.6
Social science	3.3	0	0	0	0.8
Soil science	0	0	0.3	0.3	1.6
Statistics	0	0.75	1.4	0.3	0.8
Others	0	0.65	1.5	1.5	0
Total FTE	84.5	50.8	85.2	85.4	84
Total scientists	103	76	134	130	105
Proportion FTE/total	0.82	0.67	0.64	0.66	0.80

and head bugs can result in substantial production losses. Higher allocations to entomology in both pigeonpea and sorghum are attributed to the importance of these pests. Likewise, more investment in pathology is associated with the incidence and importance of well-identified diseases that can induce catastrophic losses in production. Chickpea periodically suffers from *Ascochyta* blight, whereas ergot and downy mildew are common diseases in pearl millet.

Comparing total FTE scientists in national and international programmes by discipline

A total of about 390 FTE scientists work in the five public-sector crop improvement programmes either at the national or state level. In 2010, 44 FTE scientists worked on the five dryland crops in ICRISAT at its Headquarters in Patancheru, India. A comparison of the relative emphasis in disciplinary allocation points to the complementarities in scientific capacity between national and international agencies, even in a very large country like India (Table 14.4). National crop improvement institutes focus on applied and adaptive research; international commodity centres allocate more resources to upstream research that is less likely to be associated with a payoff in the immediate to near future. In accordance with this conventional wisdom of institutional comparative advantage, about one FTE ICRISAT scientist in six works in biotechnology, mainly in areas related to molecular biology and marker-assisted selection. A comparable ratio for national programmes is less than one scientist in 50. Proportionally, plant breeding, social science, statistics and genetic resources command significantly more resources at ICRISAT than in the Indian National Improvement Programmes on dryland crops. ICRISAT allocated no resources to agronomy in 2010–2011. This is in line with the thinking that crop management entails a high level of location specificity that is best addressed by state and national crop improvement programmes.

Since its establishment in the early 1970s, ICRISAT has allocated some programmatic resources to biotechnology-related areas but the moderately high level of investment mirrored in Table 14.4 is relatively recent, reflecting an emphasis that gained momentum in the 2000s. Earlier – in the 1970s, 1980s and on into the

Table 14.4. Comparing relative scientific capacity in Indian national and state programmes in ICRISAT for dryland crops by discipline in 2010.

	NARS	ICRISAT	Difference
Discipline	Share of FTE scientists (%)		
Biotechnology	1.7	15.9	14.2
Plant breeding	39.3	46.6	7.3
Social science	1.0	5.7	4.7
Statistics	0.9	4.5	3.7
Genetic resources	0.6	3.4	2.8
Physiology	4.0	5.7	1.7
Genetics/cytogenetics	1.0	2.3	1.3
Postharvest technology	0.2	1.1	0.9
Seed technology	0.6	1.1	0.6
Microbiology	2.4	2.3	–0.1
Agricultural engineering	0.3	0.0	–0.3
Computer application	0.3	0.0	–0.3
Soil science	0.5	0.0	–0.5
Ecobotany	0.5	0.0	–0.5
Nematology	0.7	0.0	–0.7
Others	1.0	0.0	–1.0
Biochemistry	1.0	0.0	–1.0
Entomology	13.0	5.7	–7.3
Pathology	15.5	5.7	–9.8
Agronomy	15.6	0.0	–15.6
Total	389.9	44.0	0.0

1990s – ICRISAT's research resource allocation resembled more closely that of the Indian national programmes than it does now because pathology and entomology, and to a lesser extent physiology, figured prominently in the pattern of investment in those early decades. An increasing emphasis on biotechnology was accompanied by de-emphasizing pathology, entomology and physiology as total resources contracted in the mid-1990s to early 2000s. In contrast, it is likely that the disciplinary allocation of the Indian national programmes has stayed relatively constant over time.

Comparing the educational level of FTE scientists across crops

About nine of every ten FTE scientists working on dryland genetic improvement have PhDs. This ratio is maintained across the five crops (Table 14.5). Unlike those in sub-Saharan Africa, all scientists in the Indian programmes have graduate training at least to the level of an MSc. BSc holders are viewed strictly as non-scientific, research-support staff.

With the exception of pearl millet, the numbers in Table 14.5 are synonymous with a scientific strength of more than 75 PhDs per crop. This level of educational expertise is a far cry from the very low numbers, which were quantified and discussed in Chapter 7, of PhD scientists working on sorghum, pearl millet and groundnut in West Africa.

Comparing research intensities across crops

Research intensities are compared via production and value of production critieria in Table 14.6. By either criterion, high research intensities were estimated for sorghum vis-à-vis pearl millet and for pigeonpea relative to chickpea and groundnut. As the results in Table 14.6 make abundantly clear, these differences in research intensities are driven primarily by disparities in production and value of production. They have little to do with direct investment in scientific human resources that is, for all intents and purposes, equal for four of the five crops.

Table 14.5. Educational level (%) of FTE scientists by crop.

Educational level	Sorghum	Pearl millet	Chickpea	Pigeonpea	Groundnut
PhD	93	89	91	93	92
MSc	7	11	9	7	8

Table 14.6. Estimated research intensities by crop from 2008–2009 to 2010–2011.

Estimated research intensity	Sorghum	Pearl millet	Chickpea	Pigeonpea	Groundnut
Production (FTE scientists per million tonnes)	11.6	4.6	12.6	32.8	12.1
Value of production (FTE scientists per US$100 million)[a]	8.2	3.6	3.0	7.4	3.6

[a]Prices per tonne used in calculation value of production were US$142 per tonne of sorghum, 126 for pearl millet, 416 for chickpea, 441 for pigeonpea and 337 for groundnut. These are in 2004–2006 prices and were taken from the FAOSTAT value of crop production for India.

Sorghum's high research intensity compares favourably with maize in East and Southern Africa where the private sector is very active in agricultural research (Chapter 11, this volume). Although the complex Indian Agricultural Research System is often assessed as reasonably efficient, the level of investment from the perspective of production often places India in the lower echelon of developing countries ranking behind China and developing countries in general (Pal and Byerlee, 2003). The estimate of 11.6 FTE scientists per million tonnes of production therefore seems high and atypical of the Indian context for the production of a cereal as extensively grown as sorghum. The high research intensity is partially attributed to the steeply declining area of rainy-season sorghum that has resulted in a downward trend in sorghum production. Although kharif sorghum has benefited substantially from technological change, soybean has replaced it and several other dryland crops during the monsoon season, especially in rainfall-assured zones.

Assuming that the level of FTE scientists has not changed that much over time – and this appears to be a reasonable supposition – past research intensities for sorghum were significantly lower than they are now. For example, for levels of production prevailing in the late 1960s, the estimated research intensity drops to 6.5, roughly half the estimate in Table 14.6.

In analysing the data on scientific capacity, two anomalies stand out. Both point to slowness on the part of national agricultural research in India to adjust to substantial regional shifts in the production of pulse crops. Over time, Uttar Pradesh has lost about three-quarters of its pigeonpea growing area and has dropped to fourth in state-wise importance. Yet, from the perspective of research resource allocation in terms of scientists that can be assigned to specific states, about 45% of FTE scientists are located in Uttar Pradesh. The fact that nodal research agencies are still located in Uttar Pradesh is a major explanation of why research resource allocation is incongruent with shares of production. Pigeonpea cropping systems differ markedly, however, between the North where the late-maturing pigeonpea is losing ground and Central and South India where medium-duration pigeonpea reigns as the dominant pigeonpea cropping system.

The same remarks about congruence of resource allocation and production shares apply to chickpea. Of a total of 24 research centres, only three serve the south zone, which has been one of the primary beneficiaries in the shift of production area from the north. The south zone is extensive and also includes large parts of East India.

Varietal Output

Parity in scientific capacity also applies to varietal output. The incidence and pattern of released varieties over time is broadly similar across the five dryland crops (Fig. 14.1). A total of 1013 varieties were released from the beginning of the 20th century to 2010. The number of releases ranges from a low of 159 in pigeonpea to a high of 253 in sorghum. Pearl millet, chickpea and groundnut are characterized by total releases in the narrow interval of 190 to 210. Each improvement programme also displayed a remarkable record of stability of releases over time. Since the 1970s, each programme has released at least 20 cultivars by decade until 2010. Most programmes have released at least one variety every year between 1971 and 2010. The sorghum programme epitomizes this pattern of consistency in releases over time. Between 1961 and 2010, there were only three years when the All-India Sorghum Improvement Programme did not release a variety at the central or state level.

The five improvement programmes also share a history of varietal release that predates independence in 1947. Some of the old cultivars released prior to 1971 are still widely grown today. TMV 2, released by Tamil Nadu in 1940, is the leading groundnut variety in Karnataka and Andhra Pradesh. The sorghum variety M35-1, selected from a local landrace in the late 1930s, is the dominant variety in post-rainy season production. Later releases in the 1960s also still account for large chunks of cultivated area. For example, TMV 7 released in 1968 is the most widely grown variety of groundnut in Tamil Nadu. Across the five crops, releases before 1971 make up about 10% of varietal output. Even though early genetic improvement research started in the early 1920s in chickpea, systematic efforts date only from the late 1950s. Significant momentum in releases can be observed from the 1970s.

The incidence of releases over time varies somewhat by crop. By decade, chickpea, groundnut and pearl millet display a pattern of increasing releases over time (Fig. 14.1). Sorghum adheres to the same increasing tendency with the exception of the last decade when releases declined from their peaks in the 1980s and 1990s. The declining area and production of rainy-season sorghum probably has had a dampening effect on the incidence of releases, especially in states where post-rainy season production is negligible.

In Fig. 14.1, central-level releases (centre releases) represent the difference between total and state-level releases. The importance of state-level versus central-level releases ranges from high in sorghum to very low in pearl millet. This

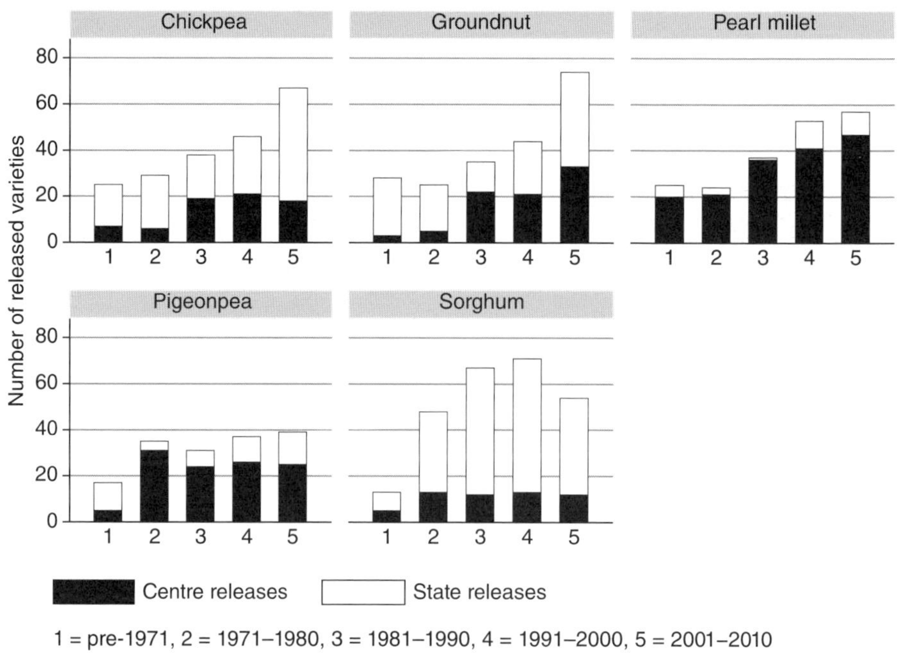

Fig. 14.1. Total and state-released varieties in India from 1971 to 2010 by crop.

variation reflects differences in the level of decentralization between these two crop improvement programmes. Pearl millet research centres in the All-India Coordinated Programme are heavily concentrated in western India in the states of Rajasthan, Haryana and Gujarat, with a sprinkling of locations in the western region of the southern states. Sorghum research centres are distributed in a dispersed pattern across more states with several states having more than one centre to cover different agroclimatic zones. A decentralized distribution of research stations is suited to more subregional adaptation and subsequent state-wise varietal output tailored to varying conditions in each agroclimatic zone. Centre-level releases from the pearl millet improvement programme should be more readily applicable to most of the subregions and zones located in Western India.

Centre- and state-level releases are also qualitatively different in the sorghum improvement programme. Centre releases are about evenly split between new hybrids and improved varieties. Most state releases are improved varieties. This distinction suggests that hybrids are more widely adapted than improved varieties or that they are more difficult to develop to meet release standards. In contrast to crop improvement programmes in sub-Saharan Africa, most improved varieties are the product of crossing parental lines. Only a small minority are selections from landrace materials or elite varieties selected by institutions outside of India.

Parental lines also feature quite prominently in the list of notified materials for release in sorghum and pearl millet. For example, parental lines constitute about 20% of sorghum releases. Their relative importance has not changed appreciably over time. Neither has the share of hybrids in total releases at the state or national level. The absence of trends in relative importance in parental lines and hybrids in total releases is puzzling because the overwhelming majority of modern cultivars in farmers' fields are hybrids. For sorghum, part of the puzzle is explained by the increasing emphasis given to the post-rainy season where most releases are improved varieties.

About 20% of the total of more than 1000 releases was related to ICRISAT materials (Fig. 14.2). From a small beginning of one sorghum and one chickpea variety, related to ICRISAT and released in the 1970s, the total number of ICRISAT-related releases increased to 197 by 2010. Broadly speaking, ICRISAT-related releases

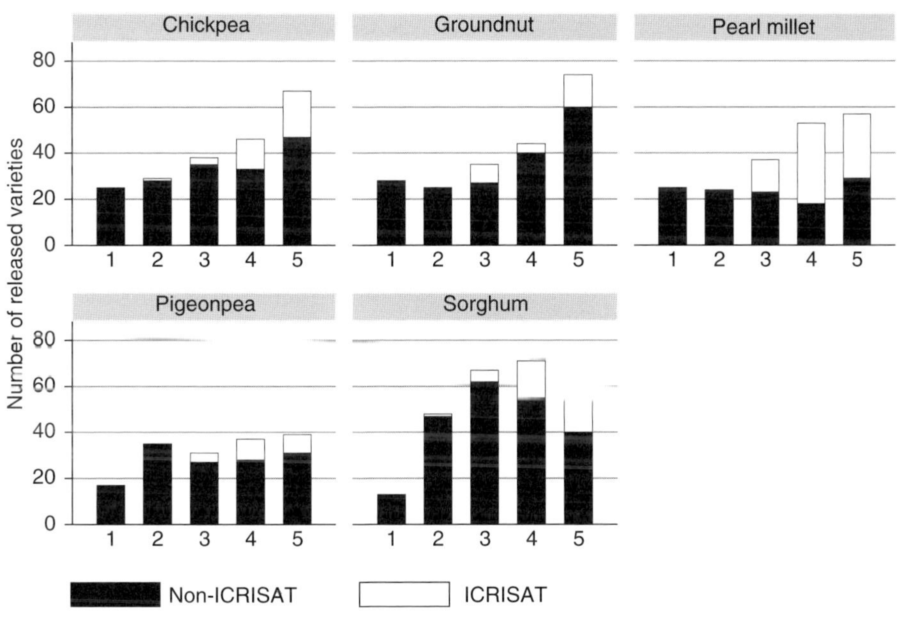

Fig. 14.2. Total- and ICRISAT-released varieties in India from 1971 to 2010 by crop.

have been increasing in all five dryland crops over time. ICRISAT has contributed to at least 20 released cultivars in each of the five crops. Four of the contributions refer to the first pigeonpea hybrids available for commercial production. Most of the recent releases in pearl millet and sorghum are parental lines. The sorghum data in Fig. 14.2 include 14 new cultivars marketed (as truthfully labelled seed) by private seed companies but they have not yet been officially released.

In terms of varietal output, ICRISAT's contribution has been more pronounced in pearl millet than in any of its other mandated commodities. Nearly 40% of ICRISAT-related releases are in this coarse cereal. Historically, pearl millet was one of ICRISAT's stronger crop improvement programmes and, arguably, pearl millet was one of India's national programmes that benefited the most from collaboration with ICRISAT. Differential strengths and weaknesses established a solid basis for sustained collaboration that has nurtured and stimulated varietal output during the past three decades.

Adoption and Varietal Turnover

The level of adoption of improved cultivars and the velocity of varietal turnover are discussed in this section, which is organized by dryland crop. Much of that discussion focuses on the leading improved varieties and hybrids in each of the main-producing states in 2010. Before cultivar-specific estimates are presented, we briefly survey the level of aggregate adoption.

Aggregate adoption of modern varieties

Consistent with the other chapters in this volume, the adoption estimates in Table 14.7 and in the rest of this section refer to modern varieties released since 1970. Estimates of the current popularity of earlier releases were also generated in the expert opinion panels and that information is referred to where it is appropriate.

Some of the adoption estimates in this section are taken mainly from the first-round expert elicitations. These scientist estimates provided a sharper definition of cultivar-specific adoption than later estimates that incorporated more information from various sources. Methodologically, these estimates are also broadly equivalent to the expert elicitation that was carried out for the majority of crop and country observations in sub-Saharan Africa. Survey estimates from ICRISAT's TRIVSA Project for rainy-season sorghum in Maharashtra and chickpea in Andhra Pradesh and from IFPRI's HarvestPlus comprehensive inquiry on pearl millet in Maharashtra and in Rajasthan are used for these four crop-by-state observations. Time-series information from the Government of India on the uptake of sorghum and pearl millet high-yielding varieties (HYVs) complements the expert and survey estimates.

Across the five dryland crops, the simple area-weighted adoption level of modern varieties is 65% in Table 14.7. Between 1970 and 2010, this estimated level is equivalent to an average increase of 1.45% per annum. After 40 years of sustained varietal output in all five crops, there are large tracts where farmers are planting third- and fourth-generation improved varieties. There are also widespread areas, usually of low production potential, where the majority of producers still cultivate desi (local) varieties.

The crops in Table 14.7 can be split into two groups: (i) sorghum and groundnut with moderate adoption levels slightly over 50%; and (ii) pearl millet, chickpea and pigeonpea with appreciably higher adoption performance ranging from 67% to 79%. Relatively low rates of adoption in the former group are attributed to specific states or seasonal production environments where the uptake of improved varieties markedly lags behind other producing-areas in India. Post-rainy season sorghum production in western Maharashtra and Andhra Pradesh and in northern Karnataka epitomizes an environment of low production potential that is almost always associated with terminal drought stress. For groundnut, the problem of lagging adoption finds its greatest expression in the southern state of Karnataka where about 90% of cultivated area is planted to TMV-2, a bold-seeded variety released in 1940 that is widely adaptable to South Indian conditions. The difficulty in replacing well-established, old commercial groundnut varieties is a recurring theme that was discussed in Chapter 7 in the context of West Africa.

Table 14.7. Adoption (%) of modern cultivars across major-producing states and seasons by crop in 2010.

State	Sorghum[a]	Pearl millet	Chickpea	Pigeonpea	Groundnut
Andhra Pradesh	–	–	95	70	40
Rainy (kharif) season	40	–	–	–	–
Post-rainy (rabi) season	40	–	–	–	–
Gujarat	–	95	–	–	–
Rainy (kharif) season	–	–	–	–	90
Irrigated summer season	–	–	–	–	100
Haryana	–	85	–	–	–
Karnataka	–	–	100	60	10
Rainy (kharif) season	90	–	–	–	–
Post-rainy (rabi) season	20	–	–	–	–
Madhya Pradesh	77	–	84	65	–
Maharashtra	–	80	70	70	85
Rainy (kharif) season	100	–	–	–	–
Post-rainy (rabi) season	20	–	–	–	–
Rajasthan	35	52	68	–	64
Tamil Nadu	–	–	–	70	60
Uttar Pradesh	–	30	65	85	–
All-India area weighted adoption (%)	53	67	79	68	54

'–' denotes minor-producing states and seasons that are not covered in the study. [a]Aggregate adoption rate for rainy season was 82%, whereas it was 21% in post-rainy season during 2010.

Sorghum's rather modest aggregate adoption outcome in Table 14.7 is exacerbated by the sharply declining trend in area of rainy-season sorghum that was characterized by 80% level of adoption in 2010. In the late 1960s, rainy-season area accounted for about two-thirds of sorghum hectareage. If that relative importance had been maintained and realized in 2010 instead of a 40% area share, the modern variety (MV) adoption level in Table 14.7 would have exceeded 60%.

Sorghum

Since CSH-1 was released in 1965, graphing the GOI (Government of India) adoption estimates of modern sorghum hybrids and varieties in the rainy season shows a consistent linear pattern of uptake in the principal producing states. In general, adoption at the state level was slower for sorghum than for pearl millet, which is characterized by a typical s-shaped diffusion path. By 2008, Maharashtra, Karnataka, Madhya Pradesh and Tamil Nadu had exceeded or were approaching 80% adoption. Gujarat and Rajasthan lagged behind in MV adoption but both states have recently made substantial progress after very slow early adoption of hybrids. Andhra Pradesh is the only state where MV adoption declined in the past decade. With a steep decline of rainy-season growing area for sorghum in Andhra Pradesh, it is likely that farmers are substituting other crops for sorghum in small subregions where modern cultivars had previously been adopted and are continuing to plant sorghum in other subregions where traditional varieties were not replaced by modern cultivars.

Going from aggregate to variety-specific adoption, several findings stand out in Table 14.8. Adoption outcomes in the rainy season are markedly superior to those in the post-rainy season. In India, sorghum improvement research from all stakeholders during the past 50 years was skewed towards development of the rainy season crop. Very little emphasis was given to the post-rainy season characterized by substantially lower production potential. However, this trend has changed during the last decade. New improved cultivars are slowly replacing the dominant landraces.

Hybrids are more extensively grown than improved open-pollinated varieties (OPVs) in the rainy season. Indeed, in Maharashtra, the largest producing state, the survey results suggested that adoption of improved sorghum varieties was negligible in 2010. Although hybrids have been indicted for poor grain quality at harvest,

Table 14.8. Adoption of modern varieties in % of sorghum-growing area from expert opinion/survey data by major-producing state and season in India.

Andhra Pradesh		Maharashtra		Karnataka		Madhya Pradesh		Rajasthan	
Cultivar	Area (%)	Cultivar[a]	Area (%)	Cultivar	Area (%)	Cultivar	Area (%)	Cultivar	Area (%)
Rainy (kharif) season									
SPV-462 (PSV-1) (1996)	20	MLSH-296 (1995)	22	CSH-14 (1992)	40	CSH-15 (1995)	13.9	CSV-15 (1996)	10.9
CSV-15 (1996)	2.5	CSH-9 (1978)	14	DSV-2 (1986)	18	CSH-18 (1999)	12.3	JKSH-592	4.4
CSV-20 (2009)	2.5	Pro-Agro 8340 (2001)	13	DSV-16 (2009)	15	Ajeet 997 (2002)	10.7	SSG-593 (1978)	2.9
NTJ-2 (1990)	2.5	Mahyco-51 (1982)	10	CSV-16 (1997)	15	Pradhan	10.0	CSV-10 (1986)	2.4
NTJ-4 (1992)	2.5	JK 22 (1999)	10	Others	2	CSH-14 (1992)	8.9	KJH-6363	2.2
Others	10	PAC 537 (2003)	4	All MVs	90	GK-4010	6.5	Others	12.2
All MVs	40	CSH-14 (1992)	3			CSH-16 (1997)	5.8	All MVs	35
		Nirmal-40 (1999)	3			Others	8.9		
		HARITA-540	2			All MVs	77		
		Ajeet-997 (2002)	2						
		MAHABEEJ-7-7A (2000)	1						
		Others hybrids	16						
		All MVs	100						
Post-rainy (rabi) season									
C-43 (1997)	10	Phule Vasudha (2008)	5	DSV-4 (1997)					
CSH-9 (1983)	10	Parbahanimoti (2005)	3	DSV-5 (1998)					
Others	20	RSLG-262 Maulee (2000)	3	CSV-216R (2000)					
All MVs	40	Phule Yashoda (2000)	3	CSV-22 (2007)					
		Phule Chitra (2008)	3	BJV-44 (2012)					
		CSV-18 (2005)	3						
		All MVs	20	All MVs	20				

[a]From ICRISAT survey data.

susceptibility to disease, especially grain mould, and low fodder production, relatively few improved sorghum varieties have found a home in many farmers' fields in rainy season production.

The low popularity of state-level releases in sorghum cultivation in the rainy season is a variation on this theme. With the exception of the DSV (Dharwar sorghum varieties) series selected at the University of Agricultural Sciences at Dharwad, few state-level varietal releases account for sizeable acreages in Table 14.8. Most adopted entries from the public sector in Table 14.8 come from the CS (coordinated sorghum) series that denotes national releases.

Private-sector hybrids are also well represented in Table 14.8, especially in Maharashtra where several larger seed companies have sited their main operations. The evidence in the next section also suggests that the adoption estimates for the private sector in Table 14.8 are likely to be underestimated and estimates for the public sector overstated because the expert panels consist primarily of public-sector scientists who are not current with demand for private-sector hybrids. Underestimation of private-sector participation seems to be more of a problem in Andhra Pradesh and Karnataka than in the other three states in Table 14.8. Additionally, farmers in Andhra Pradesh prefer to grow a local cultivar called 'yellow jowar' for its medicinal properties.

None of the adopted entries in Table 14.8 could be called a mega hybrid or variety, but there are several instances of spill-overs across states. CSH-14, CSH-15 and MLSH 296 (Dev Gen seeds) are adopted cultivars with wider adaptability across three or more states.

Hybrids are conspicuous for their absence in the rabi (post-rainy) season in Table 14.8. Few if any are recommended for the post-rainy season. In general, the estimates of adoption in the post-rainy season are higher than expected. A few of the listed adopted releases, such as Phule Vasudha, are derived from local landrace materials. Hence, they do not represent the level of qualitative change that one usually associates with modern varieties.

Much is known about varietal change and the velocity of varietal turnover in rainy season sorghum in India. Deb and Bantilan (2003) in the context of the 1998 Initiative present information on the composition of modern varieties over eight 5-year intervals from 1966 to 1999. During this timespan, adoption of MVs rose from about 1% of area in 1966, following the release of CSH-1 in 1964, to 69% in 1999. For the first two periods, only CSH-1 appears as an adopted modern cultivar in their graphical analysis. In 1976, CSH-5 joins CSH-1 in the group of modern cultivars. By 1981, CSH-1 is no longer in production because it is replaced in the diffusion of CSH-5 and a new entry, CSH-6. By 1986, CSH-9 has made its debut. CSH-5 and CSH-6 maintain their area shares from the previous period. In 1991, CSH-9 is the dominant MV accounting for slightly over 40% of area. It has replaced the earlier CSH hybrids. The private sector is now also contributing to varietal change via hybrids such as Mahyco-51 and JK 22. During the mid-to-late 1990s, public-sector hybrids CSH-13 and CSH-14 join the set of adopted cultivars together with an expanded group of private-sector hybrids. These new entrants largely replace CSH-9 and they penetrate into some regions still growing local varieties.

The transition in dominance from CSH-1 to CSH-5 to CSH-9 to a larger group of public and private sector hybrids each claiming a relatively small share of MV growing area is consistent with rapid varietal turnover, which illustrates the high productivity of the Indian sorghum improvement programme in generating genetic materials that farmers used in rainy season production. Weighted average age of modern varieties probably fell in the range of 5–10 years throughout much of this period. This low age estimate represents quite an accomplishment for a rainfed crop that does not rely on the breakdown of varietal disease resistance as an incentive for cultivar replacement.

In 2010, most of the entries in Table 14.8 were released in the 1990s. A few were released in the 1980s and the 2000s. Varietal age in 2010 was therefore probably around 15 years. Some but not many of the cultivars in Table 14.8 were already in farmers' fields by the late 1990s when Deb and Bantilan (2003) carried out their research.

Pearl millet

As with sorghum hybrid technologies and their commercialization that was pioneered by researchers at Texas A&M University and the United States Department of Agriculture (USDA) in the 1950s, India was quick to capitalize on the innovations made by Glenn Burton on the hybridization of pearl millet at the University of Georgia. Since their introduction in the mid-1960s, the uptake of pearl millet HYVs as a group has steadily climbed at the all-India level from 3% in 1966–1968 to 67% in 2006–2008. Gujarat, Haryana and Maharashtra have reached or are close to attaining full adoption of high-yielding hybrids and varieties (Table 14.9). Recently, arid Rajasthan has crossed the 50% threshold in the adoption of improved cultivars. Adoption lags behind in Uttar Pradesh. Diffusion of improved cultivars, especially hybrids, was very rapid in institutionally well-developed Gujarat. By 1977, 7 of every 10 hectares of pearl millet in Gujarat were planted to a hybrid. The higher production potential of irrigated summer cultivation was probably a favourable influence in accelerating the speed of adoption in Gujarat.

Recent large-scale national surveys are a basis for the estimates of specific MV adoption in Rajasthan and Maharashtra (Asare-Marfo *et al.*, 2013). In Gujarat, national crop improvement scientists could not assign well-defined areas to specific improved cultivars. They could name five to six of the cultivars that they believed were widely adopted but they could not distinguish among them in terms of areal importance. In other words, experts did not have well-founded prior knowledge on the extent of cultivar coverage. In contrast, in Haryana and Uttar Pradesh, experts were able to rank varieties and assign relative areas to their cultivation. Many private-sector and public-sector hybrids are available for use by farmers in all five states. Both expert assessment and the survey results coincided with estimates from the GOI on aggregate adoption.

With the exception of ICTP-8203, all the cultivars listed in Table 14.9 are hybrids. Most of the cultivars are from the private sector. The public sector is, however, well represented with ICTP-8203, HHB-67 improved, HHB-197 and GHB 558. Most of the private-sector hybrids are derived from public-sector materials (Pray and Nagarajan, 2009). Pro Agro-9444 is an apt example of private-sector collaboration with the public sector, which in this case is ICRISAT. Indeed, numerous pearl millet hybrids commercially marketed in India have made intensive use of ICRISAT-developed male sterile lines and restorers. Without a liberalized seed policy featuring open access to basic research materials, the dominance of private-sector hybrids in varietal change in pearl millet would not have been realised to the depth and extent that it has (Pray and Nagarajan, 2009).

The results in Table 14.9 also confirm some cases of spill-over varieties, namely Pioneer 86M32, the leading hybrid in Rajasthan and the second leading modern cultivar in Maharashtra. In IFPRI's HarvestPlus survey conducted by the Institute of Development Studies in Jaipur, the 'other hybrids' entry for Rajasthan in Table 14.9 sum to a total of 55 distinct names, mostly hybrids that were identified from their seed packaging. The majority of these were adopted by only 1–3 farmers in the sample of 2144 households.

The very small production areas of pearl millet in Rajasthan are one of the most relevant findings from the HarvestPlus survey (Asare-Marfo *et al.*, 2013). The average sown area per hybrid per household was only about 0.1 hectare. With an average cultivation area of 0.2 hectares, Eknath 301 was characterized by the largest growing area per household. In contrast, mean planted areas in Maharasthra were 5–10 times larger, but they still averaged less than 1 hectare. The fact that farmers who each plant such limited areas to the crop have access to such a wide array of pearl millet hybrids is impressive. Some of the diversity of hybrids in Rajasthan is attributed to different emphases in end uses among households and in varying subregional production conditions. Of the popular hybrids, heat-tolerant Pro Agro-9444 has penetrated into several of the arid districts of western Rajasthan. The Pioneer hybrids are mainly found in central and eastern Rajasthan. In general, the hybrids seem to be competitive with local varieties in all districts except Barmer and Jaisalmer, which represent the most arid production environment in the Rajasthan.

The velocity of varietal turnover of pearl millet hybrids in farmers' fields is rapid in India. The simple average varietal age across the five states in Table 14.8 is only 10 years. About 70% of pearl millet cropped area in modern varieties is occupied by cultivars released in the 2000s. The predominance of recent releases is especially marked in Gujarat and Haryana. Among the five ICRISAT mandate crops in India, the decadal-age profile of pearl millet adopted releases is consistent with the fastest rate of varietal change (Fig. 14.3).

Table 14.9. Adoption of modern varieties in % of pearl-millet-growing area from expert opinion/survey data by major-producing state in India.

Rajasthan[a]		Maharashtra[a]		Gujarat		Uttar Pradesh		Haryana state	
Cultivar	Area (%)	Cultivar	Area (%)	Cultivar	Area (%)	Cultivar	Area (%)	Cultivar	Area (%)
Pioneer 86M32 (2002)	7	Mahyco 204 (1995)	22	GHB 558/568 (2002)	95	Kaveri Super Boss (2007)	6	Pro-Agro-9444 (2004)	40
Pioneer 86M52	6	Pioneer 86M32 (2002)	14	Pioneer 86-M-86		ICTP-8203 (1988)	5	HHB-67 Improved (2005)	30
Bayer Proagro 9444 (2004)	6	Mahyco 2210 (2010)	9	MLBH 1012		Pioneer hybrids	4	HHB-197 (2008)	10
Eknath 301 (1991)	3	Nirma 9	7	Sagarlaxmi (2008)		Others	15	Others	5
Nandi 42	3	Mahalaxmi 308 (1998)	7	Pro Agro-9444 (2004)		All MVs	30	All MVs	85
HHB-67 Improved (2005)	2	Mayhco 167	6	Ratan 666					
HHB-67 (1990)	2	Dhanya 7870	6	Others					
Guhu MH 169 (1987)	2	Mahabeej ICTP 8203 (1988)	4	All MVs	95				
Nandi 52 (2004)	2	Ganga Kaveri 1044 (1997)	3						
Other hybrids	19	Nirmal 40 (2002)	3						
All MVs	52	Other hybrids	18						
		All MVs	99						

[a]From survey data from Asare-Marfo *et al.*, 2013.

A very competitive private sector coupled with the need for new sources of downy mildew resistance are two forces that drive the rapid replacement rate of improved pearl millet cultivars by farmers in India.

Chickpea

The state-wise cultivar specific adoption estimates elicited through expert consultations are summarized in Table 14.10 where the high MV

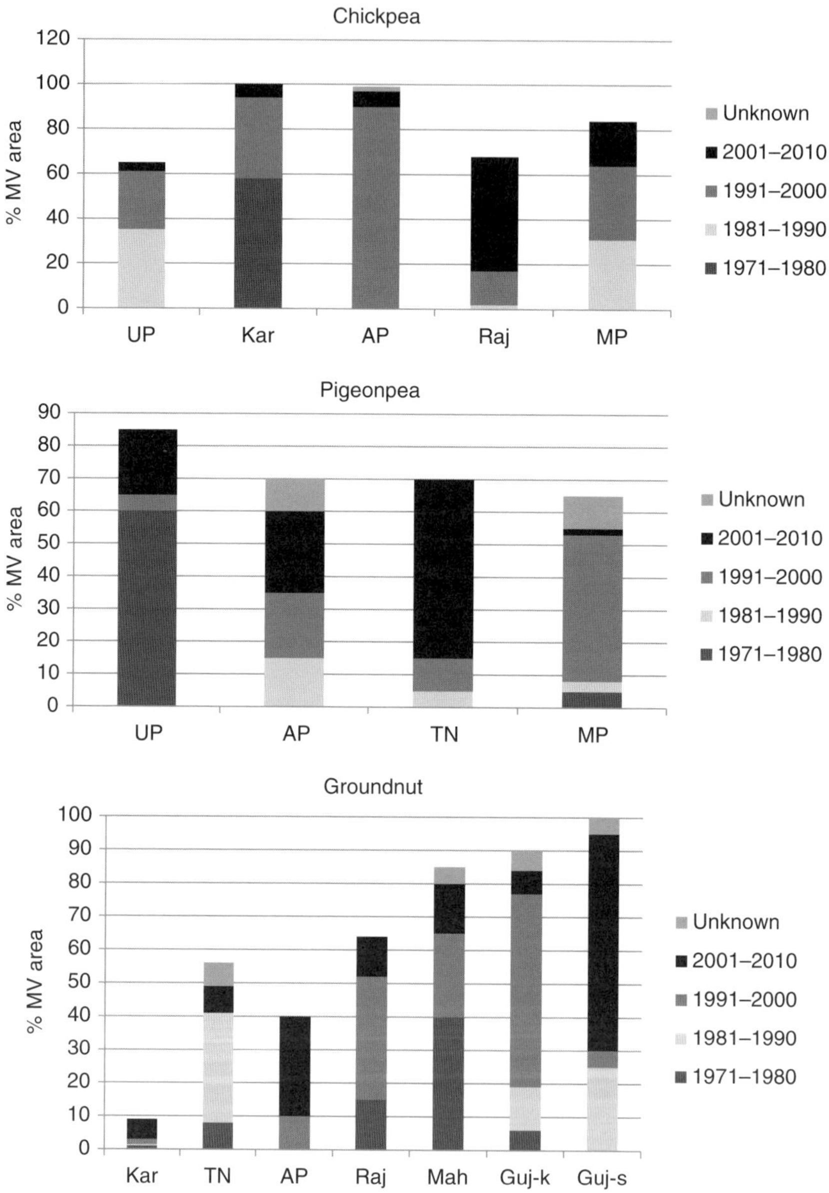

Fig. 14.3. Proportion of MV area by varietal age. AP, Andhra Pradesh; Guj, Gujarat; Har, Haryana; Kar, Karnataka; Mah, Maharastra; MP, Mahdra Pradesh; Raj, Rajahastan; TN, Tamil Nadu; UP, Uttar Pradesh; k, kharif (rainy season); r, rabi (post-rainy season); s, summer. Source: Expert elicitation surveys conducted from 2010 to 2012.

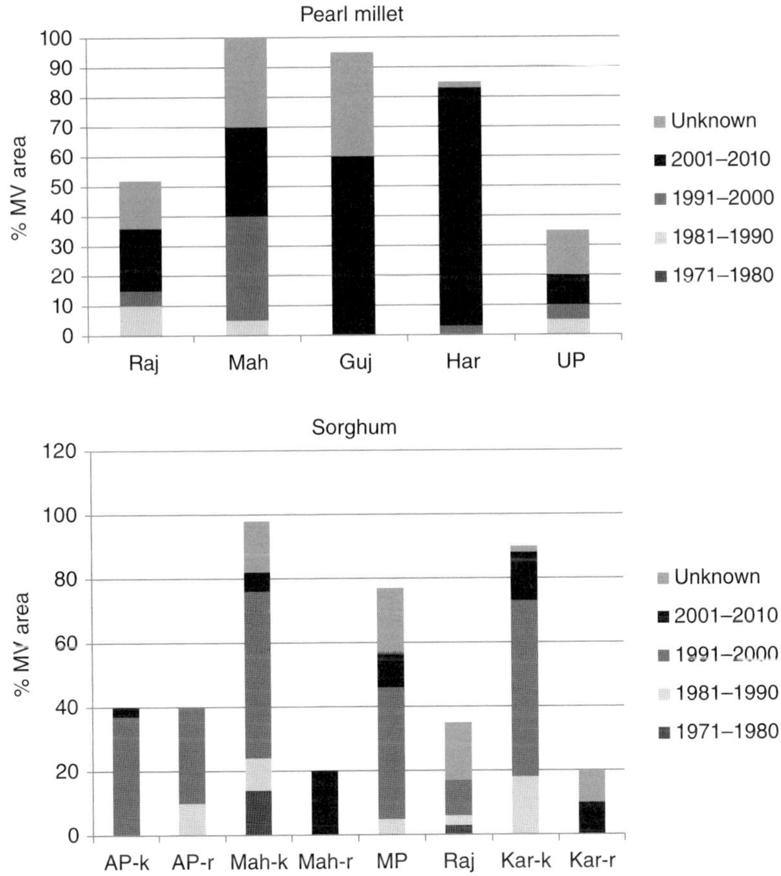

Fig. 14.3. Continued.

Table 14.10. Adoption of modern varieties in % of chickpea-growing area from expert opinion/survey data by major-producing state in India.

Andhra Pradesh[a]		Karnataka		Uttar Pradesh		Rajasthan		Madhya Pradesh	
Cultivar	Area (%)	Cultivar	Area (%)	Cultivar	Area (%)	Cultivar	Area (%)	Cultivar	Area (%)
JG-11 (1999)	84	Annegiri-1 (1978)	58	Avarodhi (1987)	25	RSG-888 (2002)	10	JG315 (1981)	27
		JG-11 (1999)	34						
Vihar (2002)	7	BGD 103 (2000)	4	KWR 108 (1996)	10	GNG-663 (1995)	15	JG130 (2002)	13
KAK-2 (1999)	6	MNK-1 (2010)	2	DCP 92-3 (1998)	7	RSG-973 (2004)	14	JG322 (1997)	13
Others	2	Others	2	Pusa 256 (1985)	5	RSG 963 (2005)	5	Vijay (1994)	7
All MVs	99	All MVs	100	Others	18	Others	15	Others	24
				All MVs	65	All MVs	68	All MVs	84

[a]From ICRISAT survey data.

adoption level in Andhra Pradesh stands out. The lion's share of cropped area is occupied by a single dominant cultivar, JG 11 (a desi type released in 1999). JG 11 was developed by ICRISAT and JNKVV University in Madhya Pradesh. Its strengths are high yield, early maturity, large attractive seed and resistance to fusarium wilt. It is replacing the old landrace cultivar Annigeri that dominated the southern states for several decades. The rapid adoption of JG 11, Vihar and KAK-2 has been described as the 'silent chickpea revolution' in Andhra Pradesh (see Bantilan *et al.*, 2013, for more details).

Madhya Pradesh has also exhibited tremendous growth in area and production of chickpea during the last five decades. The bulk of the cropped area in the state is under desi types, whereas the remainder is sown to kabuli types. Nearly 95% of desi-type area is covered with improved cultivars. In contrast, only 5% of kabuli area was planted to improved cultivars.

The extent of adoption of MVs is lower in the other states. Annigeri, released in 1978, still accounts for more than half of the chickpea-growing area in Karnataka. Farmers' fields in Uttar Pradesh are also the home to some rather old released cultivars. Radhey, released in 1968, was believed by the expert panel to comprise 25% of chickpea plantings in Uttar Pradesh. If we ignore the cultivars released before the 1980s or exclude landrace varieties such as Annigeri, the aggregated weighted adoption level estimated at the all-India level was around 70%.

The proportion of chickpea cropped area under recent releases (2000–2010) is quite high in Rajasthan and Madhya Pradesh. Most of chickpea cropped area in Andhra Pradesh and Karnataka is also cultivated in recent releases because JG-11 was only 11 years old in 2010. The weighted average varietal age of 10–15 years indicates a reasonable speed for varietal turnover in a pulse crop.

Pigeonpea

Maharashtra, Karnataka, Andhra Pradesh, Uttar Pradesh, Madhya Pradesh and Gujarat are the major pigeonpea growing states, which together represent more than 90% of cropped area and production in the country. In the two leading producing states, expert estimates were not that informative. In Maharashtra, experts could estimate an aggregate level of adoption and name a few of what they believed to be the leading improved cultivars. In Karnataka, the information was coarser as the expert panel could only venture an estimate that improved varieties covered 60% of pigeonpea-growing area. More precision was obtained in the other states where released varieties appear with estimated areas in Table 14.11.

In compiling Table 14.11, we did not include one old long-duration variety, Bahar, released in

Table 14.11. Adoption of modern varieties in % of pigeonpea-growing area from expert opinion by major-producing state in India.

Uttar Pradesh		Tamil Nadu		Andhra Pradesh		Madhya Pradesh		Maharashtra	
Cultivar	Area (%)	Cultivar	Area (%)	Cultivar	Area (%)	Cultivar	Area (%)	Cultivar	Area (%)
MAL 13 (2003)	25	LRG 41 (2007)	29	LRG-41 (2007)	15	ICPL87119 (1993)	37	BSMR-786 (1996)	70
NDA 1 (1996)		CORG 9701 (2004)	26	LRG-30 (1982)	10	No. 148 (1975)	4	BSMR-853 (2001)	
NDA-2 (2008)		Co 6 (1993)	10	ICPL-85063 (1997)	10	JA4 (1991)	4	BSMR-708	
All MVs	25	TTB 7 (1987)	5	ICPL-87119 (1993)	10	Others	20	ICPL87119 (1993)	
		All MVs	70	PRG-158 (2007)	10	All MVs	65	Others	
				PRG-100				All MVs	70
				Others	15				
				All MVs	70				

1980. Bahar is believed to still account for 60% of growing area in Uttar Pradesh. In spite of the fuzziness of the information in Table 14.11, it is apparent that several varieties are characterized by a wide adaptability because they are popular in multiple states. For example, ICPL 87119 is the first medium-duration variety with combined resistance to two of the most devastating diseases of pigeonpea, sterility mosaic and *Fusarium* wilt.

Pigeonpea is characterized by a mix of younger and older releases in farmers' fields but most were notified between 1980 and 1999. About 30% of MV area is made up of varieties released since 2000. The velocity of varietal turnover for pigeonpea is somewhat slower than that for chickpea because varietal age averages 15–20 years for the varieties in Table 14.11 that were released after 1980.

Groundnut

Cultivar-specific adoption across major groundnut-growing states is described in Table 14.12. In spite of a solid and improving performance in varietal output, recent groundnut releases have not been widely adopted by farmers. In Gujarat (kharif), the single most dominant variety is GG 20 released in 1991. GG 2 (released in 1984) is the leading cultivar in summer cultivation in Gujarat. JL 24 (1978), TAG 24 (1991) and TMV 10 (1970) are the most widely grown cultivars in Maharashtra; they are 20–40 years old. TMV 2 (1940), which is not listed in Table 14.12 because of its age, still occupies nearly 90% of the cropped area in Karnataka and 60% in Andhra Pradesh. TMV 7 (1967, not listed in Table 14.12) and VRI 2 (1989) are dominant cultivars in Tamil Nadu. This research highlights the problem of the permanency of old vintages and the lack of significant dynamism in varietal replacement across states. If we ignore cultivars released before the 1980s, weighted aggregate adoption at the all-India level is estimated at 45%. Weighted average varietal age exceeds 25 years. Massive systematic efforts, coupled with both institutional and policy support, are required to enhance adoption. In general, the lack of varietal change in groundnut in peninsular India has a lot in common with the adoption experience for the crop in West Africa that was discussed in Chapter 7.

Unlike the other four crops, for groundnut relatively few ICRISAT-related varieties are listed in the cultivar-specific adoption table. ICGV 91114 in Andhra Pradesh is one of the exceptions. It is suited to the difficult production conditions in Anantapur in the dry semi-arid Rayalaseema region where groundnut is one of the few cash crops available to farmers in rainy-season production.

Validating Expert Opinion on Cultivar Adoption

Comparisons among different methods for generating adoption estimates are highlighted in this section. In particular, estimates from village focus-group meetings and representative household surveys are used to validate estimates from expert opinion. The recent and relevant experience of HarvestPlus in eliciting cultivar-specific adoption for pearl millet in Rajasthan and Maharashtra is also reviewed. Three crop- and state- specific adoption and diffusion contexts are presented to deepen understanding about any systematic differences that could emerge between expert elicitation and focus group and survey methods.

Adoption of rainy-season sorghum improved cultivars in Maharashtra

All five of the ICRISAT mandate crops are grown extensively in Maharashtra but the spatial distribution of production is concentrated in different agroclimatic zones by crop and growing season in this very large state in central India. Initially, ICRISAT tried to develop an integrated sampling framework to address adoption and diffusion of several crops in the state. But the uneven pattern of sown area of these five dryland crops in Maharashtra was not conducive to a multi-crop adoption survey. After several iterations and interactions with various sampling experts, ICRISAT decided to conduct an independent survey for two of the crop-by-state observations discussed in this exercise. ICRISAT selected rainy-season sorghum as the first case and conducted a state-level survey

Table 14.12. Adoption of modern varieties in % of groundnut-growing area from expert opinion by major-producing state and season in India.

Gujarat (rainy)		Maharashtra		Karnataka		Tamil Nadu		Andhra Pradesh		Gujarat (summer)		Rajasthan	
Cultivar	Area (%)	Cultivar	Area (%)	Cultivar	Area (%)	Cultivar	Area (%)	Cultivar	Area (%)	Cultivar	Area (%)	Cultivar	Area (%)
GG 20 (1991)	50	JL24 (1978)	30	GPBD-4 (2004)	4	VRI 2 (1989)	25	Kadiri 6 (2005)	18	GG 2 (1984)	25	GG-20 (1991)	30
GG 11 (1987)	7	TAG 24 (1991)	25	TAG-24 (1991)	2	VRI 3 (1990)	8	TAG 24 (1991)	10	TG 37A (2004)	35	M-13 (1978)	15
GAUG 10 (1973)	6	TMV- 10 (1970)	10	Others	3	JL 24 (1978)	8	ICGV 91114 (2007)	10	TPG 41 (2004)	30	TG 37A (2004)	5
GG 5 (1996)	8	SBXI/JL 11 (1965)	5	All MVs	9	Others	15	Others	2	Others	10	Others	14
GG 2 (1984)	6	JL501 (2009)	5			All MVs	56	All MVs	40	All MVs	100	All MVs	64
Others	13	Others	25										
All MVs	90	All MVs	100										

covering 13 districts, 20 tehsils (blocks equivalent to townships), 60 villages and 360 sample households.[1]

Maharashtra is the leading sorghum-producing state, contributing a share of 55% in the country's total acreage and 49% of production. The performance of kharif sorghum is dominated by hybrids, whereas the post-rainy (rabi) crop is still sown to varieties and landraces only. Nearly 45 state-specific sorghum improved cultivars have been developed and released during the past 50 years. The vast majority of these have been released for the rainy-season crop, which is at or approaching full adoption. The private sector dominates the sorghum seed market in the state. Nearly 70–75% of total kharif sorghum seed is marketed by private seed companies; the remaining 25% is supplied by the public sector. Because of the adoption of hybrids, more than 95% of farmers buy seed from the market every year. Therefore, the focus of this case study is not on the level of aggregate adoption but on the estimated level of cultivar-specific adoption in the context of widespread diffusion of hybrids developed and multiplied by the private sector and, to a lesser extent, by the public sector.

The comparative results in Table 14.13 suggest good agreement between the focus-group and survey estimates, but poor concurrence between the expert estimates and focus-group estimates and the expert and survey estimates. Because of the relatively large number of improved cultivars available in the market in the state and private-sector dominance in the crop, experts, who were public-sector crop improvement scientists, were unable to provide estimates for all cultivars during the two rounds of expert elicitation process. They provided estimates only for some of the public-bred cultivars that they were familiar with and were not that cognizant about the uptake of specific private-sector hybrids that account for 70% of cultivated area according to the village focus groups and household surveys. Indeed, the public-sector experts substantially overstated the importance of improved OPVs, which, for all intents and purposes, were not mentioned in the 60 village focus groups and the 360 household interviews.

Between the community focus-group interviews and the household surveys, noticeable differences were observed for only one or two cultivars. However, like expert opinion, the focus-group participants placed more importance on public-sector hybrids than the household- survey respondents. The focus-group participants estimated the adoption level of the four public-sector hybrids (indicated by superscript 'a' in

Table 14.13. Comparison of estimates of adoption of modern sorghum varieties by source.

Cultivar name	Expert elicitation (% area)	Community level (% area)	Household survey (% area)
CSH-9[a]	40	19.0	13.9
MLSH-296 (Dev Gen)	–	18.2	22.2
Mahyco-51	–	10.4	10.1
Mahabeej-7[a]	–	9.4	1.0
Pro Agro-8340	–	7.4	13.2
JK-22	–	6.0	9.8
CSH-14[a]	30	3.4	2.5
MSH-296[a]		3.4	0.0
NSH-18	–	2.3	0.0
PAC-537	–	2.3	3.8
Nirmal-40 (NJH-40)	–	0.0	3.4
HARITA-540	–	0.0	2.4
Ajeet-997 (Ajeet company)	–	0.0	2.3
Other hybrids	10	18.2	16.1
Other OPVs	20	0.0	0.0
Area under total MVs	100.0	100.0	99.7
Area under locals	0.0	0.0	0.3

[a]Public-bred cultivars.

Table 14.13) at about 35%; the estimate for the household survey was only 17.5%. One can speculate that because the public-sector hybrids were somewhat older than their private-sector counterparts, the focus-group participants, some of whom may not have planted rainy-season sorghum in the last cropping season or even in the very recent past, may not have been up to date with the newer private-sector hybrids.

In the context of widespread adoption and annual market purchase of rainy-season sorghum hybrids, the household survey seems to have provided the most reliable results of the three sources of information. Survey responses were not plagued by the endemic problem of unknown improved varieties, the names of which vary from place to place. Because of their commercial importance and the prevailing tendency of farmers to purchase hybrids each year, almost all interview responses could be readily identified and correctly tagged with a cultivar name.

Adoption of pearl millet improved cultivars in Maharashtra

Pearl millet is the second crop in Maharashtra state chosen for understanding the cultivar specific adoption estimates across three different methods (expert versus community versus household level). A similar sampling framework was adapted by using block-level data collected from Maharashtra Department of Agriculture. The primary household survey collected data from nine districts, 20 thesils, 60 villages and 360 sample households in the state. Similarly, 60 focus group meetings were also organized in each sample village.[2]

Maharashtra stands third in pearl millet production in India, with an 11% share both in area and in production. Pearl millet is mainly produced in western Mahrashtra where rainfall is low and erratic in the kharif season. Around 30 improved cultivars have been released and made available to farmers in the state during the past 50 years.

Hybrids have penetrated profusely into the markets and the fields of farmers in Maharashtra. Since the 1980s, private-sector seed companies have had a higher market share of the seed market in pearl millet than in sorghum. In general, an improved hybrid cultivar produces nearly 30–40% yield advantage than any OPV grown in that particular location; however, OPVs are still preferred by farmers that have shallow soils and low rainfall regimes.

Cultivar-specific adoption estimates by method are compared in Table 14.14. The estimates from community and household surveys (more or less) coincide; the mean differences between them are insignificant. In this context with an overwhelming dominance of private-sector materials, representative and well-conducted focus-group meetings may be competitive with household surveys in lowering research costs. It is important, however, that constituents of the focus groups are pearl millet producers from the most recent cropping season. Even in this case, it is likely that many minor hybrids will be missed in the interview process.

There is a large gap in information between expert elicitations and the survey results. Experts assessed the aggregate level of adoption at or near 80% but they could name only five leading pearl millet cultivars – Pioneer 86-M-32, Pioneer 86-M-64, Mahyco 2240, Mahyco 2210 and Pro-Agro (XL-51). From the perspective of the survey results in Table 14.14, they underestimated aggregate adoption by 20 percentage points. Only one of their five leading varieties appears in the top three cultivars in Table 14.14 with more than 10% adoption.

Table 14.14. Comparison of pearl millet adoption estimates, community level (focus group) versus household level.

Cultivar	Community level (% area)	Household level (% area)
ICTP-8203[a]	25.86	27.80
Pioneer 86-M-32	17.15	15.40
Mahyco-204	15.71	18.40
GK1044	6.62	2.70
MDBH-318	6.17	0.00
Dev Gen 308	4.69	3.90
Nirmal-9	4.04	2.00
Dhanyaa 7872	2.68	4.50
Mahyco-163	2.57	2.50
Mahyco-2210	1.58	0.00
Other hybrids	12.52	22.8
Area under total MVs	99.59	100.00
Area under locals	0.41	0.00

[a]Public bred cultivars.

In general, experts were good at providing the estimates of adoption when the incidence of releases was low and well known. Because most pearl millet improved cultivars are developed and marketed by private seed companies, awareness of public-sector experts on field-level adoption was limited. Very few public-sector varieties and hybrids were present in farmers' fields in 2010.

Expert opinion on the adoption of improved pearl millet cultivars can also be validated from the perspective of the recent HarvestPlus surveys in Maharashtra and Rajasthan (Asare-Marfo *et al.*, 2013). Based on a large-scale representative survey of more than 2000 households, the HarvestPlus survey results were presented in Table 14.9. From the perspective of the HarvestPlus survey, experts were able to name the top two leading hybrids but two of their top five were not in the top ten from HarvestPlus.

The results from the HarvestPlus and ICRISAT survey differed markedly over the uptake of the improved OPV ICTP-8203. It fell from the top-ranking cultivar with 28% of area in the ICRISAT survey to eighth position with only 4% of area in the HarvestPlus survey, which pointed to its lower yield than hybrids and its importance as stover for livestock.

HarvestPlus also conducted a mail/interview survey of 58 block agricultural extension officers and 789 seed suppliers in Maharashtra. Interviewees were asked to name the three leading improved cultivars on the basis of area sown or on their own seed sales information. ICTP-8203 ranked first among the extension officers but dropped to sixth position in seed sales. In general, seed sales information was a better match to the survey results than the responses of the extension officers.

Compared to the smaller ICRISAT survey, the HarvestPlus survey sampled twice as many blocks, three times as many villages and six times as many farmers. Both surveys were conducted in the nine most important pearl-millet-producing districts in the state. The fact that the two surveys show such a large discrepancy in the estimated leading variety warrants more comparative analysis of sampling design and results, especially at the block level.

In Rajasthan, where none of the adopted hybrids exceeded 7% of area coverage in 2010, experts faced a more formidable challenge than those in Maharashtra (Table 14.9). The expert elicitation resulted in the following position of hybrids with their area shares: Pro-Agro 9444 (11%); HHB-67 Improved (10%); MH 169 (7%); JKBH 26 (4%); and others (20%). Experts correctly assessed aggregate adoption and they correctly perceived the importance of Pro-Agro 9444 and HHB-67. MH 169 and JK 26 also rank in the top 15 varieties. They did not perceive the importance of the top-ranked hybrids from Pioneer in Table 14.9 but that may because those hybrids are not notified, that is, officially released. They did about as well as agricultural officers at the block level and agricultural input suppliers in the private sector who were also surveyed in the HarvestPlus research. Overall, scientists in Rajasthan did better than their peers in Maharashtra.

Adoption of chickpea improved cultivars in Andhra Pradesh[3]

Chickpea was not even a minor crop in Andhra Pradesh until 1985. Short winters, terminal moisture stress, wilt disease and pod borer were the major constraints for growing chickpea in this southern state of India. Offsetting these disadvantages were two major advantages: it was easy to grow and it was characterized by a higher harvest index, indicative of a shorter growing period. Until 1985, late-maturing varieties namely Gulabi and Jyoti (selections from landraces) were under cultivation in Andhra Pradesh. Research collaboration between NARS partners and ICRISAT on crop improvement and management addressed the above constraints and harnessed opportunities to develop new cultivars that could make chickpea a most suitable crop for the region. The close and sustained collaborative efforts led to the development of short-duration chickpea varieties that were introduced in late 1990s and have been widely adopted by farmers in the state. All local cultivars have been replaced by these improved short-duration cultivars which resulted in what is now referred to as the 'AP chickpea silent revolution' with a fivefold increase in area, doubling productivity and a tenfold increase in production in Andhra Pradesh.

A representative sampling framework was developed based on Objective-2 DIIVA guidelines

(Walker and Adam, 2011). A total of 810 sample chickpea-growing households were interviewed with well-structured survey instruments from 30 mandals, seven districts and 90 villages of Andhra Pradesh. For enhancing the proper identification of improved cultivars at the farm-level, a varietal identification protocol was developed and administered for the chickpea adoption study in Andhra Pradesh. This has increased efficiency in proper identification of chickpea cultivars through a systematic validation process. A well-designed protocol not only minimizes the misidentification of improved cultivars but also reduces outliers.

The comparisons of cultivar-specific adoption estimates are summarized in Table 14.15. The estimates are much closer in community (focus group) and household level surveys. Compared to the sorghum and pearl millet validation discussed above, expert perceptions on chickpea varietal adoption were more precise; however, a few significant differences emerged between their perceptions and the survey results. Experts overstated the importance of KAK-2, an improved kabuli variety, and underestimated the importance of Vihar, also an improved kabuli variety. They correctly perceived that JG-11 was a truly dominant variety and that adoption of modern varieties approached 100%. Because of the low incidence of released chickpea improved cultivars in Andhra Pradesh, experts were able to provide a reasonable picture of the varietal reality in farmers' fields. Moreover, the role of the private sector in crop improvement, as well as in seed multiplication, is almost negligible.

Table 14.15. Comparison of chickpea adoption estimates, expert versus community level (focus group) versus household level.

Cultivar	Expert elicitation (% area)	Community level (% area)	Household level (% area)
JG-11	70.00	84.57	84.19
Vihar	0.00	8.35	7.39
KAK-2	20.00	4.03	5.92
Bold/Dollar	2.00	1.04	0.57
JAKI-9218	0.00	0.44	0.31
JG-130	0.00	0.02	0.08
Divijay	0.00	0.01	0.0
Annigeri	3.00	1.55	1.52
Total	95.00	100.00	100.0

Summary and Conclusions

This assessment of performance in crop improvement for five of the most important dryland crops in South Asia from the perspectives of scientific capacity, varietal output and adoption has shed light on many strengths and accomplishments and has also uncovered some areas for improvement. Although most of the findings in this chapter are not new, they are worth repeating.

Stability in making more varieties available via increasing releases over time is arguably the most impressive achievement of the dryland crop improvement programmes in India. There are few dry spells in output because the five programmes have consistently been able to release varieties annually during the past 50 years. Only pigeonpea is characterized by stagnant varietal output over time. Most of the programmes show a diversified mix of central- and state-level releases. Private-sector participation is also increasing over time in the provision of sorghum and pearl millet hybrids. For pearl millet and sorghum, recent survey results cited here show that more than 50 well-identified, notified cultivars from the public and private sector and unreleased but commercialized private-sector cultivars have been adopted by at least one farm household in large-scale surveys in Maharashtra and Rajasthan. An abundant provision of improved cultivars for adoption is especially noteworthy in western and central Rajasthan where the network of input-supply stores is sparse and where growing areas of pearl millet only average about one-tenth of a hectare.

The finding that aggregate adoption is still increasing at a rate exceeding 1% per annum in many major-producing states is also a laudable outcome that speaks to the stability of crop improvement programmes over time. Although the area-weighted average level of MV adoption across the five crops was estimated at less than 70% in several producing states, MVs are now at or approaching full adoption in several large-producing states. In pearl millet, the diffusion experience with the bajra hybrids since the mid-1960s in some states, like Gujarat, rivals the speed of adoption in irrigated wheat and rice. In terms of adoption, the conventional wisdom that dryland farmers would be bypassed by the Green Revolution does not hold for the ICRISAT-mandated crops. Diffusion of the public- and private-sector hybrids in rainy season sorghum has also been rapid and efficient in most major-producing states, especially in Maharashtra.

The high estimated velocity of varietal turnover in both the dryland cereals is another impressive finding. The weighted average age of pearl millet improved cultivars in farmers' fields is only about 10 years from their date of notification. Early adopters are planting their fourth or fifth different hybrid since HB-1 was released in 1964.

Some areas for improvement are transparent and well known. For example, the uptake of modern groundnut cultivars has lagged behind the other crops in MV adoption. The 'permanency' of old released varieties in farmers' fields has also translated into very slow varietal turnover. In particular, the dominance of TMV-2 in South India since the 1950s has eroded returns to groundnut improvement. The absence of varietal change in post-rainy season sorghum production is another formidable challenge that requires some out-of-the-box thinking because past research has not resulted in practical impact.

Sustained progress in MV adoption does not require new thinking for all lagging areas. For example, pearl millet hybrids are increasingly penetrating into central and even western Rajasthan. Following the same course with private-sector hybrids supported by public-sector parental lines should continue to lead to more positive adoption outcomes.

Other findings point to areas for improvement that are more subtle. Since the founding of ICRISAT in 1972, the dryland crops have not changed their rainfed character. But their locus of production has shifted substantially since then. Chickpea was displaced by the Green Revolution in irrigated wheat in northern India; it has shifted to central India, mainly Madhya Pradesh, and to southern India, primarily Andhra Pradesh and Karnataka. Likewise, Uttar Pradesh in the North has lost three-quarters of its pigeonpea area. As a result, long duration pigeonpea is no longer as relevant as it once was. Increasingly, medium-duration pigeonpea is the dominant maturity type. Rainy-season sorghum has secularly declined over time, especially in the wet semi-arid tropics where it is being displaced by soybean and Bt cotton. Sorghum is increasingly a crop produced and consumed in the very large state of Maharashtra. Post-rainy-season sorghum is not declining as fast because few alternatives compete with it in an environment of terminal drought stress. In India, post-rainy-season sorghum is produced in a compact production region spanning three states. Hence, sorghum-growing area is declining and its spatial concentration is increasing. Pearl millet is increasingly being relegated to the drier regions in the states in which it is produced. In the mid-1960s, the weighted mean annual rainfall of the districts in which it was produced was 900 mm; by the early 2000s, the mean had declined to 600 mm. Effectively, it was losing area in the wetter districts of higher production potential. Anecdotal evidence suggests that groundnut is also increasingly produced in droughty environments, although there is no solid empirical basis to support this conjecture. These shifts are not short-term phenomena. They reflect longer-term agronomic and economic trends.

Like the federal-state agricultural research system in several countries, such as the USA, the centre-state system in India imparts stability to agricultural research. It is difficult to understate the importance of stability as an attribute for productive agricultural research. But a two-tier integrated system can be an unwieldy institutional structure to respond to longer-term change such as geographic shifts in production. Several of the findings in the section on scientific capacity suggest frictions and impediments in responding to change. Largely because of its declining importance, the estimated research intensity of sorghum is rising and it has now reached a level that is high for a cereal with several million hectares of growing area. Both chickpea and pigeonpea are still characterized by a relatively high deployment of FTE scientists in the North relative to Central and South India. The regional location of research stations, centres and sub-centres seems highly appropriate for the 1970s but not for the 2000s. There may be good reasons to maintain the status quo; however, a priority-setting exercise would seem to be in order to take an analytical and critical look at research resource allocation in these dryland crops.

Over the past 50 years, the major institutional change has been private-sector participation in varietal development and multiplication in pearl millet and sorghum. When the private sector becomes very active in the generation and the multiplication of hybrids, the public sector moves upstream to support the activities of the private sector. The character of public-sector research qualitatively changes from adaptive to more applied and even strategic. Variation in the disciplinary composition of the five improvement programmes was seen as responding to different constraints in each crop. However, the disciplinary composition did not vary that much between the grain legume programmes on one hand and the cereal programmes on the other. In other words, there was not much evidence to indicate that the sorghum and millet improvement programmes had moved or were allocating more resources for upstream research that exploited the comparative advantage of each sector engaged in improvement research.

The adoption data also are indicative of what worked and what did not work in terms of varietal types. For example, in pigeonpea we did not see any adoption of extra-early short-statured pigeonpea varieties that were actively promoted by ICRISAT in the 1980s. These high-yielding materials were sole-cropped, highly regarded by scientists and economists, and widely tested in peninsular India. However, they were severely attacked by pod borer because they matured at a time when few other host plants were available.

We also found that improved varieties have only played a minor role in sorghum and are largely absent in varietal change in pearl millet. Improved varieties may have been competitive with hybrids in the 1970s and the 1980s but their window of opportunity seems to have closed in the production of rainy-season sorghum and pearl millet. Negligible adoption in turn suggests a low rate of return on varietal improvement compared to hybrid development. That improved sorghum varieties and improved pearl millet composites and OPVs are still being released for rainy-season production is puzzling given their limited uptake in the recent past.

Methodologically, the validation results confirmed that expert elicitation is not an effective means of generating adoption estimates when the private sector is actively engaged in hybrid development. In this context, there is no substitute for household surveys and complementary enquiries at the level of agricultural supply stores. In the case of pulses and oilseeds without private-sector participation in varietal development, the example of chickpea in Andhra Pradesh shows that responses on cultivar-specific adoption were very similar for village-focus groups and household surveys. In the chickpea example, expert panel estimates were also reasonably congruent with the results from the focus groups and household surveys when the number of improved varieties was low. But expert panels also gave fuzzy and diffuse information for pigeonpea in the major-producing states of Maharashtra and Karnataka. The lack of clarity in cultivar-specific adoption for pigeonpea in these two states was, arguably, the most disappointing outcome of this research. The uncertainty attached to cultivar adoption in pigeonpea underscores the importance attached to future survey research in Maharashtra and Karnataka. Less is still known about adoption of MVs in pigeonpea than about any of the other dryland crops in ICRISAT's mandate.

The TRIVSA Project has updated and enriched ICRISAT's knowledge on the adoption and diffusion of these five dryland crops in India. It also helped the team to better understand the adoption process as it affects different crops. Familiarization with the crop and its market players (public or private) is an important step before undertaking any adoption and diffusion study. Enlisting a senior breeder throughout the entire process helps enhance understanding. In general, expert opinion seems to be more accurate when a particular cultivar is in its early stages of adoption or at its peak stage. Disaggregated (district/mandal) estimates lead to greater precision in the adoption estimates. In the long run, institutionalization of a monitoring and adoption process is important for bringing more credibility to information on adoption.

Several of the findings also have implications for the prospects for varietal change in sub-Saharan Africa. Those implications are discussed in Chapter 19 of this volume.

Notes

[1] The sample design and the survey questionnaire are discussed in Kumara Charyulu *et al.* (2014a), a rainy-season sorghum technology adoption and impact study in Maharashtra State.
[2] The sample design and the survey questionnaire are discussed in Kumara Charyulu *et al.* (2014b), a pearl millet technology adoption and impact study in Maharashtra State.
[3] This particular activity was additionally co-funded by the Standing Panel on Impact Assessment (SPIA) to understand the cultivar specific adoption pattern of chickpea improved cultivars in Andhra Pradesh through a state-level representative survey.

References

Asare-Marfo, D., Birol, E., Karandikar, B., Roy, D. and Singh, S. (2013) *Varietal Adoption of Pearl Millet (Bajra) in Maharashtra and Rajasthan, India: A Summary Paper*. HarvestPlus, International Food Policy Research Institute, Washington, DC.

Bantilan, M.C.S. and Deb, U.K. (2003) Impacts of genetic enhancement in pearl millet. In: Evenson, R.E. and Gollin, D (eds) *Crop Variety Improvement and Its Effect on Productivity. The Impact of International Agricultural Research*. CAB International, Wallingford, UK, pp. 215–240.

Bantilan, M.S.C., Deb, U.K. and Nigam, S.N. (2003) Impacts of genetic improvement in groundnut. In: Evenson, R.E. and Gollin, D (eds) *Crop Variety Improvement and Its Effect on Productivity. The Impact of International Agricultural Research*. CAB International, Wallingford, UK, pp. 293–313.

Bantilan, M.C.S., Kumara Charyulu, D., Gaur, P.M., Shyam Moses, D. and Davis, J. (2013) *Short Duration Chickpea Technology: Enabling Legumes Revolution in Andhra Pradesh, India*. ICRISAT, Patancheru, India.

Deb, U.K. and Bantilan, M.C.S. (2003) Impacts of genetic improvements in sorghum. In: Evenson, R.E. and Gollin, D (eds) *Crop Variety Improvement and Its Effect on Productivity. The Impact of International Agricultural Research*. CAB International, Wallingford, UK, pp. 183–214.

Evenson, R.E. and Gollin, D. (2003) *Crop Variety Improvement and Its Effect on Productivity. The Impact of International Agricultural Research*. CAB International, Wallingford, UK.

Kumara Charyulu, D., Davala, M., Bantilan M.C.S., Rao K.P.C., Borikar S.T. *et al.* (2014a) *Rainy Season Sorghum Technology Adoption and Impact Study in Maharashtra.* ICRISAT, Patancheru, India.

Kumara Charyulu D., Davala, M., Bantilan, M.C.S., Borikar, S.T., Mohan Rao, Y. *et al.* (2014b) *Pearl Millet Technology Adoption and Impact Study in Maharashtra.* ICRISAT, Patancheru, India.

Pal, S. and Byerlee, D. (2003) *The Funding and Organization of Agricultural Research in India: Evolution and Emerging Policy Issues.* Policy Paper 16, National Centre for Agricultural Economics and Policy Research, ICAR, New Delhi.

Pray, C.E. and Nagarajan, L. (2009) *Pearl Millet and Sorghum Improvement in India*. IFPRI Discussion Paper 919. IFPRI, Washington, DC.

Walker, T.S. (2009) *Updating and Reviewing Future Challenges and Opportunities for Agricultural R&D in the Semi-Arid Tropics for ICRISAT Strategic Planning to 2020*. ICRISAT, Patancheru, India.

Walker, T.S. and Adam, A. (2011) *Guidelines for Data Collection for Objective 2 of the DIIVA Project.* Fletcher, North Carolina.

World Bank (2012) *World Bank Annual Report 2012*. www.worldbank.org (Accessed 1 October 2013).

15 Maize Technologies and Rural Poverty Reduction in Ethiopia

D. Zeng,[1*] J. Alwang,[2] G.W. Norton,[2] B. Shiferaw,[3] M. Jaleta[4] and C. Yirga[5]
[1]*The University of Adelaide (formerly of Virginia Tech), Australia;* [2]*Department of Agricultural and Applied Economics, Virginia Tech;* [3]*Partnership for Economic Policy (formerly of CIMMYT-Nairobi), Nairobi, Kenya;* [4]*CIMMYT, Addis Ababa, Ethiopia;* [5]*Ethiopian Institute of Agricultural Research, Addis Ababa, Ethiopia*

Introduction

Maize is a widely grown food and cash crop in many environments in sub-Saharan Africa. In Ethiopia, maize accounts for the largest share of production by volume and is produced by more farms than any other crop (Chamberlin and Schmidt, 2012). Over time, area planted to maize has increased, especially in highland areas with reliable moisture; these are the same areas where the majority of the Ethiopian population is found (Taffesse *et al.*, 2012). From the 1960s to 2009, the dietary calorie and protein contributions of maize to total consumption in Ethiopia have doubled to around 20% and 16%, respectively (Shiferaw *et al.*, 2013). Maize, like other major cereals such as teff, wheat and sorghum, is mainly grown by smallholder farmers in Ethiopia.[1] These farmers are often poor and efforts to improve productivity in maize-producing areas are likely to have important impacts on poverty reduction.[2]

Partly as a result of the importance of maize production, substantial resources have been devoted to crop genetic improvement (CGI) for maize in Ethiopia. In the past four decades, more than 40 improved maize varieties, including hybrids and open-pollinated varieties (OPVs), have been developed and released in Ethiopia by the Ethiopian Institute of Agricultural Research (EIAR) in collaboration with the International Maize and Wheat Improvement Center (CIMMYT). Most improved varieties were released after the mid-1990s.

The objective of this chapter is to investigate the ex-post impact of these improved maize varieties on household well-being and on aggregate rural poverty. Information on economic impacts of maize research is needed because the last comprehensive assessment of the impacts of crop research was completed more than a decade ago (Evenson and Gollin, 2003). Studies of impacts of research on poverty and distribution are even scarcer and, given that poverty reduction has been a high priority for Ethiopia's government, it is important to understand how all publicly supported endeavours contribute to poverty reduction.

The empirical analysis uses rural household survey data collected in four regions of Ethiopia in 2010. Plot-level estimates of yield changes are obtained and suggest a yield advantage of 47.6–63.3%

* E-mail: d.zeng@uark.edu

for improved maize varieties over traditional varieties. Input costs following adoption increase between 22.8 and 29.4%. These estimates are incorporated into an economic surplus framework to measure population-level impacts and examine market-level changes in prices and economic surplus. The maize research is associated with a US$175.13–195.60 million annual gain in economic surplus at the national level. Household incomes increased by US$18.82–24.50 per year for an average adopting household. Poverty impact analysis shows that improved maize varieties have led to a 0.8–1.3 percentage point reduction in the rural poverty headcount ratio, and proportional reductions in poverty depth and severity. Poor producers are, however, found to benefit the least from adoption because their land areas are limited, leaving room for policies to enhance access to resources among the poor.

This chapter is organized as follows. First, we briefly review the literature on impacts of improved crop varieties with a focus on maize in Africa. Next, we discuss the conceptual framework and how the analysis addresses challenges associated with measuring heterogeneous and disaggregated impacts when adoption is endogenous using observational data. The model is presented and discussed. This discussion is followed by a description of data collection and summary statistics on maize production in the study areas. Next, results are presented and discussed. A concluding section discusses implications.

Background

Improved crop varieties have led to substantial increases in food production worldwide (Evenson and Gollin, 2003). New varieties can increase or maintain yield and may contribute to poverty reduction by increasing incomes and food consumption of adopting farmers (de Janvry and Sadoulet, 2002). At the farm-household level, benefits of improved crop varieties are obtained via adoption, a decision made by the farmer. Econometric models of microeconomic behaviour help identify effects of technology adoption (the treatment) on field and household-level direct outcomes such as yields (e.g. Suri, 2011), per-unit cost of production, or farm incomes. When farmers decide to adopt, net income gains probably overcome perceived risks from unknown new technologies and a large volume of literature links agricultural variety research to high positive economic impacts in Africa (Alston *et al.*, 2000; Maredia *et al.*, 2000; Renkow and Byerlee, 2010).

New maize varieties may also affect welfare of non-adopters. As household-level yield changes are aggregated over many adopters, market-level impacts emerge. Market price declines due to increased supply benefit consumers but these declines can adversely affect producers, especially those that fail to adopt. Assessments of market-level impacts usually employ partial equilibrium multi-market models (e.g. Mills, 1997) or computable general equilibrium (CGE) frameworks (e.g. de Janvry and Sadoulet, 2002).

A few empirical studies of the impacts of improved maize varieties in Ethiopia and other sub-Saharan countries are found in the literature. Seyoum *et al.* (1998) find that higher maize yields are obtained by farmers in eastern Ethiopia who participate in the Sasakawa-Global maize project, which aims to disseminate proven maize technologies to smallholders through enhanced extension services. Suri (2011) uses panel data from Kenya and estimates a mean gross return of 60% for hybrid maize but returns vary among households. Maredia *et al.* (2000) provide a review of the impacts of the agricultural research of multiple food crops across sub-Saharan Africa, where the development and diffusion of improved maize varieties is considered as a major success. Manyong *et al.* (2003) investigate the impact of International Institute of Tropical Agriculture (IITA) research on maize germplasm improvement in 11 countries in Western and Central Africa, where improved maize varieties obtain a gross yield advantage of 45.3% compared to local varieties. Overall, however, little empirical work has investigated poverty impacts of improved maize varieties in Ethiopia.

Modelling Framework

The study in this chapter adopts a 'treatment effect' approach to estimate the impacts of adoption (the treatment) on crop yields and input costs. Direct effects of adoption are felt at the household level and, because adoption is chosen by the decision maker, possible selection bias must be accounted for during estimation. Measuring indirect effects involves linking the household-level treatment to market-level changes. Indirect effects occur when the diffusion of improved crop varieties leads to a shift in supply of maize, causing market price changes; producers (adopters and non-adopters) and consumers are affected by these changes. Yield and cost impacts can be estimated via a production function (e.g. Suri, 2011) and cost function, respectively. The most widely used means of aggregate welfare impact assessment is economic surplus analysis (Alston *et al.*, 1995). In an ex-post setting, the appropriate counterfactual is what would have been observed in the absence of the technology, e.g. the distribution of well-being without the new technology. The counterfactual can be found in a surplus framework using a backward derivation procedure.

Possible differences in treatment effects in terms of yield gains across crop plots and households need to be considered when measuring direct impacts. These differences, or treatment effect heterogeneity, can vary with all observed and unobserved characteristics at the plot (such as soil productivity) and household (such as managerial capability) levels. Differences among observed and unobserved characteristics may be related to the yield gain from adoption and household crop income. This heterogeneity may affect estimates of population-wide poverty impacts.

Modelling consists of four steps. First, plot-level treatment effects are estimated as yield and cost changes due to adoption. Second, direct effects on household income are estimated from plot-level treatment effects. Third, indirect impacts on well-being, caused by the technology-induced shift in maize supply and associated market price reductions, are measured and disaggregated to different households. The counterfactual income distributions are derived by combining the direct and indirect impacts on household well-being. Finally, poverty impacts are evaluated by comparing poverty measures computed using observed and counterfactual income distributions.

Treatment effect specification

The household is the basic unit of analysis and households need to be classified according to adoption status. Most studies classify households as either adopters or non-adopters (e.g. Mendola, 2007; Becerril and Abdulai, 2010). This grouping rules out the possibility of partial adoption. Although partial adopters are widely observed in Ethiopia, most maize plots contain either improved or local varieties, but not both (see below for a discussion of data). A probit model, which specifies adoption as a binary outcome affected by individual and household characteristics is appropriate under such circumstances.

For each maize plot, the farm household expects a profit by selecting a maize variety on that plot, either improved ($T = 1$) or local ($T = 0$):

$$E\left[\pi^T\right] = PY^T - C^T \tag{15.1}$$

where Y^T and C^T represent the yield and input cost of maize variety T, and P is the maize market price. In an ex-ante situation when the farm household makes the varietal decision, risks cannot be observed and so the household selects the variety with higher expected profit level. The plot-level adoption rule is written as:

$$T = \begin{cases} 1, & \text{if } E\left[\pi^1\right] > E\left[\pi^0\right] \\ 0, & \text{if } E\left[\pi^1\right] \leq E\left[\pi^0\right] \end{cases} \tag{15.2}$$

On the production side, a suitable function allows plot-level maize yields to be specified in logarithmic form:

$$y^1 = \alpha^1 + \varphi + X\beta^1 + u^1 \tag{15.3a}$$

$$y^0 = \alpha^0 + X\beta^0 + u^0 \tag{15.3b}$$

where φ is the plot-specific percentage yield gain with adoption; X is the input vector with coefficients β,[3] and u denotes unobservable variables. Eqns 15.1–15.3b jointly specify the Generalized Roy Model in linear form (Heckman and Vytlacil, 2001). Production can be expressed as $y = Ty^1 + (1+T)y^0$, or more specifically:

$$\begin{aligned} y = {} & \alpha^0 + T\left(\alpha^1 - \alpha^0\right) + T\varphi + X\beta^0 \\ & + TX\left(\beta^1 - \beta^0\right) + u \end{aligned} \tag{15.4}$$

where $u = Tu^1 + (1-T)u^0$. Estimation of Eqn 15.4 quantifies the yield advantage of improved maize varieties as the coefficient φ of the treatment indicator T. Eqn 15.4 allows for possible unobserved heterogeneity in the error term.

Cost changes due to adoption are estimated using a cost function approach, empirically specified in a similar manner:

$$C^1 = \lambda^1 + \delta + P\gamma^1 + v^1 \quad (15.5a)$$

$$C^0 = \lambda^0 + P\gamma^0 + v^0 \quad (15.5b)$$

where C^T is the production cost of maize per hectare and P is the vector that includes input prices, plot area and the level of maize output. The parameters λ, δ, and γ are estimated and v represents a random error term. Capital cost is not accounted for in the short-run analysis. Since $C = TC^1 + (1-T)C^0$, the Generalized Roy Model can be similarly expressed as:

$$C = \lambda^0 + T(\lambda^1 - \lambda^0) + T\delta + P\gamma^0 + TP(\gamma^1 - \gamma^0) + v \quad (15.6)$$

where $v = Tv^1 + (1-T)v^0$. The parameter δ is interpreted as the plot-specific treatment effect in terms of percentage cost increase due to adoption.

Estimation methods

Eqns 15.5 and 15.6 are the main specifications to be estimated. Because treatment is self-determined by farmers, instrumental variable (IV) techniques are used to account for potential endogeneity. Homogeneous treatment effects are initially assumed. This assumption implies that all adopting farmers increase their yields and costs by the same proportion. Several alternatives are used to identify the average treatment effect on the treated (ATT)[4] in the homogeneous treatment effect model. The alternative estimates help assess the robustness of the findings.[5]

To assess how the yield effect varies among heterogeneous farmers, treatment effect heterogeneity is also considered. Local instrumental variable (LIV) estimation of the marginal treatment effect (MTE) is employed to deal with the heterogeneity (Heckman *et al.*, 2006). The MTE provides treatment effect estimates at each propensity-score level[6] (see Zeng *et al.*, 2013, for details).

Computing direct changes in household income

The direct income effects can be computed using the estimated treatment effects, i.e. yield and cost increases due to adoption.[7] For household *i*'s plot *k* planted with the improved maize variety, the income change $\Delta\hat{I}_{ik}$ is computed as:

$$\Delta\hat{I}_{ik} = \left(PY_{ik}^{obs} - C_{ik}^{obs}\right) - \left(PY_{ik}^{ct} - C_{ik}^{ct}\right) = P\Delta\hat{Y}_{ik} - \Delta\hat{C}_{ik} \quad (15.7)$$

where P is the maize market price; (Y^{obs}, C^{obs}) and (Y^{ct}, C^{ct}) are observed and counterfactual yield and cost pairs of plot k and $\Delta\hat{Y}_{ik}$ and $\Delta\hat{C}_{ik}$ denote the differences in per-hectare yield and per-hectare cost due to adoption, computed using the estimated treatment effects. Household income changes are the sum of plot-level income changes across all plots with improved maize:

$$\Delta\hat{I}_i = \sum_k \left(P\Delta\hat{Y}_{ik} - \Delta\hat{C}_{ik}\right). \quad (15.8)$$

The counterfactual income for each adopting household is obtained by subtracting the estimated income change due to adoption from observed income.

Accounting for indirect effects: changes in markets

The next step is to relate plot-level outcomes to market outcomes. Income changes shown in Eqn 15.8 assume that maize market price does not change but increased market supply due to diffusion of the new varieties may lower the market price received by producers and paid by consumers. These price changes may impact the net change in well-being and its distribution. Welfare impacts of supply shifts on maize-market participants depend on the nature of supply and demand.

In a small open economy, market price is fixed in the short run as the country is a price taker in the world market. Welfare changes occur only to adopters who increase their incomes due to the reduction in per unit cost of production. Non-adopters and pure consumers experience no change in welfare (Fig. 15.1) because only producer surplus changes. In a closed economy, the market price decreases as total output increases, and all producers and consumers are affected (Fig. 15.2). Ethiopia was not a member state of the World Trade Organization (WTO) in 2010 and maize exports are occasionally restricted by cereal export bans. Ethiopia can be considered a relatively closed economy

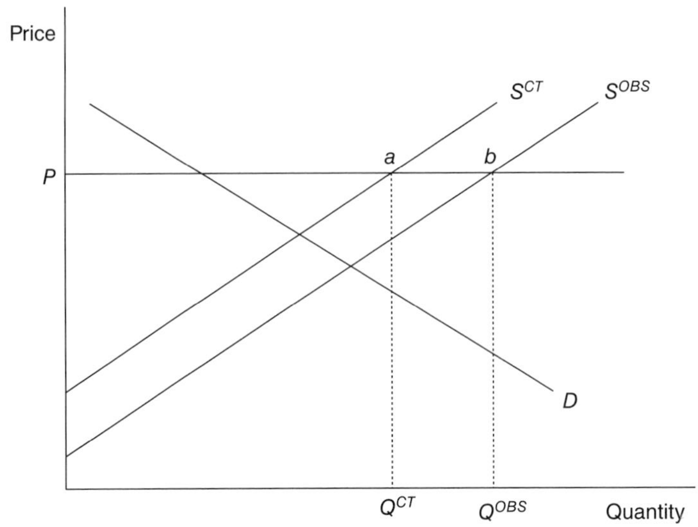

Fig. 15.1. Economic surplus change in a small open economy. The counterfactual supply (S^{CT}) reflects the supply that would have existed in the absence of research. The observed supply is reflected by S^{obs}.

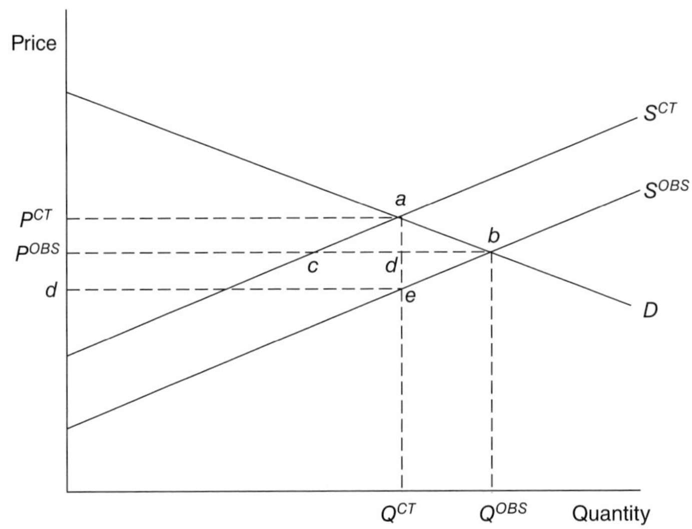

Fig. 15.2. Economic surplus change in a closed economy. The counterfactual supply (S^{CT}) reflects the supply that would have existed in the absence of research. The observed supply is reflected by S^{obs}.

for maize. Cross-border trade with neighbouring countries occurs, however, even when cereal export bans are in effect. As a result, poverty impacts are assessed under assumptions of both a small open and a closed economy. The true impacts will fall within the bounds of the estimates from these two cases.

In a closed economy case, it is necessary to estimate market-level counterfactual prices and counterfactual economic surplus. The key parameter affecting price and economic surplus differences due to the technology is the cost reduction per unit of output due to improved varieties, or the *k*-shift (Alston *et al.*, 1995). This is calculated as:

$$K = \left(\frac{\hat{\varphi}}{\varepsilon} - \frac{\hat{\delta}}{1+\hat{\varphi}} \right) \times \text{Adoption rate} \qquad (15.9)$$

where ε is the supply elasticity; $\hat{\varphi}$ and $\hat{\delta}$ are the estimated yield and cost treatment effects, respectively (see Zeng *et al.*, 2013). Using the estimated *k*-shift, the counterfactual output price level is retrieved. As shown in Fig. 15.2, the idea is to derive backwardly from the observed equilibrium price at *b* (P^{obs}, Q^{obs}) the counterfactual equilibrium price at *a* (P^{ct}, Q^{ct}). It can be shown that the counterfactual equilibrium price can be obtained using Eqn 15.9:

$$P^{ct} = P^{obs}(\varepsilon + \eta)/(\varepsilon + \eta - K\varepsilon) \qquad (15.10)$$

where η is the absolute value of the demand elasticity. Q^{ct} is computed by subtracting aggregate yield gains from Q^{obs}. The actual and counterfactual producer and consumer surpluses are computed using standard formulae (Alston *et al.*, 1995).

Producer and consumer surplus changes are allocated to individual households. On the demand side, only maize buyers experience consumer surplus changes. Thus, we allocate ΔCS to surveyed households (using appropriate sample weights) according to their purchased quantities as a share of total market supply. Allocation of producer surplus change is more complicated. Welfare impacts vary by household net sales position. The aggregate producer surplus change can be decomposed into yield and price effects:

$$\Delta PS = \Delta PS_{yield} + \Delta PS_{price} \qquad (15.11)$$

where ΔPS_{price} is equal to $-K\varepsilon P^{obs}Q^{ct} / (\varepsilon + \eta - K\varepsilon)$, and ΔPS_{yield} is the difference between ΔPS and ΔPS_{price}. ΔPS_{price} is allocated to all maize sellers based on their market shares because only sellers suffer from the price drop. All adopting plots, however, observe productivity and cost changes. Thus, ΔPS_{yield} is first allocated to all adopting plots (which may have different yield and cost MTEs) based on their shares of yield gains weighted by plot-level profitability, and then aggregated to the household level. This procedure accounts for partial adoption, and includes measurement of the direct benefits from adoption and indirect effects from market price changes. The counterfactual household income is computed by subtracting the income change from observed income and a counterfactual income distribution is created.

Foster-Greer-Thorbecke (FGT) poverty indices (Foster *et al.*, 1984) are then calculated for a given poverty line using the observed and counterfactual incomes. The poverty impacts, in terms of reductions in poverty headcount ratio, depth and severity are measured as the differences of the respective poverty indices with and without the improved varieties.

Data and Summary Statistics

Data are from a household survey conducted jointly by CIMMYT and EIAR during 2009–2010. Four major regions for maize production are covered: Oromia, Amhara, Tigray, and Southern Nations, Nationalities and People's Region (SNNPR). The survey uses a three-stage stratified and proportionate random sampling strategy that includes interviews with farmers from 30 woredas[8] across regions and yields nationally representative data.[9] A total of 1396 farm households were surveyed, of which 1359 grew maize on 2496 plots. Plot areas were reported by farmers and details of crop production such as varieties, yields and inputs were gathered. Household demographic and socioeconomic information as well as characteristics of household heads are recorded.

Maize varieties can be grouped into three categories: hybrids, improved open-pollinated varieties (OPVs) and local OPVs. Hybrid maize has the highest yield but requires the purchase of new seeds for each cropping season to restore hybrid vigour and the seeds cost more than OPVs. OPVs generally have lower yields than hybrids (still higher than local varieties) but the seeds may be recycled for up to three seasons. Many OPVs are developed for challenging conditions (i.e. droughts, pests) and under circumstances where seed markets are underdeveloped or missing. Whatever varieties farmers grow, inbred lines are crossed through open pollination. Thus, varieties are only differentiated as being either improved or local. Any hybrid that has been ever recycled or OPV that has been recycled for more than three seasons is categorized as local.[10] After accounting for sampling weights, the data suggest an adoption rate of 39.1% of improved maize varieties by area. Of the 1359 households, there are 503 adopters, 583 non-adopters and 273 partial adopters

(Table 15.1). Most regions observe more than half of their areas planted with improved maize varieties, although such a portion is extremely low in Tigray.

Farmers tend to adopt improved varieties on plots that are larger, flatter and closer to their homes (Table 15.2). These differences appear rather small even if they are statistically significant. Wealthier households with larger maize areas[11] and more family members tend to adopt improved maize varieties, whereas partial adopters seem to have the largest total cultivated area, maize area and household size (Table 15.3). Heads of full- and partial-adopting households are more likely to be male, married and better educated than those of non-adopting households.

Farmers tend to grow improved varieties during the long rainy season (mid-June to mid-September) more frequently than during the short rainy season (February to April) (Table 15.4). Oxen power, fertilizer and other inputs reported in monetary terms, such as purchased seeds, pesticides and herbicides, are significantly higher for plots with improved varieties, whereas labour does vary significantly between plots with improved and local varieties. Improved varieties yield about 1275 kg more per hectare than local varieties, about a 59.0% yield difference. This difference is larger than earlier estimates, namely 30–40% for hybrids and 14–25% for OPVs (see Maredia *et al.*, 2000 for a synthesis of previous research).

Empirical Results

Modelling adoption

Prior to turning to adoption, it is necessary to discuss the instrumental variables (IVs) used to identify the treatment effects. These IVs should affect the adoption decision, but only affect the outcome (yield or cost) through their impacts on adoption. Although there are statistical tests of the appropriateness of IVs, it is first necessary to justify them on the basis of conditions in maize growing areas. Several alternative IVs were discussed with project experts and CG scientists at a Diffusion and Impact of Improved Varieties in Africa (DIIVA) Project meeting held in January 2011 in Bamako, Mali. Further discussions were held at a meeting with project participants and experts at the Ethiopian Ministry of Agriculture during May 2011 in Addis Ababa. The universe of potential IVs was determined following these discussions.

Table 15.1. Adoption of improved maize varieties across regions.

	Tigray	Amhara	Oromia	SNNPR
Percentage of land area under improved maize varieties	4.84	67.54	52.50	67.39
Percentage of full adopters	6.78	26.80	39.80	24.91
Percentage of non-adopters	93.22	38.31	28.36	50.92
Percentage of partial adopters	0.00	34.89	31.84	24.18

Table 15.2. Descriptive statistics of plot characteristics.

	Improved[a] (n = 1214)	Local[a] (n = 1282)	Difference[b]
Altitude (metres)	1832.5 (304.5)	1830.1 (255.4)	2.4 (.832)
Walking minutes from home	9.73 (18.43)	14.26 (28.87)	–4.53 (0.000)
Plot area (ha)[c]	0.453 (0.416)	0.334 (0.357)	0.119 (0.000)
Soil slope (1–3: gentle–medium–steep)	1.43 (0.65)	1.52 (0.70)	–0.11 (0.002)
Soil depth (1–3: shallow–medium–deep)	2.21 (0.84)	2.17 (0.85)	0.05 (0.162)
Soil fertility (1–3: good–average–poor)	2.45 (0.62)	2.47 (0.60)	–0.02 (0.359)

[a]Standard deviations are reported in parentheses. [b]*P*-values of *t*-tests are represented in parentheses.
[c]Originally recorded in kert. 1 kert = 0.25 ha for most areas in Ethiopia, whereas in a few areas 1 kert = 0.125 ha, which has been adjusted accordingly.

Table 15.3. Descriptive statistics of maize households by adoption type.[a]

	Adopters (n = 503)	Non-adopters (n = 583)	Partial-adopters (n = 273)	Statistical significance level of pairwise *t*-tests		
				Adopters and non-adopters	Adopters and partial-adopters	Non-adopters and partial-adopters
Total cultivated area (ha)	2.02 (1.51)	1.86 (1.33)	2.37 (1.89)		5%	1%
Total maize area (ha)	0.709 (0.674)	0.553 (.545)	1.090 (1.173)	1%	1%	1%
Household size	6.58 (2.46)	6.29 (2.21)	6.91 (2.40)	5%	10%	1%
Total household wealth[b] (thousand ETB)	18.83 (35.31)	13.18 (29.51)	22.69 (61.[illegible]8)	1%		1%
Head gender (% of male)	95.0 (21.8)	91.3 (28.3)	98.1 (13.4)	5%	5%	1%
Head age (years)	42.01 (12.95)	43.90 (12.52)	43.15 (11.34)	5%		
Head marital status (% married)	94.6 (22.6)	90.6 (29.3)	96.7 (17.9)	5%		1%
Head education (years)	2.92 (3.36)	2.48 (2.99)	2.99 (3.32)	5%		5%
Head illiteracy rate[c]	0.549 (0.492)	0.592 (0.498)	0.582 (0.494)			
Poverty rate by household (< US$1.25/person/day)	40.36%	46.14%	39.93%	–	–	–

[a]Standard deviations are reported in parentheses. [b]Computed as the sum of the self-reported values of all household assets. The daily average exchange rate in 2010 is US$1 = 14.38 ETB (Ethiopian Birr). [c]Defined as no education at all, as opposed to at least some education.

Table 15.4. Descriptive statistics of maize cropping practice.

	Improved[a] (n = 1214)	Local[a] (n = 1282)	Difference[b]
Season (1 = long; 0 = short)	0.945 (0.228)	0.915 (0.279)	0.030 (0.003)
Intercropping (1 = yes; 0 = no)	0.129 (0.266)	0.173 (0.384)	–0.044 (0.135)
Labour days per ha	105.0 (115.4)	102.9 (78.5)	2.1 (.588)
Ox plough days per ha	8.01 (7.87)	4.92 (4.63)	3.09 (.000)
Fertilizer (kg per ha)	150.6 (243.3)	56.3 (305.8)	94.3 (.000)
Other inputs per ha[c] (ETB[d])	299.1 (398.9)	67.7 (210.8)	231.4 (.000)
Yield (kg per ha)	3434.9 (2176.2)	2159.6 (1610.8)	1275.2 (.000)

[a]Standard deviations are reported in parentheses. [b]*P*-value of *t*-tests are reported in parentheses. [c]Including cost for seeds purchased and pesticides. [d]The daily average exchange rate in 2010 is US$1 = 14.38 ETB.

Five potential IVs are considered: the distances to the nearest seed dealer, agricultural extension office, farmer cooperative and main market, and the quality of roads to the main market. These reflect the accessibility of improved seeds and markets, extension efforts, availability of credit and business services, and degree of commercialization.

To gain a better understanding of the adoption decision and the correlation between potential IVs and adoption, a series of probit models is estimated: the baseline probit model, the baseline model with regional dummies, and the baseline model with regional dummies and potential IVs (Table 15.5). All specifications are significant at the 1% level. Factors significantly associated with technology adoption include cropping season, plot area and the age of household head. Adoption decisions are systematically different across regions and the five potential IVs are jointly correlated with adoption at the 1% level (columns 2 and 3 of Table 15.5). Farmers in Tigray are far less likely to adopt improved maize compared with the other three surveyed regions (see Table 15.1). Farmers are more likely to adopt improved varieties for planting in the long rainy season (with probabilities 0.134–0.154 higher) than in the short rainy season.

Yield and cost impacts

Production and cost functions are estimated with endogenous technology choice using IV techniques. All five IVs are employed to identify the yield ATT. These IVs affect adoption but only affect yield through their influence on adoption. Further justification of the IVs is necessary. Although it may be argued that access to markets and other distance variables might affect other input uses (such as fertilizer and labour) and thus have a direct effect on yield, this is not the case in a production function model. In a well-specified production function where levels of inputs and soil quality are already controlled for (see below for details on variables included in the estimation), measures of access to resources such as distances to nearest seed dealer, extension office and farm cooperative should not affect maize yield other than through their impact on varietal adoption. Similar logic holds for the other IVs. For example, roads might be placed in areas with higher fertility and, hence, there might be correlation between road access and the outcome. Under such conditions, it would not be appropriate to exclude road access from the outcome equation. However, the production function includes variables reflecting soil fertility and the exclusion of road quality in the outcome equation is logical because the variable mediating the relationship between road quality and the outcome is already controlled for during estimation. A similar IV (distance to the nearest fertilizer dealer) was used in production function estimates by Suri (2011). The IVs underwent tests of endogeneity, under-identification, over-identification and weak identification.[12]

Maize yield is measured on a per-hectare basis. Besides the instrumented adoption decision, other explanatory variables included are per-hectare inputs in logarithmic form: labour days (L), ox ploughing days (O), fertilizer (F) and other capital inputs (K), human capital indicators (total household size and wealth, characteristics of household head such as gender, age,

Table 15.5. Probit estimation of the plot-level adoption decision (n = 2,496).[a, b]

	Baseline		With regional dummies		With regional dummies and IV	
	Estimate	Marginal effect	Estimate	Marginal effect	Estimate	Marginal effect
Season (1 = long; 0 = short)	0.337***	0.134	0.368***	0.134	0.385***	0.154
	(0.101)		(0.104)		(0.110)	
Walking minutes to home	–0.005***	0.002	–0.003**	0.001	–0.003***	0.001
	(0.001)		(0.001)		(0.001)	
Soil slope	–0.158	–0.063	–0.041	–0.016	0.013	0.005
(1–3: gentle–medium–steep)	(0.040)		(0.042)		(0.044)	
Soil depth	0.071	0.028	0.042	0.017	0.068**	0.027
(1–3: shallow–medium–deep)	(0.031)		(0.032)		(0.033)	
Soil fertility	–0.065	–0.026	–0.082*	–0.033	–0.082*	–0.033
(1–3: good–average–poor)	(0.043)		(0.044)		(0.045)	
Plot area (ha)	0.130***	0.052	0.144***	0.058	0.124***	0.049
	(0.018)		(0.018)		(0.019)	
Village altitude (m)	0.000**	0.000	0.000**	0.000	0.000	0.000
	(0.000)		(0.000)		(0.000)	
Total household size	0.026**	0.011	0.020	0.008	0.023*	0.009
	(0.013)		(0.013)		(0.013)	
Reported total assets (ETB)	0.000*	0.000	0.000	0.000	0.000	0.000
	(0.000)		(0.000)		(0.000)	
Head gender (1 = M; 0 = F)	0.165	0.065	0.153	0.061	0.154	0.061
	(0.186)		(0.189)		(0.192)	
Head marital status	0.171	0.068	0.148	0.059	0.186	0.073
(1 = married; 0 = other)	(0.164)		(0.164)		(0.167)	
Head age (years)	–0.044**	–0.018	–0.039*	–0.015	–0.039***	–0.016
	(0.014)		(0.014)		(0.015)	
Age square of household head	0.000**	0.000	0.000**	0.000	0.000**	0.000
	(0.000)		(0.000)		(0.000)	
Head literacy	0.037	0.015	0.072	0.029	0.029	0.012
(1 = literate; 0 = illiterate)	(0.056)		(0.057)		(0.059)	
Regional dummy: Amhara			1.476***	0.510	1.512***	0.520
			(0.262)		(0.271)	
Regional dummy: Oromia			1.086***	0.406	0.956***	0.361
			(0.258)		(0.265)	

Continued

Table 15.5. Continued.

	Baseline		With regional dummies		With regional dummies and IV	
	Estimate	Marginal effect	Estimate	Marginal effect	Estimate	Marginal effect
Regional dummy: SNNPR			1.656*** (0.269)	0.512	1.450*** (0.277)	0.475
Walking minutes to the nearest main market					–0.007*** (0.001)	–0.003
Quality of roads to the nearest main market					0.031 (0.020)	0.012
Walking minutes to the nearest seed dealer					–0.000 (0.001)	–0.001
Walking minutes to the nearest farm corporative					–0.001* (0.001)	–0.003
Walking minutes to the nearest extension office					0.001 (0.001)	–0.003
% of correct prediction	61.14		62.46		67.67	
LR χ^2	142.82***		239.97***		441.46***	

[a]*, **, *** indicate significance at 1%, 5% and 10% level, respectively. [b]Standard errors are clustered at woreda level, the primary sampling unit.

marital status and education), maize area,[13] soil conditions (slope, depth and fertility, all on a discrete poor–average–good scale), season (short or long), village altitude and regional dummies.

A Cobb–Douglas production function is estimated via 2SLS, probit-2SLS, GMM and Heckit procedures to estimate the yield ATT, or $\hat{\varphi}$. Alternative estimates under assumed treatment heterogeneity are obtained by taking weighted averages of LIV-estimated MTEs, as described previously. As reported in the upper panel of Table 15.6, these results are numerically close. Across different models, $\hat{\varphi}$ is estimated to be between 47.6 and 63.3% and is highly significant in all the estimates. Estimated MTEs are graphed in Fig. 15.3. As a robustness check, a flexible translog functional form is used, and $\hat{\varphi}$ is estimated as 55.1–61.6%, with high significance. Additional robustness checks confirm these findings (see Zeng *et al.*, 2013, for details).

Estimated yield MTEs are highest for households with mid-low propensity scores, as observed using both the Cobb–Douglas (Fig. 15.3a) and translog function form (Fig. 15.3b). These results indicate negative selection: farmers are less likely to grow improved varieties on plots that are more likely to observe a higher yield gain, a pattern also found in Suri (2011). About half of the households surveyed grow maize only on a single plot, and negative selection indicates that farmers planting maize on plots with higher yield potential may be more conservative. As a test for heterogeneity in $\hat{\varphi}$, ordinary least squares (OLS) regressions were run of the estimated MTEs on propensity scores, with the null hypothesis being a zero slope. Similar to Suri (2011), the slopes were negative and significant at the 1% level, confirming the existence of heterogeneity.

The cost ATT, or $\hat{\delta}$, is estimated in a similar manner to the production model. Only three of the five IVs are included in the cost effect estimation: distances to the nearest extension office, farm corporative and seed dealer, which are not supposed to correlate with total cost per hectare other than through adoption. Distance and quality of road to the main market are excluded because they reflect the degree of commercialization and may be correlated with input prices, such as those of improved seeds and fertilizer. The IVs passed a battery of tests. Other variables include the prices of inputs (e.g. labour, fertilizer, ox plough, pesticides), maize yield per hectare, maize area, plot and household characteristics, and regional dummies.

Following Jacoby (1993), shadow prices of labour (P_L) and ox plough (P_O) are computed from the production function estimates and employed here.[14] The shadow price of oxen ploughing is computed similarly. Other explanatory variables include: prices of other inputs such as fertilizer (P_F), seeds (P_S) and other inputs in monetary terms (P_K); cropping season, characteristics of plot, household and household head; and regional dummy variables to control for heterogeneity across areas.

Assuming a homogeneous treatment effect, $\hat{\delta}$ is estimated to be 22.8–29.4% under a Cobb–Douglas specification and 23.1–27.4% with a translog cost function (lower line, Table 15.6). The LIV estimates are similar, with 27.8 and 25.3% cost increases due to additional inputs.[15] Estimated cost MTEs generally decrease

Table 15.6. ATT estimation of yield and cost effects.[a, b]

ATT	Model specification	2SLS[c]	probit-2SLS[c]	GMM	Heckit	LIV
Yield ATT	Cobb–Douglas	0.588**	0.476***	0.561***	0.496***	0.633***
		(0.230)	(0.128)	(0.145)	(0.129)	(0.242)
	Translog	0.616***	0.564***	0.594***	0.551***	0.535***
		(0.221)	(0.126)	(0.146)	(0.128)	(0.203)
Cost ATT	Cobb–Douglas	0.276**	0.228***	0.261**	0.294***	0.278**
		(0.122)	(0.084)	(0.102)	(0.093)	(0.110)
	Translog	0.243**	0.231***	0.239**	0.274***	0.253***
		(0.102)	(0.089)	(0.110)	(0.095)	(0.098)

[a]*, **, *** indicate significance at 1%, 5% and 10% level, respectively. [b]Standard errors of the treatment effects are reported in parentheses; LIV standard errors are obtained by bootstrapping 100 times. [c]Standard errors are clustered at woreda level, which is the primary sampling unit.

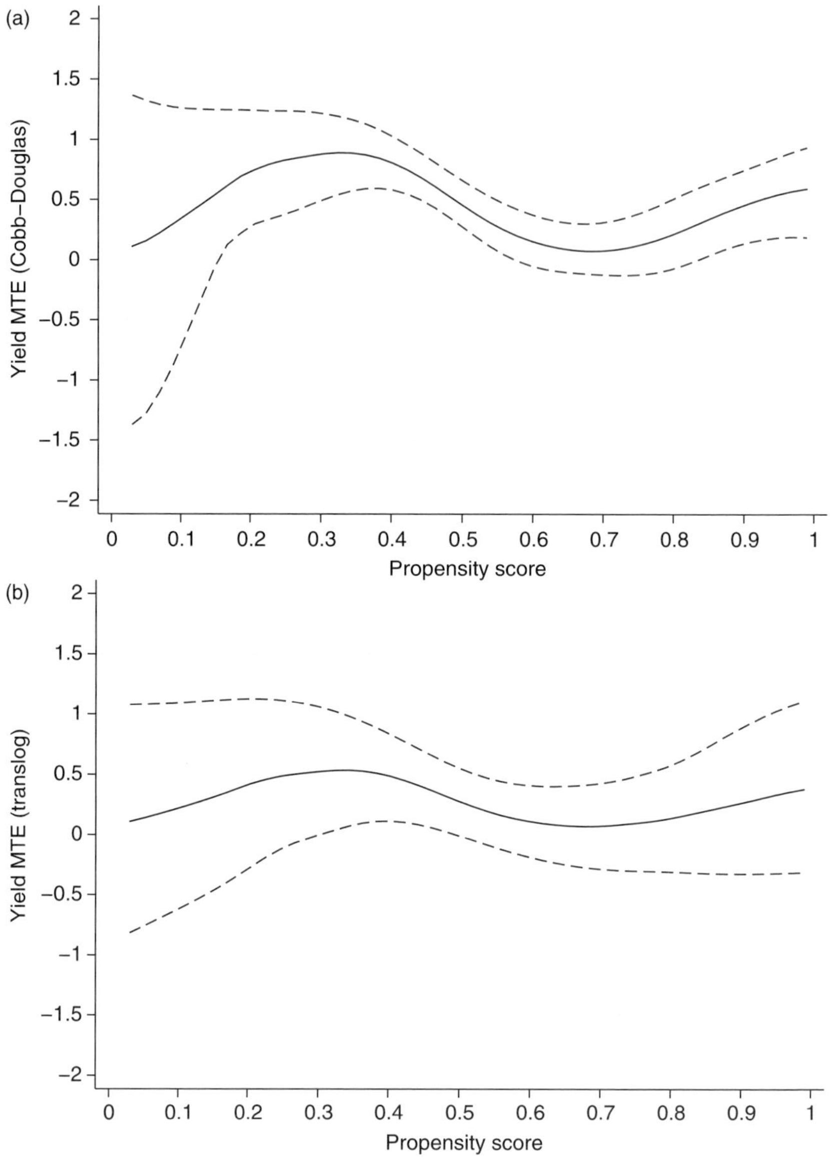

Fig. 15.3. Yield MTE using semiparametric LIV estimator. Yield MTEs were estimated using (a) the Cobb–Douglas and (b) the translog functional form. (Estimated using local polynomial regression. Solid line shows the estimated MTE; dashed lines are 95% confidence intervals obtained via bootstrapping.)

as propensity scores increase (see Figs 15.4a and 15.4b). OLS regressions of MTE on propensity scores yield a negative slope with 1% significance, confirming the existence of heterogeneity in cost treatment effects as well. These results offer a possible explanation for the negative selection observed in yield MTEs: farmers are less likely to adopt improved maize varieties given high additional costs even if the yield potential is high.

Aggregate impacts in a closed economy

In the small open economy, the maize market price does not change as supply is shifted outward.

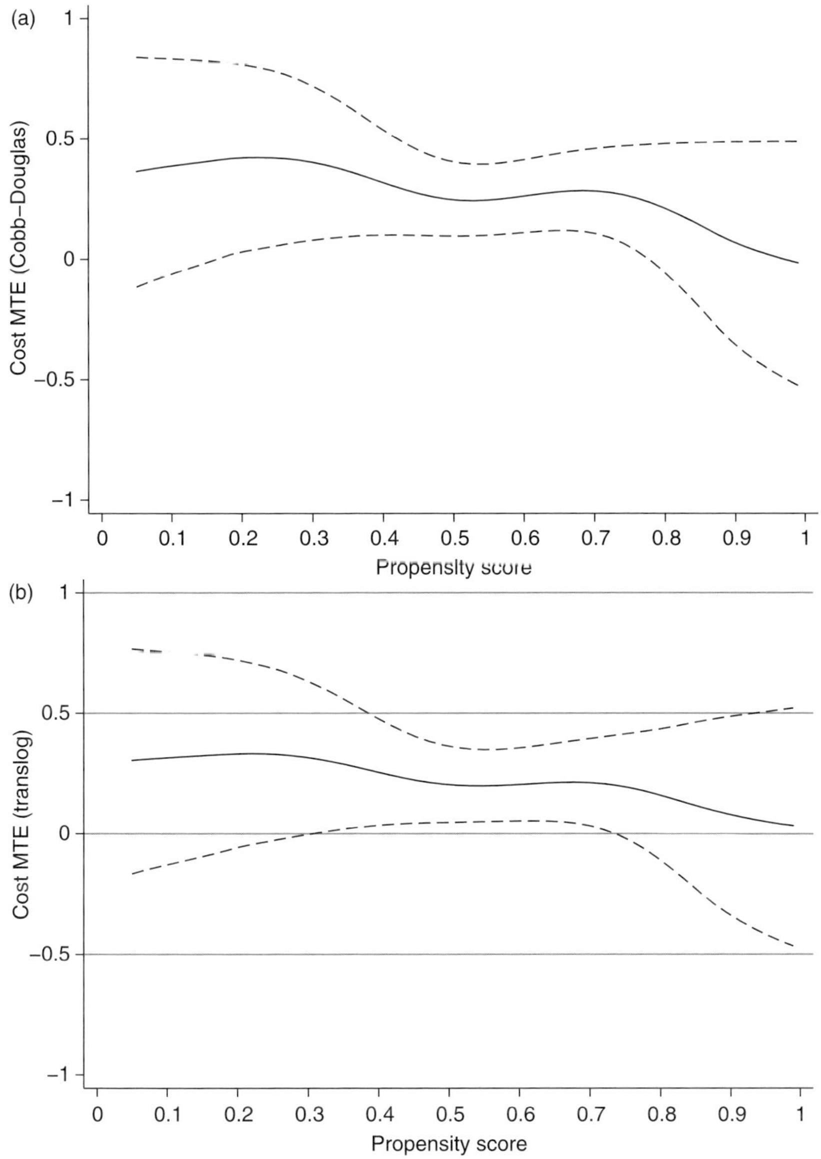

Fig. 15.4. Cost MTE using semiparametric LIV estimator. Cost MTEs were estimated using (a) the Cobb–Douglas and (b) the translog functional form. (Estimated using local polynomial regression. Solid line shows the estimated MTE; dashed lines are 95% confidence intervals obtained via bootstrapping.)

For the closed economy case, a natural next step is to obtain estimates of price elasticities of supply and demand to derive the counterfactual price level. Given the cross-sectional nature of our data and the lack of demand-side information, the elasticities of both maize supply and demand were obtained from existing literature.[16] Previous literature suggests a wide variation of price elasticity of maize supply in sub-Saharan Africa.[17] We assume the supply elasticity to be 0.5 in our short-run analysis because maize is commonly intercropped with other crops and it is comparatively easy to switch in and out of maize production in the long run. Sensitivity analysis was conducted on this parameter and other estimated parameters, as detailed below.

Given that the variation of demand elasticity has no effect in the small open economy case, and only minor effects on the economic surplus change at household level, the analysis conservatively assumes a unit absolute value of demand elasticity in the poverty analysis, which is followed by sensitivity analysis.

The market price P^{obs} is obtained as an 11-year average (2000–2010) of national-level annual producer prices from FAOSTAT, which is US$0.166 per kilogram.[18] With P^{obs}, and sample-level Q^{obs}, the k-shift is computed as a 37.4% cost reduction per kilogram of maize. A P^{ct} of US$0.191 per kilogram is obtained by averaging the LIV estimates from the Cobb–Douglas or translog technologies.[19] The aggregate producer surplus and consumer surplus changes are US$75,118 and US$37,559 among the surveyed households, respectively; 6.37% of the latter is allocated to these households according to their maize consumption share of total supply. Plot-level yield and cost MTEs are used to derive counterfactual incomes. At the national level (with 3.897 million tonnes of total maize production in 2010, FAOSTAT), the total changes in producer surplus and consumer surplus are US$130.40 million and US$65.20 million, respectively. For comparison purposes, we also compute aggregate impacts in the small open economy.

Aggregate impacts in a small open economy

In a small open economy, the total surplus change is equal to producer surplus change. The welfare improvement at the market level can be computed as (Alston *et al.*, 1995):

$$\Delta PS = KPQ^{ct}(1+0.5K\varepsilon) \quad (12)$$

where K is the k-shift computed previously; P is the fixed output price level; ε is the supply elasticity and Q^{ct} is the counterfactual national-level output level. The latter is computed using the yield ATT.

The national producer surplus change in the small open economy at the national is computed as US$175.13 million. The impacts of maize research in Ethiopia are substantial. At the national level, 7.149 million maize-cropping households hold a total of 1.772 million hectares (Central Statistical Agency of Ethiopia, 2010). Along with the closed economy estimates, these aggregate impacts translate into an income increase of US$18.82–24.50 for an average adopting household.

Assessing poverty impacts

Before assessing the poverty impacts from maize technology adoption, it is helpful to understand the poverty situation using information on household incomes from our data. Poor rural households in maize-growing areas of Ethiopia have more members, lower asset values and less-educated household heads (Table 15.7). Adoption rates are lower among poor households. The difference in adoption rates by household poverty status narrows as the poverty threshold increases; the poorest of the poor are least likely to adopt new maize varieties and the distribution of well-being for adopting households falls to the right of non-adopters (Fig. 15.5).

The poverty headcount, depth and severity indices are computed using MTEs and LIV-estimated ATTs. All three show higher poverty rates in the absence of the technology; adoption of improved maize varieties has led to lower rural poverty in Ethiopia (Table 15.8). Impacts on the poverty headcount reduction are larger under the assumption of a small open economy, where the poverty headcount ratio drops by 0.9–1.3 percentage points, as compared to 0.8–0.9 percentage points in the closed economy. This makes sense because the profitability of maize decreases as market price drops and only a small portion of total consumer surplus is enjoyed by surveyed households in the closed economy. These numbers further imply that 1.7–3.1% of the rural poor have escaped poverty in the current year owing to the adoption of improved maize.[20] The depth and severity estimates show similar patterns; a 2.3–3.1% decrease in poverty depth and a 3.1–4.0% decrease in poverty severity are observed. Results are robust across all poverty lines.

To explore the distribution of impacts, the variations in producer benefits from adoption of improved maize varieties along the counterfactual income levels are presented in Fig. 15.6. Poor adopters benefit the least from the new technologies. Analysis shows that the poor are

Table 15.7. Descriptive statistics for household characteristics by poverty status.[a, b]

Poverty line	US$1		US$1.25		US$1.45	
Poverty status	Non-poor (n = 955)	Poor (n = 404)	Non-poor (n = 778)	Poor (n = 581)	Non-poor (n = 667)	Poor (n = 692)
Household size	6.191	7.317***	6.057	7.153***	5.954	7.077***
Total assets (ETB)[2]	20.220	10.010***	22.076	10.635***	23.680	10.924***
Head gender (1 = M; 0 = F)	0.934	0.955	0.936	0.947	0.934	0.947
Head age (years)	42.80	43.63	42.70	43.52	42.83	43.27
Head marital status (1 = married; 0 = other)	0.924	0.955**	0.923	0.947*	0.922	0.945
Head education (years)	2.902	2.376***	2.960	2.458***	2.988	2.512***
Head literate (1 = yes; 0 = no)	0.480	0.406**	0.472	0.396***	0.463	0.393***
Adopter (%)	37.82	35.05	37.56	36.41	37.33	36.75
Non-adopter (%)	41.66	47.86	41.76	45.46	41.90	45.01
Partial-adopter (%)	20.52	18.09	20.68	18.13	20.77	18.24

[a]***, **, * denote that the difference between non-poor and poor is significant at the 1%, 5% and 10% level via *t*-test, respectively. [b]Computed as the sum of the self-reported values of all household assets.

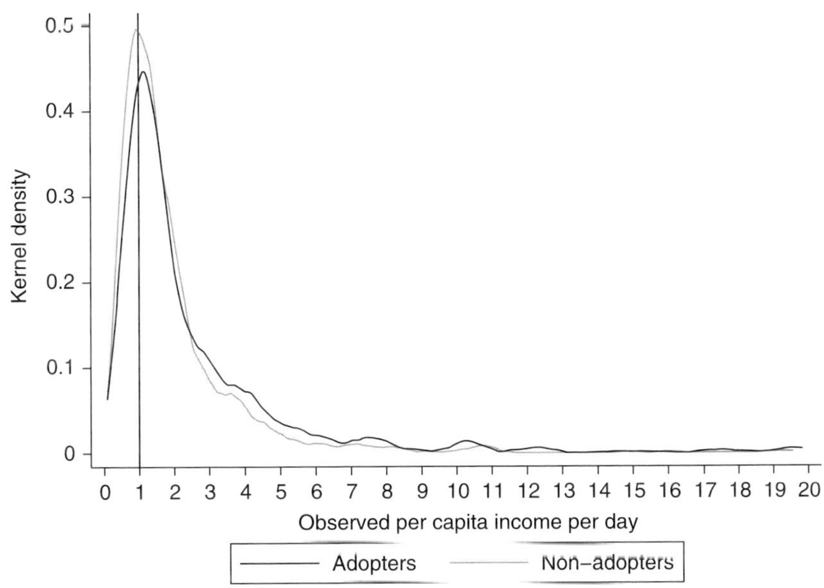

Fig. 15.5. Kernel density estimates of per capita income per day by adoption. (Partial adopters are omitted in this estimation. The US$1 poverty line is added as a reference line.)

as likely to adopt as the non-poor and their yield and cost MTEs are generally similar. Limited land area, rather than an inability to adopt, explains why the poor receive relatively few producer benefits.[21] Sensitivity analyses of the poverty impact estimates are shown in Zeng *et al.* (2013) and confirm that these findings are robust to the assumptions made. The estimate of a 0.6–1.2 percentage point reduction in poverty implies that 47.8–95.6 thousand households in rural Ethiopia have escaped poverty because of the adoption of improved maize varieties.

Table 15.8. Poverty impacts of improved maize varieties.

Poverty line		Observed	Counterfactual: small open economy	Poverty impact[a]	Counterfactual: closed Economy	Poverty impact[a]
$1	Headcount	0.2894	0.2987	0.0093	0.2973	0.0079
	Depth	0.0963	0.0994	0.0031	0.0991	0.0028
	Severity	0.0435	0.0453	0.0018	0.0449	0.0014
$1.25	Headcount	0.4162	0.4291	0.0129	0.4255	0.0093
	Depth	0.1496	0.1534	0.0038	0.1537	0.0041
	Severity	0.0724	0.0748	0.0024	0.0749	0.0025
$1.45	Headcount	0.4957	0.5057	0.0100	0.5043	0.0086
	Depth	0.1947	0.1996	0.0049	0.1992	0.0045
	Severity	0.0983	0.1021	0.0038	0.1020	0.0037

[a]Computed as the difference in percentage point change between the observed and counterfactual measures.

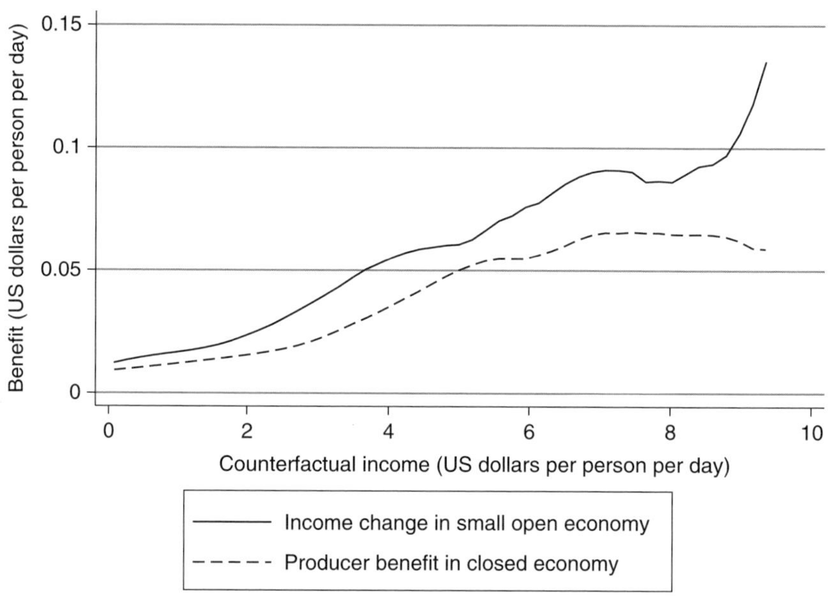

Fig. 15.6. Benefits due to adoption across counterfactual income levels. Counterfactual incomes of 95% households are less than US$5 per person per day. About 1% households who have counterfactual incomes above US$10 per person per day are excluded.

Concluding Remarks

Crop genetic improvement in maize has had substantial impacts on poverty in rural Ethiopia. A 1% reduction in overall poverty owing to maize varieties alone is a major achievement. With research investments in multiple crops, agricultural research as a whole is likely to have a correspondingly higher poverty-reducing impact. The distribution of benefits from CGI is uneven: the poor benefit the least from adoption owing to limited resources such as land holdings. Policies to increase benefit flows to limited resource farmers might be further explored but the analysis finds no evidence that poorer farmers are inhibited from adopting improved maize varieties.

This study employs cross-sectional household survey data, and may not reflect the dynamics of the poverty impact of improved maize varieties. In fact, the poverty impacts of maize research should grow over time given the expanding maize area in Ethiopia. However, with detailed information on maize varieties at a single point of time, we are able to provide a snapshot of the poverty impact following decades of maize diffusion and adoption.

Estimation procedures account for treatment effect heterogeneity and provide a creative means of estimating the economic surplus change at household level. The impact of adopting improved maize varieties on each household is differentiated, whether the household is an adopter, a non-adopter or a partial-adopter. Results are consistent across alternative poverty lines. Multiple robustness checks and sensitivity analyses have been implemented that consistently support the findings.

We combined this estimated treatment effect with an economic surplus model to understand how maize price changes from widespread adoption affect maize producers and consumers. With field-level treatment effect estimation, a backward derivation procedure was developed and yielded the counterfactual income distribution. The method may serve as a standard procedure of ex-post impact assessment of agricultural technologies where randomized controlled trials are not applicable. It can easily be adapted to general cases where distributional impacts are of interest.

Notes

[1] Of the farms in the cereal producing highlands 80% are 1.52 hectares and smaller and account for about 43% of total area planted in these cereal-dominant areas (Taffesse *et al.*, 2012).

[2] More than 62% of the population resides in areas appropriate for maize production (moisture-reliable highlands) even though these areas represent less than 25% of production area.

[3] X can include nonlinear combinations of individual inputs (e.g. a translog production function), while the model is linear in coefficients β. The same applies to cost function specification.

[4] In a homogeneous treatment effect model, the ATT is the same for all treated units and is assumed to be the same as the average effect on the untreated (ATUT). With heterogeneous treatment effects, ATT does not necessarily equal ATUT and marginal treatment effects may differ across observations.

[5] A simple 2SLS procedure is consistent, but additional econometric techniques are implemented to check for robustness. One alternative is to use the probit-estimate of the propensity score as the IV in the 2SLS procedure (probit-2SLS). The estimator is efficient and robust for mis-specifications in the probit model (Wooldridge, 2002). To allow arbitrary heteroskedasticity, Eqn 15.5 is also estimated using generalized method of moments (GMM, Hansen 1982). Finally, a generalized selection model (Heckit) is estimated via Heckman's two-stage procedure (Heckman, 1979). The latter provides consistent and efficient treatment effect estimates assuming joint normality of error terms.

[6] The propensity score, or the probability of adopting the new variety, is introduced at different evaluation levels of the LIV estimator. This approach differs from the propensity score matching literature in which the treatment effect, a scalar value, is defined as the average difference of the variable of interest between treated and untreated observations matched by propensity score.

[7] Here maize market price is held fixed. This assumption is relaxed below.

[8] A woreda is an administrative district, comparable to a US county.

[9] See Zeng *et al.* (2013) for sampling details.

[10] This cut-off is suggested by local experts.

[11] Wealth is computed as the sum of the self-reported values of all household assets.

[12] Standard errors are clustered at the village level. See Zeng *et al.* (2013) for further details on specification testing.

[13] Although the yield and all inputs are measured on a per-hectare basis, maize area is also included because empirically smaller plots tend to report higher per-hectare yields. An expansive literature has emerged on this topic; see Benjamin (1995), for example.

[14] See Zeng *et al.* (2013) for details.

[15] Propensity score matching as another robustness check was implemented in cost ATT estimation. Nearest neighbour matching, radius matching and kernel matching were all employed. Results suggests a per hectare cost reduction of 22.1–25.6%. These and additional robustness checks are reported in Zeng *et al.* (2013).

[16] We conduct a sensitivity analysis on these and other parameters, as detailed below.

[17] Details of decision-making about supply and demand elasticities are presented in Zeng *et al.* (2013).

[18] Producer price is used here, as retail price has larger variation across areas. The FAOSTAT producer prices are 2004–2006 International Dollar prices.
[19] The two counterfactual prices are numerically very close; thus the average should be acceptable.
[20] Computed as the percentage reduction divided by the counterfactual poverty headcount ratio. For example, in the small open economy, the counterfactual poverty headcount ratio and poverty impact under the $1 poverty line are 0.2987 and 0.0093, respectively. Thus, the percentage of the originally poor who have escaped poverty is computed as 0.0093 / 0.2987 = 0.0311, or 3.1%. Similar computations with respect to poverty depth and severity are applied.
[21] Further computation shows that the mean differences of maize areas between the poor and non-poor are significantly different under each of the three poverty lines.

References

Alston, J.M., Norton, G.W. and. Pardey, P.G. (1995) *Science under Scarcity: Principles and Practice for Agricultural Research Evaluation and Priority Setting.* Cornell University Press, Ithaca, NY.

Alston, J.M., Chan-Kang, C., Marra, M.C., Pardey, P.G. and Wyatt, T.J. (2000) *A Meta-Analysis of Rates of Return to Agricultural R&D.* Research Report 113. International Food Policy Research Institute, Washington DC.

Becerril, J. and Abdulai, A. (2010) The impact of improved maize varieties on poverty in Mexico: a propensity score-matching approach. *World Development* 38, 1024–1035.

Benjamin, D. (1995) Can unobserved land quality explain the inverse productivity relationship? *Journal of Development Economics* 46, 51–84.

Central Statistical Agency of Ethiopia (2010) Agricultural Sample Survey 2009/2010. Available online at http://www.csa.gov.et (Accessed 11 May 2015)

Chamberlin, J. and Schmidt, E. (2012) Ethiopian agriculture: a dynamic geographic perspective. In: Dorosh, P. and Rashid, S. (eds) *Food and Agriculture in Ethiopia: Progress and Policy Challenges.* University of Pennsylvania Press, Philadelphia, PA, pp. 21–52.

de Janvry, A. and Sadoulet, E. (2002) World poverty and the role of agricultural technology: direct and indirect effects. *Journal of Development Studies* 38, 1–26.

Evenson, R.E. and Gollin, D. (2003) Assessing the impact of the green revolution. *Science* 300, 758–762.

FAOSTAT (2010) Food and Agriculture Organization of the United Nations, Rome. http://faostat3.fao.org/faostat-gateway/go/to/home/E (Accessed 15 May 2015).

Foster, J., Greer, J. and Thorbecke, E. (1984) A class of decomposable poverty measures. *Econometrica* 52, 761–766.

Hansen, L.P. (1982) Large sample properties of generalized method of moments estimators. *Econometrica* 50, 1029–1054.

Heckman, J.J. (1979) Sample selection bias as a specification error. *Econometrica* 47, 153–161.

Heckman, J.J. and Vytlacil, E.J. (2001) Local instrumental variables. In: Hsiao C., Morimune K. and Powell, J.L. (eds) *Nonlinear Statistical Modeling: Proceedings of the Thirteenth International Symposium in Economic Theory and Econometrics: Essays in Honor of Takeshi Amemiya.* Cambridge University Press, New York, pp. 1–46.

Heckman, J.J., Urzua, S. and Vytlacil, E. (2006) Understanding instrumental variables in models with essential heterogeneity. *The Review of Economics and Statistics* 88, 389–432.

Jacoby, H.G. (1993) Shadow wages and peasant family labour supply: an econometric application to the Peruvian Sierra. *Review of Economic Studies* 60, 903–921.

Manyong, V.M., Kling, J.G., Makinde, K.O., Ajala, S.O. and Menkir, A. (2003) Impact of IITA germplasm improvement on maize production in West and Central Africa. In: Evenson, R.E. and Gollin, D. (eds) *Crop Variety Improvement and its Effect on Productivity: The Impact of International Agricultural Research.* CAB International, Wallingford, UK, pp. 159–182.

Maredia, M.K., Byerlee, D. and Pee, P. (2000) Impacts of food crop improvement research: evidence from sub-Saharan Africa. *Food Policy* 25(5), 531–559.

Mendola, M. (2007) Agricultural technology adoption and poverty reduction: a propensity-score matching analysis for rural Bangladesh. *Food Policy* 32, 372–393.

Mills, B. (1997) Ex-ante agricultural research evaluation with site-specific technology generation: the case of sorghum in Kenya. *Agricultural Economics* 16, 125–138.

Renkow, M. and Byerlee, D. (2010) The impacts of CGIAR research: a review of recent evidence. *Food Policy* 35, 391–402.

Seyoum, E.T., Battese, G.E. and Fleming, E.M. (1998) Technical efficiency and productivity of maize producers in eastern Ethiopia: a study of farmers within and outside the Sasakawa-Global 2000 project. *Agricultural Economics* 19(3), 341–348.

Shiferaw, B., Smale, M., Braun, H.J., Duveiller, E., Reynolds, M. and Muricho, G. (2013) Crops that feed the world 10. Past successes and future challenges to the role played by wheat in global food security. *Food Security* 5, 291–317.

Suri, T. (2011) Selection and comparative advantage in technology adoption. *Econometrica* 79, 159–209.

Taffesse, A.S., Dorosh, P. and Asrat, S. (2012) *Crop Production in Ethiopia: Regional Patterns and Trends.* International Food Policy Research Institute (IFPRI) and Ethiopian Development Research Institute (EDRI). ESSP Working Paper 16. IFPRI, Washington, DC.

Wooldridge, J. (2002) *Econometric Analysis of Cross Section and Panel Data*. MIT Press, Cambridge, MA.

Zeng, D., Alwang, J., Norton, G.W., Shiferaw, B., Jaleta, M. and Yirga, C. (2013) *Ex-Post Impacts of Improved Maize Varieties on Poverty in Rural Ethiopia.* Diffusion and Impact of Improved Varieties in Africa (DIIVA) – Objective 3 Report. Virginia Tech, Blacksburg, VA and CIMMYT, Nairobi.

16 Impacts of Improved Bean Varieties on Poverty and Food Security in Uganda and Rwanda

C. Larochelle,[1]* J. Alwang,[1] G.W. Norton,[1] E. Katungi[2] and R.A. Labarta[3]

[1]Department of Agricultural and Applied Economics, Virginia Tech, Virginia, USA; [2]Pan Africa Bean Research Alliance, CIAT, Kampala, Uganda; [3]CIAT (formerly of CIP), Cali, Colombia

Introduction

A major objective of crop genetic improvement (CGI) research is to enhance the productivity and quality of food crops and contribute to poverty reduction and food security. The common bean (*Phaseolus vulgaris*) is an important subsistence crop for smallholding farmers in Rwanda and Uganda and elsewhere in sub-Saharan Africa (SSA). In countries where consumption is high, beans are a major source of dietary protein and provide other nutrients such as iron. Rwanda has the highest per capita bean consumption in the world and consumption in Uganda is significant in areas where beans are part of the average diet (Kalyebara and Buruchara, 2008). According to the Food and Agriculture Organization (FAO) 2009, bean consumption is about 29 kg per capita per year[1] in Rwanda and 11 in Uganda.

Widespread adoption of improved varieties with their higher yields, accompanied by a shift from bush to climbing beans that produce more per unit of land, allowed Rwanda to move from being a net importer to a net exporter of beans in 2005. Through the 1980s and 1990s, publicly supported research in the country focused on improving bean yield and most varieties released during this period were high yielding.[2] However, as yields improved, selection criteria shifted and improved bean varieties are now released for attributes such as marketability, short production cycle, high iron content, disease-resistance, seed size and others.

According to the International Center for Tropical Agriculture (CIAT) database of improved varieties, 37 (18) CIAT-improved bean varieties were released in Rwanda (Uganda) since 1985, with 20 (11) of these being released in or following 1998 or later.[3] The national agricultural research systems (NARS) in both countries also release their own varieties, which are more likely to be landraces (Johnson *et al.*, 2003). Released landraces follow a process of evaluation, cleaning and careful selection; they are improved but improvement involves purification of existing genetic material, not introduction of new material. Bean is the crop that receives the most research attention by the Research Agriculture Bureau (RAB) in Rwanda, followed by sweetpotatoes and bananas (Karangwa, 2007).

* E-mail: claroche@vt.edu

This chapter has three objectives. First, it documents the spread and determinants of adoption of improved bean varieties in Uganda and Rwanda. Second, it assesses the causal impact of CGI in beans on household welfare using a treatment-effect estimation approach. Household-specific impacts are then aggregated to market levels to understand impacts on aggregate poverty. Third, the study examines the impact of adoption of improved beans on household food security. Food security impacts are aggregated to national levels to understand how bean-related CGI has affected aggregate national food security in both countries.

This study focuses on bean varieties released since 1998 because adoption and benefits of varieties released prior to 1998 are documented in Johnson *et al.* (2003). About 15% of the 360,000 hectares planted to beans in Uganda in 1998 were CIAT-related varieties. Their yield gain over local varieties averaged 200 kg per hectare, and corresponded to an annual production increment of 8830 tonnes valued at US$2.6 million.[4] In Rwanda, CIAT-related varieties also represented 15% of the area planted to beans in 1998, but the yield gain over local varieties was higher, an estimated 900 kg per hectare. The higher gain in Rwanda is partially due to a shift from bush to climbing beans in Northern Rwanda. In Rwanda, the yield gain contributed to an annual incremental production of 28,888 tonnes of a gross annual value of US$8.7 million (Johnson *et al.*, 2003).

Findings from this study indicate that improved bean varieties released since 1998 are associated with substantial yield gains in both countries, varying from 43% to 82%. Poverty impacts, obtained by comparing the observed distribution of income to that which would have existed in the absence of the new varieties, are relatively small. In the absence of crop varietal improvements, poverty in 2011 would have been about 0.4 percentage points higher in Rwanda and 0.1 in Uganda. The small size of the poverty impacts is due to the small share of land planted to beans and the small share of total household income coming from bean production even among bean-producing families. Improved beans, however, contribute substantially to enhanced food security. Food insecurity would have been 16 percentage points higher in Rwanda without the improved beans and about 2 percentage points higher in Uganda.

The remainder of the chapter is divided as follows. Data are described in the next section and descriptive statistics are briefly discussed. Section three highlights the conceptual framework used to estimate yield gain, profit, poverty and food security impacts associated with bean CGI. The empirical specifications and results are given in sections four and five, respectively. The chapter ends with concluding remarks.

Data and Summary Statistics

Because secondary data on variety spread and yield impacts of beans in Rwanda and Uganda are incomplete, this study included a major effort to collect data on household- and community-level variables associated with adoption and technology impacts. Comprehensive household surveys were conducted during the 2011–2012 cropping seasons in both countries, which encompass 1440 households (1298 bean producing) in Rwanda and 1908 (1722) in Uganda. The surveys covered: household and housing characteristics; productive and household assets; social networks and farmer knowledge; adoption of improved varieties of beans; production activities including land areas, input use and yields; market participation and access; food security; and access to agricultural inputs. Information was collected for 1963 household bean plots in Rwanda and 2957 in Uganda. The household survey included a consumption questionnaire that was administered to about half of the sample in each country.[5] Respondents were asked to recall household food consumption over a 7-day reference period. Food consumption includes food purchased, home-produced, and received as a gift or in kind payment by any household member. The surveys, administered over two rounds, are nationally representative.

The surveys also included community questionnaires to gather information on village characteristics, market access, crop prices and agriculture services, such as access to extension specialists, input distributors, the presence of seed distribution programmes, seed availability, cultivar changes and agroclimatic shocks.

Eighty communities were surveyed in Rwanda and 108 in Uganda. These community-level data were used to help identify the causal effect of technology adoption on outcomes of interest.

Variety identification

A critical issue in measurement of impacts of CGI is linking reported variety names to the database of improved varieties. Variety identification presented numerous problems in both countries. In Rwanda, surveyed farmers reported over 400 unique bean variety names to describe planted varieties. With the help of RAB and CIAT, these names were grouped into 165 unique varieties. Varieties were classified as improved, selected (purified local varieties), local, uncertain (6% of observations) and unknown (9%). In other words, neither farmers nor experts could exactly identify about 15% of the bean varieties named by farmers during surveying.

In Uganda, survey respondents reported more than 500 bean names to describe the varieties in their bean plots. With inputs from the Uganda National Agricultural Research Organization (NARO) and CIAT, the names were grouped together using the same scheme as in Rwanda, yielding 278 unique named bean varieties. Varietal uncertainty in Uganda occurs for about 14% of observations and unknown varieties represent 16% of observations; for the country as a whole about 30% of bean varieties encountered could not be identified with certainty. For the purpose of the analysis, unknown varieties were grouped with local varieties and varietal uncertainty was disregarded. Improved and selected varieties were combined together and considered as improved if released after 1997. For the remainder of the report, the term 'improved' refers to selected and improved varieties released in 1998 and afterwards, whereas 'local' refers to the remainder.

Adoption of improved bean varieties and bean plot characteristics

Rwanda

About 34% of bean-producing households in Rwanda reported planting improved bean varieties, but nearly 18% are partial adopters (Table 16.1). The latter include households with parts of their bean land planted to improved varieties. Full-adopters represent about 16% of bean producers. In addition to partial adoption at the household-level, partial adoption is observed at the plot level. About 6% of bean plots were sown with a mixture of improved and local seeds, 22% were sown with improved seeds only and the remaining plots were planted with local seeds only. Adoption can also be measured as the share of bean land under improved varieties, which reached 23.1% for Rwanda.

The average bean plot in Rwanda is 0.13 hectares.[6] Land under beans is allocated almost equally between climbing and bush beans. The mean number of bean varieties per plot is 1.2. Plots with improved bean varieties are the smallest, on average, followed by plots with local varieties; mixed-variety plots tend to be considerably larger. Yield is significantly greater for plots under improved varieties (782 kg per hectare) compared to local varieties (688 kg per hectare). Intercropping is less common for plots under local varieties than under mixed and improved varieties (Table 16.2).

Uganda

In Uganda, about 26% of bean-producing households plant improved bean varieties; 22.6% are partial adopters; and only 3.9% are full adopters. Partial adoption at the plot level is more common in Uganda than Rwanda. About 10% of bean plots are mixed, whereas about 80% are planted with local varieties and 10% with improved seeds only. The adoption rate in terms of land area under improved varieties is 13.2% in Uganda. The lower share of bean land planted to improved bean varieties in Uganda is due to the lower overall rate of adoption and the greater frequency of partial adoption.

Intercropping bean varieties is more frequent in Uganda than in Rwanda, taking place on about two-thirds of the plots. A plot includes on average 1.5 bean varieties. Climbing beans are far less common in Uganda (Table 16.2). Plots with a combination of local and improved seeds are, on average, larger than those with either only local or improved varieties. Intercropping is significantly more frequent when mixed varieties are planted. Climbing beans are, however, more likely

Table 16.1. Production and household (HH) characteristics among non-adopters, partial adopters and full adopters, second growing season, Rwanda and Uganda.

	Rwanda				Uganda			
	Adoption				Adoption			
Variables	Non	Partial	Full	Sample	Non	Partial	Full	Sample
Avg. bean plot (ha)[b,c,d,e]	0.14	0.16	0.11	0.14	0.15	0.13	0.11	0.15
Number of bean plots[a,c,d,f]	1.5	1.91	1.42	1.56	1.7	2.32	1.6	1.83
Total bean area (ha)[a,b,c,e,f]	0.19	0.26	0.16	0.19	0.24	0.28	0.18	0.24
Bean production (kg)[a,c,d,f]	106.71	145.66	93.09	111.38	152.23	210.22	130.09	164.46
Land cropped (ha)[a,c,d]	0.88	1.18	0.82	0.92	1	1.2	1.01	1.05
HH size[c,e,f]	5.13	5.32	4.94	5.13	6.42	6.67	5.69	6.44
HH head age[c]	44.74	46.23	42.77	44.69	45.88	46.84	48.87	46.21
HH age gender	0.74	0.72	0.79	0.74	0.8	0.82	0.81	0.8
Household head education								
None	0.28	0.3	0.26	0.28	0.17	0.14	0.12	0.16
Primary	0.68	0.67	0.7	0.68	0.58	0.58	0.51	0.58
Secondary and higher[e]	0.04	0.03	0.04	0.04	0.25	0.28	0.37	0.26
Wealth index[b,c,d,e]	–0.18	–0.07	–0.34	–0.18	–0.04	0.23	0.44	0.04
N	863	228	207	1298	1265	389	68	1722
%	66.49	17.57	15.95	100	97.46	29.97	5.24	100

[a]Mean is statistically different at the 5% level between non-adopters and partial adopters in Rwanda; [b]Mean is statistically different at the 5% level between non-adopters and full adopters in Rwanda; [c]Mean is statistically different at the 5% level between partial adopters and full adopters in Rwanda; [d]Mean is statistically different at the 5% level between non-adopters and partial adopters in Uganda; [e]Mean is statistically different at the 5% level between non-adopters and full adopters in Uganda; and [f]Mean is statistically different at the 5% level between partial adopters and full adopters in Uganda.

to be found in plots with improved varieties. This finding is unlike Rwanda, where climbing beans are more common in plots with local varieties.[7]

Household summary statistics

Rwanda

A typical bean-producing household in Rwanda plants 0.92 hectares, or about 0.18 hectares per person (see Table 16.1). Land devoted to beans averages 0.2 hectares per household, corresponding to 1.56 plots of 0.14 hectares each. Thus, beans represent only slightly more than one-fifth of the average household's rather small land under production. Bean harvest, expressed in dry bean equivalent,[8] averages 111 kg per household per season. Subtle differences are evident between full, partial and non-adopters of improved beans. Full adopters cultivate slightly smaller bean plots compared to non- or partial adopters (see Table 16.1). Partial adopters have more bean plots, more land area devoted to bean production and more harvested bean weight than non- and full adopters. Bean harvest during the survey season averaged 146 kg per household for those who partially adopted improved varieties compared to 93 and 107 kg for full and non-adopters.

Uganda

Bean-producing households in Uganda plant, on average, 1.05 hectares of land (Table 16.1). Household sizes in Uganda are larger than in Rwanda resulting in 0.16 hectares of land cropped per capita. Ugandan households cultivate on average 1.83 plots of beans, with a plot size of 0.15 hectares, corresponding to 0.24 hectares of beans cultivated per season (slightly less than one-quarter of the total land). Bean harvest averages 164 kg per household per season.

Table 16.2. Plot characteristics per local, mixed and improved bean variety plots, Rwanda and Uganda.

	Rwanda				Uganda			
	Variety types				Variety types			
Variables	Local	Mixed	Imp.	Sample	Local	Mixed	Imp.	Sample
Plot size (ha)[a,b,c,d,e,f]	0.13	0.22	0.1	0.13	0.13	0.16	0.09	0.13
Yield (kg/ha)[b,d]	688	704	782	710	965	1205	1067	999
Distance to HH (walking minutes)	16.58	17.5	17.92	16.93	22.24	22.02	22.22	22.22
Intercropped (1 = yes)[a,b,d,f]	0.43	0.64	0.55	0.47	0.64	0.77	0.63	0.65
Climbing seeds (%)[a,b,e,f]	54.48	37.58	31.49	48.37	0.09	0.12	0.27	0.11
Nb. of varieties[a,b,c,d,e,f]	1.16	2.32	1.03	1.2	1.4	2.54	1.03	1.47
Soil fertility								
Good[e,f]	0.41	0.36	0.4	0.41	0.46	0.41	0.58	0.46
Medium[e,f]	0.49	0.49	0.51	0.49	0.43	0.49	0.33	0.43
Poor	0.1	0.15	0.09	0.1	0.11	0.1	0.09	0.11
Elevation (m)[a,b,e,f]	1714	1642	1626	1690	1358	1373	1517	1375
N	1410	118	435	1963	2379	286	292	2957
%	71.83	6.01	22.16	100.00	80.45	9.67	9.87	100.00

[a]Mean is statistically different at the 5% level between non-adopters and partial adopters in Rwanda; [b]Mean is statistically different at the 5% level between non-adopters and full adopters in Rwanda; [c]Mean is statistically different at the 5% level between partial adopters and full adopters in Rwanda; [d]Mean is statistically different at the 5% level between non-adopters and partial adopters in Uganda; [e]Mean is statistically different at the 5% level between non-adopters and full adopters in Uganda; and [f]Mean is statistically different at the 5% level between partial adopters and full adopters in Uganda.

Similar patterns are observed to those in Rwanda among Ugandan full, partial and non-adopters of improved bean varieties. Partial adopters cultivate more land compared to non- and full adopters (see Table 16.1). The amount of land planted to improved bean varieties is significantly smaller for full adopters compared to non- and partial adopters. Bean plots of non-adopters are larger compared to those of partial and full adopters. Partial-adopters also have more bean plots and a greater bean harvest than non- and full adopters. In Uganda, the average quantity of beans harvested per household member is 25 kg, compared to 22 kg in Rwanda.[9]

Conceptual Framework

The conceptual framework introduced in de Janvry *et al.* (2011) is followed to estimate the impact of adoption on different outcomes including yield, household income, aggregate poverty, and household and aggregate food security. In this framework, agricultural technology adoption is driven by the expected profitability of the technology compared to alternatives. Farmers are assumed to be heterogeneous agents making adoption decisions on the basis of a constrained optimization process. Potential constraints to adoption include insufficient finance, limited credit access, or lack of knowledge or information about the new variety. The adoption decision is assumed to be binary:[10] either the farmer adopts or not. Adoption depends on observable and unobservable factors and a random error component. Farmers are assumed to adopt the new technology if the expected profitability of adoption exceeds that of not adopting.

The study, however, is ultimately interested in the impact of technology adoption on outcomes such as household income, poverty status and food insecurity. If unobservable factors affecting the decision to adopt also influence the outcome of interest, selection bias will occur. Selection bias is a common problem in studies using observational data. The magnitude of the bias depends on the importance of the unobservable factors and their correlations across decisions and outcomes. The potential bias complicates identification of the treatment effect – the causal impact of adoption on the outcome. Identification of the treatment effect in the presence of potential selection bias requires careful statistical modelling.

When estimating the impact of technology adoption on farm profitably, selection bias can arise for various reasons. Farmers with greater abilities will probably be able to make better use of the new technology, making adoption more profitable and thus more likely. These same abilities positively affect farm profit regardless of the adoption decision.[11] A similar logic holds for plot characteristics. The new technology might perform better in good compared to bad soil, making profitability higher and the adoption decision more likely in the former. If the farmer recognizes these differences in soil productivity, but they are not observed by the statistician, selection bias occurs. It can therefore be argued that adopters and plots where adoption occurs differ from non-adopters in terms of observable and unobservable characteristics. Under such circumstances, it is necessary to control for the endogenous nature of the adoption decision. Similar arguments can be made about how selectivity can confound causal identification of impacts on other outcomes such as yields and household food security.

Poverty impacts

Poverty can be measured at the household or population level. As individual households adopt a more profitable technology, their specific position relative to a socially defined poverty cutoff may change. This change depends on the magnitude of income gains from adoption and the household's position relative to the poverty line in the absence of the technology. Policy makers are interested in aggregate poverty outcomes: how has the technology affected population poverty? Poverty outcomes depend on the nature of the market; if bean prices fall owing to widespread adoption of a new technology, all participants in the bean market will be affected. Because dry beans are an exportable commodity in both countries, a small open-economy framework is employed. Under this assumption, all benefits related to adoption of improved beans accrue to adopters.

The change in income among adopters is the additional profit (revenues minus cost) resulting from adoption of improved varieties. Additional revenues are computed using the estimated yield gain assuming no change in output prices. Additional production costs are assumed to come from input adjustments associated with adoption and the cost of the new technology (the seed) itself. Household incomes with and without the technology are compared to a poverty line to provide estimates of the change in the prevalence of poverty.

Food security

Adoption of improved beans can affect household food security directly through its impact on income and indirectly through changes in production patterns. Greater income allows the household to increase and diversify its consumption. Income increases can enable precautionary savings and allow the household to take steps to insure itself against food-related shocks (Alwang *et al.*, 2001). As a result of these factors, adoption of improved bean varieties is likely to have a greater impact on food security than on poverty.

Food security is a multidimensional concept that includes the quantity, quality and social acceptability of the food consumed, and variability in food consumption. The result of this multidimensionality is that competing measures of food security are found in the literature, each of which has strengths and liabilities. For example, food security measures based on anthropometric techniques identify the malnourished but do not reflect the root causes of the insecurity. Other measures such as caloric intake require extensive surveying that can be costly and cumbersome to conduct. Owing to its dynamic nature, proper measurement of food insecurity requires panel data. These can be costly and cost considerations can reduce their representativeness due to limited sample sizes and potential attrition. The need for a simple, low-cost and easily administered proxy that accurately reflects nutrient intake and food security has led to widespread use of dietary diversity scores (Kennedy *et al.*, 2011).

Numerous studies have validated the association between dietary diversity and household food security (Swindale and Bilinsky, 2006; Kennedy *et al.*, 2011). Hoddinott and Yohannes (2002) found that increased dietary diversity is associated with an increase in dietary energy availability and improvement in household socio-economic status across many countries. Other studies show that dietary diversity is positively associated with macronutrient and micronutrient adequacy across all age groups (Steyn *et al.*, 2006; Kennedy *et al.*, 2007, 2011; Arimond *et al.*, 2010). As a result of these findings, a dietary diversity measure is used to reflect household food security and the causal impact of bean technology adoption on this measure is identified using similar techniques as for plot-level productivity and household incomes.

Empirical Specifications

The study econometrically estimates a treatment effect (TE), or the difference between yields on plots planted to improved beans compared to what they would have yielded had they been planted to local varieties. The TE is estimated at the plot level to help avoid the problem of partial adoption at the household level. The study considers plots planted with 50% or more improved seeds to be treated and those with less than 50% to be non-treated.[12] In Rwanda, partial adopting plots represent 6% of observations; under the 50% assumption, 42% of mixed plots are considered to be non-treated (Fig. 16.1a). In Uganda, mixed plots correspond to 9.6% of observations (Fig. 16.1b). Under the 50% assumption, 60% of partially adopting plots are considered to be non-treated.

Three econometric models are estimated. A homogenous TE model assumes that the yield gain from adoption is identical across observations, regardless of whether the household adopted or failed to adopt the technology. The second model, an observed heterogeneous TE model, relaxes this assumption and allows the yield impact of improved varieties to vary over a set of observed variables. Impacts might differ by education or agronomic factors. The final model, the unobserved and observed heterogeneous TE model, allows the treatment to vary with unobservable characteristics and observable variables. Estimation of the three models provides insights into how adoption impacts differ across households in the sample. A critical step in this

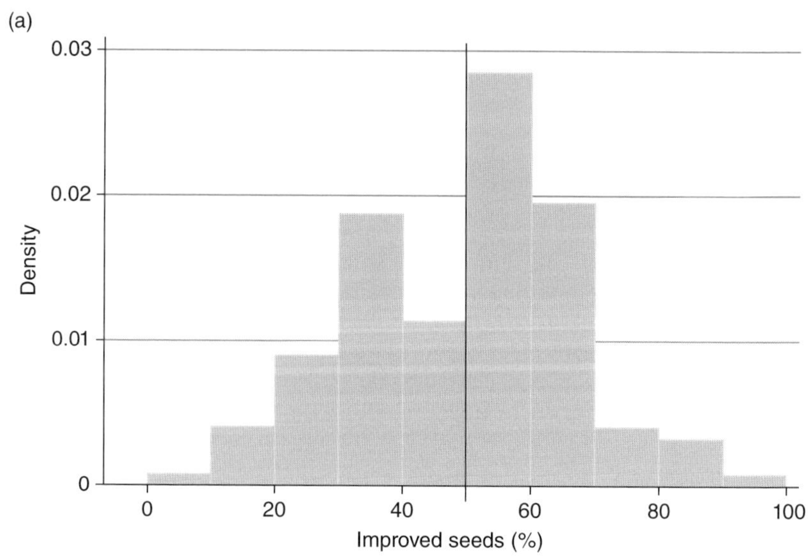

Fig. 16.1. Area (%) of improved bean seed in mixed plots in (a) Rwanda and (b) Uganda.

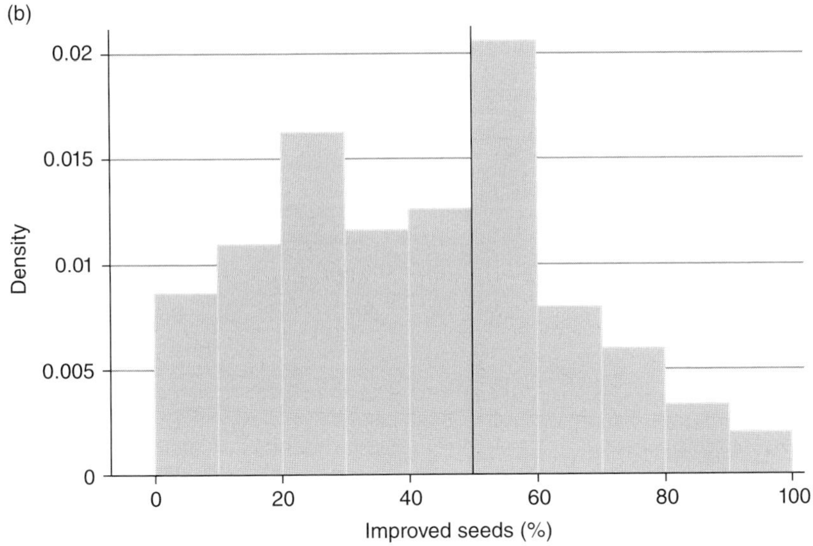

Fig. 16.1. Continued.

estimation is identification of the TE; instrumental variables (IVs) are needed that reflect factors inducing variation in technology adoption without directly affecting the outcome. This identification is discussed in more detail below.

Production functions

Plot-level production functions are estimated assuming a Cobb–Douglas functional form, as in Suri (2011). Under the homogenous TE[13] assumption, the production function is expressed as:

$$\ln(y_{ij}) = \beta_0 + \alpha T_{ij} + \beta_1 \ln(x_{ij}) + \beta_2 \ln(pc_{ij}) + \beta_3 \ln(hc_i) + \beta_4 D_i u_{ij} \quad (16.1)$$

The dependent variable, y_{ij}, is the yield (kg/ha) of household i for plot j. The coefficients to be estimated are the TE (α) and the βs. Yield is assumed to be function of the adoption of improved varieties (T_{ij}), agricultural inputs (x_{ij}), plot characteristics (pc_{ij}), household characteristics (hc_i), and geographical dummy variables (D_i). Inputs comprise the number of pesticide applications, quantity of seeds[14] (kg/ha), labour (person day/ha), organic fertilizer (compost and manure) application[15] (kg/ha), chemical fertilizer[16] and agricultural equipment.[17] Inputs and outputs are divided by plot size and the specification can also be called a yield function. *Plot characteristics* refer to soil fertility (represented by dummy variables distinguishing between good, medium and poor soil fertility as reported by the farmers), distance between the plot and household residence, whether the plot is intercropped, number of bean varieties in the plot, percentage of climbing seeds and plot elevation. *Household characteristics* include age, gender and education of the household head. *Geographical dummy variables* represent the 10 agroecological zones in Rwanda and the four main regions in Uganda, providing a way of distinguishing across production systems. Summary statistics for variables included in the production function are given in Table 16.3.

Under TE heterogeneity on observables,[18] the production function takes the following form:

$$\ln(y_{ij}) = \beta_0 + \alpha T_{ij} + \beta_1 \ln(x_{ij}) + \beta_2 \ln(pc_{ij}) + \beta_3 \ln h(hc_1) + \beta_4 D_i + \gamma_1 T_{ij}\left(x_{ij} - \bar{x}\right) + \gamma_2 T_{ij}(pc_{ij} - \overline{pc}) + \gamma_3 T_{ij}\left(hhhc_1 - hh\overline{hc}\right) + u_{ij} \quad (16.2)$$

With this model specification, three additional vectors of coefficients, γ_1, γ_2 and γ_3, must be estimated, in addition to those defined in Eqn 16.1. These coefficients are those

Table 16.3. Summary statistics of variables included in the production function, Rwanda and Uganda.

	Rwanda			Uganda		
	Local	Improved	Sample	Local	Improved	Sample
Yield (kg/ha)	689.94	789.98	713.44	964.01	1,159.36	991.52
Seeds (kg/ha)				176.96	207.72	181.3
No. of pesticide application	0.01	0.01	0.01	0.04	0.04	0.04
Labour (days/ha)	231.4	260.05	238.13	321.94	350.23	325.92
Organic fertilizer (kg/ha and 1 = yes)	6,377.84	7,220.96	6,594.43	0.03	0.04	0.03
Chemical fertilizer (1 = yes)	0.09	0.09	0.09	0.01	0.01	0.01
Agr. equipment (UGX)	0.12	−0.03	0.08	51,656.75	48,113	51,168.99
Soil fertility						
Good	0.41	0.41	0.41	0.45	0.55	0.46
Medium	0.49	0.51	0.49	0.44	0.36	0.43
Poor	0.1	0.09	0.1	0.11	0.09	0.11
Distance to HH (walking minutes)	16.6	18.84	17.13	22.27	21.89	22.22
Plot is intercropped (1 = yes)	0.43	0.54	0.46	0.66	0.66	0.66
Number of bean varieties planted						
1	0.84	0.88	0.85	0.66	0.69	0.67
2	0.11	0.11	0.11	0.23	0.26	0.23
3	0.05	0.01	0.04	0.11	0.05	0.10
Percent of climbing seeds	53.96	33.05	49.05	0.09	0.24	0.11
Elevation (m)	1,712.17	1,620.74	1,690.69	1,358.03	1,497.11	1,377.61
Agroecological zones and regions						
1	0.05	0.17	0.08	0.32	0.35	0.32
2	0.14	0.05	0.12	0.14	0.03	0.13
3	0.03	0.06	0.04	0.1	0.01	0.09
4	0.07	0.1	0.08	0.44	0.62	0.46
5	0.08	0.11	0.09			
6	0.26	0.16	0.24			
7	0.05	0.05	0.05			
8	0.16	0.15	0.15			
9	0.12	0.04	0.1			
10	0.05	0.11	0.07			
Household head age	44.94	43.67	44.59	45.92	47.61	46.16
Household head gender (1 = male)	0.73	0.78	0.74	0.8	0.81	0.81
Household head education						
No education	0.28	0.27	0.28	0.15	0.11	0.14
Primary education	0.67	0.7	0.68	0.59	0.6	0.59
Secondary education and higher	0.05	0.03	0.04	0.26	0.29	0.27
N (plots)	1,460	503	1,963	2,550	407	2,957

Note: UGX, Uganda Shillings.

capturing observed heterogeneity in treatment, obtained by interacting the treatment (T) with $(x_{ij}-\bar{x}).(pc_{ij}-\overline{pc})$ and $(hc_i-\overline{hc})$. The first term in parentheses is the plot and/or household specific variable and the second term, the sample mean for that given variable. Heterogeneity is assumed to come from two agricultural inputs – organic and chemical fertilizer application; two variables for plot characteristics – soil fertility and share of climbing beans; and three household characteristics – household head age, gender and education. The choice of variables affecting observed treatment heterogeneity is guided by the theory of varietal technology impacts and knowledge of the potential variability in treatment effects. Good soil fertility and fertilizer application practices might increase the differential yield gain of planting improved varieties compared to the situation when improved varieties are cultivated on poor soil and/or without fertilizer application. Treatment heterogeneity in bean types permits testing for differences in the impact of improved seeds between climbing and bush beans. Household head characteristics are included to control for the possibility that farmer knowledge and learning abilities influence gains from adopting bean varietal technology.

The third model specification allows treatment heterogeneity on observable and unobservable characteristics. This specification is a variant of the Heckman selection model, where the source of inconsistency associated with selectivity bias is eliminated by including two Mills ratio-like terms, the last two terms, ρ_1 and ρ_2, in Eqn 16.3. The model is:

$$\begin{aligned} In(y_{ij}) &= \beta_0 + \alpha T_{ij} + \beta_1 In(x_{ij}) + \beta_2 In(pc_{ij}) \\ &+ \beta_3 In(hc_i) + \beta_4 D_1 + \gamma_1 T_{ij}(x_{ij} - \bar{x}) \\ &+ \gamma_2 T_{ij}(pc_{ij} - \overline{pc}) + \gamma_3 T_{ij}(hc_i - \overline{hc}) \\ &+ p_1 T_{ij} \frac{\hat{\varphi}_{ij}}{\hat{\phi}_{ij}} + \frac{p_2(1-T_{ij})\hat{\varphi}_{ij}}{1-\hat{\phi}_{ij}} + u_{ij} \end{aligned} \quad (16.3)$$

In Eqn 16.3, y_{ij} corresponds to bean yield (kg/ha) of household *i* for plot *j* and is assumed to be a function of adoption of improved varieties (T_{ij}), variable inputs (x_{ij}), plot characteristics (pc_{ij}), household characteristics (hc_i), geographical dummy variables (D_i) and an error term, μ_{ij}. Variables considered influencing observed treatment heterogeneity are identical to Eqn 16.2. The predicted probabilities, $\hat{\phi}_{ij}$, and densities, $\hat{\varphi}_{ij}$, of adopting improved varieties are included (estimated using a Probit model).

Changes in farm profit

A measure of income changes associated with adoption of the new bean technologies requires estimating the change in revenue associated with the yield effect, adjusting for changes in input costs, and cost of the new seed. Since yield gains are estimated using the production function at the plot-level, the results must be aggregated to the household-level.

Revenues

In order to obtain the additional revenue due to adoption of the improved varieties, the counterfactual yield (y^0_{ij}) is calculated for each plot *j* cultivated by household *i*. The TE reflects the difference between the actual yield (y^1_{ij}) and that which would have existed without the new bean variety. Under the homogeneity assumption, the TE is constant across all observations but it becomes plot specific if the treatment is heterogeneous. Multiplying the observed minus counterfactual yield ($y^1_{ij} - y^0_{ij}$), by plot size (a_{ij}) gives the change in production (Δq_{ij}) resulting from adoption (see Larochelle *et al.*, 2015 for details). The additional value of production from adopting improved varieties is computed by summing the additional production across all *j* plots, and valuing it at village average output sale price.[19] The village bean sale price reflects between-village variability in price, which is reasonably assumed to be exogenous to household decisions because household bean production is quite small relative to the quantities of bean sales transacted in survey villages.

Input expenditures

Use of improved varieties is likely to be accompanied by input adjustments. For example, applications of chemical fertilizer might be necessary to obtain the expected yield gain from adoption. Adoption might also be linked to increased labour requirements, such as at harvest. To test whether input adjustments take place when planting improved varieties, input use is calculated on a per-hectare basis for all plots and a test for the equality of the means is performed between plots, with and without improved varieties, for the entire sample. Changes in production costs (ΔC^a_i) resulting from input adjustments are computed as:

$$\begin{aligned} \Delta C^a_i &= \sum_{j=1}^{J} \Delta C^a_{ij} = \sum_{j=1}^{J}\sum_{k=1}^{K} \left(x^1_{ijk} - x^0_{ijk}\right) w_{ik} . a_{ij} \\ &= \sum_{j=1}^{J}\sum_{k=1}^{K} \left(x^1_{ijk} - \frac{x^1_{ijk}}{1+\Delta x_k} \right) w_{ik} . a_{ij} \end{aligned} \quad (16.4)$$

where x_{ijk} represents the use of input *k* on a per-hectare basis, and the subscripts 1 and 0 distinguish between observed and counterfactual input usage respectively. The counterfactual input usage, x^0_{ijk}, depends on the current input use, x^1_{ijk}, and Δx_k, the per-hectare average percentage change in use of input k resulting from adoption of improved varieties. In Eqn 16.4, w_{ik} is the village-level unit price of input k, which is assumed to be the same for all inputs regardless of the variety used (seed prices are discussed below). Therefore, by summing first over inputs $k = 1$ to K, and then over plots $j = 1$ to J, a measure is obtained of changes in input expenditures induced by technology adoption for household i.

Seed costs and profit

Seed quantities used on a per-hectare basis do not vary with technology choice, but seed unit price does. Therefore, the cost of adopting improved varieties is simply the difference in input prices between improved and unimproved seeds $(w^1_{ij} - w^0_{ij} = \Delta w_{ij})$, multiplied by the quantity of seeds used on a per-hectare basis, expressed as in Eqn 16.5 and by plot size.

$$\Delta C^b_i = \sum_{j=1}^{J} r_{ij}\left(w^1_{ij} - w^0_{ij}\right) * a_{ij} = \sum_{j=1}^{J} r_{ij}\Delta w_{ij} * a_{ij} \tag{16.5}$$

Summing the additional costs of improved seeds over J plots gives the cost of the technology for household i, expressed as ΔC^b_i in Eqn 16.5. Adding changes in production costs, ΔC^a_i, due to input adjustments to the cost of the technology provides a measure of total input expenditures, ΔC_i, associated with adopting improved varieties, $\Delta C_i = \Delta C^a_i + \Delta C^b_i$.

Subtracting the additional costs of production of improved variety crops from the additional revenues from the varieties provides a measure of profitability.[20]

Poverty impacts

To determine the causal impact of adoption on poverty, household per capita consumption expenditures are used as a measure of household well-being. These are compared to a predetermined poverty line. In developing countries, expenditures are a better measure of well-being than income because incomes are seasonal, difficult to measure and may be under-reported in household surveys (Deaton, 1997). Well-being is also generated through the flow of services derived from durable goods and housing ownership. All these components of well-being are included in an expenditure measure. Per-capita expenditures are compared to a poverty line and differences in poverty prevalence associated with technology adoption are estimated.

Food security and technology adoption

Dietary diversity, the measure of food security, is the simple count of food items or food groups consumed by household members over a predetermined period (Ruel, 2003). Measures of dietary diversity based on the number of food *groups* consumed, rather than food *items*, are likely to reflect more accurately the diversity of macronutrient and micronutrient intakes. When constructing a dietary diversity index, different methods are available to classify food items into food groups. The FAO employs the Household Dietary Diversity Score (HDDS), which is based on 12 food groups and is an indicator of household economic access to food (Kennedy *et al.*, 2011). This study employs the HDDS. Computation of the HDDS requires a recall of all the food items consumed by the household over a reference period (here 7 days is used). Each food is classified into one of 12 food groups defined by the FAO. The HDDS is the number of food groups consumed by the household over the recall period. Each food group is counted only once. A high HDDS reflects a diverse diet and indicates that the household is not food insecure. A low HDDS suggests food insecurity.

There is no international consensus for HDDS guidelines or target values (Swindale and Bilinsky, 2006; Smith and Subandoro, 2007). On the basis of the assumption that food expenditure diversity increases with income, Swindale and Bilinsky (2006) suggest setting the average score for the richest third of the population or the average score for households ranking in the top 33% of dietary diversity as the target goal to be achieved. Using the latter criterion, literature on food security indicators and country specific food insecurity statistics from the World Food Program (WFP), the following benchmarks are set for this study. Households consuming less than 6 food groups over the 7-day recall period are considered as food insecure. Households consuming 6 to 9 food groups are considered moderately food secure. Households consuming more than 9 food groups are considered food secure.

Because the HDDS is a count of the number of food groups consumed by household members, a Poisson model is the most appropriate econometric specification. Selectivity is also likely to be an issue because, as in the case for yield, unobservable characteristics affecting adoption are also likely to influence HDDS. An IV approach is employed. A Generalized Method of Moments (GMM)-IV Poisson model is used to address adoption endogeneity of adoption and the count nature of the dependent variable (see Larochelle *et al.*, 2015, for details).

Coefficient estimates of the (endogenous) decision to adopt improved beans are used to compute the counterfactual HDDS for adopting households. Food insecurity incidences, based on the benchmarks specified above, are calculated for the observed and counterfactual HDDS. Differences in the food insecurity incidences between the two measures reflect the food security impacts associated with technology adoption. The impacts on food security are expected to be more pronounced than those on poverty because improved varieties influence food consumption through channels in addition to the farm profitability channel. For example, improved varieties with shorter production cycles can free up labour, allowing household members to be engaged in additional income-generating activities. Higher productivity can also allow households to reallocate resources to other crops, which can increase food consumption diversity.

Instrumental variables

Various steps were taken to identify the most relevant IVs for this study. At a Mali DIIVA meeting in January 2011, a large portion of time was devoted to discussions with experts about potential IVs. Later, workshops were held in Rwanda and Uganda; these workshops included academics and agricultural extension agents. The workshops helped refine the list of potential IVs and the community questionnaires were developed accordingly. During estimation of the statistical models, appropriate tests were conducted to establish validity of the IVs.

The IVs used to address endogenous technology adoption and identify the different TEs are localized events that contribute to discontinuity in seed availability (either at the household or village level) and proxies for transaction costs of accessing improved seeds. Discontinuity in seed availability following a natural disaster such as droughts or floods might result in households receiving seed aid (where seeds distributed under the seed aid programme are commonly improved varieties) or being forced to buy new seeds (instead of the custom of recycling seeds[21]), favouring adoption. There should be no current independent correlation between yield and previous nature disasters. The probability of adoption is expected to be greater among households whose transaction costs of accessing improved seeds are low. However, low transaction costs should not have a direct impact on yield other than through their influence on adoption of improved seeds.[22] Examples of factors that could influence transaction costs are distance to paved road, distance to input distribution centre and village population. With many potential IVs, tests are performed of under-identification, weak identification and over-identification to identify the most appropriate ones.

Results

Rwanda: Yield and income gains

Yield gain

For each data set and causal relation, tests are necessary to identify the most relevant IVs. These variables are used to identify the effect of technology adoption and correct for potential selectivity in the outcome (Eqns 16.1, 16.2 and 16.3). The tests, performed while adjusting for potential heteroskedasticity, support the following choice of IVs for yield gain in Rwanda: whether there was a flood or drought in the village during the last 10 years, existence of marketing services for agricultural crops in the village and existence of credit service in the village (see Larochelle *et al.*, 2015, for details). These IVs were also identified by experts on Rwandan agriculture as possibly affecting seed distribution but not having an independent effect on productivity (given properly specified models).

The three models – homogenous treatment, heterogeneous treatment on observables, and heterogeneous treatment on observables and unobservables – all performed well. Most variables have significant coefficients, with the expected signs, and are of similar magnitude across the three models (Table 16.4). All inputs (pesticide, organic and chemical fertilizer, labour and agricultural equipment) are found to significantly increase yield. Focusing on model 3 (treatment heterogeneity on observable and unobservable characteristics), an additional application of pesticide would boost yield by about 27%. Increasing organic fertilizer application and labour by 10% would lead to a 1% and 3.3% increase in bean yield, respectively. Application of chemical fertilizer is associated with a yield gain of 17% while increasing the value of agricultural equipment index by one point raises productivity by about 3%. Intercropping reduces yield by about 14% while varietal diversity contributes significantly and positively to yield. The climbing bean type is associated with a yield advantage of about 28% over bush beans.

Household head education has a strong and positive impact on yield. Bean productivity is about 13% and 27% higher for households whose head has some primary and secondary education, respectively, compared to households whose head has no formal education. The agro-ecological variables are strong and significant determinants of yield, highlighting the importance of environmental production constraints. A test for the joint significance of the variables explaining heterogeneity in TE supports their inclusion. However, few variables explaining treatment heterogeneity are statistically significant.

Under the assumption of homogenous treatment, the estimated average TE is 0.803 (*P*-value = 0.004), meaning that adopting improved bean varieties increases yield on average by 80% controlling for all other factors. Under the assumption of treatment heterogeneity based on observable variables, the estimated average TE is 0.539 (*P*-value = 0.048). Since this model allows the estimated effect to vary, it produces separate estimates of TEs on the treated (ATET) and the non-treated (ATENT). These are 0.567 and 0.529, respectively. Similarly, under the assumption of treatment heterogeneity based on observable and unobservable characteristics, the estimated average TE is 0.527 (*P*-value = 0.023). Assuming treatment heterogeneity on observable and unobservable variables leads to greater differences between the ATET (0.821) and ATENT (0.426).

Both treatment heterogeneity models suggest that the average gain from improved varieties is smaller for non-adopters than adopters, indicating positive selection bias. Adopters experience greater yield gains from adopting than non-adopters would have if they had adopted. This is a plausible finding. Results also suggest that not accounting for unobservable heterogeneity could understate returns from adoption for adopters and overstate them for non-adopters. A Wald test supports the assumption of unobserved heterogeneity in returns from adopting improved varieties. Thus, the remainder of the results for income gains in Rwanda is discussed in light of observed and unobserved heterogeneity in TE.

Profitability

Using plot-specific yield gain estimates, under observed and unobserved heterogeneity, the expected additional bean production from planting improved varieties is computed for each plot and then summed to the household level. On average, households that planted improved bean varieties obtained an additional 42 kg of beans compared to what they would have had they planted local varieties. Additional production is valued at the village average bean sale price, and, on average, planting improved bean varieties increases household bean revenues by 11,971 Rwanda Francs (Rwf) per season, or US$49.88 in Purchasing Power Parity Dollars (PPP).[23]

Average additional labour cost per household associated with adoption is estimated to be 1347 Rwf, or $5.61 at PPP, per agricultural season. This estimate is based on the assumption that proportionally more labour is needed given increased yield at harvest. Pesticide and chemical fertilizer applications do not differ statistically between plots planted with improved and local varieties, leading to no cost adjustments. Improved seeds cost an additional 200 Rwf per kg and, with a seeding rate of 80 kg per hectare, the cost of the new seeds is 16,000 Rwf per hectare. However, this is an upper bound of the seed cost because bean seeds can be recycled from season to season. The average additional expenditures per household from planting improved varieties are 3572 Rwf (US$14.88 at PPP) per season; about 62% of this increment is due to higher seed price.

Table 16.4. Results of the homogenous and heterogeneous TE models on yield, Season B 2011, Rwanda.

	Homogenous model		Heterogeneous model 1		Heterogeneous model 2	
Yield	coef	se	coef	se	coef	se
ATE (1 = adoption)	0.8026***	0.2758	0.5390**	0.2719	0.5271**	0.2311
No. of pesticide application	0.2793**	0.1381	0.2605**	0.1261	0.2696***	0.1042
Organic fertilizer (kg/ha)	0.1264***	0.0200	0.0564	0.0359	0.0999***	0.0213
Labour (days/ha)	0.3245***	0.0248	0.3236***	0.0247	0.3268***	0.0221
Plot is intercropped (1 = yes)	–0.1441***	0.0391	–0.1571***	0.0379	–0.1471***	0.0352
Chemical fertilizer (1 = yes)	0.0758	0.0685	0.3729***	0.0944	0.1693**	0.0694
Agr. equipment (UGX)	0.0314**	0.0130	0.0328***	0.0126	0.0312***	0.0117
Soil fertility (Base = good)						
Medium	–0.0439	0.0358	–0.0200	0.0651	–0.0137	0.0369
Poor	–0.1291**	0.0613	–0.1516	0.1073	–0.1084*	0.0644
Head education (Base = none)						
Primary education	0.1735***	0.0437	0.0089	0.0891	0.1317***	0.0451
Secondary education	0.2760***	0.0899	0.1313	0.1302	0.2735***	0.0872
Number of varieties (Base = 1)						
Two varieties	0.0692	0.0583	0.1157**	0.0568	0.0874*	0.0504
Three varieties	0.3476***	0.0866	0.3596***	0.0861	0.3512***	0.0816
Distance to HH (walking minutes)	–0.0104	0.0130	–0.0109	0.0127	–0.0112	0.0116
Plot elevation (m)	–0.3412*	0.2024	–0.2473	0.2113	–0.3275*	0.1894
Percentage of climbing seeds	0.0028***	0.0006	0.0030***	0.0008	0.0028***	0.0006
HH head age	–0.0266	0.0598	–0.0916	0.1302	–0.0451	0.0594
HH head gender (1 = male)	0.0141	0.0418	0.1754**	0.0810	0.0574	0.0431
Agroecological zone (Base = 1)						
2	0.1693	0.1482	0.0316	0.1443	0.1006	0.1370
3	0.0473	0.1510	0.0192	0.1507	0.0092	0.1350
4	0.0220	0.1067	–0.0157	0.1078	–0.0151	0.0923
5	0.5511***	0.1277	0.4549***	0.1244	0.4873***	0.1140
6	0.2936**	0.1263	0.2127*	0.1234	0.2396**	0.1152
7	0.5297***	0.1383	0.4525***	0.1358	0.4911***	0.1254
8	0.6847***	0.1237	0.5774***	0.1265	0.6177***	0.1127
9	0.2041	0.1494	0.0651	0.1463	0.1379	0.1376
10	0.6820***	0.1077	0.6328***	0.1056	0.6507***	0.0923
Head education (Base = none)						
Primary education * T			0.5045*	0.2960	0.1342	0.0874
Secondary education * T			0.3681	0.4800	–0.1153	0.2025
HH head gender (1 = male) * T			–0.5625*	0.2917	–0.1378*	0.0809
HH head age * T			0.0209	0.4661	–0.0280	0.1214
Soil fertility (Base = good)						
Medium soil fertility * T			–0.0867	0.2330	0.1120	0.0729
Poor soil fertility * T			0.0040	0.4016	–0.0845	0.1300
Organic fertilizer (kg/ha) * T			0.2871**	0.1161	0.1104***	0.0366
Percentage of climbing seeds * T			–0.0023	0.0028	–0.0006	0.0008
Dummy if NPK app. = 1 *T			–1.2126***	0.3683	–0.3777**	0.1618
ρ_1					–0.2179	0.1377
ρ_2					0.4745***	0.1751
Constant	5.5611***	1.5237	6.0435***	1.7214	5.8347***	1.4206
Number of observations	1963		1963		1963	

Note: *** *P*<0.01, ** *P*<0.05, * *P*<0.1. Standard errors (se) are heteroskedasticity robust. ATE, average treatment effect; coef, coefficient.

*T denote interactions between the variable in question and the treatment (adoption). These interactions are necessary in the presence of heterogeneous treatments.

Net profit from adoption, obtained by subtracting from the additional revenue the additional costs of improved varieties, is on average 8399 Rwf (US$35.00 at PPP) per household per agricultural season. As there are two bean cropping seasons, the additional annual profit from the technology is estimated. According to the most recent Agricultural Census of Rwanda (2008) bean production is about 10% higher in season A than in season B (the surveyed season) resulting in an estimated average additional yearly profit of 17,638 Rwf (US$73.49 at PPP) for households adopting improved varieties. Although improved beans are associated with a statistically significant increase in production and household income, the increase is relatively small (about US$0.20 PPP per day). As a result of this relatively small income effect, poverty impacts are likely to be relatively small.

Impact on poverty in Rwanda

The counterfactual well-being, i.e. the well-being in the absence of technology for households who adopted improved varieties, is obtained by subtracting the additional farm profit resulting from cultivating improved bean varieties from household consumption expenditures. Then the observed and counterfactual well-beings are adjusted for household size and compared to the poverty line provided in the poverty report of the National Institute of Statistics of Rwanda (NISR) (2008).[24] The observed poverty head count is 47.2%, which is close to NISR estimates of rural poverty, even though different measures of well-being and sampling procedures were used.[25] Applying this poverty line suggests that poverty among bean producers would be 47.5%, or about 0.4 percentage points higher, in the absence of improved bean varieties. Although this change is modest, it indicates that some of the poor producers in Rwanda are able to escape poverty by adopting improved bean varieties.

Adopter households have mean observed per capita expenditures of US$697 at PPP compared to US$659 at PPP for non-adopters. Corresponding poverty incidences are 42.4% and 49.1%, respectively. The counterfactual per capita expenditures for adopters is US$679 at PPP on average, corresponding to a poverty level of 43.6%. This indicates that, within a 1-year time period, poverty prevalence among adopters of improved bean varieties decreased by 1.2 percentage points owing to adoption. The actual and counterfactual distributions are illustrated in Figs 16.2 and 16.3, where the vertical dashed line represents the poverty line.

The small magnitude of the poverty impacts can be explained by small areas farmed and generally low productivity of beans. The average household in Rwanda produces only

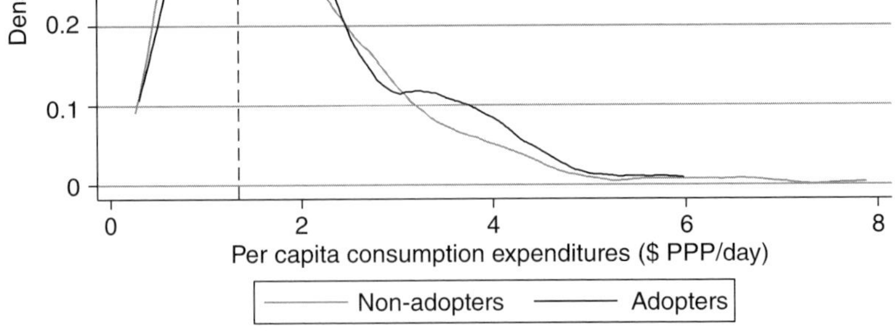

Fig. 16.2. Actual distribution of well-being over 1 year in Rwanda. Note: The Epanechnikov kernel was used. The vertical dashed line reflects the poverty line.

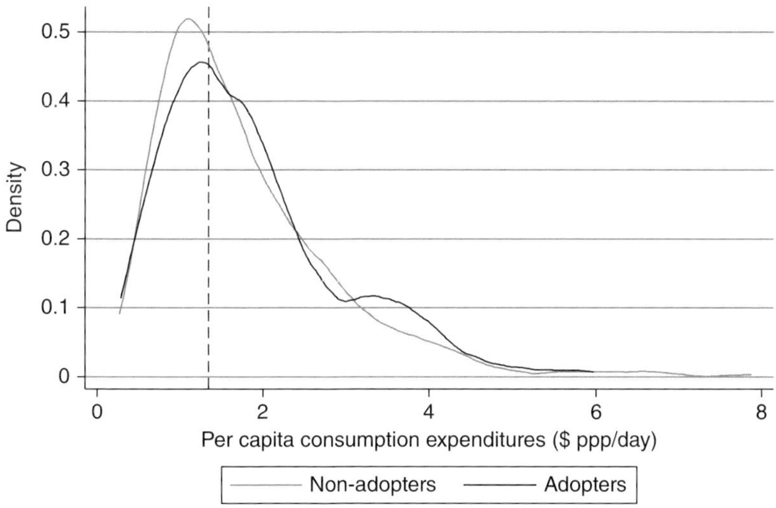

Fig. 16.3. Counterfactual distribution of well-being over 1 year in Rwanda. Note: The Epanechnikov kernel was used. The vertical dashed line reflects the poverty line.

111 kg of dry beans per agricultural season. Low sale prices also attenuate the poverty impact of technology adoption. A kilogram of beans sells for about 300 Rwf, approximately US$1.25 PPP. As a result, farm profits from bean production represent only a small share of total household consumption expenditures. In addition, the poverty impacts are computed for adopters of improved varieties released in 1998 and afterwards as opposed to all adopters of improved varieties. By making this distinction in release date, the measured adoption rate drops by half, reducing potential impacts of crop varietal technology. Moreover, the estimated returns from improved varieties might be understated because the counterfactual, which reflects conditions in the absence of the new varieties, includes older improved varieties and unknown varieties.

Uganda: Yield and income gains

Yield Gain

The three models[26] of causal impacts of adoption of improved varieties on yield for Uganda provide consistent results (Table 16.5). All inputs with the exception of chemical fertilizer, which is applied to only 1% of the bean plots, are statistically significantly in explaining variation in yield. Compared to good soil fertility, planting in poor-quality soil reduces bean yields by about 22%. As for Rwanda, varietal diversity and growing climbing beans instead of bush beans have a positive impact on yield. Cultivating two bean varieties within the same plot would raise yield by about 9%, whereas planting three varieties is associated with a yield gain of 20%. The yield advantage of climbers over bush beans is estimated to be 26% in Uganda, compared to 28% in Rwanda. Elevation has a significant and negative impact on yield in Uganda, whereas this variable is not statistically significant in Rwanda. In contrast to Rwanda, where intercropping reduces productivity, intercropping does not have a statistically significant impact on yield in Uganda. The age of the household head has a significant (negative) impact on yield, whereas his/her education has a weak and small effect. As in Rwanda, variation in yield closely related to environmental production constraints, captured by the regional dummy variables included in this model.

Under assumptions of homogenous treatment and heterogeneity based on observable and unobservable characteristics, the average treatment effect (ATE) in Uganda is estimated to be 0.603. Under the assumption of treatment

Table 16.5. Results of the homogenous and heterogeneous TE models on yield, Second season 2011, Uganda.

	Homogenous model		Heterogeneous model 1		Heterogeneous model 2	
Yield	coef	se	coef	se	coef	se
ATE (1 = adoption)	0.60*	0.32	0.56	0.52	0.60**	0.29
Seed (kg/ha)	0.40***	0.02	0.38***	0.03	0.40***	0.02
Labour (days/ha)	0.29***	0.02	0.29***	0.02	0.29***	0.02
No. of pesticide application	0.22***	0.08	0.24***	0.09	0.22***	0.08
Chemical fertilizer (1 = Yes)	–0.16	0.14	–0.13	0.20	–0.18	0.16
Organic fertilizer (1 = Yes)	0.22***	0.08	0.48**	0.24	0.22***	0.08
Agr. equipment (UGX)	0.04***	0.02	0.05***	0.02	0.04***	0.02
Soil fertility (Base = good)						
Medium	–0.04	0.03	-0.05	0.07	–0.06	0.04
Poor	–0.19***	0.06	-0.21	0.20	–0.22***	0.06
Number of varieties (Base = 1)						
Two varieties	0.09**	0.04	0.08*	0.04	0.09**	0.04
Three varieties	0.21***	0.06	0.19***	0.07	0.20***	0.06
Distance to HH (walking minutes)	–0.04***	0.01	–0.04***	0.01	–0.04***	0.01
Plot elevation (m)	–1.73***	0.19	–1.76***	0.22	–1.72***	0.18
Percentage of climbing seeds	0.19**	0.09	–0.31	0.46	0.26**	0.10
Plot is intercropped (1 = yes)	–0.05	0.04	–0.04	0.05	–0.05	0.04
HH head age	–0.18***	0.06	–0.10	0.11	–0.18***	0.06
HH head gender (1 = male)	–0.01	0.04	0.02	0.09	–0.02	0.04
Head education (Base = none)						
Primary education	0.11**	0.05	–0.02	0.10	0.10*	0.06
Secondary education	0.05	0.06	0.03	0.11	0.07	0.07
Regions (Base = 1)						
2	–0.50***	0.07	–0.53***	0.08	–0.50***	0.08
3	–0.33***	0.08	–0.38***	0.09	–0.32***	0.09
4	0.25***	0.05	0.25***	0.06	0.25***	0.05
Head education (Base = none)						
Primary education * T			1.18	0.91	0.11	0.17
Secondary education * T			0.39	0.99	–0.12	0.19
HH head gender (1 = male) * T			–0.34	0.72	0.11	0.12
HH head age * T			–0.52	0.70	–0.02	0.15
Soil fertility (Base = good)						
Medium soil fertility * T			0.19	0.57	0.12	0.10
Poor soil fertility * T			0.66	1.95	0.20	0.16
Chemical fertilizer * T			–0.21	1.17	0.30	0.26
Organic fertilizer * T			–1.60	1.63	0.03	0.21
Percentage of climbing seeds * T			2.19	1.80	–0.23	0.16
ρ_1					–0.30*	0.17
ρ_2					0.35	0.27
Constant	15.54***	1.37	15.56***	1.70	15.48***	1.33
Number of observations	2957		2957		2957	

Note: *** *P*<0.01, ** *P*<0.05, * *P*<0.1. Standard errors (se) are heteroskedasticity robust. ATE, average treatment effect; coef, coefficient; UDX, Uganda Shillings.

* T denote interactions between the variable in question and the treatment (adoption). These interactions are necessary in the presence of heterogeneous treatments.

heterogeneity based on observables, the estimated ATE is 0.563. While assuming treatment heterogeneity on observables, the ATET and ATENT are 0.805 and 0.525, respectively, suggesting that adopters gain on average more from the technology than would non-adopters if they were to adopt (again, positive selection bias). The remaining results are discussed in light of treatment heterogeneity on the basis of observable and unobservable characteristics. This focus is appropriate because differences between models in Uganda are not great and it provides a consistent analysis across the countries.

Profitability

Profitability is computed by first evaluating the additional bean revenue from planting improved varieties. The yield gain associated with adopting improved varieties corresponds to an increment of 40 kg of beans on average per household for one cropping season. This additional production is valued at the village average sale price, which corresponds to an additional household revenue per agricultural season of 44,787 Uganda Shillings (UGX) or US$46.65 in PPP.[27] The average additional labour costs per household owing to adoption of improved varieties is estimated at 8379 UGX (US$8.73 at PPP). The average technology cost of improved seeds compared to local ones is 1671 UGX (US$1.74 at PPP) per adopting-household per season. Adoption of improved varieties generates average additional bean profits of 34,783 UGX (US$36.23 at PPP) per season per household. Based on the 2011 Agricultural Census of Uganda, bean production in the first agricultural season is estimated to about 72% of the production of the second season. This leads to an estimated average additional yearly profit of US$62.32 at PPP among adopting households compared to what they would have had without the improved beans.

Impact on poverty

Using household observed and counterfactual consumption expenditures per capita and a poverty line of US$2 at PPP per day yields an observed poverty head count of 34.9% compared to a counterfactual of 35.0%. Poverty among all bean producers would have been 0.1 percentage points higher in the absence of improved bean varieties in Uganda, a minor difference. Similar factors explain the small impact of improved bean varieties on poverty in Rwanda and Uganda. These include small land holding, low productivity and low sale prices. The decision to include only improved varieties released after 1998 may also contribute to the low poverty impacts. Exclusion of pre-1998 varieties will naturally lower the estimated overall impact of CGI research and it also contaminates the counterfactual (making 'traditional' varieties appear to be more productive than they actually are). The average household in Uganda produces 164 kg of dry beans per agricultural season, where a kilogram is worth approximately US$1.13 PPP. In addition, the lower adoption rate in Uganda helps explain smaller poverty impacts.

Differences in income and poverty status between adopters and non-adopters are, however, greater in Uganda than Rwanda. The observed per capita expenditures for adopters at PPP is US$1727.37 versus US$1353.89 for non-adopters, and poverty incidences are 25.4% and 36.3%, respectively, for the two groups. The expenditure distributions for adopters and non-adopters are statistically different, providing clear evidence that better-off (ex-ante) households are more able to adopt in Uganda (Figs 16.4 and 16.5). In contrast, the equality of observed and counterfactual well-being distributions could not be rejected in Rwanda, suggesting that the household economic status does not influence adoption. For adopters, the average counterfactual and observed per capita consumption expenditures are US$1718.53 and US$1727.37, corresponding to poverty rates of 25.9% and 25.4%, respectively. Adoption of improved beans is associated with a reduction in poverty prevalence among adopting households of 0.5 percentage points.

Food security

The average HDDS in Rwanda is 7.37 food groups, with a minimum of three and a maximum of 12 food groups. Based on the HDDS benchmarks discussed above, 13% of bean-producing households are considered food insecure, 75% moderately food secure and 12% are food secure. The HDDS is significantly greater for adopting households (7.70) compared to non-adopters (7.24). As a result, food insecurity is significantly more prevalent among non-adopters (15.47%) than adopters (7.95%).

Households in Uganda consumed on average 8.73 food groups over the recall period, meaning that they have a greater dietary diversity

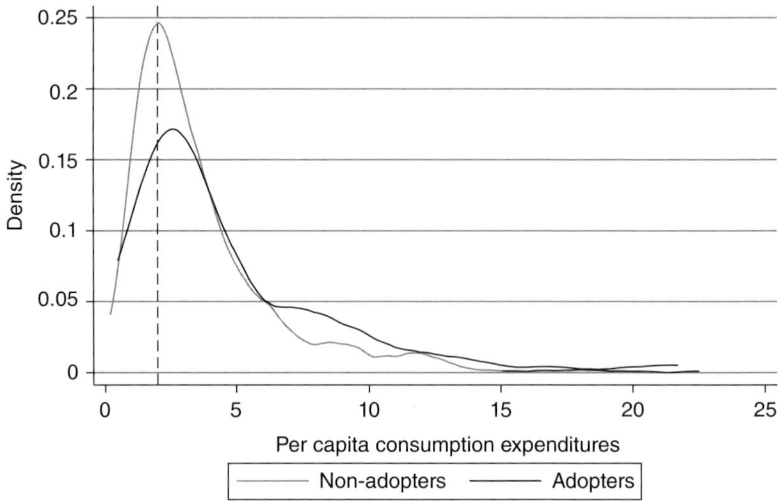

Fig. 16.4. Actual distribution of well-being over 1 year in Uganda. Note: The Epanechnikov kernel was used. The vertical dashed line reflects the poverty line.

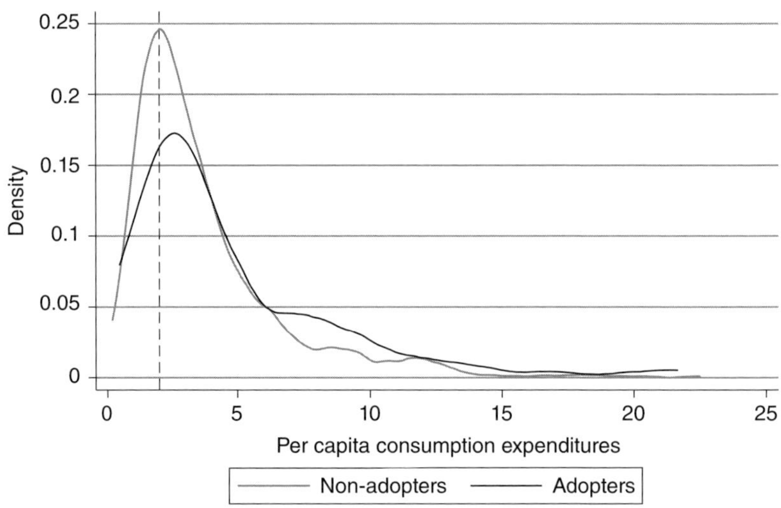

Fig. 16.5. Counterfactual distribution of well-being over 1 year in Uganda. Note: The Epanechnikov kernel was used. The vertical dashed line reflects the poverty line.

than those in Rwanda. The gap in HDDS between adopters and non-adopters is smaller in Uganda; the HDDS is 8.88 for adopters compared to 8.69 for non-adopters. Food security prevalence for the Uganda sample is 39.12%, with 45.45% and 37.56% among adopters and non-adopters, respectively.

Technology adoption and food security[28]

Rwanda

Estimates of the causal impact of improved bean adoption on household food security in

Rwanda indicate that the HDDS would be about 44% lower in the absence of improved bean varieties (Table 16.6). The counterfactual average HDDS is 6.74 (compared to the actual of 7.37) and the percentage of food insecure households increases from 13.43% to 29.32% without adoption. Without improved bean varieties 63% and 7% of households would be moderately food secure and fully food secure compared to the current situation (75% and 12% are moderately food secure and food secure). Key variables explaining variation in HDDS are education of the household head and household wealth. Household composition also influences household diet diversity but its impact is relatively small.

Table 16.6. Results of GMM IV Poisson model on HDDS, Rwanda and Uganda.

	Rwanda		Uganda	
HDDS	Coefficients	*P*-values	Coefficients	*P*-values
Adoption (1 = yes)	0.4438	0.005	0.3236	0.043
HH head gender (1 = male)	–0.016	0.569	0.0213	0.392
HH head age	–0.0015	0.199	–0.0006	0.488
Head education (Base = none)				
Primary education	0.0735	0.007	0.0073	0.843
Secondary education	0.1779	0.002	–0.03	0.467
Number of elderly	0.1707	0.072	0.0378	0.541
Number of elderly squared	–0.0848	0.007	–0.0224	0.484
Number of adults	–0.0122	0.732	0.0527	0.044
Number of adults squared	0.0029	0.47	–0.0057	0.073
Number of children (6–14 years old)	0.0015	0.939	–0.0147	0.094
Number of children squared (6–14 years old)	0.0024	0.539	0.0022	0.001
Number of children (0–5 years old)	0.0664	0.076	0.0228	0.074
Number of children squared (0–5 years old)	–0.0254	0.063	–0.0027	0.172
Land cropped (ha)	–0.0252	0.182	–0.0271	0.041
Land cropped squared (ha)	0.0022	0.075	0.0023	0.14
Wealth index (Base = poorest quintile)				
Quintile 2	0.0512	0.112	0.1068	0.001
Quintile 3	0.0961	0.013	0.1184	0
Quintile 4	0.1244	0.005	0.0986	0.003
Quintile 5	0.1549	0	0.1343	0
Tropical livestock unit (TLU)	0.0101	0.343	0.0009	0.838
Agr. equipment (UGX) (Base = quintile 1)				
Quintile 2	0.0436	0.273	0.0409	0.191
Quintile 3	0.0191	0.655	0.0866	0.006
Quintile 4	0.0377	0.323	0.1158	0
Quintile 5	–0.0111	0.801	0.136	0
Agroecological and ecological zone (Base = 1)				
2	0.0026	0.978	0.0415	0.688
3	0.0451	0.085	–0.0572	0.599
4	0.0172	0.798		
5	–0.0203	0.776		
6	0.0855	0.356		
7	–0.0494	0.700		
8	0.0545	0.435		
9	0.0047	0.964		
10	–0.0046	0.933		
Constant	1.7242	0	1.8497	0
Number of observations	641		736	

Note: *P*-values are estimated from heteroskedasticity-robust standard errors clustered at the village level.

Uganda

Adoption of improved beans also causes reduced household food insecurity in Uganda (Table 16.6). The mean HDDS would be about 32% lower in the absence of improved bean varieties. Without the improved beans, the counterfactual average HDDS would fall to 8.35 from its actual value of 8.73, meaning that the prevalence of household food insecurity would be 8.15% compared to the observed 6.05%. Without improved bean varieties the proportion of food-secure households would be reduced; 31.66% of households would be food secure compared 39.12% under the actual situation.

Significant determinants of HDDS in Uganda are household wealth, captured both by an asset index and by the value of agriculture equipment. As in Rwanda, the influence of household composition is significant but of a lesser magnitude than the other determinants.

Conclusions

Bean production is important to household income in many areas of Rwanda and Uganda. This chapter reports on estimated causal impacts of adoption of improved bean varieties on: (i) productivity and income at the field and household level; (ii) aggregate poverty; and (iii) food security. Food security is important because many NARS and CG Centers are moving toward breeding beans to mitigate the adverse effects of climate change and are now developing shorter-season and drought-tolerant varieties. Recent advances in the econometrics of TEs allow us to create quasi-experimental cases that rely on exogenous variation in IVs to identify the adoption TE. These approaches are convincing because they allow TEs to vary over heterogeneous households, agroecologies and field conditions. They take advantage of information on impacts generated from a long process of CGI diffusion and, given a properly designed sample, reflect country-level impacts on bean producers.

The study identified significant impacts on field-level yields and on household food security for adopting producers. These findings are consistent across countries: yield TEs are consistent with expectations and show that adopters are better off than they would have been in the absence of the technologies. However, given the relatively small areas planted to beans and their low sale prices, the magnitude of the poverty impacts is small.

Average yield gain from adopting improved bean varieties is similar in the two countries; the yield gain from improved varieties is 53% in Rwanda and 60% in Uganda. Treatment heterogeneity due to observable and unobservable factors is highly significant in Rwanda but insignificant in Uganda. This implies that in Uganda non-adopters would have gains similar to those of adopters if they were to adopt. Further, evidence from Uganda shows that household wealth is associated with adoption – poorer bean-producing households are less likely to adopt the new bean varieties compared to the non-poor. This evidence shows that there are potentially important poverty reductions that could occur in Uganda if poorer producers can gain access to bean technologies. In Rwanda, results indicate that higher yield gains occur among adopters (82%) compared to non-adopters (43%).

Because data were collected over only one agricultural season, it is difficult to confidently assert dynamic effects on poverty reduction resulting from the adoption of improved bean varieties. It could be argued, however, that poverty changes reported in this study correspond to annual poverty reduction rates, as additional yearly profits from adoption are assessed against annual consumption expenditures. Improved bean varieties could therefore have had substantial cumulative impacts on poverty given that some of these varieties have been released for 15 years.

Impacts of improved varieties on food security were assessed using a dietary diversity measure of household food consumption. Results were consistent across multiple model specifications and IV choices in Rwanda, whereas slightly more noise was found in Uganda. In both countries, adoption of improved bean varieties was found to have a strong and positive effect on the HDDS. The average HDDS of adopting households would have been 43% and 32% lower in the absence of improved varieties, meaning that household food insecurity would have been about 16 and 2 percentage points higher in Rwanda and Uganda, respectively, in the absence of the bean technology. Consistent with expectations, food security impacts are stronger than the poverty impacts because the influence of improved varieties on food consumption is likely to occur through channels other than bean profitability. Alternative measures of food security, such as percentage of expenditure on food or percentage of food expenditures on staples, could be computed and their corresponding food security estimates derived to assess whether the conclusions hold for alternative measures.

Notes

[1] Statistics reported for 2009.

[2] Before the 1990s, few resources were invested in bean breeding across Africa (Johnson *et al.*, 2003). Public investments grew during the 1990s but by 1998 Rwanda had only one full-time equivalent (FTE) scientist devoted to bean breeding, whereas Uganda had two.

[3] Legume Innovation Lab – Michigan State University.

[4] In 1990 US dollars.

[5] Consumption expenditure information was gathered for 704 and 852 households in Rwanda and Uganda, respectively.

[6] Due to the challenges of obtaining an accurate measure of land area using farmers' estimates of plot size and plot share devoted to bean cultivation (when plot is intercropped), planting density was used to estimate land area in Rwanda. Recall bias regarding quantity of seeds planted is likely to be smaller than the measurement errors associated with estimation of land area and share devoted to beans when the plot is intercropped. Using planting density rather than farmer estimations to compute plot size gives more coherent yield estimates.

[7] 'Local' varieties include improved and selected varieties released prior to 1998, which encompass various climbing bean varieties in Rwanda.

[8] Dry bean equivalent includes bean harvested in green, fresh and dry forms, where a weight loss of 12.5% is assumed for green and fresh harvest.

[9] The cropping season under consideration is different in Uganda and Rwanda. Both countries have two cropping seasons per year, and one season is more important for bean production than the other one. In Uganda, the cropping season under consideration corresponds to the most important one for bean production but not in Rwanda.

[10] Means of addressing partial adoption are discussed below.

[11] Negative selection bias is also possible and has been found in two recent studies of maize technology impacts (Suri, 2011; Zeng *et al.*, 2013).

[12] This assumption is made to simplify estimation. Sensitivity analysis reveals that the findings do not vary under different assumptions about what percentage of bean acreage constitutes adoption (Larochelle *et al.*, 2015).

[13] The homogenous TE is estimated using Probit-Two-Stage-Least-Squares (Probit-2SLS). This method consists of instrumenting the treatment using the predicted probability of being treated. This method, while requiring stronger assumptions than standard IV methods, leads to a more efficient estimator of the TE. The procedure is also robust to misspecification in the data generating process of the treatment. Please see Larochelle *et al.* (2015) for assumptions underlying identification of the TE using this model.

[14] In Rwanda, plot size is estimated on the basis of the quantity of seed planted, assuming a planting density of 80kg/ha, leading to the exclusion of seeds as an input in the production function.

[15] When estimating the production function for Rwanda, observations with zero organic fertilizer application are handled as suggested in Battese (1997) to avoid dropping these observations or creating bias. In Uganda, where few households apply organic fertilizer, a dummy variable for fertilizer application is used instead of quantity.

[16] Because chemical fertilizer is rarely applied, a dummy variable is used to indicate its application.

[17] In Uganda, agricultural equipment is measured as the total value of agricultural tools owned by a household, where tool values are estimated by the farmers and reported in Uganda Shillings (UGX). Due to the lack of price information on agricultural equipment in Rwanda, an agricultural equipment index is derived using polychoric PCA.

[18] Please see Larochelle *et al.* (2015) for a discussion of the assumptions made during estimation of this and other models presented in this chapter.

[19]The prices of the main agricultural crops, including beans, were recorded in the community questionnaire for periods of high and low availability.
[20] Note that the term 'profitability' used here does not necessarily require that the household sells its entire output. If the household chooses to consume the additional output, the increased value of this consumption affects household well-being.
[21] Bean seeds do not lose their potency as a result of being recycled, meaning there is no time limit to the number of seasons over which seeds can be recycled. This is because bean seeds are self-pollinated and do not commonly cross-pollinate.
[22] One can argue that the lower transaction costs of obtaining improved seeds can also apply to other agricultural inputs, such as fertilizer, impacting yield. However, any problems are avoided since agricultural inputs are included in the production function, controlling for their influence on yield; these will be correlated with the instrumented variable, but the correlation will not imply bias. Similar IVs were employed by Suri (2011).
[23] Using 240 Rwf = 1 International dollar at PPP in 2011.
[24] The poverty line is 118,000 Rwf (US$491.67 at PPP), which is about US$1.35 per person per day.
[25] In the report, 48.7% of the rural population is identified as poor.
[26] The most relevant IVs were identified: (i) whether the household reported drought within the last 5 years; (ii) presence of a distribution centre for agricultural inputs in the village; and (iii) village population.
[27] Using 960 UGX = 1 International dollar at PPP in 2011.
[28] Potential IVs are identical to those considered in estimation of the production functions. Similar tests were run. IVs for Rwanda were: (i) whether a village was affected by flood in the past 10 years; and (ii) village population, captured by the number of households within the village. For Uganda, they are: (i) whether a village was affected by drought in the past 10 years; (ii) village population, captured by the number of households within the village; and (iii) distance from village centre to a paved road in kilometres.

References

Alwang, J., Siegel, P.B. and Jorgensen, S.L. (2001) *Vulnerability: A View from Different Disciplines*. Social Protection Discussion Paper Series. The World Bank, Washington, DC.

Arimond, M., Wiesmann, D., Becquey, E., Carriquiry, A., Daniels, M.C. *et al.* (2010) Simple food group diversity indicators predict micronutrient adequacy of women's diets in 5 diverse, resource-poor settings. *The Journal of Nutrition* 140(11), 2059S-2069S. doi: 10.3945/jn. 110.123414.

Battese, G.E. (1997) A note on the estimation of Cobb-Douglas production functions when some explanatory variables have zero values. *Journal of Agricultural Economics* 48(2), 250–252.

Deaton, A. (1997) *The analysis of household surveys : a microeconometric approach to development policy*. Washington, D.C.: The World Bank. http://documents.worldbank.org/curated/en/1997/07/694690/analysis-household-surveys-microeconometric-approach-development-policy (Accessed 9 October 2015).

de Janvry, A., Dunstan, A. and Sadoulet, E. (2011) *Recent Advances in Impact Analysis Methods for Ex-post Impact Assessments of Agricultural Technology: Options for the CGIAR*. Report prepared for the workshop: Increasing the rigor of ex-post impact assessment of agricultural research. A discussion on estimating treatment effects, organized by the CGIAR Standing Panel on Impact Assessment (SPIA), 2 2010, Berkeley, California. Independent Science and Partnership Council Secretariat, Rome.

Food and Agricultural Organization of the United Nations (FAO). (2009) Food Balance Sheets. http://faostat3.fao.org (Accessed 12 May 2015).

Hoddinott, J. and Yohannes, Y. (2002) *Dietary Diversity as a Food Security Indicator*. Food Consumption and Nutrition Division. Retrieved from http://www.ifpri.org/sites/default/files/publications/fcndp136.pdf (Accessed 12 May 2015).

Johnson, N.L., Pachico, D. and Wortmann, C.S. (2003) The impact of CIAT's genetic improvement research on beans. In Evenson, R.E. and Gollin, D. (eds) *Crop Variety Improvement and Its Effect on Productivity: The Impact of International Agricultural Research*. CAB International, Wallingford, UK.

Kalyebara, R. and Buruchara, R. (2008) *Farm Level Impacts of Improved Bean Varieties and Agronomic Technologies in Rwanda*. Highlights: CIAT in Africa no. 41. http://ciat-library.ciat.cgiar.org:8080/jspui/handle/123456789/5398 (Accessed 12 May 2015).

Karangwa, P. (2007) *Plant Breeding and Related Biotechology Capacity, Rwanda*. Global Partnership Initiative for Plant Breeding Capacity Building, Kigali, Uganda.

Kennedy, G.L., Pedro, M.R., Seghieri, C., Nantal, G. and Brouwer, I. (2007) Dietary diversity score is a useful indicator of micronutrient intake in non-breast-feeding Filipino children. *The Journal of Nutrition* 137(2), 472–477.

Kennedy, G., Ballard, T., Dop, M.C. and Union, E. (2011) *Guidelines for Measuring Household and Individual Dietary Diversity*. FAO, Rome.

Larochelle, C., Alwang, J., Norton, G.W., Katungi, E. and Labarta, R.A. (2015) Ex-post impact of adopting improved bean varieties on poverty and food security in Rwanda and Uganda. ISPC Secretariat, Rome, in press.

National Institute of Statistics of Rwanda (NISR). (2008) *National Agricultural Survey Report*. http://statistics.gov.rw/publications/national-agricultural-survey-report-nas-2008 (Accessed 12 May 2015)

Ruel, M.T. (2003) Operationalizing dietary diversity: a review of measurement issues and research priorities. *The Journal of Nutrition* 133(11), 3911S–3926S.

Smith, L.C. and Subandoro, A. (2007) *Measuring Food Security Using Household Expenditure Surveys. Food Security in Practice*. IFPRI, Washington, DC.

Steyn, N., Nel, J., Nantel, G., Kennedy, G. and Labardarios, D. (2006) Food variety and dietary diversity scores in children: are they good indicators of dietary adequacy? *Public Health Nutrition* 9(5), 644–650.

Suri, T. (2011) Selection and comparative advantage in technology adoption. *Econometrica* 79(1), 159–209.

Swindale, A. and Bilinsky, P. (2006) *Household Dietary Diversity Score (HDDS) for Measurement of Household Food Access: Indicator Guide (v. 2).* Food and Nutrition Technical Assistance, USAID, Washington, DC. http://www.fantaproject.org/sites/default/files/resources/HDDS_v2_Sep06_0.pdf (Accessed 12 May 2015).

Zeng, D., Alwang, J., Norton, G.W., Shiferaw, B., Jaleta, M. and Yirga, C. (2013) *Ex-Post Impacts of Improved Maize Varieties on Poverty in Rural Ethiopia*. Diffusion and Impact of Improved Varieties in Africa (DIIVA) – Objective 3 Report. Virginia Tech, Blacksburg, VA and CIMMYT, Nairobi.

17 The Diffusion and Impact of Improved Food Crop Varieties in Sub-Saharan Africa

K. Fuglie* and J. Marder
Economic Research Service, US Department of Agriculture and Department of Agricultural and Resource Economics, University of California, Davis, USA

Introduction

The Diffusion and Impact of Improved Varieties in Africa (DIIVA) data on the adoption of improved crop varieties in sub-Saharan Africa (SSA)[1] tell a confounding story. On the one hand, there has been significant progress over the past decade in disseminating improved crop cultivars to farmers. By 2010, total area sown to improved varieties of food crops exceeded 37 million hectares (mha), more than double the estimated area in 2000 (Walker *et al.*, 2014). On the other hand, even this achievement represented only 35% of area planted to these crops in the countries included in the DIIVA surveys.[2] In most cases, the rate of diffusion of new crop varieties appears to have been quite slow. Moreover, the impact of crop variety adoption on agricultural productivity in SSA is not well documented. Because the speed of diffusion of new technology is likely to be correlated with its profitability, the slow pace of diffusion in SSA suggests that the productivity impact of improved crop varieties may be limited.

Our objectives in this chapter are twofold. First, drawing on the DIIVA data and supplementing these with information from other studies, we examine the pattern of diffusion on improved crop varieties in SSA. Second, we evaluate the impact of improved varieties on agricultural productivity. We use a panel of country-level observations to see how adoption may have affected growth in agricultural total factor productivity (TFP).[3] Agricultural TFP provides a national-level indicator of the rate of technical change in the farm sector. It captures improvements in crop and livestock yield net of change in the use of production inputs. While crop variety improvement represents only one type of technical change occurring in African agriculture, the fairly widespread use of improved varieties implies that they should contribute to TFP if they are improving productivity in any significant way. Our methodological approach provides an aggregate estimate of the combined productivity impact of improved varietal adoption across 21 food crops; it does not differentiate impacts for specific crops. The information on the productivity impact of aggregate variety diffusion provides evidence of the overall contribution of crop variety improvement to agricultural growth in the region.

Diffusion of Improved Crop Varieties

As Dalrymple (1986a,b) documented for the Green Revolution of the 1960s and 1970s, diffusion of improved crop varieties can be quite rapid when their impacts are large. He estimated

* E-mail: kfuglie@ers.usda.gov

that, in the 18 years between 1965 and 1983, high-yielding varieties of rice and wheat were adopted on more than 120 million hectares in Asia and Latin America (but only about 0.7 million hectares in SSA). Diffusion was especially rapid and thorough on irrigated cropland where yield impacts were relatively large. In rainfed and marginal production environments, diffusion was slower and spotty. After 1980, improved varieties were successfully extended to more regions and environments, and to more crops, including coarse grains, roots and tubers, oilseeds and pulses. Areas already at full adoption continued to reap productivity gains as new generations of improved varieties replaced earlier ones (Evenson and Gollin, 2003a).

On the basis of observations on national adoption for 1995–1998 and 2009–2010, the DIIVA data provide an opportunity to compare and contrast the pace of diffusion of improved varieties across crops and regions. A sigmoid-shaped diffusion curve like the logistic function provides a convenient framework for modelling the diffusion process (Griliches, 1957). Over a specified geographic area, say a country, let IV_{ct} be the proportion of the crop area planted to improved varieties of crop *c* in year *t*. The logistic diffusion curve is given by:

$$IV_{ct} = \frac{K}{1+e^{-(\alpha+\beta t)}} \quad (17.1)$$

where *K* is the adoption ceiling, β is the rate of diffusion (sometimes called the 'rate of acceptance') and α is the constant of integration related to the adoption rate in the early years of diffusion. The original motivation behind such a diffusion process was the assumption that adoption works like an epidemic – farmers adopt a new technology after they have come into contact with others who have successfully adopted it (Rogers, 2003). However, modern interpretations of the diffusion process acknowledge that simply being aware of a new technology is not sufficient for adoption to occur. Potential adopters must become convinced of its advantages and have access to the technology and complementary services that facilitate its profitable use. Furthermore, early adopters assume the risk of verifying its profitability and stability. They may expend resources to adapt local management practices to better suit the new technology. Other farmers may wait to benefit from this verification and adaptation. Who adopts first and how quickly others follow will be influenced by the institutional and market environment, farm size, heterogeneity in agroecological conditions of farms, farmers' education and health, the quality of agricultural extension and advisory services, availability of credit and inputs, marketing infrastructure, economic policies, prices, security of property rights, and the rule of law, among other factors (Feder *et al.*, 1985). Early adopters will tend to be those with greater access to information, services and markets, who are more able to assume risks and who possess farmland similar to the conditions under which the varieties were first selected and adapted.

To reflect the various stages of the diffusion process, we identify a few threshold points along the diffusion curve: when diffusion reaches 20% of the target population,[4] we call this the 'year of origin'. Before this point, we assume early adopters are experimenting with the technology and it may yet prove unsuccessful. Once it has reached this level of adoption, the technology has a high likelihood for rapid uptake by the majority of the target population. When diffusion reaches 80% we designate this as the 'year of saturation'. At this point we assume the technology is approaching its peak adoption and environmental or sociological constraints may significantly slow further spread of the technology.

Transforming the diffusion curve in Eqn 17.1 into linear form gives:

$$ln\left(\frac{IV_{tc}}{(K-IV_{tc})}\right) = \alpha + \beta t \quad (17.2)$$

Lacking information on the adoption ceiling (unobserved because adoption is not yet complete), we let *K* be 1.00 (or 100% of crop area). Then, so long as we observe IV_{tc} for at least two points in time (denoted t_1 and t_2), the slope parameter β is given by:

$$\hat{\beta} = \frac{ln\left(\frac{IV_{2c}}{(1-IV_{2c})}\right) - ln\left(\frac{IV_{1c}}{(1-IV_{1c})}\right)}{t_2 - t_1} \quad (17.3)$$

The constant of integration can be derived as:

$$\hat{\alpha} = ln\left(IV_{tc}/(1-IV_{tc})\right) - \hat{\beta}t \quad (17.4)$$

If more than two observations of IV_{tc} exist then α and β can be estimated by linear regression.

Selected evidence from around the world on the rate of diffusion of improved crop varieties in rainfed cropping systems suggests that the value of β typically ranges between 0.2 and 0.8 (Table 17.1). A low value of β, say 0.2, implies it takes about 14 years for diffusion in a population of farmers to increase from 20% to 80%, whereas a value of β = 0.8 implies this diffusion would only take 3.5 years. Observations of β outside of the 0.2–0.8 range would generally appear to be exceptions. Hybrid corn, an extremely successful innovation that was introduced in the USA in the 1930s, diffused very rapidly in the Corn Belt (with β as high as 0.95 in Iowa) but at rates more typical of improved crop varieties in the south-eastern USA (with β as low as 0.24 in Tennessee). Griliches (1957) found that the value of β was positively correlated with the average yield gain from hybrid corn obtained in those states, with higher rates of diffusion in regions where the new hybrids offered higher productivity advantages. It is interesting to note from the selective review in Table 17.1 that rates of crop variety diffusion in developing countries do not appear to be all that dissimilar to rates of diffusion of hybrid corn and wheat in the USA. A successful variety (or class of improved varieties) usually takes from 4 to 14 years to diffuse through a farm population. If a variety is taking longer than this to diffuse, it may be because it offers only slim yield advantages (or is only narrowly adapted to the agroecological conditions under which the crop is grown) or other institutional constraints are constraining adoption.

The rate of acceptance of improved varieties

The DIIVA data provide 97 crop-by-country observations on improved varietal adoption in SSA for 1995–1998, 151 observations for 2009–2010 and 61 observations for the same crop-by-country combinations in both periods. For these

Table 17.1. Typical rates of variety acceptance in rainfed agriculture.

Study	Country	Crop	Rate of acceptance (β)[a]	Years from 20% to 80% adoption
Dixon (1980)	USA, Corn Belt states	Hybrid corn, 1932–1960	0.75–0.95	3–4
	USA, south-eastern states	Hybrid corn, 1932–1960	0.24–0.70	4–12
Knudson (1991)	USA, Plain states	Semi-dwarf wheat, 1959–1984	0.67	4
	USA, eastern states	Semi-dwarf wheat, 1959–1984	0.18	15
Frisvold (2004)	USA	Bt cotton, 1996–2003	0.44	6
Byerlee and Hesse de Polanco (1986)	Mexico altiplano wet zone	Barley, 1960–1980	0.14	20
	Mexico altiplano dry zone	Barley, 1970–1980	0.26	11
Fuglie (1989)	Thailand, north-east region	Rice, rainfed paddy, 1975–1988	0.55	5
Bera and Kelley (1990)	Bangladesh	Rice, deepwater and rainfed paddy, 1970–1985	0.20	14
Jarvis (1981)	Uruguay	Improved pastures, 1960–1978	0.28	10
Jansen (1988)	India	Hybrid corn, 1956–1984	0.13	22
	India	Hybrid sorghum, 1956–1984	0.23	12
	India	Hybrid millet, 1956–1984	0.30	9

[a]The rate of acceptance is the β or slope parameter from the logistic diffusion curve (see Eqn 17.1 in the text).

61 observations, it is possible to estimate α and β of the logistic diffusion curve for these crops and countries. We supplement the DIIVA observations with information from other studies to generate a total of 73 cases in which national-level adoption is observed for a crop for at least two points in time. Estimates of the diffusion parameters for these cases and the sources of supplementary information on variety adoption are listed in Appendix Table 17.A1.

These estimates confirm that crop variety diffusion in SSA has been exceptionally slow. The average value of β from these 73 cases is only 0.115 (Table 17.2), considerably outside the usual range of 0.2–0.8. The slow rate of diffusion holds across most crops and countries. In fact, of the 73 country–crop estimates of the rate of acceptance of improved varieties, only 21 have values of β greater than 0.2, and only two greater than 0.5 (see Table 17.A1). For some crops, including rice and potatoes, diffusion stagnated or regressed in some major-producing countries.[5] Among crop types, the average diffusion rate for vegetatively propagated crops such as roots and tubers is somewhat slower than for cereal grains or legumes. Multiplication rates for the planting material of clonal crops like potato (seed tubers) and cassava (stem cuttings) are much lower than crops grown from seed and this may be a factor in the relatively slow diffusion of new varieties of clonal crops. The results also confirm that breeding programmes historically exhibited a preference for the major cereal staples. Improved varieties of legumes and root and tuber crops have been relative latecomers to SSA. The diffusion curve for 'all crops' in Table 17.2 gives an adoption rate of 45% in 2010, very close to the 48% that Walker *et al.* (2014) estimated from the DIIVA data using multiperiod observations on variety adoption.[6] It is important to recall, however, that these observations represent only a subset of the crop area in the region, and may be biased toward countries that have released more improved varieties over a longer period. The value

Table 17.2. Summary of crop variety diffusion estimates for sub-Saharan Africa.

Crop	No. of countries in estimation	Average estimate of parameters[a] α	β	Year of origin (20% adoption)	Year of saturation (80% adoption)	Years from origin to saturation
Cereal grains	40	–3.131	0.101	1997	2025	27
Barley	1	–4.386	0.128	2004	2025	22
Maize	19	–3.545	0.133	1996	2017	21
Pearl millet	1	–1.735	0.031	1991	2080	89
Rice	9	–1.646	0.050	1985	2041	56
Sorghum	4	–2.359	0.052	1999	2052	54
Teff	1	–6.778	0.206	2006	2020	13
Wheat	5	–3.405	0.138	1995	2015	20
Legumes and oilseeds	12	–5.826	0.189	2003	2018	15
Beans	6	–5.470	0.168	2004	2021	17
Cowpea	1	6.002	0.191	2004	2019	15
Groundnut	3	–6.161	0.219	2002	2014	13
Pigeonpea	1	–12.088	0.417	2006	2012	7
Soybean	1	–5.467	0.230	1998	2010	12
Clonally propagated crops	21	–4.194	0.120	2003	2027	23
Banana	1	–11.293	0.297	2013	2023	9
Cassava	15	–3.221	0.098	1999	2027	28
Potato	5	1.733	–0.082	na	na	na
All crops	73	–3.632	0.115	2000	2024	24

na = not applicable. The years to origin and saturation could not be estimated because of negative value of β. [a]For the full set of parameter estimates of crop variety diffusion curves, see Table 17.A1 in the Appendix. The averages reported here are weighted averages of the estimates in Table 17.A1, where the weights are area harvested of that crop in a country.

of this analysis is that it exploits the time series available in the DIIVA database to quantity the typical rate of diffusion once improved varieties have been successfully developed and released. In the next section, we assemble a more complete picture of variety diffusion among all food crops and countries in SSA.

Assembling a picture of the aggregate diffusion of all improved varieties

In order to examine how the diffusion of improved varieties affected agricultural TFP growth in SSA, it is necessary to have a more complete picture of aggregate diffusion over time. With such data, it is possible to examine whether countries that achieved greater aggregate diffusion also experienced greater growth in their agricultural TFP. Such evidence could enable us to quantify the economic impact of improved-variety adoption in the region, as long as we control for the influence of other factors that may affect both variables.

A starting point for this analysis is the historical interpolations derived by Evenson (2003, p. 450, Table 22.8). Based on findings from the 1995–1998 DIIVA adoption data, Evenson (2003) interpolated the rates of variety adoption by decade back to 1970 (Douglas Gollin kindly provided us with his adoption estimates by crop and by country). Although the assumptions Evenson used for deriving these estimates are not entirely clear, they seem to be based on linear interpolations of from when improved varieties first became available in a country to 1995–1998. It is also apparent he drew from Dalrymple's (1986a,b) estimates of adoption of modern rice and wheat high-yielding varieties in 1982–1983. Evenson's adoption estimates for 1980 correspond with Dalrymple's for all cases except rice in Madagascar.[7]

Our approach is to take any available data on initial and observed adoption (from DIIVA or other studies) and interpolate adoption estimates by crop and country for each 5-year period from 1971–1975 to 2006–2010. Besides the information from the DIIVA data for the late 1990s and late 2000s and Dalrymple's estimates for rice and wheat in the early 1980s, we add adoption estimates for various years for maize from Alene *et al.* (2009), for rice from WARDA (Africa Rice Center, 2008), for potato from Theile *et al.* (2008), for soybean from Sanginga *et al.* (1999), for teff from Minten *et al.* (2013) and for pigeonpea from Shiferaw *et al.* (2008). For the 71 crop–country combinations in Table 17.A1, we use these estimates of diffusion curves to interpolate adoption rates for other periods. For cases where DIIVA data were unavailable in 2009–2010 but were available from other sources for an earlier period, we simply extended the latest available adoption rate to the more recent periods. For wheat and potato, several countries in SSA were already at full adoption in 1995–1998 and were not included in the 2009–2010 round so we assumed that full adoption continued to 2010. For several cases involving sorghum, pearl millet, cassava and groundnut, however, the latest available adoption rates are from the 1995–1998 survey and were well below 100%. Because we also assume no further increase in the adoption rates after 2000 for these cases, it is likely our estimates of diffusion for these crops in later periods are low.

Two important food crops for which we could find no evidence of variety adoption in any country in SSA are sesame, an important oilseed in semi-arid regions, and cocoyam (taro), an important subsistence crop in the humid tropics.[8] For countries or crops for which we could find no data on variety adoption, we assume these rates are zero.

After constructing adoption estimates for each food crop in a country, we summed the adoption areas to generate the total area harvested in modern varieties and the aggregate adoption rate for each 5-year period between 1970 and 2010. Thus, our estimate of total improved-variety adoption for food crops in SSA is a composite of 23 food crops (21 of which had at least some adoption). In 2006–2010, these 23 crops were harvested on an average 147 million hectares per year in SSA, or about 80% of the total area of all crops in the region. The aggregate rate of improved-variety adoption for a country in a year is the total adopted area divided by total food crop area that year. Total food crop area is total area planted to the 23 food crops (including sesame and taro, even though no adoption of improved varieties for these crops was observed).

Although not a central focus of our study, for the sake of comparison we also reviewed evidence on the adoption of improved varieties of non-staple and plantation crops. Available estimates of variety adoption in smallholder non-staples include:

- Cocoa hybrids in Ghana for 2001 (Edwin and Masters, 2005);
- Improved tea clones in Kenya for 1999 (Nzuma, 2011); and
- Cotton variety diffusion in Senegal during 1980–1997 (Seck *et al.*, 1998).

In addition, Tschirley *et al.* (2009) report that in countries with cotton ginning monopolies, the ginning companies select varieties and provide seed to farmers, so that diffusion of replacement varieties is very rapid. A similar role exists in countries where sugar mills support breeding research (although sugar is generally not a smallholder crop in SSA). The industry-supported South African Sugarcane Research Institute has been particularly influential, with its cultivars being widely disseminated in southern and eastern African countries (Zhou, 2013). We could find no evidence on adoption of improved varieties or clones of rubber, oil palm, coffee, cashew or coconut, which are other important non-staple tree crops in the SSA region. Of these, only the area under cashew saw substantial growth during 1970–2010, whereas areas planted to rubber, oil palm, coffee and coconut remained stable or declined. Because these are all long-lived perennials, the likelihood of substantial new plantings of improved clones when area is stagnant is low.

Taking food and non-food crops together, we estimate that during 2006–2010, the area under improved varieties of food and plantation crops in the SSA region averaged 43.2 million hectares, or about 23.3% of the total cropland harvested in the region (Table 17.3). Considering only the 23 most significant food crops (cereal grains, root and tubers, legumes and oilseeds, and banana, in Table 17.3), there were about 40.5 million hectares under improved varieties, or about 27.5% of average annual area of these crops in 2006–2010.[9] This represented more than a doubling of the area under improved food crop varieties from a decade earlier (during 1996–2000), when the adoption area averaged 18.3 million hectares. Cereal grains represented by far the largest share of area under improved varieties in 2006–2010, at about 27.5 million hectares, for an aggregate adoption rate of nearly 32%. For other crop types, the areas under improved varieties in 2006–2010 were: 6.7 million hectares for legumes and oilseeds (21% adoption rate), 6.3 million hectares for clonally propagated crops (22% adoption rate), and 2.7 million hectares for non-food or plantation crops (12.4%).

The estimates of adoption in Table 17.3 are generally lower than those Evenson (2003, p. 450, Table 22.1) reported for the decades from 1970 to 2000. He estimated that by 2000, 27% of the total area planted in food crops was in improved varieties, whereas we find about 15% of food crop area was planted with improved varieties (or 18% considering just the nine crops in Evenson's total). Our estimates of adoption rates are similar to Evenson's for rice and cassava, higher for wheat and maize, but substantially lower for sorghum, pearl millet, beans, groundnuts and potatoes. One explanation is that Evenson's estimates are based only on countries included in the 1995–1998 DIIVA survey, whereas the new 2009–2010 DIIVA data have a much more complete coverage of improved variety adoption in the region. Essentially, Evenson assumed similar adoption rates for these crops in non-reporting countries, which the 2009–2010 DIIVA data showed was in many cases an over-estimate.

Another representation of these data is provided in Table 17.4, where adoption rates of improved varieties are given by country (i.e. aggregated across all crops grown in that country). In 2006–2010, Nigeria had by far the largest area under improved varieties of any SSA country, nearly 11.5 million hectares out of 42.6 million hectares in all crops (for a 27% adoption rate). Countries with the largest share of their cropland in improved varieties were in the southern part of the continent: Swaziland, Zimbabwe and Zambia had more than 50% adoption rates across all crops. Outside of southern Africa, Kenya and Senegal had the highest overall adoption rates at more than 40% of total cropland. Kenya achieved relatively early success with the uptake of improved varieties of wheat and maize as far back as the 1970s, whereas Senegal's success was more recent and mostly involved groundnut, a major export commodity for that country. One of the surprises in these data is the Democratic Republic of the Congo (DR Congo, formerly Zaire), which despite poor governance and significant armed conflict, was able to achieve impressive diffusion of new crop varieties, especially in cassava.

At the other end of the spectrum are several large agricultural countries with very low adoption rates: Niger, Cote D'Ivoire, Chad, Guinea and Angola all have at least 3 million hectares of cropland but less than 15% of that area under improved

Table 17.3. Adoption of improved crop varieties in sub-Saharan Africa.

	Total crop area				Share of crop area in improved varieties				Area in improved varieties			
	('000 ha)				(% of area)				('000 ha)			
Commodity	1976–1980	1986–1990	1996–2000	2006–2010	1976–1980	1986–1990	1996–2000	2006–2010	1976–1980	1986–1990	1996–2000	2006–2010
Cereals												
Sorghum	13,518	17,573	22,348	25,826	0.3	5.4	14.5	22.7	46	941	3,236	5,868
Maize	12,426	18,215	20,152	25,470	5.9	15.8	30.9	51.8	739	2,881	6,229	13,191
Pearl millet	12,268	14,938	19,507	21,207	0.3	1.3	4.4	13.5	33	194	866	2,873
Rice	4,112	5,291	6,802	8,365	3.2	12.7	37.7	35.6	132	671	2,567	2,974
Teff	1,397	1,307	2,128	2,625			6.5	35.2	0	0	138	924
Wheat	1,082	1,234	1,582	2,188	36.6	55.5	65.0	61.6	396	684	1,029	1,347
Barley	822	981	1,033	1,105			10.6	32.3	0	0	110	357
Root and tubers												
Cassava	6,983	8,164	10,623	12,021		6.6	17.3	35.9	0	542	1,840	4,316
Yam	1,275	1,826	3,575	4,585			1.3	29.7	0	0	46	1,362
Sweet potato	1,109	1,281	2,161	3,242			0.5	3.1	0	0	11	99
Potato	290	367	684	1,271	13.4	23.3	32.0	21.9	39	85	219	278
Cocoyam (taro)	709	759	1,089	1,259					0	0	0	0
Legumes and oilseeds												
Cowpea	3,394	4,765	8,840	10,537		2.7	8.8	25.7	0	127	775	2,710
Groundnut	6,251	5,521	8,417	10,106	1.2	4.3	9.3	19.8	77	238	782	2,000
Beans (*Phaseolus*)	2,560	3,059	3,979	5,667		1.9	7.4	20.8	0	58	296	1,181
Sesame	1,511	1,566	2,640	3,001					0	0	0	0
Soybean	311	632	765	1,000		2.1	15.3	50.9	0	13	117	509
Faba bean (*Vicia*)	306	281	402	563		0.5	2.8	13.7	0	1	11	77
Field pea	312	309	400	488				0.7	0	0	0	3
Pigeonpea	241	308	423	515			2.2	33.6	0	0	9	173
Chickpea	213	263	377	411				8.6	0	0	0	35
Lentil	55	49	75	104			0.9	9.5	0	0	1	10

Continued

Table 17.3. Continued.

Commodity	Total crop area ('000 ha)				Share of crop area in improved varieties (% of area)				Area in improved varieties ('000 ha)			
	1976–1980	1986–1990	1996–2000	2006–2010	1976–1980	1986–1990	1996–2000	2006–2010	1976–1980	1986–1990	1996–2000	2006–2010
Other food staples												
Bananas and plantains	3,793	4,235	5,033	5,682			0.5	4.0	0	0	23	227
Non-staple crops												
Cocoa	3,306	3,534	4,473	6,119		1.7	9.0	13.7	0	60	403	840
Oilpalm	2,992	3,133	4,009	4,426					0	0	0	0
Cotton	3,346	3,065	4,104	3,839		18.2	25.4	34.4	0	557	1,042	1,321
Coffee	3,145	3,218	2,620	2,231					0	0	0	0
Cashew	411	303	929	1,879					0	0	0	0
Sugarcane	579	785	801	1,093		24.6	25.5	36.9	0	193	204	404
Coconut	526	615	800	1,052					0	0	0	0
Rubber	291	371	545	718					0	0	0	0
Tea	165	170	211	275			44.3	45.6	0	0	94	125
All food staples	74,937	92,924	123,033	147,236	2.0	6.9	14.9	27.5	1,462	6,436	18,303	40,514
Non-staple crops	14,761	15,193	18,492	21,634	0.0	5.3	9.4	12.4	0	810	1,743	2,690
All crops	97,993	118,828	155,553	185,535	1.5	6.1	12.9	23.3	1,462	7,246	20,046	43,205

Sources: Total crop area from FAOSTAT. 'All crops' includes crops listed plus area planted to fruits and vegetables. Area in improved varieties from DIIVA data set and other studies (see text).

Table 17.4. Adoption of improved crop varieties in sub-Saharan Africa.

	Total crop area ('000 ha)				Share of crop area in improved varieties (% of area)				Area in improved varieties ('000 ha)			
Commodity	1976–1980	1986–1990	1996–2000	2006–2010	1976–1980	1986–1990	1996–2000	2006–2010	1976–1980	1986–1990	1996–2000	2006–2010
Nigeria	17,701	25,423	40,888	42,624	0.4	5.1	12.7	26.9	77	1,298	5,208	11,446
Benin	937	1,272	2,132	2,371	1.0	4.8	8.1	27.7	10	61	172	657
Côte d'Ivoire	4,323	5,701	6,197	7,260	1.4	3.3	4.7	13.4	61	187	293	973
Ghana	3,277	3,378	4,894	6,449	0.3	4.5	20.6	27.6	9	152	1,008	1,777
Guinea	1,349	1,572	2,436	3,226	0.0	2.1	9.9	14.1	0	33	242	456
Guinea-Bissau	271	244	377	465	0.0	0.0	0.0	0.0	0	0	0	0
Liberia	465	506	420	533	0.0	0.0	0.0	0.0	0	0	0	0
Sierra Leone	630	720	646	1,112	0.0	8.2	12.3	10.0	0	59	79	111
Togo	688	1,048	1,498	1,602	0.3	0.7	1.2	7.8	2	7	18	125
Other West Africa	11,940	14,441	18,599	23,019	0.7	3.5	9.8	17.8	83	499	1,813	4,100
Burkina Faso	2,629	3,711	4,039	5,557	0.1	5.8	10.8	19.1	4	214	436	1,060
Chad	1,817	1,673	2,831	3,641	0.0	3.2	9.7	10.9	0	54	275	398
The Gambia	163	200	230	411	0.0	0.0	1.3	6.2	0	0	3	25
Mali	1,940	2,717	3,325	4,840	0.5	14.2	35.8	36.3	10	386	1,192	1,756
Mauritania	179	260	373	385	0.0	0.0	0.0	0.0	0	0	0	0
Niger	4,728	7,380	10,973	15,234	0.0	1.4	6.0	13.7	0	101	654	2,090
Senegal	2,428	2,252	2,349	2,645	2.6	9.4	19.4	43.1	64	213	456	1,141
Sahel	13,909	18,274	24,192	32,775	0.6	5.3	12.5	19.7	78	967	3,016	6,471
Cameroon	3,334	3,084	3,636	4,797	1.3	5.2	12.5	19.1	43	159	453	916
Central African Republic	891	649	794	947	0.0	0.0	0.0	1.8	0	0	0	17
Congo	213	235	219	306	0.0	0.0	0.0	0.0	0	0	0	0
DR Congo	4,924	6,335	6,162	5,920	0.1	4.7	17.8	29.4	3	298	1,094	1,739
Gabon	146	190	195	228	0.0	1.0	3.5	3.5	0	2	7	8
Central Africa	9,670	10,662	11,160	12,335	0.5	4.3	13.9	21.7	46	459	1,554	2,679
Burundi	951	1,169	1,140	1,076	0.5	0.7	2.6	7.3	5	9	30	79
Kenya	3,687	4,057	4,156	4,781	4.5	14.2	32.6	37.2	164	576	1,356	1,780
Rwanda	989	1,284	1,316	1,790	1.5	3.2	6.9	7.1	15	41	91	127

Continued

Table 17.4. Continued.

	Total crop area				Share of crop area in improved varieties				Area in improved varieties			
	('000 ha)				(% of area)				('000 ha)			
Commodity	1976–1980	1986–1990	1996–2000	2006–2010	1976–1980	1986–1990	1996–2000	2006–2010	1976–1980	1986–1990	1996–2000	2006–2010
Tanzania	5,915	3,763	7,152	11,313	0.6	3.5	7.4	23.7	35	239	531	2,687
Uganda	4,884	4,549	5,984	7,388	0.5	1.9	5.6	22.3	23	89	336	1,648
East Africa	16,435	17,826	19,754	26,352	1.5	5.3	11.9	24.0	242	952	2,343	6,321
Ethiopia and Erteria	7,313	7,798	11,024	14,029	1.4	4.4	8.7	19.9	105	346	964	2,788
Somalia	739	976	735	788	0.0	0.0	0.0	0.0	0	0	0	0
Sudan	6,895	8,121	12,319	12,586	2.0	6.6	11.7	23.0	137	538	1,440	2,892
Horn of Africa	14,948	16,901	24,091	27,413	1.6	5.2	10.0	20.7	242	884	2,404	5,680
Angola	1,673	1,871	1,925	3,579	0.0	0.0	7.9	11.2	0	0	151	400
Botswana	205	248	205	128	0.0	9.8	14.3	9.6	0	24	29	12
Lesotho	220	263	217	223	8.9	22.0	44.5	45.4	20	58	97	101
Madagascar	2,344	2,600	2,785	3,053	2.8	9.6	21.9	16.0	65	248	609	487
Malawi	2,081	2,290	2,709	3,561	0.0	3.9	8.1	35.3	0	88	218	1,256
Mozambique	3,095	3,575	4,129	5,058	0.0	6.4	12.4	23.2	0	230	510	1,172
Namibia	218	244	364	349	6.2	19.3	34.7	32.2	14	47	127	112
Swaziland	140	184	173	148	18.1	56.1	56.4	65.9	25	103	97	98
Zambia	1,176	1,232	1,252	1,617	0.6	9.0	17.4	54.5	6	111	217	882
Zimbabwe	2,059	2,586	2,888	3,069	27.5	44.9	53.9	63.5	565	1,162	1,557	1,950
Southern Africa	13,391	15,301	16,869	21,018	5.2	14.3	22.0	31.2	696	2,186	3,712	6,558
Sub-Saharan Africa	97,993	118,828	155,553	185,535	1.5	6.1	12.9	23.3	1,462	7,247	20,051	43,255

Total crop area is the area harvested for all crops. Area under improved varieties includes food staples and non-food or plantation crops. Sources: Total crop area from FAOSTAT. Area in improved varieties from DIIVA data set and other studies (see text).

varieties. Several of these countries have recent histories of political instability and civil unrest that undermined public institutions and private commerce. If these constraints could be addressed, these countries offer significant opportunities for rapidly expanding the area sown with improved varieties.

Finally, there is a group of ten countries with very limited data on adoption of crop varieties (Botswana, Burundi, Central African Republic, Congo Republic, Gabon, The Gambia, Guinea-Bissau, Liberia, Mauritania and Somalia). Because we are assuming zero adoption rates in cases where data on diffusion are missing, we may be underestimating aggregate adoption rates for these countries. However, by and large these tend to be small agricultural producers with less than 1 million hectares of cropland. Altogether these countries have about 5 million hectares under crops with only about 140,000 hectares classified as being sown to improved varieties. Adoption may be higher in some of these countries than we have estimated but it will not significantly alter adoption statistics for the region as a whole.

Impact of Adoption of Improved Varieties on Agricultural Productivity

As a first step to assessing how the adoption of improved varieties affected productivity, we consider whether there is any correlation between observed growth of the share of a crop's area under improved varieties and the change in its average yield. Using FAOSTAT (2013) data for SSA as a whole, Fig. 17.1 shows that between 1980 and 2010 there was a higher yield gain for crops experiencing more growth in the adoption of improved varieties. On average, each percentage increase in the share of crop area under improved varieties was associated with a 0.71% increase in crop yield. The relation is somewhat higher for soybean, cowpea, lentil and chickpea, a bit lower for maize, yam, sorghum, beans and sweetpotato, but fairly consistent for wheat, cassava, rice, groundnut, pearl millet, potato and others. The intercept of the dotted line (0.13) indicates that there was a general upward trend in yield of about 13% during this period, independent from variety adoption.[10]

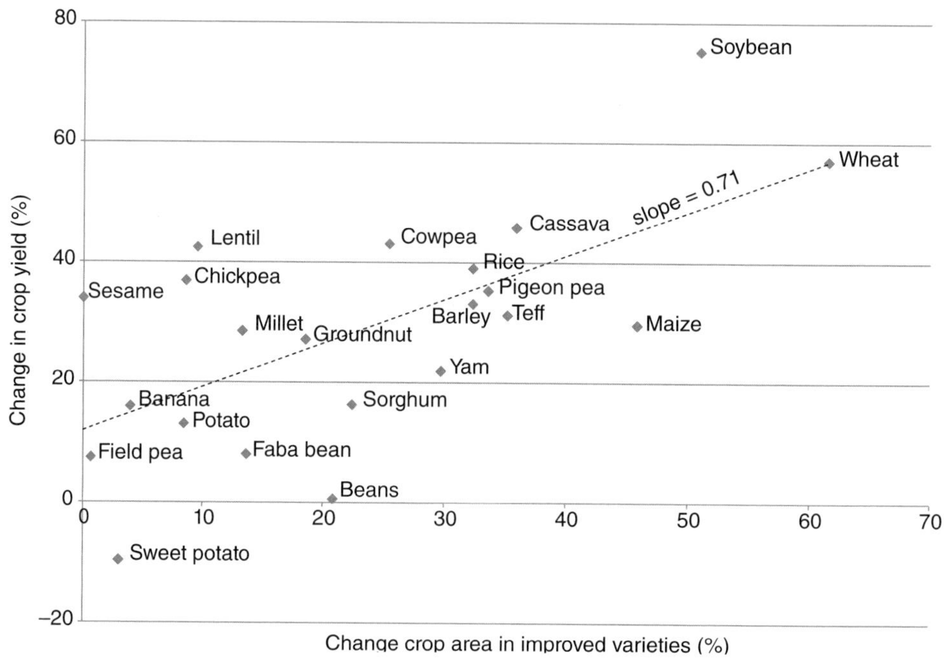

Fig. 17.1. Adoption of improved varieties and rate of yield improvement in sub-Saharan Africa between 1980 and 2010. Source: Crop yield change is the average crop yield in 2006–2010 relative to the average crop yield during 1976–1980 for the whole sub-Saharan Africa region, estimated from FAOSTAT data. The change in percentage area in improved varieties estimated by authors from DIIVA data and other sources (see text).

Conceptual limitations to using the information in Fig. 17.1 to draw inferences on productivity impact of improved varieties include (i) the attribution problem and (ii) the way productivity is measured. The attribution problem arises because other factors could be driving both higher yields and higher adoption. For example, countries that significantly increase their investments in agricultural research, extension and farmer schooling may achieve more rapid rates of adoption of a wide range of technologies, not just improved varieties, all of which could be contributing to higher yields. In econometric terms, the attribution problem refers to the endogeneity of both variety adoption and productivity growth: they both represent farmers' responses to external (exogenous) factors. Moreover, even if adoption leads to higher yield, it may be accompanied by greater use of other complementary inputs like fertilizer, irrigation, or labour, which increase costs. In principle, a welfare measure of adoption impact should subtract additional input costs to obtain the change in 'net yield', which is analogous to an improvement in TFP.

Our empirical strategies for dealing with these conceptual issues are to: (i) use a recursive econometric model (described below) to address the endogeneity problem to isolate the effect of variety adoption on productivity while controlling for the influence of other factors; and (ii) identify aggregate agricultural TFP growth as the metric of productivity change. By evaluating how adoption of improved varieties of food crops affected aggregate TFP, our model should capture the direct effects of adoption on crop yield net of changes in the use of other inputs in production. In addition, if the higher productivity from varietal adoption induces farmers to plant a greater share of their land to these crops or improves livestock productivity attributed to greater availability of animal feedstuffs, sector-level TFP growth will capture these effects.[11]

For TFP growth we use Fuglie (2011), who estimated indices of agricultural TFP for each SSA country from 1961 to 2008 (which was later extended to 2010 by Fuglie and Rada, 2013). These TFP indices show that before the mid-1980s there was essentially no agricultural TFP growth in the SSA region as a whole – in other words, the rate of output growth was proportional to the increases in land, labour and capital, and not from improvements in the productivity of these resources. From the mid-1980s onward, however, average agricultural TFP growth in SSA accelerated to about 1% per year, though with significant variation across countries. This suggests that around the time that improved varieties began to be widely disseminated, at least some countries in the region started experiencing improvements in agricultural productivity. Fuglie and Rada (2013) found significant correlations between growth in national agricultural R&D capital stock, dissemination of 'CGIAR-related technologies'[12] and economic reforms with the acceleration of TFP growth in African countries during this period. Our purpose here is to determine to what extent this growth in TFP can be specifically associated with the diffusion of improved food crop varieties.

Empirical model for determining impact

To determine the impact of improved varieties on agricultural productivity, we use a recursive system of equations to control for the endogeneity of variety adoption:

$$\begin{aligned}&(\mathrm{a})\quad IV = f(X_1)+\varepsilon_1\\&(\mathrm{b})\quad TFP = g(IV, X_2)+\varepsilon_2 \qquad (17.5)\end{aligned}$$

IV is the share of total cropland in a country where improved varieties of food crops have been adopted, *TFP* is an index of agricultural total factor productivity, X_1 and X_2 are sets of exogenous variables affecting adoption and productivity, respectively, and the error terms ε_1 and ε_2 include measurement error and omitted variables. The problem with estimating Eqn 17.5b directly is that there may be unobserved factors affecting both *IV* and *TFP* (which would cause *IV* to be correlated with the error term ε_2), and bias the parameter estimates. We address this problem using a two-stage estimation method. In the first stage we estimate Eqn 17.5a. We then use this estimation to create a predicted rate of technology adoption $\widehat{IV}$. In the second state we estimate TFP growth as a function of $\widehat{IV}$ and X_2. Because $\widehat{IV}$ should display less correlation with ε_2, the parameters of this estimation should no longer be biased. Thus, the coefficient on $\widehat{IV}$ will give an unbiased estimate of the productivity impact of *IV* adoption. The sets of exogenous explanatory variables X_1 and X_2 may contain some of the same variables, but to identify the model X_1 should contain at least one variable not found in X_2.

In equation Eqn 17.5a, the dependent variable IV_{ct} (where the subscript *c* represents country and *t* represents the time period), is the proportion of total cropland planted to the 23 major food crops with improved varieties of crops in period *t*. Its construction was described in the previous section of the paper. For the functional form of Eqn 17.5a, we use a dynamic specification of the logistic diffusion curve which allows the rate of acceptance (β) to vary among countries and over time.[13] In other words, Eqn 17.1 is written as:

$$IV_{ct} = \frac{K}{1+e^{-(\alpha+\beta t(X_{1c}))}} \qquad (17.6)$$

In its linear form (and setting K = 1) Eqn 17.6 becomes:

$$ln\left(\frac{IV_{ct}}{(1-IV_{ct})}\right) = \alpha + \beta_0 t + \beta_1(tX_{11ct}) + \beta_2(tX_{12ct}) + \cdots + \beta_j(tX_{1jct}) + \varepsilon_{1ct} \qquad (17.7)$$

where *j* is the number of variables in vector X_1 and an error term ε_{1ct} has been added to the right-hand-side. With estimates of the parameters in Eqn 17.7, the predicted adoption rate $\widehat{IV_t}$ can then be determined. The second stage of the model is specified as:

$$ln(TFP_{ct}) = \gamma_0 + \phi\left(\widehat{IV}_{ct}\right) + \gamma_1(X_{21ct}) + \gamma_2(X_{21ct}) + \cdots + \gamma_k(X_{2kct}) + u_c + \varepsilon_{2ct} \qquad (17.8)$$

where *k* is the number of variables in vector X_2 and ϕ, $\gamma_0, \ldots \gamma_k$ are parameters to be estimated. If the assumptions of the model are satisfied, ϕ provides an unbiased estimate of how a change in the share of cropland under improved varieties affects TFP. Since $\widehat{IV_t}$ varies in value from 0 to 1, ϕ indicates the percentage change in TFP if all cropland were planted to improved varieties. Equivalently, it indicates the average percentage change in net yield per hectare due to the adoption of an improved variety. The error terms ε_{1ct} and ε_{2ct} are assumed to be independently distributed, and a random error term u_c has been added to Eqn 17.8 to account for other country-specific effects not included in the model.

Equation 17.7 is estimated with ordinary least squares (OLS) and Eqn 17.8 is estimated using the random effects (RE) model. The RE model helps to control for unobserved differences across countries that remain constant over time. The fact that the dependent variable TFP_{tk} is set to a value of 100 for every country in the base year of the index should control for systematic differences among countries in factors such as agroecological conditions or cropland quality. Still, it is possible that these differences could systematically influence the growth in TFP. The RE model should control for these effects; however, it is vulnerable to omitted variable bias if any unobserved factors that influence productivity are correlated with any of the explanatory variables in the regression. In the Appendix to this chapter, we report estimates using the fixed effects (FE) method, which should be free of omitted variable bias. The FE estimates of Eqn 17.8 are similar to the RE estimates, and a Hausman test suggests that the RE model is the appropriate choice for these data. See the Appendix for more discussion of RE versus FE models.

The endogenous variables in the model are defined as:

- IV_{ct}: The proportion of total crop land planted with improved food crop varieties *t* years since improved food crop varieties first became available in country *c*. We generally assume 1970 as the 'start date' for introduction of improved varieties of food crops in SSA (i.e. t_0 = 1970). From the DIIVA data on crop variety releases and adoption, it is clear, however, that a number of countries had a relatively late start on crop variety improvement. We set t_0 = 1980 for Botswana, Central African Republic, Congo Republic, Gabon, The Gambia, Guinea, Mozambique, Malawi, Mauritania and Niger.
- TFP_{ct}: An index of agricultural TFP for country *c* in time *t* with a base year value of 100 in 1977 for each country (Fuglie and Rada, 2013). To filter out annual fluctuations in TFP owing to weather and other short-term shocks, the series was smoothed using the Hodrick–Prescott filter (setting λ = 6.25).

Exogenous variables included in the model are variables that influence the supply and demand for technology and the ability of farmers to evaluate and access new technologies and complementary inputs and services. Whereas many of the factors influencing adoption decisions are micro in nature (i.e. specific to the farm household), in this model we are confined to macro variables (specific to the country). We include:

- Agricultural research investment: The stock of knowledge from past spending on agricultural research. Fuglie and Rada (2013) developed national agricultural research stocks for each SSA country and the CGIAR research stock for the region as a whole using a 17-year research lag structure.
- Labour force schooling: The average years of schooling of an adult in the labour force (Barro and Lee, 2010).
- Economic policy distortions: The nominal rate of assistance (NRA) to agriculture or the per cent deviation of producer prices from prices that would have prevailed without policy interventions (Anderson and Masters, 2009, with updates from Anderson and Nelgen, 2013).
- Density of paved roads: Kilometres of paved roads per km^2 of a country's land area (International Road Federation, 2012).
- HIV-AIDS infection: The proportion of the national population infected with the HIV-AIDS virus (World Development Indicators, 2012).
- Armed conflict: The cumulative number of years since 1980 that the country has experienced significant armed conflict (Uppsala Conflict Data Program, 2012).

Note that we only include variables in which values change over time. The influence of fixed factors (like agroenvironmental heterogeneity within a country) are accounted for by setting the value of TFP to 100 for each country in the base year (1977) and by the RE error structure of the method.

The research stock variables represent 'technology capital' and are available for 30 countries in SSA for the period of our study. The other variables, on economic policy, marketing infrastructure, farmer education and health, and governance, represent the 'enabling environment' for the diffusion of new technology. Measures of these factors are available for these 30 countries over 1980–2010 except for the nominal rate of assistance to agriculture. NRA is available for only 17 of the 30 countries in our sample. To expand the data to the full sample of countries, we impute values of NRA for the 13 countries with missing values by using regional averages of NRA.[14] To test the sensitivity of the model to this assumption, we also estimate the model using only the 17-country sample.

The panel data consists of observations on 30 countries during 1977–2010 but we use only every 5th year of data beginning in 1980 (i.e. 1980, 1985, ... 2010). Some variables, such as farmer schooling and road density, are reported from their original sources in 5-year intervals. Using a 5-year interval between observations on improved varietal adoption also gives a better representation of what we actually know about crop variety diffusion in SSA because it is less dependent on interpolating adoption rates between survey observations. To account for farmer expectations regarding economic policy, we use a 4-year ARIMA forecast of NRA in the model (i.e. a weighted average of the past NRA values). With up to seven observations per country, the 30-country sample has potentially 210 observations and the 17-country sample 119 observations. However, owing to missing values on some variables, the panel is slightly unbalanced, with a total of 191 observations available for the 30-country and 111 for the 17-country samples.

The technology capital and enabling environment variables are included in both X_1 and X_2. In other words, we assume they affect both the rate of diffusion of improved varieties and agricultural productivity through other pathways. If they affect productivity primarily through the variety adoption pathway, their impacts will be embedded in the estimate of ϕ, the coefficient on $\widehat{IV}$. If they influence agricultural productivity through other pathways (i.e. by providing other types of technologies or improving conditions and incentives for farmers), then these impacts are captured in the estimates of $\gamma_1 \ldots \gamma_k$ and will not be attributed to variety adoption. In addition to these terms, we add to X_1 the cumulative number of improved crop varieties released in a country (available from the DIIVA data set) per 1000 hectares of total cropland harvested. The number of varieties released should be strongly correlated with adoption area but not with unobserved factors influencing national TFP growth (i.e. uncorrelated with ε_2). To further identify the model, we include higher-order terms of the exogenous variables in X_1.

Figure 17.2 depicts the countries included in the 30-country sample and 17-country sample, as well as seven other countries for which we have information on improved variety adoption and TFP growth (but not the explanatory variables). The larger sample covers about 90% of the improved variety adoption area in SSA, whereas the smaller sample covers about 80%. Countries not included in the econometric model are island states except Madagascar, very small countries with under one million in population (Guinea-Bissau, Djibouti and Namibia) and some larger countries with poor quality or incomplete data (Angola, Chad, DR Congo, Liberia, Sierra Leone and Somalia). The most significant omitted country is the DR Congo, which in 2006–2010 accounted for about 3% of the region's cropland and 4% of adoption area of improved varieties. Whereas the 30-country sample provides a more complete and representative picture of agricultural change in SSA, the 17-country sample allows us to test whether imputation procedures for missing data may bias the estimate of the impact of improved varieties on productivity.

Findings: The Average Impact of Crop Variety Adoption on Total Factor Productivity

Although our main interest is the impact of food crop variety diffusion on agricultural productivity, we first comment briefly on the results of our stage-one regression on the factors affecting the rate of diffusion of improved crop varieties (Eqn 17.7 above). Appendix Table 17.A2 reports the estimates from this regression. The excluded instrument in our model (varieties released per hectare of cropland) appears to be a strong instrument for variety adoption. The F-statistic in both the 30-country and 17-country samples is greater than 10, which Stager and Stock (1997) suggest is a valid screen against weak instruments. In addition to the number of crop varieties released, the stock of agricultural research is also positively associated with more rapid rates of diffusion. Obviously these are closely related factors. But greater research investment, controlling for the number of varieties released, could indicate higher quality or greater adaptability of the released varieties that have enabled them to spread more quickly and more widely in a country. Overall, the model appears to capture much of the cross-country variation in the rate of improved variety adoption of food crops, with an R-squared of 58% (and an R-squared of 68% in the 17-country sample). More discussion of these results, including what the estimated coefficients imply about the rate of improved variety acceptance (value of β in the diffusion curve) across countries and over time, can be found in the Appendix.

Our findings on the impact of food crop variety diffusion and other factors on agricultural productivity growth in sub-Saharan Africa are given in Table 17.5. The second column reports a simple regression of the predicted share of cropland under improved varieties and TFP growth, excluding other factors. The coefficient on improved variety adoption from this regression is 0.74, which is close to the correlation observed between the regional improved variety adoption rate and crop yield growth of 0.71 shown in Fig. 17.1 (even though one is computed from observations on national cropland and the other from regional crop-specific data). The upward bias of this estimate is evident from the regression results, which control for the influence of sources of technology, farmer education and health, economic policies, market infrastructure and governance. When these variables are included in the model the estimated impact of improved variety adoption on TFP growth is 0.47 (Table 17.5, columns 3 and 4) (both the large country sample and the small country sample give the same result). The difference between this estimate of 0.47 and the simple correlation of 0.74 indicates the size of the bias from not controlling for the effects on productivity of the other variables in the model.

The significance of CGIAR agricultural research stock in the TFP impact model suggests that CGIAR research is raising agricultural productivity through means other than just supplying improved varieties. One such example is the highly effective biological control programme for cassava insect pests (Zeddies *et al.*, 2001). Maredia and Raitzer (2006) estimated that, prior to the year 2000, these biological control programmes accounted for as much as 80% of the economic impact of the CGIAR in SSA. National agricultural research is influencing agricultural TFP through

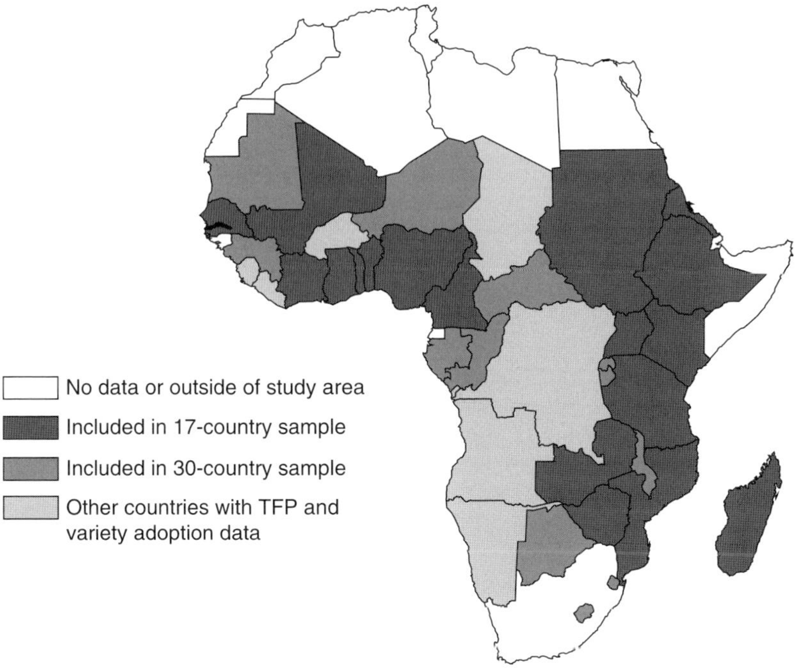

Countries included in small sample model (c = 17)		Additional countries for large sample model (c = 30)		Other countries with TFP and improved variety adoption information (c = 37)		Countries not included in analysis	
1	Benin	18	Botswana	31	Angola	38	Cape Verde
2	Cameroon	19	Burundi	32	Burkina Faso	39	Comoros
3	Côte d'Ivoire	20	Cent Afr Rep	33	Chad	40	Djibouti
4	Ethiopia	21	Congo Rep	34	DR Congo	41	Eq Guinea
5	Ghana	22	Gabon	35	Liberia	42	Eritrea*
6	Kenya	23	The Gambia	36	Namibia	43	Guinea-Bissau
7	Madagascar	24	Guinea	37	Sierra Leone	44	Mauritius
8	Mali	25	Lesotho			45	Reunion
9	Mozambique	26	Malawi			46	Sao Tome & Principe
10	Nigeria	27	Mauritania			47	Seychelles
11	Senegal	28	Niger			48	Somalia
12	Sudan	29	Rwanda			49	South Africa
13	Tanzania	30	Swaziland			50	South Sudan*
14	Togo						
15	Uganda						
16	Zambia						
17	Zimbabwe						

* Information on Eritrea included with Ethiopia, and information on South Sudan included with Sudan.

Fig. 17.2. Area of study and country-samples for econometric models.

Table 17.5. Impact of adoption of improved food crop varieties and other factors on agricultural total factor productivity (TFP), random effects (RE) model.

Dep. variable = TFP Index (1977 = 100 for each country) Explanatory variables	Large sample of countries		Large sample of countries		Small sample of countries	
Improved variety adoption (per cent of crop area harvested)	0.736	***	0.469	***	0.473	***
	(6.810)		(3.009)		(2.628)	
CGIAR agricultural research (log of R&D stock)			0.109	***	0.146	***
			(4.262)		(4.588)	
National agricultural research (log of R&D stock)			0.0205		0.0233	
			(1.224)		(0.902)	
Combined agricultural research (log of R&D stock)						
Labor force schooling (years)			0.0218	**	0.0138	
			(2.013)		(0.779)	
Nominal rate of assistance to agriculture (per cent change in farm prices due to policies)			0.0856		0.180	
			(0.770)		(1.538)	
Density of paved roads (log of km paved roads/km^2 country area)			−0.0155		−0.00292	
			(−1.014)		(−0.111)	
HIV-AIDS infection (per cent of adult population)			−0.812	***	−1.107	***
			(−2.991)		(−2.783)	
Armed conflict (cumulative years of armed conflict)			−0.00457	*	−0.00592	**
			(−1.786)		(−2.064)	
Constant	4.626	***	3.995	***	3.944	***
	(202.20)		(28.43)		(19.81)	
Observations:	212		191		111	
Number of countries	34		30		17	
R-squared: Within	0.2047		0.3575		0.5140	
Between	0.0645		0.2822		0.3970	
Overall	0.1206		0.3209		0.4671	
σ_u (std dev within-country error)	0.1074		0.1053		0.1090	
σ_e (std dev between country error)	0.1308		0.1224		0.1166	
rho (fraction of variance due to u_i)	0.4027		0.4256		0.4664	
Wald $\chi 2$	46.33	***	97.14	***	103.45	***
Prob > $\chi 2$	0.0000		0.0000		0.0000	

T-statistics in parentheses: ***, **, and * indicate 1%, 5% and 10% significance levels, respectively. Note: Estimates of NRA are available for only some SSA countries, and including this variable limits the sample to 17 countries (the small sample estimates). To estimate the model with the large sample of 30 countries, average regional values of NRA were used for missing observations on NRA. See text for further details.

its contribution to the development and dissemination of improved varieties but the results in Table 17.5 show only weak evidence of impact through other pathways. This result only applies to national agricultural research systems in SSA on average; it is certainly possible that some national systems are having significant impacts on agricultural productivity through technologies other than improved varieties. Better farmer education and health are also associated with higher TFP growth, whereas armed conflict has suppressed it.

The estimate in Table 17.5 on improved variety adoption implies that if the share of cropland planted with improved food crop varieties increased from 0 to 100%, then TFP would increase by 47%. This estimate (from the large sample of countries) has a 95% confidence interval of 0.164 to 0.775. An increase in TFP is equivalent to an increase in net production (growth in output minus any growth in inputs, including land area, which may accompany variety adoption). This can be interpreted as the average increase in net yield attributable to adoption of an improved food crop variety. It is an average effect for the region as a whole across all 21 food crops reporting some improved variety adoption.

This estimated impact of variety adoption on net crop yield is generally consistent with what has been reported in other, crop-specific studies. In Table 17.6 we show average yield impacts of improved varieties in SSA and India from the case studies in Evenson and Gollin (2003b), summarized in Evenson (2003, p. 455, Table 22.3). Also included in Table 17.6 are estimates from other more recent studies that report improved variety impacts for cowpea, potato and banana in SSA. From this literature, the average yield impacts of improved varieties in SSA ranged from 22% for banana to 66% for cowpea, with an average across commodities of 41%. Our estimate of 47% TFP improvement is close to this average.

With this information on variety adoption and impact, we can draw some inferences about the aggregate economic impact from adoption of improved varieties of food crops in SSA. For this exercise we use average crop yields from the period 1976–1980 as the 'base yield' and then take 47% of this yield to be the average yield growth on area harvested from improved varieties of that crop in 2010. Multiplying this average yield growth by the change in improved

Table 17.6. Estimates of the productivity impact of improved crop varieties.

Crop	% yield or TFP impact from adoption: Sub-Saharan Africa	India	Source
Maize	45	65	Evenson (2003, p. 455, Table 22.3)
Pearl millet	38	45	Evenson (2003, p. 455, Table 22.3)
Rice	24	33–65	Evenson (2003, p. 455, Table 22.3)
Beans	55	–	Evenson (2003, p. 455, Table 22.3)
Cassava	48	–	Evenson (2003, p. 455, Table 22.3)
Sorghum	34	37–40	Deb and Bantilan (2003)[a]
Cowpea	66	–	Kristjanson *et al.* (2002)[b]
Potato	40	–	Kaguongo *et al.* (2008), Rueda *et al.* (1996)[c]
Banana	22	–	Edmeades *et al.* (2007)[d]
Average across the nine studies listed above	41	–	See above
Average across 20 food crops in DIIVA survey plus teff	38	–	This study

[a]Deb and Bantilan (2003, p. 205, Table 9.7) report estimates of yield impacts of improved sorghum varieties for various regions of Nigeria, Cameroon and Chad. These estimates range from 7% to 63%, with weighted average (weighted by sorghum area planted in these regions) of 34%. [b]Kristjanson *et al.* (2002, p. 29, Table 7) used a crop simulation model to derive estimates of yield impact of improved cowpea varieties in the dry savannah regions of West Africa. They estimated an average 66% increase in grain yield and a 23% reduction in fodder yield from improved varieties. [c]From farm survey data from Kenya and Uganda, Kaguongo *et al.* (2008) estimated yield differences of 40% between farms using improved and traditional potato varieties. While survey estimates may be subject to sample self-selection bias, Rueda *et al.* (1996) also found average yield gains of 40% using data from 75 on-farm potato variety trials in the East African highlands of Rwanda, eastern Uganda, Burundi and western DR Congo. [d]Using farm survey data from Tanzania, Edmeades *et al.* (2007, p. 138, Table 9.4) found that households that adopted banana hybrids had 22% higher average yields than non-adopters; however, some of these differences could be due to sample self-selection.

variety adoption area between 1980 and 2010 gives an estimate of the production increase due to improved variety adoption over these three decades for each of the 21 food crops with some improved variety adoption. We then value the production increase for each food crop using constant 2005 prices and sum up these values to get the total value of improved productivity from all improved variety adoption in SSA. We use FAOSTAT world average farm-gate prices for valuing output.

In volume terms, the increase in improved variety adoption area from 1.45 million hectares to 40.46 million hectares between 1980 and 2010 produced an increase in crop output totalling 36 million tonnes (fresh weight) or 21 million tonnes per year in grain-equivalent weight (grain-equivalent weight discounts for the water content of roots, tubers and bananas).[15] In value terms, by 2010 improved variety adoption in SSA added US$6.16 billion to the annual value of agricultural output in the region. This amounts to 15.3% of the total growth in food crop production and 7.6% of the growth in total agricultural output in SSA between 1980 and 2010.

On cropland harvested using improved varieties, grain-equivalent yield rose on average by 0.55 tonnes per hectare, raising the gross value of yield by US$156 per hectare. If farmers were capturing the entire value of these yield improvements, a gain of this size would seem to imply fairly rapid adoption. The slow rate of improved variety uptake suggests that many farmers in SSA are not able to realize gains of this magnitude or face constraints to adoption such as lack of access to improved variety seed or markets for surplus production. If varieties are only narrowly adapted to the agroecological conditions under which these crops are grown, then these yield benefits could fall off significantly with adoption area. Therefore, even though average yield gains are significant the marginal yield gains from additional adoption may not be. Policies that suppressed trade or lowered agricultural prices (to the benefit of consumers) for these commodities would also have reduced incentives for farmers to adopt improved varieties. The profitability of adoption to farmers and the welfare distribution of the social gains from technology adoption among producers and consumers are likely to be location-specific, depending on market, policy and institutional conditions.

Other results from the regression analysis are generally consistent with Fuglie and Rada (2013), and the reader is referred to that paper for discussion of the quantitative effects of how R&D investment and the enabling environment have affected TFP growth in SSA agriculture. These variables affect TFP growth by influencing the rate of improved variety adoption (which then affects TFP) or by enabling improvements to other farming practices or technologies. Our estimate of the average impact of improved variety adoption of 0.47 is at the lower end of the 0.46 to 0.82 range that Fuglie and Rada (2013) estimated as the average impact of 'CGIAR-related technologies' (which included impacts of biological control as well as improved varieties that CGIAR Centers helped to develop).

Summary and Conclusions

Since improved varieties first made their appearance in much of SSA in the 1970s, it took about three decades to extend them to 20 million hectares, but only one more decade to double that to over 40 million hectares. This was achieved by deepening the pool of improved varieties available to farmers, both in terms of their adaptability to more environments but especially to a wider set of crops beyond major cereal grains, including oilseeds, legumes, roots, tubers and bananas. These genetic improvements to food crops appear to have raised aggregate food crop output in SSA by more than 15%. At FAOSTAT global average prices, this amounted to over US$6 billion per year by 2010. If present adoption rates and per hectare impacts continue, the added-value from improved varieties could approach US$12 billion per year by 2020. As these first-generation improved varieties approach full adoption, the challenge will increasingly turn to introducing replacement-generation improved varieties with new traits that can sustain growth in productivity.

Despite this improving picture, diffusion rates in SSA for improved varieties remain significantly below historical experiences of rainfed farming systems in other parts of the world. Whereas our analysis suggests countries that have invested more in agricultural research and released more new varieties have achieved more rapid rates of diffusion, our understanding of the technology diffusion process in SSA agriculture is far from complete. Much could be learned from micro-level studies of the institutional, environmental and economic factors that influence decisions by farm households to adopt (or not) improved varieties and other agricultural innovations. At the local level, the role of agricultural extension, seed and fertilizer supply systems, farm credit, crop insurance, farm size, land tenure rights, gender, crop marketing systems, price policies, the heterogeneity of agroecologies and other factors are likely to be important determinants conditioning acceptance and adoption by farmers of new agricultural technologies. More micro-level studies on farm adoption could provide helpful insights for national efforts to increase the pace of technical change in African agriculture.

Our estimates suggest that the impact of improved variety adoption on farm productivity in SSA has been significant, raising average net crop yield on adopting areas by around 0.55 tonnes per hectare, or by 47%, from 1976–1980 average levels. If farmers received global average farm prices for their crops, this would translate into gains of about US$156 per hectare, likely to be enough to generate widespread interest in, if not rapid adoption of, the new improved varieties. The fact that improved variety diffusion has been so slow suggests that institutional and/or environmental barriers may be constraining adoption. Three possible reasons are: (i) farmers receive significantly less than global average prices for their surplus production; (ii) there may be real constraints in accessing sufficient quantities of quality improved variety seed and other complementary inputs and services; and (iii) due to ecological factors, these average impacts may diminish as adoption area increases. New research emphasizing micro-level studies of farm household adoption behaviour are needed to better understand the roles of markets, policies and prices in incentivizing or constraining adoption.

Notes

[1] For the purposes of this study, we define 'sub-Saharan Africa' to include the 50 nations south of the Sahara. We combine information on Ethiopia and Eritrea to create a single consistent geographic region. 'Sudan' is defined as Sudan as it existed until 2011, when it separated into Sudan and South Sudan. The Republic of South Africa is excluded from the statistical analysis because of its reliance on large, modern commercial farms.

[2] In this chapter, we refer to the DIIVA surveys or DIIVA data as the combined information on variety releases and adoption area in SSA countries reported in the volume by Evenson and Gollin (2003b) and updated information collected as part of the Diffusion and Impact of Improved Varieties in Africa Project. Walker *et al.* (2014) have combined this baseline and updated surveys into one DIIVA database (Chapters 3 and 4, this volume). Consistent with Walker *et al.* (2014), we consider a variety to be improved if it was released after 1970.

[3] Fuglie (2011) developed indexes of agricultural TFP for each SSA country since 1961. These were updated in Fuglie and Rada (2013).

[4] By 'target population' we refer to the population of potential adopters. In practice, what we typically observe is not the share of farmers who have adopted but the share of total crop area in a country where improved varieties are grown.

[5] Here, diffusion 'stagnation or regression' is relative to the proportion of a crop's total area sown with improved varieties. In much of SSA total crop area has been rising, so it is possible that area grown with new varieties could be rising even as its proportion falls.

[6] These adoption rates refer only to crops and countries where estimates of adoption exist for at least two periods: 61 cases in Walker *et al.* (2014, Table 4.5) and for our estimate, the 73 cases listed in Table 17.A1.

[7] Dalrymple did not report an estimate for adoption of rice improved varieties in Madagascar, while Evenson assumed a 5% adoption rate (180,000 hectares) for 1980. While we have not found direct evidence of improved rice variety adoption in Madagascar other than the DIIVA 2009–2010 survey (which showed about 450,000 hectares, or 35% of total rice area, in improved varieties that year), the DIIVA survey indicated that pre-1980 releases were still sown on about 90,000 hectares in 2009–2010. In fact, the second-most widely grown rice variety in Madagascar in 2009–2010 was released in 1970. Allowing for variety replacement, it seems highly plausible that improved varieties of rice had been adopted in Madagascar by the early 1980s.

[8] We include cocoyam and sesame (also known as beniseed) in our estimate of total area planted to food crops in a country. Both cocoyam and sesame are planted to substantially over 1 million hectares annually in SSA, according to FAO.

[9] Our estimates of crop and areas under improved varieties differ somewhat from Walker *et al.* (2014) because of differences in coverage. We include improved variety adoption and crop area of 23 major food crops in all countries of SSA except South Africa, whereas Walker *et al.* (2014) include 20 crops and only the areas planted to these crops in the countries covered in the 2009–2010 DIIVA survey. Also, we estimate total crop area using a 5-year average over 2006–2010 from FAOSTAT data. Walker *et al.* (2014) use 2010 FAOSTAT data for crop area. Thus, whereas Walker *et al.* report 37.5 million hectares planted to improved varieties from a total of 108 million hectares in 2010, we estimate 40.5 million hectares of adoption over 147 million hectares on average over 2006–2010.

[10] Not included in Fig. 17.1 is cocoyam, which recorded a phenomenal yield gain of 128% over these 30 years, though we have found no record of an improved variety (or other technical improvement, for that matter) being developed or adopted for this crop in Africa. But apparent yield gains are only as real as the quality of the data. Measuring production and yield of crops that are largely grown for home consumption, harvested piecemeal, and often intercropped, are exceptionally difficult, especially for statistical agencies with limited capacities and resources. The questionable quality of the yield data affects all crops in SSA but is probably most serious for roots, tubers and beans.

[11] McKinsey and Evenson (2003) modelled separately the ‘direct’ impact of variety adoption on crop yield and the ‘indirect’ effects on area planted and investment in irrigation. By focusing on the relationship between improved-variety adoption and sector-wide agricultural TFP, we consider these impacts together.

[12] CGIAR-related technologies in the Fuglie and Rada (2013) study included improved crop varieties, biological control of crop pests, and improved methods of natural resource management where there were significant contributions from CGIAR Centers. In this chapter, we consider improved varieties from all sources (Walker *et al.*, 2014, note that only about two-thirds of the region’s crop area under modern cultivars is planted to CGIAR-related varieties) but exclude other types of improved technology.

[13] Frisvold (2004) describes the general specification of a dynamic logistic function as $IV_t = \frac{K(W)}{1+e^{-(\alpha(Z)+\beta t(X))}}$, where the adoption ceiling, origin and rate of acceptance are functions of the exogenous variables *W, Z* and *X*, respectively. In our application, we simplify this model by assuming $K = 1.00$ (or that the potential adoption area is 100% of crop area) and hold α constant across countries. The assumption of $K = 1.00$ is probably the best choice for examining aggregate trends over the long term. Even if currently available improved varieties are not well adapted to all growing conditions or crops in a country, it is probably only a matter of time before they are developed for these conditions and crops. In Griliches (1957) study of hybrid corn adoption in the USA, he estimated values of $K \leq 1$ for each state based on data available at the time of his study, but in just a few years after his study adoption reached virtually 100% in all states (Dixon, 1980). Similarly, Jansen (1988) derived estimates of $K \leq 1$ for improved variety adoption in Indian districts, only to find that in subsequent years adoption surpassed these ceilings (Walker, 2014, personal communication). The assumption that α is constant could be a limitation on the goodness of fit of the model, but allowing the year when improved varieties were first introduced ($t = 0$) to vary by country allows diffusion curves to vary during early years of the diffusion process. Examples of the use of dynamic diffusion curves to model agricultural technology adoption can be found in Jarvis (1981), Knudson (1991) and Frisvold (2004).

[14] For Lesotho and Swaziland, we use the NRA values for the Republic of South Africa. Lesotho and Swaziland have been in a customs union with South Africa and their currencies are pegged to the Rand, so it is reasonable to assume that farmers in these countries face similar prices as South African farmers. For Niger, we use the average NRA of Mali, Burkina Faso, Togo and Benin, which all share the same currency with Niger. Moreover, the NRA in all four of these countries has tended to be near zero over 1970–2010. For Gabon, Congo Republic and the Central African Republic, we use the average NRA from Chad and Cameroon, which also share the same currency and have historically similar NRAs. For the other seven countries with missing NRA values, we use the regional average NRA for Africa excluding Nigeria, Republic of South Africa and Egypt (Anderson and Neglen, 2013).

[15] To adjust for the high water content of roots, tubers and bananas, a ‘grain equivalent’ weight for these crops is found by multiplying their fresh weight production by 30%.

[16] Adoption rates of improved cultivars in maize in East and Southern Africa for 1990 reported in Byerlee and Heisey (1996) and in the early 1980s by Timothy *et al.* (1988) were not used: these were in most cases considerable higher than the DIIVA estimates for 1995–1998 and appear to include pre-1970 releases (and therefore are not consistent with the notion of an ‘improved variety’ we have adopted here).

References

Africa Rice Center (2008) *NERICA Adoption and Impact: Summary of Findings from Four Countries*. Research and Development Brief August 2008. Africa Rice Center (WARDA), Cotonou, Benin.

Alene, A., Menkir, A., Ajala, S., Badu-Apraku, B., Olanrewaju, A. *et al.* (2009) The economic and poverty impacts of maize research in West and Central Africa. *Agricultural Economics* 40, 535–550.

Anderson, K. and Masters, W. (eds) (2009) *Distortions to Agricultural Incentives in Africa*. World Bank, Washington, DC.

Anderson, K. and Nelgen, S. (2013) *Updated National and Global Estimates of Distortions to Agricultural Incentives, 1955 to 2011*. World Bank, Washington, DC. Available at: www.worldbank.org/agdistortions (Accessed 13 May 2015).

Barro, R. and Lee, J. (2010) *A New Data Set of Educational Attainment in the World, 1950 – 2010*. Working Paper No. 15902, April, National Bureau of Economic Research (NBER), Cambridge, Massachusetts.

Bera, A. and Kelley, T. (1990) Adoption of high yielding rice varieties in Bangladesh: an econometric analysis. *Journal of Development Economics* 33(2), 263–285.

Byerlee, D. and Heisey, P. (1996) Past and potential impacts of maize research in sub-Saharan Africa: 'a critical assessment' *Food Policy* 21, 255–277.

Byerlee, D. and Hesse de Polanco, E. (1986) Farmers' stepwise adoption of technological packages: evidence from the Mexican Altiplano. *American Journal of Agricultural Economics* 68(3), 519–527.

Dalrymple, D. (1986a) *Development and Spread of High-Yielding Rice Varieties in Developing Countries*. USAID, Washington, DC.

Dalrymple, D. (1986b) *Development and Spread of High-Yielding Wheat Varieties in Developing Countries*. USAID, Washington, DC.

Deb, U. and Bantilan, C. (2003) Impacts of genetic improvement in sorghum. In: Evenson, R. and Gollin, D. (eds) *Crop Variety Improvement and Its Effects on Productivity: The impact of International Agricultural Research*. CAB International, Wallingford, UK, pp. 183–214.

Dixon, R. (1980) Hybrid corn revisited. *Econometrica* 48, 1451–1461.

Edmeades, S., Nkuba, J. and Smale, M. (2007) Use of hybrid cultivars in Kagera Region, Tanzania, and their impact. In: Smale, M. and Tushemereirwe, W. (eds) *An Economic Assessment of Banana Genetic Improvement and Innovation in the Lake Victoria Region of Uganda and Tanzania.* Research Report 155, International Food Policy Research Institute, Washington, DC, pp. 129–140.

Edwin, J. and Masters, W. (2005) Genetic improvement and cocoa yields in Ghana. *Experimental Agriculture* 41, 491–503.

Eicher, C. (1995) Zimbabwe's maize-based Green Revolution: preconditions for replication. *Food Policy* 23, 805–818.

Evenson, R. (2003) Production impacts of crop genetic improvement. In: Evenson, R. and Gollin, D. (eds) *Crop Variety Improvement and Its Effects on Productivity: The impact of International Agricultural Research*. CAB International, Wallingford, UK, pp. 447–471.

Evenson, R. and Gollin, D. (2003a) Assessing the impact of the Green Revolution, 1960 to 2000. *Science* 300, 758. doi: 10.1126/science.1078710.

Evenson, R. and Gollin, D. (eds) (2003b) *Crop Variety Improvement and Its Effects on Productivity: The impact of International Agricultural Research*. CAB International, Wallingford, UK.

FAOSTAT (Date of retrieval: 10/01/2013) Food and Agriculture Organization of the United Nations, Rome. http://faostat3.fao.org/faostat-gateway/go/to/home/E (Accessed 15 May 2015).

Feder, G., Just, R. and Zilberman, D. (1985) Adoption of agricultural innovations in developing countries: a survey. *Economic Development and Cultural Change* 33, 255–298.

Frisvold, G. (2004) *Diffusion of Bt Cotton and Insecticide Use*. Paper presented at the Annual Meeting of Western Agricultural Economics Association, Honolulu.

Fuglie, K. (1989) *The Adoption of New Agricultural Technology in a Rainfed Rice Farming Systems in Northeast Thailand*. PhD thesis. University of Minnesota, St Paul, Minneapolis.

Fuglie, K. (2011) Agricultural productivity in Sub-Saharan Africa. In: Lee, D.R. and Ndulo, M.B. (eds) *The Food and Financial Crises in Sub-Saharan Africa: Origins, Impacts and Policy Implications*. CAB International, Wallingford, UK, 122–153.

Fuglie, K. and Rada, N. (2013) *Resources, Policy and Agricultural Productivity in Sub-Saharan Africa*. Economic Research Report 145, Economic Research Service, USDA, Washington, DC.

Griliches, Z. (1957) Hybrid corn: an exploration in the economics of technological change. *Econometrica* 25, 501–522.

International Road Federation (2012) *World Road Statistics*. Geneva, Switzerland, International Road Federation. Available at: http://www.irfnet.org/ (Accessed 15 May 2015).

Jansen, H. (1988) *Adoption of Modern Cereal Cultivars in India: Determinants and Implications of Interregional Variation in the Speed and Ceiling of Diffusion*. PhD thesis. Cornell University, Ithaca, New York.

Jarvis, L. (1981) Predicting the diffusion of improved pastures in Uruguay. *American Journal of Agricultural Economics* 63, 495–502.

Kaguongo, W., Gildemacher, P., Demo, P., Wagoire, W., Kinyae, P. *et al.* (2008) *Farmer Practices and Adoption of Improved Potato Varieties in Kenya and Uganda*. Social Sciences Working Paper 2008–5, October, International Potato Center, Lima.

Knudson, M. (1991) Incorporating Technological Change in Diffusion Models. *American Journal of Agricultural Economics* 72, 724–733.

Kristjanson, P., Tarawali, S., Okike, I., Singh, B., Thornton, P., *et al.* (2002) *Genetically Improved Dual-Purpose Cowpea: Assessment of Adoption and Impact in the Dry Savannah of West Africa*. ILRI Impact Assessment Series No. 9. September, International Livestock Research Institute, Nairobi.

Maredia, M. and Raitzer, D. (2006) *CGIAR and NARS Partner Research in Sub-Saharan Africa: Evidence of Impact To Date*. Commissioned Paper, CGIAR Science Council, Rome.

McKinsey, J. and Evenson, R. (2003) Crop genetic improvement impacts on Indian agriculture. In: Evenson, R. and Gollin, D. (eds) *Crop Variety Improvement and Its Effects on Productivity: The impact of International Agricultural Research*. CAB International Wallingford, UK, pp. 387–408.

Minten, B., Tamru, S., Engida, E. and Kuma, T. (2013) *Ethiopia's Value Chains on the Move: The Case of Teff*. ESSP Working Paper 52. International Food Policy Research Institute, Washington, DC.

Nzuma, J. (2011) *Producer Funding of Agricultural Research: The Case of Kenya's Tea Industry*. Background paper prepared for the ASTI-IFPRI/FARA Conference, Accra, Ghana, 5–7 2011. Agricultural Science and Technology Indicators Project, International Food Policy Research Institute, Rome.

Rogers, E. (2003) *Diffusion of Innovations*. 5th edn. Free Press, New York.

Rueda, J. Ewell, P., Walker, T., Soto, M., Bicamumpaka, M. and Berrios, D. (1996) Economic impact of high-yielding, late-blight resistant varieties in the Eastern and Central African Highlands. In: Walker, T. and Crissman, C. (eds) *Case Studies of the Economic Impact of CIP-Related Technologies*. International Potato Center, Lima, pp. 15–30.

Sanginga, P., Adesina, A., Manyong, V. and Dashiell, K. (1999) *Social Impact of Soybean in Nigeria's Southern Guinea Savanna*. Mimeo. International Institute for Tropical Agriculture (IITA), Ibadan, Nigeria.

Seck, P., Sibide, M., Moustapha, A. and Moustapha, K. (1998) *Impact Social de la Recherche et du Transfert de Technologies sur le Coton au Sénégal*. Institut Sénégalais de Recherches Agricoles, Dakar.

Shiferaw, B., Kebede, T. and You, L. (2008) Technology adoption under seed access constraints and the economic impacts of improved pigeon pea varieties in Tanzania. *Agricultural Economics* 39, 309–323.

Stager, D. and Stock, J. (1997) Instrumental variables regression with weak instruments. *Econometrica* 65, 557–586.

Theile, G., Hareau, G., Suarez, V., Chujoy, E., Bonierbale, M. and Maldonado, L. (2008) *Varietal Change in Potatoes in Developing Countries and the Contribution of the International Potato Center: 1972–2007*. Social Science Working Paper 2008-6, International Potato Center (CIP), Lima.

Timothy, D., Harvey, P. and Dowswell, C. (1988) *Development and Spread of Improved Maize Varieties and Hybrids in Developing Countries*. United States Agency for International Development, Washington, DC.

Tschirley, D., Poulton, C. and Labaste, P. (eds) (2009) *Organization and Performance of Cotton Sectors in Africa: Learning from Reform Experience.* World Bank, Washington, DC.

Uppsala Conflict Data Program (Date of retrieval: 05/31/2012) *UCDP Conflict Encyclopedia*: www.ucdp.uu.se/database, Uppsala University, Sweden.

Walker, T., Alene, A., Ndjuenga, J., Labarta, R., Yegezu, Y. *et al.* (2014) *Measuring the Effectiveness of Crop Improvement Research in Sub-Saharan Africa from the Perspectives of Varietal Output, Adoption and Change: 20 Crops, 30 Countries, and 1150 Cultivars in Farmers' Fields*. Report of the Standing Panel on Impact Assessment (SPIA), CGIAR Independent Science and Partnership Council (ISPC) Secretariat, Rome.

World Development Indicators (Date of retrieval: 05/01/2012) World Bank, Washington, DC. Available at: http://data.worldbank.org/data-catalog/world-development-indicators (Accessed 18 May 2015).

Zeddies, J., Schaab, R., Neuenschwander, P. and Herren, H. (2001) Economics of biological control of cassava mealybug in Africa. *Agricultural Economics* 24, 209–219.

Zhou, M. (2013) Conventional sugarcane breeding in South Africa: progress and future prospects. *American Journal of Plant Sciences* 4, 189–197.

Appendix 17.1. Supplemental Material on Diffusion and Impact of Improved Varieties in Africa

Diffusion of improved crop varieties: Country-level findings

Table 17.A1 shows the estimates of 73 logistic diffusion curves for improved varieties of food crops in different SSA countries. Estimating the parameters of the diffusion curve requires information on adoption rates for at least two points in time. For the cases listing DIIVA as a data source, these observations were from 1995–1998 (reported in the chapters in Evenson and Gollin) and 2009–2010 (Walker *et al.*, 2014). Estimates of adoption of improved varieties of wheat and rice for the early 1980s from Dalrymple provide a longer profile of the diffusion curve for these crops. The estimates of improved maize varietal adoption in West Africa in Alene *et al.* provide additional observations from 1981 and 2005.[16] We were also able to supplement DIIVA data with additional observations on adoption from a number of other studies, such as potato in Tanzania (Theile *et al.*, 2008), soybean in Nigeria (Sanginga *et al.*, 1999), rice in West Africa (Africa Rice Center, 2008) and pigeonpea in Tanzania (Shiferwa *et al.*, 2008). Finally, Minten *et al.* (2013) provides figures on the diffusion on improved varieties of teff in Ethiopia, an important staple in that country not included in the DIIVA surveys.

Because the diffusion curves reported in Table 17.A1 are derived from so few observations on adoption (just two in most cases), they should be considered as first approximations only. Obtaining more observations on adoption is necessary to improve confidence in the pattern of diffusion of improved varieties in SSA. It will be especially important to obtain more information on adoption rates during the next decade to clarify trends and prospects, as adoption of improved varieties is still at an early stage in much of SSA.

Adding up the estimates of improved varietal adoption by crop establishes a profile of the aggregate adoption of improved varieties as a share of total crop area in a country. This provides the basis for the econometric estimation of the 'dynamic diffusion curve' reported in Table 17.A2. In this model, the overall rate of acceptance, or the slope of the diffusion curve, varies among countries and over time as a function of the enabling environment for technology dissemination. In this model, the enabling environment consists of country and regional factors that measure the accumulated stock of research capital, the education and health of farmers, policies affecting agricultural terms of trade, market infrastructure and the rule of law. Whereas these variables account for only some of the factors that make up the enabling environment (and imperfectly measured, no doubt), they account for most of the variation within and across countries in adoption of improved genotypes. Figure 17.A1 compares observed diffusion curves with those predicted from the model for all of the 37 countries included in this analysis (including three – Congo Republic, Liberia and Mauritania – where adoption was not observed for any crop and assumed to be zero). In nearly all cases, predicted rates of adoption track actual rates closely. The enabling environment variables included in the model appear to capture the differences in improved variety diffusion rates observed among the countries of the region. But they do not explain why diffusion of improved varieties in SSA has been so slow overall.

The estimates of the dynamic diffusion curve shown in Table 17.A2 suggest that multiple factors affect varietal adoption. Clearly, the supply of new technology, represented by the stock of research capital and the number of varieties released, significantly influences adoption. But demand factors – farmers' education and health and the economic policies they face – are also quite significant. The significance (and opposite sign) of the squared terms of the variables suggests that each variable by itself (holding other factors unchanged) faces rapidly diminishing returns. In other words, the conditions for agricultural technology dissemination among African smallholders are complex and require attention to both supply and demand factors to accelerate and sustain rapid diffusion. Twenty years ago close observers of technology dissemination among African smallholders were noting the same multidimensional constraints (Eicher, 1995; Byerlee and Heisey, 1996); since then this situation has only improved marginally.

Table 17.A1. Rates of diffusion of improved crop varieties in sub-Saharan Africa.

Crop	Country	Data sources	No. of observations on adoption	Estimated parameters of the logistic diffusion curve		Year of origin (20% adoption)	Year of saturation (80% adoption)	Years from origin to saturation	Total crop area ('000 ha, 2006–2010 annual avg.)
				α	β				
Cereal grains									
Barley	Ethiopia	DIIVA	2	–4.386	0.128	2004	2025	22	1,068
Maize – ESA	Angola	DIIVA	2	–1.708	–0.017	na	na	na	1,249
Maize – ESA	Ethiopia	DIIVA	2	–4.711	0.130	2006	2027	21	1,727
Maize – ESA	Kenya	DIIVA	2	1.064	–0.009	na	na	na	1,819
Maize – ESA	Malawi	DIIVA	2	–4.369	0.141	2001	2021	20	1,568
Maize – ESA	Mozambique	DIIVA	2	–2.459	0.009	2099	2405	307	1,520
Maize – ESA	Tanzania	DIIVA	2	–7.260	0.230	2006	2018	12	2,816
Maize – ESA	Uganda	DIIVA	2	–7.418	0.283	2001	2011	10	860
Maize – ESA	Zambia	DIIVA	2	–5.573	0.242	1997	2009	11	774
Maize – ESA	Zimbabwe	DIIVA	2	–0.306	0.100	1969	1997	28	1,552
Maize – WCA	Benin	DIIVA, Alene *et al.* (2009)	4	–3.620	0.135	1997	2017	21	796
Maize – WCA	Burkina Faso	DIIVA, Alene *et al.* (2009)	4	–3.108	0.139	1992	2012	20	580
Maize – WCA	Cameroon	DIIVA, Alene *et al.* (2009)	4	–2.529	0.094	1992	2022	30	674
Maize – WCA	Cote D'Ivoire	DIIVA, Alene *et al.* (2009)	4	–3.227	0.128	1994	2016	22	304
Maize – WCA	Ghana	DIIVA, Alene *et al.* (2009)	4	–4.016	0.199	1993	2007	14	875
Maize – WCA	Guinea	DIIVA	2	–4.313	0.181	1996	2011	15	431
Maize – WCA	Mali	DIIVA, Alene *et al.* (2009)	4	–3.742	0.148	1996	2015	19	432
Maize – WCA	Nigeria	DIIVA, Alene *et al.* (2009)	4	–3.151	0.179	1990	2005	16	3,673
Maize – WCA	Senegal	DIIVA, Alene *et al.* (2009)	4	–3.034	0.243	1987	1998	11	165
Maize – WCA	Togo	DIIVA, Alene *et al.* (2009)	4	–4.295	0.041	2051	2118	67	497
Pearl millet	Mali	DIIVA	2	–1.735	0.031	1991	2080	89	1,529
Rice	Benin	DIIVA, WARDA (2008)	2	–13.919	0.510	2005	2010	5	34
Rice	The Gambia	DIIVA, WARDA (2008)	2	–4.100	0.149	1998	2017	19	46
Rice	Côte d'Ivoire	DIIVA, Dalrymple (1986)	3	–2.138	0.063	1992	2036	44	371
Rice	Ghana	DIIVA, Dalrymple (1986)	3	–1.949	0.105	1985	2012	26	142
Rice	Guinea	DIIVA, WARDA (2008)	3	0.568	–0.080	na	na	na	816
Rice	Mali	DIIVA, Dalrymple (1986)	3	–1.223	0.025	1973	2084	111	564
Rice	Nigeria	DIIVA, Dalrymple (1986)	3	–2.097	0.090	1988	2019	31	2,227

Continued

Table 17.A.1. Continued.

Crop	Country	Data sources	No. of observations on adoption	Estimated parameters of the logistic diffusion curve		Year of origin (20% adoption)	Year of saturation (80% adoption)	Years from origin to saturation	Total crop area ('000 ha, 2006–2010 annual avg.)
				α	β				
Rice	Senegal	DIIVA, Dalrymple (1986)	3	–2.157	0.147	1985	2004	19	115
Rice	Sierra Leone	DIIVA, Dalrymple (1986)	3	–2.066	0.024	2008	2124	116	544
Sorghum	Mali	DIIVA	2	–1.062	0.011	1951	2201	250	1,063
Sorghum	Nigeria	DIIVA	2	–1.786	0.014	2009	2209	200	6,442
Sorghum	Sudan	DIIVA	2	–2.353	0.068	1994	2035	41	6,379
Sorghum	Tanzania	DIIVA	2	–8.570	0.275	2006	2016	10	823
Teff	Ethiopia	Minten *et al.* (2013)	2	–6.778	0.206	2006	2020	13	2,625
Wheat	Ethiopia	DIIVA, Dalrymple (1986)	3	–3.670	0.119	1999	2022	23	1,523
Wheat	Sudan	DIIVA, Dalrymple (1986)	3	–3.240	0.173	1991	2007	16	277
Wheat	Kenya	DIIVA, Dalrymple (1986)	2	–1.209	0.215	1979	1992	13	135
Wheat	Tanzania	DIIVA, Dalrymple (1986)	2	–2.251	0.253	1983	1994	11	44
Wheat	Nigeria	DIIVA, Dalrymple (1986)	2	–2.930	0.279	1986	1995	10	39
Vegetatively propagated crops									
Banana	Uganda	DIIVA	2	–11.293	0.297	2013	2023	9	
Cassava	Malawi	DIIVA	2	–7.192	0.264	2002	2013	11	171
Cassava	Benin	DIIVA	2	–7.524	0.282	2002	2012	10	231
Cassava	Togo	DIIVA	2	–4.521	0.140	2002	2022	20	137
Cassava	Angola	DIIVA	2	–3.552	0.097	2002	2031	29	859
Cassava	Kenya	DIIVA	2	–3.990	0.130	2000	2021	21	62
Cassava	Nigeria	DIIVA	2	–3.527	0.115	1999	2023	24	3,543
Cassava	DR Congo	DIIVA	2	–2.960	0.100	1996	2023	28	1,857
Cassava	Côte d'Ivoire	DIIVA	2	–2.103	0.025	2009	2121	112	329
Cassava	Ghana	DIIVA	2	–1.919	0.046	1992	2053	61	838
Cassava	Guinea	DIIVA	2	–1.922	0.019	2009	2157	148	132
Cassava	Uganda	DIIVA	2	–1.221	0.021	1972	2106	134	397
Cassava	Cameroon	DIIVA	2	–1.168	0.020	1969	2105	136	215
Cassava	Tanzania	DIIVA	2	–0.482	–0.018	na	na	na	898
Cassava	Zambia	DIIVA	2	–15.502	0.526	2007	2012	5	194
Cassava	Zimbabwe	DIIVA	2	–6.570	0.229	2003	2015	12	47

Continued

Table 17.A1. Continued.

Crop	Country	Data sources	No. of observations on adoption	Estimated parameters of the logistic diffusion curve		Year of origin (20% adoption)	Year of saturation (80% adoption)	Years from origin to saturation	Total crop area ('000 ha, 2006–2010 annual avg.)
				α	β				
Potato	Kenya	DIIVA	2	–2.568	0.058	2000	2048	48	133
Potato	Tanzania	DIIVA, Theile *et al.* (2007)	2	–1.386	0.000	na	na	na	134
Potato	Ethiopia	DIIVA	2	–0.908	–0.011	na	na	na	61
Potato	Uganda	DIIVA	2	2.350	–0.045	na	na	na	97
Potato	Rwanda	DIIVA	2	9.877	–0.360	na	na	na	134
Legumes and oilseeds									
Beans	Malawi	DIIVA	2	–12.443	0.436	2005	2012	6	251
Beans	Tanzania	DIIVA	2	–8.116	0.274	2005	2015	10	1,197
Beans	Ethiopia	DIIVA	2	–5.843	0.189	2004	2018	15	227
Beans	Uganda	DIIVA	2	–3.141	0.078	2002	2038	35	894
Beans	Rwanda	DIIVA	2	–1.974	0.018	2013	2171	158	343
Beans	DR Congo	DIIVA	2	2.759	–0.158	na	na	na	209
Cowpea	Nigeria	DIIVA	2	–6.002	0.191	2004	2019	15	3,646
Groundnut	Uganda	DIIVA	2	–6.799	0.242	2002	2014	11	248
Groundnut	Malawi	DIIVA	2	–6.536	0.228	2003	2015	12	259
Groundnut	Zambia	DIIVA	2	–4.564	0.167	1999	2016	17	160
Pigeonpea	Tanzania	DIIVA, Shiferaw *et al.* (2008)	2	–12.088	0.417	2006	2012	7	71
Soybean	Nigeria	DIIVA, Sanginga *et al.* (1999)	3	–5.467	0.230	1998	2010	12	550

na = not applicable. In these cases, there was little or no increase in diffusion between the two periods and the estimated value of β is too low to generate a diffusion curve. It is likely that peak adoption will remain far below 100% unless further improvements are made to new varieties.

Table 17.A2. OLS estimates of dynamic diffusion model.

Dep. variable = % crop area in improved varieties	Large sample of countries (30)		Small sample of countries (17)	
Exogenous variables	Linear terms	Quadratic terms	Linear terms	Quadratic terms
Constant	−3.294***		−3.067***	
	(−6.821)		(−3.083)	
Time since improved varieties first released or adoptions (years)	−0.746***		−0.992***	
	(−2.522)		(−2.548)	
Number of releases of improved varieties (cumulative number per 1000 ha cropland)	0.288**	−0.819**	0.276*	−0.864**
	(2.289)	(−2.177)	(1.697)	(−1.961)
CGIAR agricultural research (log of R&D stock)	0.245**	−0.0219	0.270*	−0.0224
	(1.985)	(−1.623)	(1.871)	(−1.488)
National agricultural research (log of R&D stock)	0.0178	−0.00016	0.0964***	−0.0110***
	(1.141)	(−0.0622)	(2.916)	(−2.373)
Labour force schooling (years)	0.0150**	−0.00172**	−0.00289	0.00031
	(2.005)	(−2.303)	(−0.275)	(0.263)
Nominal rate of assistance to agriculture (per cent change in farm prices due to policies)	0.144**	0.401***	0.0843	0.237
	(2.852)	(2.537)	(1.399)	(1.507)
Density of paved roads (log of km paved roads/km² country area)	−0.00633	−0.00017	−0.0200	−0.00172
	(−0.332)	(−0.0796)	(−0.584)	(−0.471)
HIV-AIDS infection (per cent of adult population)	−0.169	1.660***	−0.310*	2.532***
	(−1.060)	(2.588)	(−1.761)	(3.223)
Armed conflict (cumulative years of armed conflict)	−0.00082	1.96e-05	−0.00192	5.10e-05
	(−0.609)	(0.396)	(−1.181)	(0.885)
Number of observations	167		111	
F-test of the regression	12.00***		11.39***	
Prob<F	0.0000		0.0000	
Root MSE	0.9165		0.8414	
R-squared	0.5778		0.6755	
Adjusted R-squared	0.5297		0.6162	

T-statistics in parentheses: ***, ** and * indicate 1%, 5% and 10% significance levels, respectively.

The estimates of the diffusion curve in Table 17.A2 provide insights on how constraints to technology adoption have varied across countries and over time in SSA. Recall that the estimated value of β (the rate of acceptance or slope parameter of the diffusion curve) indicates the speed of technology diffusion. Since β is a function of the parameter estimates and the explanatory variables, its value varies by country and over time. Using the hat (^) symbol for estimated regression coefficients, the estimated value of β for country c in year t (see Eqns 17.6 and 17.7 in the text) is given by:

$$\hat{\beta}_{ct} = \hat{\beta}_0 + \hat{\beta}_1\left(X_{11ct}\right) + \hat{\beta}_2\left(X_{12ct}\right) + \cdots + \hat{\beta}_j\left(X_{1jct}\right)$$

A higher or rising value of $\hat{\beta}_{ct}$ would imply that the conditions for technology dissemination are better or improving. We can also compare these values with the benchmark range of 0.2 to 0.8 that seemed to characterize typical diffusion rates for improved varieties in other parts of the world. Using the parameters from the 30-country regression, Fig. 17.A2 shows that β for the SSA region rose from below zero in the 1980s to about 0.05 by 2000 and then stabilized at that level. The increased availability of improved varieties,

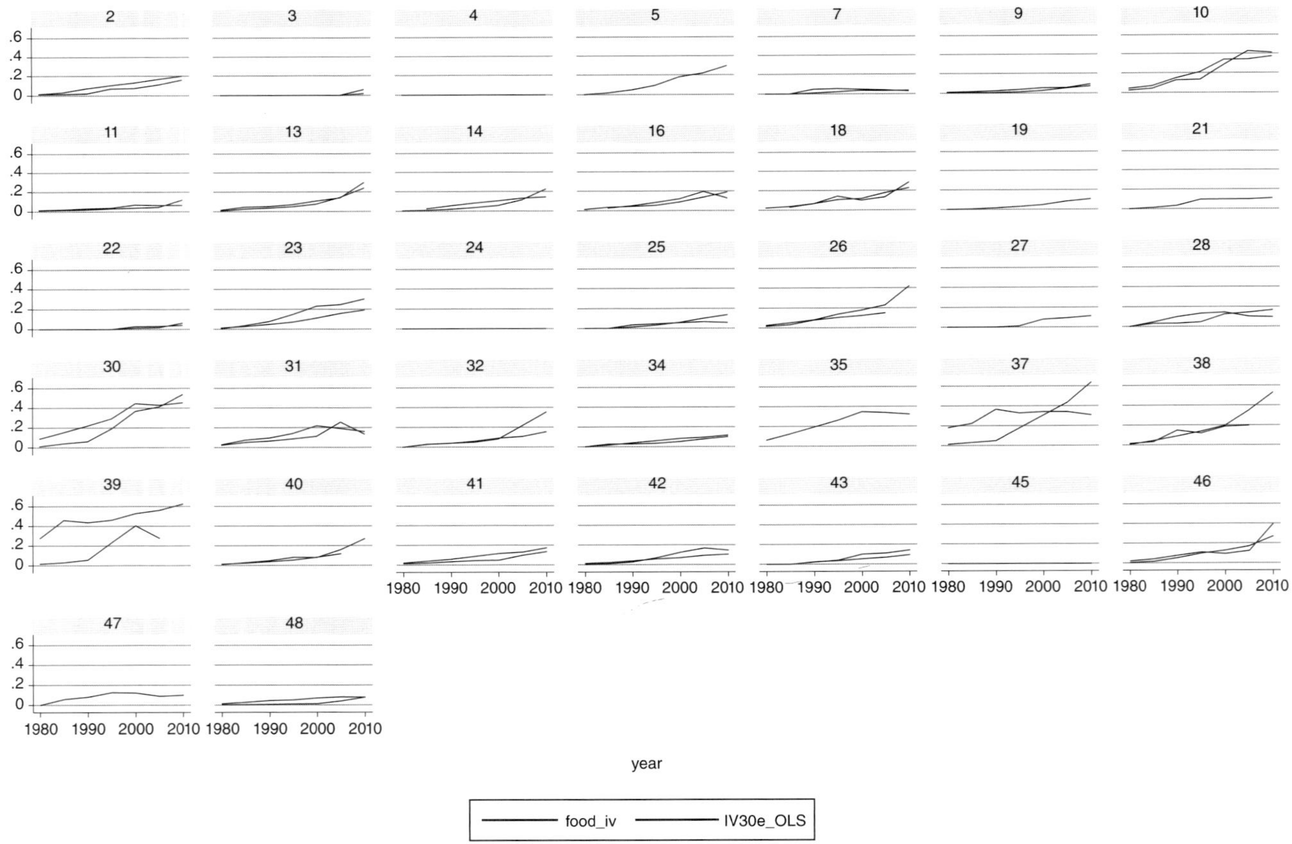

Figure 17.A1. Actual and predicted rates of adoption of improved varieties, by country. (Note: food_iv = actual improved variety adoption rate; IV30e_OLS = predicted improved variety adoption rate from dynamic diffusion model.)

Country	Code	Country	Code	Country	Code
Cameroon	2	Chad	21	Swaziland	37
Cent Aft Rep	3	The Gambia	22	Zambia	38
Congo Rep	4	Mali	23	Zimbabwe	39
DR Congo	5	Mauritania	24	Benin	40
Gabon	7	Niger	25	Cote d'Ivoire	41
Burundi	9	Senegal	26	Ghana	42
Kenya	10	Angola	27	Guinea	43
Rwanda	11	Botswana	28	Liberia	45
Tanzania	13	Lesotho	30	Nigeria	45
Uganda	14	Madagascar	31	Sierra Leone	47
Ethiopia	16	Malawi	32	Togo	48
Sudan	18	Mozambique	34		
Burkina Faso	19	Namibia	35		

Figure 17.A1. Continued.

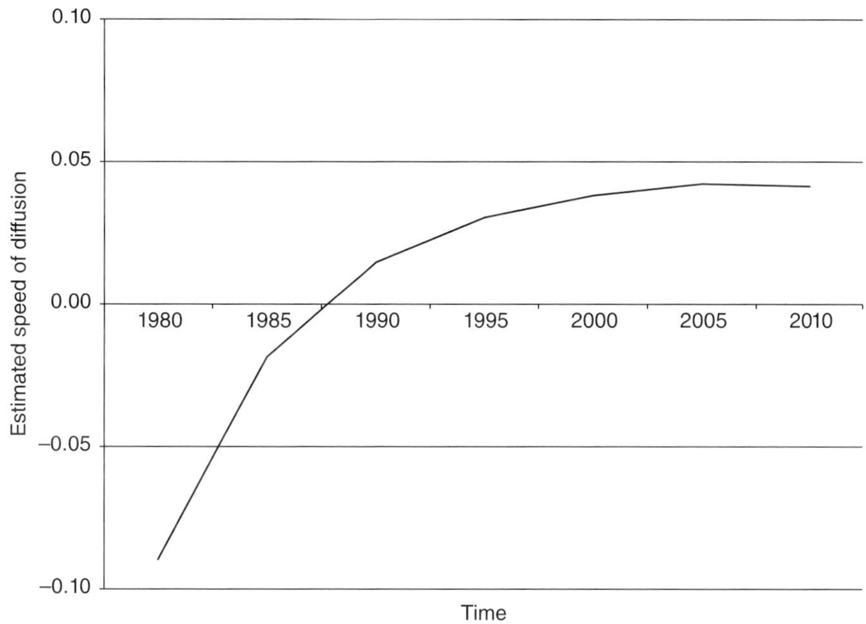

Figure 17.A2. Rate of acceptance (β) of improved varieties – average across SSA countries.

gradually rising stocks of research capital and labour force schooling levels, and higher nominal rates of assistance to agriculture all contributed to raising the value of β in the region. Together, they imply that the enabling environment for technology dissemination has markedly improved in the 1980s and 1990s, but then stagnated and still remains very low according to our benchmark range for β. Figure 17.A3 compares the average values of β in 2010 among countries in SSA. This provides a ranking of which countries seem to have the most (and least) favourable environment for crop variety dissemination, at least according to the variables included in the model. Besides Swaziland and Lesotho (small countries with a high number of varieties released per hectare of cropland due to their relatively small crop area), Kenya and Nigeria seem to have the most favourable environments for crop variety dissemination. But even in these countries the value of

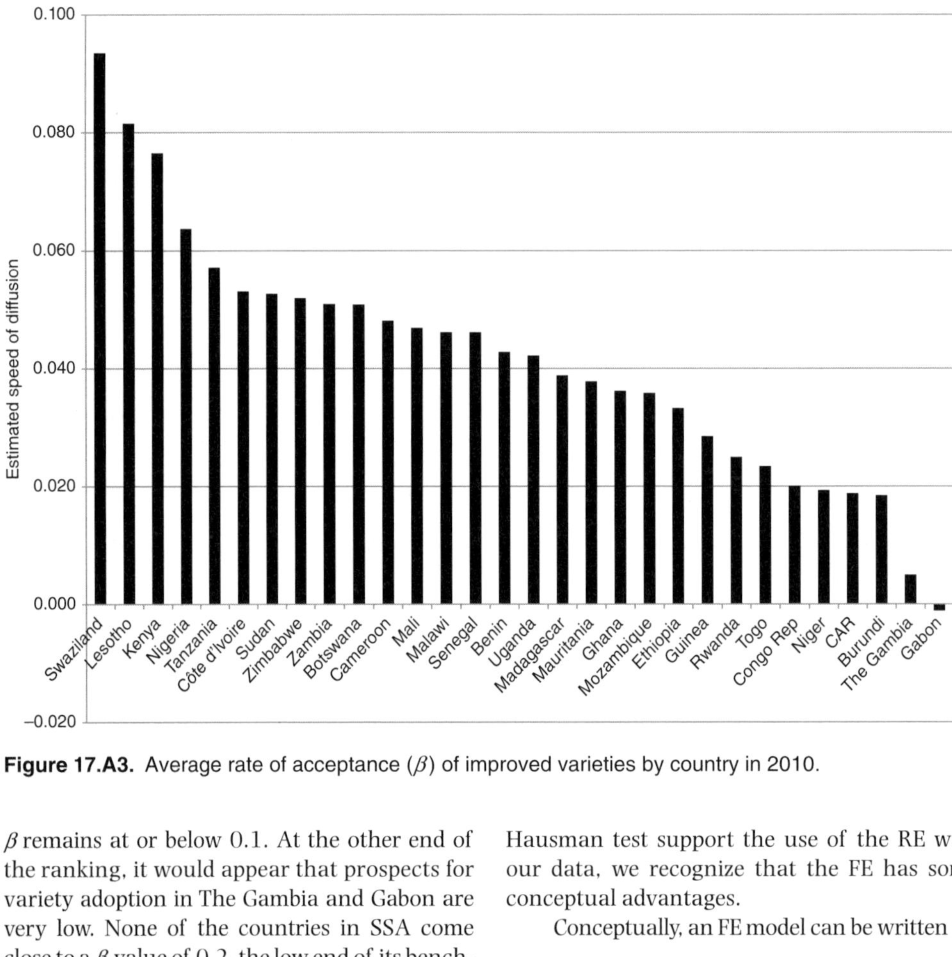

Figure 17.A3. Average rate of acceptance (β) of improved varieties by country in 2010.

β remains at or below 0.1. At the other end of the ranking, it would appear that prospects for variety adoption in The Gambia and Gabon are very low. None of the countries in SSA come close to a β value of 0.2, the low end of its benchmark value. All of the countries in the region need to substantially enhance their enabling environment for technology dissemination to achieve rates of diffusion typical of other regions of the world.

Impact of variety adoption on productivity: fixed effects model

Fixed effects (FE) and random effects (RE) models are two of the most popular methods of econometric estimation with panel data. In Table 17.A3 we report FE estimates of the impact of adoption and other factors on agricultural TFP to complement the RE estimates reported in the main text of the chapter. Table 17.A3 also reports the results of the Hausman test commonly used to select among FE and RE models given the data. Although the results of the Hausman test support the use of the RE with our data, we recognize that the FE has some conceptual advantages.

Conceptually, an FE model can be written as:

$$Y_{ct} = \beta X_{ct} + u_c + \varepsilon_{ct}$$

where u_c is a country fixed-effect that removes the influence of all time invariant characteristics of a country on the dependent variable Y. However, results of this estimation are considered valid only for the sample and not generalizable to the broader population (in this case, to other countries in SSA not included in the sample). Moreover, FE models have fewer degrees of freedom and thus tend to be more demanding of the data. The RE model, on the other hand, treats u_c as a random variable drawn from some distribution, and results can be generalized to apply to the larger population from which the sample of observations is drawn. A drawback of RE, however, is that if unobserved characteristics of a country (in the error term ε_{ct}) embody elements that are correlated with both the regressor Y and one or more of the explanatory variables in X, then the estimates of β may suffer from omitted variable bias. In other words, the model would attribute the influence of an omitted variable to a variable in X, leading to an over- or underestimation of β.

The FE estimates in Table 17.A3 are very similar to the RE estimates reported in the text. The average impact of improved varietal adoption is slightly higher in the FE model, but this difference is not statistically significant. The positive effect of agricultural research and the negative impact of HIV-AIDS infection in the population are significant and of similar magnitudes. The main difference is that the effect of schooling of the labour on TFP is no longer significant in the FE model.

Table 17.A3. Impact of adoption of improved food crop varieties and other factors on agricultural total factor productivity (TFP), fixed effects (FE) model.

Dep. variable = TFP Index (1977 = 100 for each country) Exogenous variables	Large sample of countries		Small sample of countries	
Improved variety adoption	0.596	***	0.538	***
(per cent of crop area harvested)	(3.548)		(2.640)	
Combined CGIAR and national agric. research	0.122	***	0.175	***
(log of R&D stock)	(3.183)		(3.258)	
Labour force schooling	0.00376		−0.00934	
(years)	(0.182)		(−0.278)	
Nominal rate of assistance to agriculture	0.153		0.217	
(per cent change in farm prices due to policies)	(1.218)		(1.639)	
Density of paved roads	0.0248		0.0438	
(log of km paved roads/km² country area)	(0.905)		(1.184)	
HIV–AIDS infection	−0.909	***	−0.979	***
(per cent of adult population)	(−2.822)		(−2.035)	
Armed conflict	−0.00523		−0.00533	
(cumulative years of armed conflict)	(−1.540)		(−1.376)	
Constant	4.220	***	4.125	***
	(19.41)		(14.55)	
Observations	191		111	
Number of countries	30		17	
R-squared: Within	0.3522		0.5053	
Between	0.0600		0.1555	
Overall	0.1917		0.3536	
σ_u (std dev within-country error)	0.1304		0.1178	
σ_e (std dev between country error)	0.1238		0.1179	
rho (fraction of variance due to u_i)	0.5200		0.4999	
F-Test of regression	11.96	***	12.69	***
Prob > F	0.0000		0.0000	
Test of fixed versus random effects (if Prob <0.05, then FE, otherwise RE):				
Hausman $\chi 2$	4.87		4.66	
Prob > $\chi 2$	0.6761		0.7013	

T-statistics in parentheses: ***, ** and * indicate 1%, 5% and 10% significance levels, respectively.

Note: World Bank estimates of NRA are available for only some SSA countries, and including this variable limits the sample to 17 countries (the small sample estimates). To estimate the model with the large sample of 30 countries, average regional values of NRA were used for missing observations on NRA.

18 Varietal Generation and Output

T.S. Walker,* A. Alene, J. Ndjuenga, R. Labarta, Y. Yigezu, A. Diagne, R. Andrade, R. Muthoni Andriatsitohaina, H. De Groote, K. Mausch, C. Yirga, F. Simtowe, E. Katungi, W. Jogo, M. Jaleta, S. Pandey and D. Kumara Charyulu

The substantive findings in Chapters 6–17 are synthesized and reviewed in this and the following chapter, which draw heavily on Walker *et al.*, 2014. Findings are synthesized from two perspectives: a cross-sectional analysis across the 20 crops in 2009–2011 and a before-and-after comparison with the 1998 benchmark and the 2009–2011 data. Findings in this chapter are organized from the evaluation framework of inputs and outputs that was described in Chapter 3. Hypotheses from that chapter are revisited at the end of each thematic section. Where appropriate, results from South Asia reported in Chapters 13 and 14 are cited to provide a spatial benchmark for the outputs of data analysis in sub-Saharan Africa (SSA).

Varietal Generation: Full-Time Equivalent Scientists by Crop Improvement Programme

As discussed in Chapter 3, our emphasis on inputs in varietal generation focuses on full-time equivalent (FTE) scientists in crop improvement programmes in national agricultural research systems (NARS) broadly defined as public crop improvement programmes, universities and private-sector companies.[1]

Scientist numbers, research intensities, and congruence estimates

The total number of FTE scientists across the 151 national crop improvement programmes approaches 1300 (Table 18.1). The actual number engaged in crop improvement researcher is larger. For example, the 126 FTE scientists in rice refer to the time allocated by 289 researchers (Diagne *et al.*, Chapter 10, this volume). Only about 25–30% of these scientists commit 75–100% of their time to rice research.

More scientific resources are allocated to maize than to any other crop in SSA (Table 18.1). Cassava is a distant second to the total for maize across its two major regions of production.

* E-mail: walkerts@msu.edu

Table 18.1. Full-time equivalent (FTE) scientists by crop improvement programme in SSA in 2010.

Crop	Countries	Total FTE scientists	Min.	Median	Max.
Maize (ESA)	9	243.2	12.0	17.0	62.0
Maize (WCA)	11	139.5	3.0	5.8	77.5
Cassava	17	138.8	1.0	7.2	22.5
Rice	14	125.0	0.9	8.3	15.3
Bean	10	86.5	2.6	5.9	21.4
Potato	5	57.3	3.0	4.6	30.0
Cowpea	18	76.5	0.4	2.9	16.0
Wheat	4	70.1	12.0	15.0	28.0
Soybean	14	52.2	0.8	2.4	14.6
Sweetpotato	5	32.7	2.0	4.0	15.9
Yam	8	49.5	3.0	4.6	12.1
Sorghum	7	42.3	2.4	3.0	18.2
Groundnut	10	23.9	1.15	2.1	5.0
Banana	1	40.0	40.0	40.0	40.0
Chickpea	2	27.0	8.4	13.5	18.6
Pigeonpea	3	6.9	3.9	1.2	5.0
Barley	2	22.1	1.0	11.1	21.1
Pearl millet	5	20.4	1.5	4.5	6.8
Faba bean	2	15.5	6.9	7.8	8.7
Lentil	3	11.0	2.0	3.7	5.3
Field pea	1	6.9	6.9	6.9	6.9
Total/mean	151	1,289	na	8.6	na

na, not applicable.

Maize in East and Southern Africa (ESA), with a longstanding tradition of national programmes promoting hybrids, has benefited from a sharp and sustained increase in private sector maize breeding, especially in Kenya, Malawi, Zambia and Zimbabwe (De Groote *et al.*, Chapter 11, this volume). The private sector has yet to make its presence felt in maize production in much of West and Central Africa (WCA) where national programmes have emphasized Open Pollinated Varieties (Alene *et al.*, Chapter 6, this volume). Nonetheless, relative to other crops, the public sector has allocated substantial scientific resources to maize research in several of the 11 producing countries covered in WCA. Maize in Nigeria has the largest scientific cadre equivalent to 77 FTE scientists. Some of these are university research staff who allocate part of their time to maize research.

The median programme size is 8–9 FTE scientists, which should be sufficient to get the job done for all small and most medium-sized producing countries unless the crop is produced in highly diverse agroecologies or unless changes in basic knowledge lead to a radical shift in the distribution of yield potential. In agricultural research, there are diminishing marginal returns to sampling from the same distribution when knowledge is stagnant or only increasing incrementally (Kislev, 1977). In other words, most crop improvement programmes are subject to economies of scale as we would not expect the desirable number of scientists in a programme to increase proportionally to rising production. Very large programmes will not have hundreds of scientists.

In contrast to other crops, the number of scientists engaged in all the maize improvement programmes in ESA is not a cause for concern. The nine programmes are all staffed by more than 12 FTE scientists, with Angola and Mozambique tied for the smallest programme. Even the smallest maize programmes in ESA have more scientists than the median-sized programme for 16 of the 19 other crops (Table 18.1).

A median programme size of 15 for wheat underscores the continuing commitment of governments to invest heavily in this import substitute that is grown on large farms, often with access to irrigation in Kenya, Zambia and Zimbabwe. Ethiopia, where wheat is grown by smallholders, is by far the largest wheat producer in SSA. A value of 11 for barley reflects the emphasis that Ethiopia places on agricultural research.

Pearl millet is at the other end of the human resource spectrum. Indeed, its largest country programme only has about 7 FTE scientists. With the exception of the largest-producing countries in West Africa, pearl millet is almost always a shared programme with other coarse cereals. Groundnut suffers a similar outcome (Table 18.1) and is often a member of a composite programme made up of pulses and/or oilseeds.

Saying something more conclusive about the data in Table 18.1 requires adjusting for the differences in the size of production across different countries. Attaining a critical mass of scientists is needed to make progress in large-producing countries and crossing a threshold size of production is required before resources should be committed to investing in crop improvement in very small-producing countries (Maredia and Eicher, 1995).

In Table 18.2, the size of production has been normalized across crops and countries by calculating research intensities that express FTE scientists as ratios from the perspectives of area, production and value of production. As anticipated, crops characterized by small areas and values of production are associated with higher estimated research intensities than those with very large areas, production levels and value of production.

The ranking of the crops in terms of research intensity varies somewhat across the three criteria in Table 18.2. Potato ranks high in research intensity on area but occupies a low position on production and value. Banana ranks high on area, low on production and high on value. However, there are more aspects in common than are different across the three criteria.

Table 18.2. Estimated research intensities by crop in SSA in 2010 from the perspectives of area, production and value of production.[a]

Area		Production		Value of production	
Crop	FTE scientists per million hectares of production	Crop	FTE scientists per million tonnes of production	Crop	FTE scientists per US$100 million of the crop
Chickpea	112.4	Lentil	89.1	Banana	25.2
Pigeonpea	64.2	Chickpea	83.6	Soybean	21.4
Potato	61.3	Soybean	45.6	Chickpea	18.4
Lentil	55.6	Bean	43.3	Pigeon pea	17.5
Banana	45.9	Field pea	31.4	Lentil	16.2
Soybean	44.0	Wheat	20.5	Field pea	14.0
Wheat	42.9	Faba bean	20.5	Wheat	13.7
Beans	32.5	Pigeonpea	20.3	Barley	12.8
Field pea	29.7	Barley	15.1	Maize (ESA)	8.5
Faba bean	25.3	Maize (ESA)	12.3	Sweetpotato	7.0
Rice	24.0	Cowpea	11.3	Faba bean	6.2
Barley	22.8	Rice	10.1	Beans	6.1
Sweetpotato	22.1	Maize (WCA)	8.1	Maize (WCA)	5.7
Maize (ESA)	16.5	Potato	6.5	Cowpea	5.3
Maize (WCA)	14.0	Groundnut	4.2	Rice	3.9
Cassava	12.6	Banana	4.2	Potato	3.4
Yam	10.6	Sweetpotato	3.6	Sorghum	2.2
Cowpea	6.6	Sorghum	2.9	Groundnut	1.4
Groundnut	5.3	Pearl millet	1.6	Cassava	1.2
Sorghum	2.5	Yam	1.0	Pearl millet	1.0
Pearl millet	1.4	Cassava	0.9	Yam	0.4

[a]All estimates are weighted averages of area in hectares, production in tonnes and value of production in total US$.

In general, several pulses rank high in research intensity in all three criteria. The first five crops listed in the production column of Table 18.2 are all pulse crops with relatively small areas of production. The exceptions are soybean in Nigeria and pulses that are produced in Ethiopia, which has invested substantial scientific resources in its NARS in terms of the number of scientists. Bean's high ranking speaks to the stability of the Pan-African Bean Research Alliance (PABRA) – one of the regional crop improvement associations that survived a shrinking budget for international crop improvement research in the 1990s and early 2000s (Muthoni and Andrade, Chapter 8, this volume). Cowpea, which is the lowest ranking pulse in Table 18.2, is produced almost entirely in West Africa.

Turning to the cereals in Table 18.2, barley does well because of its location in Ethiopia, which has a large and regionally decentralized national programme at the Ethiopian Institute of Agricultural Research (EIAR). Rice also displays a research intensity estimate above 10 from the perspective of production. Potato has a leading position in roots and tubers because of its high market orientation and demand in East Africa.

Cassava, yams and pearl millet appear at the bottom of Table 18.2. Relative to their area, production and value of production, all three of these semi-subsistence food crops appear to be starved of research resources. In terms of area, groundnut and sorghum are also characterized by very low research intensities.

The estimated research intensities for pearl millet and sorghum in the arid and semi-arid tropics of India in Chapter 14 (this volume) are three to four times larger than those in Table 18.2 for the same crops in SSA. Apparently, higher research intensities associated with smaller country size are not sufficient to compensate for the lack of investment in agricultural research on these coarse cereals in West Africa.

These intercontinental differences would be even greater if educational attainment was factored into the estimation of research intensity. Nine of ten scientists in pearl millet and sorghum research in India have PhDs; only slightly more than one-third of the FTE scientists in SSA are PhD holders.

The disparities in research intensity between India and SSA are also notable in groundnut. Estimated research intensities are more than twice as high in India than in SSA. If smaller country programmes in ESA were not included, the difference between research intensities would be similar to those encountered in pearl millet and sorghum.

In contrast to pearl millet, sorghum and groundnut, rice's weighted average research intensity of 24 FTE scientists per million hectares of growing area in SSA is 10 to 12 times larger than what one typically finds for predominantly rainfed rice cultivation in South Asia (Chapter 13, this volume). Part of this difference is attributable to small-producing countries in SSA. Increasing urban demand and related policies that favour import substitution are other major contributors to the position of rice as the cereal with the highest research intensity in SSA in 2010.

Specific cases of resource deprivation can be identified by counting the incidence of falling below an arbitrary but seemingly reasonable threshold of critical mass. This lower bound threshold for large programmes exceeding 2 million tonnes of production is established at nine scientists (the median-size programme as shown in Table 18.1). Ten large-producing crop-by-country combinations fall below this minimum threshold: cassava in Benin, Côte d'Ivoire, Malawi and Mozambique; cowpea in Côte d'Ivoire and Guinea; groundnut in Nigeria; pearl millet in Niger and Nigeria; and sorghum in Nigeria. From the perspective of production, the estimated research intensity of these 10 crops is in the range of 0.2–2.0 and averages 1.0.

Building on the estimated research intensities in Table 18.2, it is useful to compare the actual allocations of FTE scientists with normative allocations calculated from a congruence rule. This states that research resources should be allocated in proportion to the value of production across commodities, if all other things are considered equal (Alston *et al.*, 1995). In priority setting, 2% of value of production is a common assumption because studies have shown that research investment proportional to agricultural gross domestic product (GDP) often exceeds 2% in developed countries (Walker *et al.*, 2006). In developing countries in SSA, the 2% criterion is rarely obtained (Beintema and Stads, 2011). In large countries, such as China and India, where economies of scale and size prevail, research investments in the order of 1% of agricultural GDP are commonplace.

When comparing normative to actual allocations, we have assumed that 1% of the value of production is desirable for the size of research investment and that each scientist costs on average US$115,000 in purchasing-power parity (PPP) in 2010. The latter assumption is well within the range of comparable estimates in the ASTI (Agricultural Sciences and Technology Indicators) Initiative country studies. We also cap the maximum size of a crop-by-country programme at 80 FTE scientists, recognizing economies of size and scale in agricultural research. This admittedly arbitrarily imposed limit is slightly above the size of the largest programme – maize in Nigeria.

In order to achieve congruence or parity in research intensities across crops with a fixed budget, resources would have to be reassigned from the crops with positive estimates in Table 18.3 to the commodities with negative estimates. The sign and size of the estimates by crop are sensitive to our assumptions on a desirable target for research intensity, the cost of each FTE scientist and the limit on the size of the programme. The relative position of the crops in Table 18.3 will change somewhat as these assumptions vary but not as much as their numerical values. Assuming payoffs are the same – a very large and strong supposition – these more formal results reinforce the findings on research intensities in Table 18.2. Using the congruence rule to set priorities shows that research into cowpea, groundnut, pearl millet, sorghum and yams is underinvested in relative to other crops, from the perspective of the value of production.

Table 18.3. Comparing the actual allocation of FTE scientists in SSA to a normative allocation by crop programme.

	Simple average in FTE scientists		
Crop	Actual allocation	Normative allocation[a]	Difference
Banana	42.0	11.1	30.9
Wheat	17.5	8.5	9.0
Chickpea	13.5	4.9	8.6
Maize (ESA)	27.0	21.3	5.8
Barley	11.1	5.8	5.3
Pigeonpea	7.8	3.0	4.8
Soybean	3.9	1.2	2.7
Lentil	3.6	1.5	2.1
Field pea	6.9	5.2	1.7
Sweetpotato	6.5	6.3	0.3
Faba bean	7.8	8.4	−0.6
Beans	8.7	9.5	−0.9
Cowpea	4.5	5.6	−1.2
Maize (WCA)	12.7	14.7	−2.0
Rice	9.6	16.6	−7.0
Potato	7.6	14.7	−7.1
Groundnut	3.4	16.4	−13.0
Sorghum	6.6	20.3	−13.7
Pearl millet	4.1	27.8	−23.8
Cassava	8.2	32.1	−24.0
Yams	7.0	47.3	−40.4

[a]Assumes a research intensity of 1% of value of crop production, a cost per FTE scientist of US$115,000 and a maximum programme size of 80 FTE scientists.

Differences in scientific strength over time

Results on differences in scientific strength over time are mixed. Between 1998 and 2010 more programmes have gained scientists than have lost researchers but, because of rising production, estimates of research intensity have not improved and have even declined for the majority of the 65 country programmes with information available to carry out paired comparisons. Before addressing changes over time, we briefly examine the results of previous estimates of scientific staff strength in 1998 for SSA (Walker, Chapter 5, this volume).

1. Nigeria stood out as a country with consistently low researcher intensity. Indeed, Nigerian farmers appeared to be afflicted by some of the lowest readings on researcher intensity ever estimated anywhere in the world. Mean readings of the ratio of FTE scientists to million tonnes of production were 0.1 for cassava, 0.5 for sorghum, 1.7 for rice, 1.8 for pearl millet and 2.6 for maize, which benefited from some private sector participation in research. Nigeria ranked among the lowest in researcher intensity in each of the five commodity groups to which it was a major contributor. The country also figured prominently when the performance indicators for these same crops were aggregated.

2. Ethiopia, Kenya, South Africa and Sudan were characterized by a higher investment in scientific staff than other countries in the 1998 data set. This behaviour was reflected in positive and statistically significant estimated country

coefficients in an additive effects model regressing total scientists years on production, crops and countries.
3. Researcher intensity was lower in cassava than in other crops even when the relatively inferior output value of cassava was factored into the calculation. Rice and sorghum also had lower than expected research intensities, although not as extreme as cassava.
4. Estimates of researcher intensity declined exponentially as the size of production increased from under 50,000 tonnes to more than 5 million tonnes.

Data are available for a before-and-after analysis of the changes in scientific capacity for 65 matching crop-by-country observations that feature eight of the continuing crops (Table 18.4). Thirty of the 65 programmes had fewer FTE scientists in 2010 than in 1998. Among the 35 programmes that gained staff, two observations were unduly influential in the results: maize programmes in Nigeria and Zimbabwe both experienced increases that were equivalent to more than 40 FTE scientists.

Some of this change is undoubtedly real, but some may be attributable to an underestimation of scientific capacity in 1998, e.g. maize in Nigeria included substantially more university researchers in 2010 than in 1998. Excluding maize in Nigeria and Zimbabwe, the mean scientific strength in 1998 was 8.4 FTE scientists compared to 9.7 in 2010, resulting in a positive but statistically insignificant change at the 0.05 level. The median programme also gained 1.3 FTE scientists as the difference between the two time periods was normally distributed. Overall, these results suggest a marginal increase in scientific capacity.

Cassava appears in Table 18.4 as the largest loser of scientific capacity. Maize in ESA, potato, rice and wheat were the biggest gainers.

These gains in staff were not sufficient to translate into increased research intensity in most crops. The net decline in research intensity was about 1.7 scientists per million tonnes of production, which suggests that growth in production outstripped the smaller positive changes in staffing. Maize and wheat in ESA were the only crop categories that accrued substantial gains in researcher intensity (Table 18.4).[2]

A first-difference comparison of the bulk of the overlapping crop-by-country observations is presented in Fig. 18.1. For reasons of scale, three high-end outliers are excluded: maize in Kenya that had very large values in 1998 and 2010, and maize in Nigeria and Zimbabwe that had high values in 2010.

A small majority of the 62 remaining observations increased their numbers of scientific staff between the two periods. One of these was cassava in Nigeria which added about 6 scientists. Notably, we also see that several of the largest commodity programmes on the right-hand side of Fig. 18.1 could not sustain their staff strength. These were mainly concentrated in cassava-growing programmes. For example, Benin, Guinea and Tanzania downsized to only 2–3 scientists per programme.

For a few maize programmes in WCA, the numbers of scientific staff also declined over time. But these declines were more than compensated for by Nigeria's dramatic increase in

Table 18.4. Differences in estimated FTE scientists and research intensities between 1998 and 2010 by crop based on 65 paired comparisons.

Crop	Mean FTE scientists	Median FTE scientists	Mean research intensity	Paired observations
Bean[a]	–0.6	–0.8	1.3	8
Cassava	–2.4	–2.3	–4.7	14
Maize (ESA)	10.8	7.0	3.9	9
Maize (WCA)	4.3	–3.3	–32.4	9
Pearl millet	–1.1	–1.0	–0.9	5
Potato	6.9	3.6	–7.9	4
Rice	4.3	3.8	–4.3	6
Sorghum	1.9	1.4	–10.3	6
Wheat	7.3	8.5	110.9	4

[a]For bean, the definition of scientists applies only to breeders.

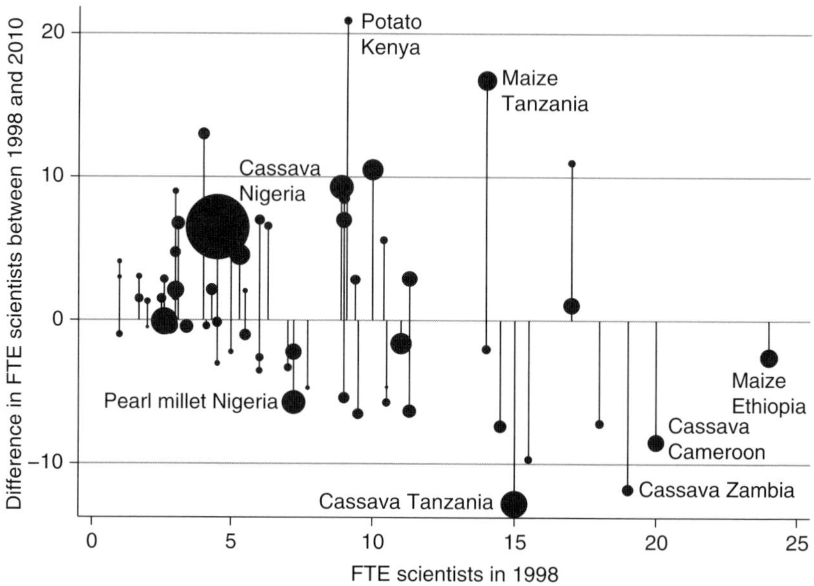

Fig. 18.1. Change in scientific staff strength in food crop improvement programmes between 1998 and 2010. The size of the circles reflects the size of production value in 2010. Note that Nigeria's observation for cassava is the largest circle in the bubble graph. (Source: DIIVA SY Database.)

scientific staff, discussed earlier. Overall, the data presented in Fig. 18.1 convey the message that larger crop improvement programmes may be highly susceptible to downsizing in times of financial crisis or when donor support ends.

Other aspects of scientific capacity: age, education and area of specialization

The problem of scientific capacity in NARS in West Africa is not only a problem of numbers but also of age. About 65% of the scientists working on groundnut, pearl millet and sorghum in the five project countries in West Africa were over 50 in 2010 (Ndjeunga *et al.*, Chapter 7, this volume).

Scientists engaged in crop improvement across WCA appear to be more highly educated than their ESA counterparts, with around 2.6 PhD holders per programme. But in future, an estimated lower number of BSc holders in WCA is a cause for concern because fewer younger scientists will be available to be mentored by, and capitalize on the experience of, older scientists (Table 18.5).

The incidence of scientists with PhDs and MSc qualifications is encouraging (Table 18.5). Only 24 of the 135 programmes did not have a PhD presence. Only four programmes had neither a PhD nor an MSc scientist involved directly in their research. More than half of the programmes have at least 1.0 FTE PhD scientist working in research. For the most part, all crops and most countries have at least one programme supported by several PhDs and MScs. Eritrea was the exception among the 30 countries in the Diffusion and Impact of Improved Varieties in Africa (DIIVA) Project. Nonetheless, it was still possible to find programmes, such as cassava in Tanzania, that were severely understaffed both numerically and educationally in 2010.

Staff stability is a primary ingredient for a recipe of sustained output from investing in crop improvement research (Eicher, 1995). Even with increasing participatory varietal selection (PVS) and marker-assisted selection (MAS) it can take, on average, about 10 years from parental crossing to progeny release in the same country. PVS is increasingly becoming a reality in rice and beans among the food crops in the DIIVA Project. MAS is still rare and newsworthy in SSA.

Table 18.5. Educational level of scientists in crop improvement programs by region in SSA.

	Number of observations	Mean number of FTE scientists by educational level			
		PhD	MSc	BSc	Total
ESA	65	1.51	3.20	2.33	7.03
WCA	70	2.61	2.84	1.66	7.12
Total	135	2.08	3.01	1.98	7.07

It has been applied to facilitate varietal development in only a few successful cases, such as sorghum in the Sudan (ICRISAT, 2013). The DIIVA Project sought to collect information on the duration of varietal generation, selection and testing. However, reliable data over time on this aspect of crop improvement performance were not forthcoming, so we cannot say whether the gestation period of new improved or modern varieties (MVs) is shortening or staying the same. We can say, though, that instability in scientific staffing levels within crop improvement programmes can severely curtail their potential. Full potential will only be reached if the routine work of varietal selection and testing takes place season after season and year after year.

Estimates on experience levels within the same area of research suggest that many scientific staff have been able to work on the same crop for an extended period of time. For example, the 289 NARS rice scientists had worked on rice improvement for an average of 12.25 years as of 2010 (Diagne *et al.*, Chapter 10, this volume). Scientists with 10 or more years' experience made up the majority of staff in five of the ten bean programmes in ESA (Muthoni and Andrade, Chapter 8, this volume). This level of experience was not anticipated because only about one scientist in six was older than 50 in 2010 across the ten bean improvement programmes.

Estimates on the allocation of scientists across specialized areas of crop improvement are presented in Tables 18.6 and 18.7 on two aspects: crop type and strength of scientific resources. We expect that relative allocations across areas of specialization will vary substantially across cereals, grain legumes, and roots and tubers. Root and tuber programmes that are based on vegetatively propagated material and on clonal selection are hypothesized to be characterized by a more diverse area allocation than cereals and grain legumes, which typically are more heavily concentrated in classical plant breeding. It was expected that increasing human resources would be accompanied by less concentration in plant breeding and agronomy, which are conventionally viewed as the core disciplinary areas of crop improvement research.

These expectations are largely confirmed in Tables 18.6 and 18.7, although the differences among programmes based on generalized crop orientation as well as small versus large programmes are not as obvious as anticipated. With regard to crop type, the main distinction focuses on roots and tubers on one hand, and cereals and grain legumes on the other. Root and tuber crop programmes invest considerably less in plant breeding and more in the biotechnological areas of molecular biology and tissue culture than cereal and grain legume programmes. With the exception of postharvest research, the other research areas are surprisingly similar across the generalized crop types. The emphases in entomology, pathology, agronomy and social science are not markedly different across the three groups of crops.

Three other findings in Table 18.6 warrant comment. First, molecular biology only accounts for 3.4% of the mean resources across the 150 programmes in the database. This level of investment is not significantly different from tissue culture, which has been a staple area in root and tuber crop improvement since the 1970s. The 3.4% is equivalent to only 40 FTE scientists, 17 of whom work on banana in Uganda. Second, the level of social science involvement in crop improvement work is much higher than 5%, which was expected. Third, postharvest work is concentrated on maize and cassava in Nigeria.

It was also anticipated that smaller programmes would have a higher concentration of disciplinary resources vis-à-vis larger programmes.

Table 18.6. Relative allocation of scientists by disciplinary specialization across roots and tubers, grain legumes and cereals in SSA in 2010 (%).

Broad areas of crop improvement work	Root and tuber crops (5)a	Grain legumes (8)	Cereals (7)	All 20 crops
Plant breeding including germplasm conservation	21.8	45.8	44.39	39.6
Plant pathology	8.3	10.9	7.80	9.2
Molecular biology and genetic engineering	11.4	0.5	1.22	3.4
Tissue culture	11.9	0.1	0.40	3.0
Entomology and nematology	5.4	6.1	7.38	6.3
Agronomy, weed science and seed production	25.2	24.6	23.68	24.4
Social science	8.7	10.3	9.36	9.6
Postharvest and food science	5.0	0.6	4.55	3.6
Other areas including soil science	1.2	0.2	0.20	0.6

[a]Numbers in parentheses refer to the number of observations in each crop category.

Table 18.7. Relative allocation of scientists by disciplinary specialization across programme-size quartiles in SSA in 2010 (%).

Broad areas of crop improvement work	Quartile 1	Quartile 2	Quartile 3	Quartile 4
Plant pathology and virology	5.9	7.6	11.19	7.1
Molecular biology and genetic engineering	1.0	2.3	2.44	3.7
Tissue culture	3.2	3.9	2.74	3.3
Entomology and nematology	3.9	7.4	11.01	5.4
Agronomy, weed science and seed production	24.3	20.4	15.51	20.5
Seed production	7.9	8.4	6.38	10.3
Social science	2.8	6.6	12.90	8.6
Postharvest and food science	1.6	3.4	5.1	5.3
Total FTE scientists	63.1	137.9	292.0	796.1

Indeed, the largest programmes in quartile 4 in Table 18.7 display a more even disciplinary allocation pattern across disciplines than the smallest programmes in quartile 1, but the differences are milder than expected. On average, even the smallest programmes from the perspective of total scientists invest about half of their resources in disciplines other than plant breeding. Nevertheless, the smallest programmes invest relatively few resources in molecular biology, entomology, social science and postharvest research compared to programmes in the quartiles with higher relative allocations. By contrast, the relative research allocations to tissue culture, pathology, agronomy and seed production do not vary systematically by size of the programme. This lack of response to programme size suggests that these areas are viewed as essential services for crop improvement.

The term 'essential' should not convey the notion that all programmes are active in these areas. Fifty of the 150 programmes do not have any representation in pathology, which historically has been one of the most productive areas in plant breeding in screening for varietal resistance and tolerance to economically important plant diseases. Investment in entomology in grain legumes was also lower than expected given the potential importance of damage from insect pests.

Comparing the disciplinary allocations in crop improvement programmes in SSA to those documented in South Asia in Chapters 13 and 14 suggests one similarity and two contrasts. Like the national programmes in SSA, biotechnology accounted for less than 5% of the total FTE scientists engaged in crop improvement by crop in both Chapters 13 and 14. However, rice research in South Asia is heavily concentrated in plant breeding and genetics (Chapter 13, this volume). A second contrast focuses on the role of pathology and entomology in dryland crop research. They figure more prominently in India than in SSA where agronomy and its related disciplines (see Table 18.6) loom larger (Chapter 14, this volume).

Revisiting the hypotheses about FTE scientists

Seven input-related hypotheses were put forth in Chapter 3. Two of these positively stated hypotheses were rejected from the empirical data on FTE scientists generated in the DIIVA Project. Most importantly, disparities in research intensities across regions and crops *were* substantial. Relative to their size of regional production, national crop improvement programmes in ESA had invested more in scientific capacity than comparable programmes in WCA. Concerns about scientific capacity in national programmes in West Africa reflect not only a problem of relative numbers but also of scientist age. About 65% of the scientists working on sorghum, pearl millet and groundnut in the five project countries in West Africa were older than 50 in 2010. Moreover, a lower incidence in BSc-holders in crop improvement research in WCA is a cause for concern because fewer younger scientists are available for mentoring by older, experienced scientists.

Of the 20 crops, cassava, yams and pearl millet consistently ranked at the bottom on research intensity. Relative to their area, production and value of production, all three of these semi-subsistence food crops appear to be losing the race for research resources. In terms of harvested area, groundnut and sorghum in West Africa are also characterized by very low research intensities.

Drought in SSA was the cause of 17 of the 100 most damaging natural disasters that occurred worldwide in the 20th century (CRED, 2014). The highest incidence of these drought-induced production shortfalls occurred in the arid and dry semi-arid tropics where pearl millet, sorghum, groundnut and cowpea are the major food crops. That these crops still lag far behind others in estimated research intensity is one of the most disappointing findings of the DIIVA Project. Without greater investments in agricultural research, the cost of relief efforts will continue to rise in the Sahel and the Horn of Africa as rural populations increase and climatic change becomes an increasing reality.

With the exception of root and tuber crops in a few country programmes, evidence for investments in biotechnology was also less visible than expected. The DIIVA studies in SSA do not show that *the disciplinary distribution of FTE scientists in crop improvement reflects an increasing capacity in biotechnology*. A similar finding was obtained at the national level in South Asia as addressed in Chapters 13 and 14: not much evidence was encountered for the increasing salience of biotechnology.

Results on the differences in scientific strength over time echoed the findings of Beintema and Stads (2006). Between 1998 and 2010, slightly more programmes have gained than have lost researchers. However, because of rising levels of crop production, mainly attributed to area expansion, estimates of research intensity have not increased and, indeed, have even declined for most of the 65 programmes that have information available to carry out paired comparisons. Therefore, we cannot state that *research intensities in national food crop improvement programmes are increasing*. We find solid support, however, for the hypothesis that *the number of FTE scientists in national food crop improvement programmes in SSA is increasing* because the gainers employed more scientists than the losers who reduced staff. The paired comparisons over time also suggest that larger public sector crop improvement programmes might be highly susceptible to downsizing in times of financial crisis or when donor support ends.

Evidence for *rising private-sector participation in research in the genetic improvement of cereal hybrids* divided along regional lines. By far, the largest increase in scientific capacity has occurred in maize across ESA, thanks largely to favourable government policies, such as relaxation on the government's exclusive right on the production of breeders' seed and the dynamism of the private sector in this region. In contrast, private-sector participation in the generation and distribution of cereal hybrids in WCA has stagnated in many large-producing countries. At best, participation seems to have stagnated or, at worst, regressed compared to what was documented in

Manyong *et al.* (2003). Unlocking constraints to greater private-sector participation in hybrid cereal production West Africa is as relevant today as it was in the late 1990s.

Some support was also uncovered for the hypothesis that *university participation in research is becoming increasingly visible from a small base.* Support was most transparent for maize improvement in Nigeria. For all crops other than maize, however, research scientists came overwhelmingly from the public-sector NARS.

Varietal Output: Released Varieties

As discussed in Chapter 3, varietal output is synonymous with released varieties broadly defined. 'Output' refers to the expansion that can be attributed to genetic improvement in the potential availability of genotypes for cultivation. Attribution is measured from a with-and-without perspective, i.e. the difference in potential availability from genetic improvement and what would have been available without an investment in plant breeding. Therefore, released varieties include many cultivars that are not officially notified including so-called informal introductions, escapes and private-sector materials that may not be officially notified but which are available to farmers. Restricting varietal releases to government-notified materials will severely understate output from crop genetic improvement that is potentially available for adoption (Alene *et al.*, Chapter 6, and De Groote *et al.*, Chapter 11, this volume).

Findings on varietal output in 1998

In the 1998 Initiative, most CGIAR (Consultative Group on International Agricultural Research) participants were successful in assembling valid release data for almost all countries, supplemented by information on so-called informal releases of suspected improved varieties. For maize in ESA, release was equated to varietal availability in the market in the late 1990s because of heavy private-sector participation in seed production and distribution. In spite of the inherent difficulties in inferring varietal output from varietal release, such data present an historical benchmark that, once consolidated carefully, can provide a firm foundation for updates over time.

In the pooled analysis of varietal release covering 1965–1998 reported in Chapter 5, we found:

1. Across all crops, annual releases increased at an accelerating rate from the 1960s to the late 1990s.
2. Political instability adversely affected varietal output in some crops in key countries in the 1990s, such as potatoes in Rwanda.
3. Some crops were characterized by high numbers of releases prior to 1975. The crop improvement programmes of the CGIAR were most likely a force that contributed to offsetting differences in initial advantage in research endowments because most CG Centers reached their full potential to generate varietal output in the 1990s.
4. Across the eight food crops in the study, the higher and more stable release rate in wheat was anticipated. In contrast, the very low release intensity for cassava was unanticipated.
5. Release profiles were often punctuated by bursts of activity sandwiched between long periods of inactivity.

Varietal output by 2010

Updating the database for the continuing crops and assembling fresh historical data on varietal output for the new crops in Table 18.8 broadly confirms the five findings cited above from the analysis of the 1998 data.

The historical data on varietal output across the 20 crops contain 3594 entries. About 90% of these have information on the year of release. The undated entries are associated with modern materials that were judged to be available to farmers or are located in countries that do not maintain a formal release registry. Many of these come from the International Institute of Tropical Agriculture (IITA; Alene *et al.*, Chapter 6, this volume) and are listed as 'informal' releases. Participants were encouraged to add escapes and other adopted materials perceived as modern to the release database so that information on their identity and characteristics was available (Walker, 2010). Most, but not all, of the dated entries in Table 18.8 imply official release.

Maize leads all crops with over 1000 entries in the cultivar-release database. Rice is second

Table 18.8. Counting the number of cultivars in the varietal release database by crop in SSA from before 1970 to 2011.[a]

Crop	Number of countries	Number of cultivars in the varietal release data	Number of released cultivars with year of release information	Output intensity (total releases/ million ha) in 2009
Banana	1	13	6	14
Barley	2	41	41	42
Bean	9	250	232	100
Cassava	17	355	207	32
Chickpea	2	27	26	108
Cowpea	17	200	157	17
Faba bean	2	28	28	46
Field pea	1	26	26	113
Groundnut	10	140	137	22
Lentil	3	15	14	158
Maize (ESA)	8	692	664	47
Maize (WCA)	11	330	271	33
Pearl millet	5	121	120	9
Pigeonpea	3	17	17	46
Potato	5	117	117	190
Rice	11	436	428	64
Sorghum	8	174	180	11
Soybean	15	201	156	170
Sweetpotato	5	89	89	60
Wheat	5	244	243	146
Yam	8	78	35	17
Total/average	148	3594	3194	68

[a]This count also includes the same cultivar released in different countries under a different name.

with over 400. Both maize in ESA and rice have had access to multiple institutional sources of modern genetic materials.

A simple index of output intensity can also be constructed for comparative analysis across crops. In Table 18.8, output intensity is expressed in terms of total releases per million hectares (ha) in 2009. Similar to research intensity, we expected the results to show that less extensively grown crops are characterized by higher levels of output intensity. Indeed, this expectation was confirmed for lentil, soybean, potato and wheat, all of which were associated with strong market demand. Additionally, during the mid-20th century both wheat and potato in SSA benefited from a strong programme of genetic improvement thanks to the Rockefeller Program in Mexico. The genetic base for many released varieties in SSA came from that early work.

At the other end of the spectrum, five crops fell under the low threshold of less than 20 cultivars released per million hectares of harvested area in 2009. Low research intensities in pearl millet and sorghum have translated into low output intensities. The same finding applies to countries producing cowpea. Relatively few varieties have been released recently (Alene and Mwalughali, 2012). A low estimated research intensity for banana is derived from the observation that hybridization is still difficult. More than all other crops in Table 18.8, low output intensity in yams is attributed to historically negligible levels of research investment.

The parity between the output intensity of cassava and maize in WCA is perhaps the most interesting finding in Table 18.8. The total number of releases and their total harvested area are almost identical for the two crops. The example of cassava suggests that low research intensity does not preordain mediocre performance in output.

Varietal output over time

Tracking cultivar release over five decades supports the anticipated finding that varietal output has been increasing over time. About 45% of the

3194 dated entries in Table 18.8 were released since 2000 (Table 18.9). The mid-point for data release was 1998. Decade by decade, the incidence of release has steadily increased over time.

Not all crops fit the pattern of a steady rise in varietal output over time. In ESA, varietal output rose exponentially in maize between the 1990s and the 2000s because of surging private-sector releases. Groundnut displays a flat trajectory in output for more than four decades and then output rises abruptly from 2000. Unfortunately, this increase in releases is confined mainly to smaller-producing countries in ESA. Meanwhile, WCA is still associated with stagnation in the incidence of released varieties, e.g. varietal output in cowpea has declined sharply from its peak in the 1990s.

Three cereals have also not been able to maintain an increase in varietal production. Varietal output in pearl millet peaked in the 1980s. Meanwhile, varietal performance in sorghum tapered off in the 2000s. Constricting resources both internationally and nationally have played a role in limiting varietal output in pearl millet and sorghum in West Africa prior to the rise in food prices in 2008.

Table 18.9. The frequency of cultivar release by decade by crop in SSA.

	Released varieties and hybrids by decade				
Crop	Pre-1970	1970s	1980s	1990s	2000s[a]
Banana	0	0	0	0	6
Barley	0	3	3	4	31
Bean	1	6	22	73	130
Cassava	0	2	31	61	113
Chickpea	0	3	2	9	12
Cowpea	3	8	49	65	32
Faba bean	0	3	2	8	15
Field pea	0	2	2	10	12
Groundnut	20	23	25	21	48
Lentil	0	0	4	5	5
Maize (ESA)	7	10	34	159	455
Maize (WCA)	12	25	75	76	82
Pearl millet	1	7	46	28	38
Pigeonpea	0	0	3	2	12
Potato	3	18	29	24	43
Rice	27	53	133	138	77
Sorghum	2	25	36	63	54
Soybean	2	13	32	52	57
Sweetpotato	0	0	9	20	60
Wheat	20	43	43	40	97
Yam	0	0	0	5	30
Total	98	244	580	863	1409

[a]The end year for the period is either 2009, 2010 or 2011, depending on the crop.

In spite of the widespread introduction of the New Rice for Africa (NERICA) varieties starting in the mid-1990s in most rice-growing countries in SSA, varietal release also slowed in rice in the 2000s. Political instability and civil war in Côte d'Ivoire and Sierra Leone severely curtailed releases caused by the closure of several rice research stations. With the exception of Senegal, West Africa shows a downturn in releases in the 2000s compared to the 1980s and 1990s. Even in Guinea, where varietal output exceeded 100 varieties in the 1980s and 1990s, rice releases are becoming increasingly rare.

Releases in the post-1998 period are described in Table 18.10. Five crops have been able to maintain a simple average annual release rate of at least one variety released per programme. Fuelled by Kenya's and Zambia's high production – with over 100 varieties released since 1998, mostly by the private sector – maize in ESA easily tops the list at five varieties released per annum per programme. Seven of the eight maize-growing countries released more than 29 varieties during this recent period.

In general, releases were unevenly distributed across countries within each crop. Thirty country programmes reported no releases, and 45% of the 148 crop–country programmes released fewer than five varieties during the 12-year period. The country with the most releases often accounted for more than one-third of the total releases and, in the case of yams in Côte d'Ivoire, the vast majority of total releases. In contrast with cowpea, none of the 17 countries in the data set released more than ten varieties in the 10-year period.

Wheat's position near the top of Table 18.10 in weighted annual release rate was anticipated. Ethiopia is by far the largest producer and recently has been prolific in varietal release, which explains why the weighted annual rate is substantially higher than the simple annual rate. The release performance of the smaller wheat-growing countries of Kenya, Tanzania, Zambia and Zimbabwe has slowed somewhat recently.

Table 18.10. Performance in varietal release from 1999 to 2011 by crop improvement programme.

		Annual release rate		Total releases[b]	
Crop[a]	Total releases	Simple	Weighted by area	Maximum	Minimum
Maize (ESA)	485	5.1	5.1	143	0
Wheat	106	1.8	4.0	53	5
Barley	31	1.3	2.2	28	3
Bean	148	1.4	1.4	27	8
Maize (WCA)	91	0.6	1.4	37	0
Yam	30	0.3	1.3	23	0
Cassava	128	0.6	1.2	20	0
Sweetpotato	66	1.1	1.1	28	1
Faba bean	15	0.6	1.0	14	1
Field pea	12	1.0	1.0	12	12
Chickpea	12	0.5	1.0	12	0
Potato	47	0.8	0.8	24	1
Sorghum	58	0.6	0.6	30	0
Banana	6	0.5	0.5	6	6
Rice	77	0.6	0.5	23	0
Cowpea	34	0.2	0.5	8	0
Soybean	61	0.3	0.4	16	0
Pearl Millet	39	0.7	0.4	17	1
Pigeonpea	12	0.3	0.4	6	2
Groundnut	46	0.4	0.4	9	0
Lentil	5	0.1	0.3	4	0

[a]The crops are ordered by annual release rate weighted by area in column 4. [b]The maximum and minimum refer to country programmes over the release period and not individual years.

Ethiopia's sustained efforts in varietal release also explain barley's ranking near the top of Table 18.10. Moreover, a decentralized regional research emphasis has reinforced release activities in Ethiopia. The buoyancy and productivity of the aforementioned PABRA network – the umbrella organization that oversees three regional genetic networks in SSA – contributed heavily to the release performance of beans in the recent period. Sweetpotato programmes also released varieties at a rate of more than 1% per annum. The fruition of a longstanding CIP (International Potato Center)-supported breeding programme in Mozambique made a substantial contribution to this output.

At the lower end of Table 18.10, there are the same crops that displayed lagging levels of human resources investment in genetic improvement programmes. The estimated release rate for cowpea, groundnut, pearl millet and sorghum indicate one release per country programme every 3–5 years.

The low position of soybean for the recent period in Table 18.10 is a surprise for an expanding commercial crop from a very small production base in most countries. Such countries are probably following a cost-effective strategy of capitalizing on finished materials from other tropical and semi-tropical countries, especially Brazil and Argentina. Nevertheless, those varieties should still appear in the varietal registries maintained by countries in SSA.

Between one-fifth and one-quarter of the 146 crop-by-country observations were characterized by more releases in the 1980s than in the 2000s. These observations are identified in Fig. 18.2 by the number of releases in the 1980s and the change in releases between the two periods. Results imply declining productivity in crop improvement varietal output over time. Some of these observations were casualties of civil war during the 1990s and early 2000s.

Civil war, as a major explanation for falling varietal output, applies to rice in Sierra Leone, potato in Rwanda and rice in Côte d'Ivoire. For other observations, the explanation for their presence in Fig. 18.2 seems to be country or region specific. Most of the observations come

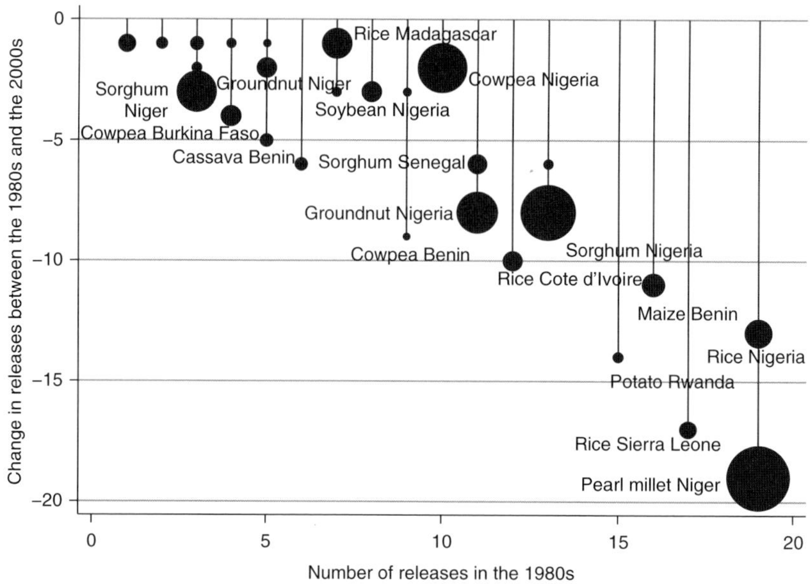

Fig. 18.2. Crop-by-country observations with more releases in the 1980s than in the 2000s.

from West Africa. As the balloons in Fig. 18.2 show, Nigeria accounts for a large share of total area of all the observations. Cowpea, groundnut, pearl millet, rice and sorghum are well represented in Fig. 18.2. With the exception of rice, these crops finished at the bottom of Table 18.2 in estimated research intensities in 2010.

The historical record on CGIAR contributions to varietal output

The commodity centres of the CGIAR can leverage varietal output through the direct distribution of elite material and their finished varieties, progenies for selection, and parents for direct crossing by NARS. About 43% of the varieties released since 1980 in Table 18.9, or some 1500 varieties, are related to the work of the CGIAR.

The CGIAR contribution is greater than 40% for the majority of crops in Table 18.11. In several cases, two or more CG Centers contribute to varietal releases of the same crop. Notable examples of joint contributions include ICRISAT and ICARDA (International Center for Agricultural Research in Dry Areas) for chickpea; IITA and CIMMYT (Centro Internacional de Mejoramiento de Maiz y Trigo) for maize in WCA; and AfricaRice, the International Rice Research Institute (IRRI), and IITA for rice (before IITA closed its rice programme).

The six crops below the 40% contribution level in Table 18.11 are suitable candidates for discussion about why their estimates are lower than those of other crops. Barley and field pea are primarily grown in Ethiopia and are researched in a strong NARS setting where the crops have considerable genetic diversity as a locus of domestication.

Other institutional suppliers play a large part in the reported estimates for banana and maize in ESA. The Honduras Foundation for Agricultural Research (FHIA) has contributed significantly to the improvement of banana in SSA, especially in finding cultivars resistant to *Fusarium* wilt – a soil-borne fungal disease – in the brewing, cooking and dessert types of banana.

Between 1958 and 2010, the private sector – without direct participation from other institutions – was responsible for 56% of maize releases in ESA (De Groote *et al.*, 2011). In Chapter 11, the CGIAR is credited with a 23% share of improved maize variety releases, together with NARS and the private sector. This estimate is substantially higher than what is currently shown in the DIIVA database but even a 23% contribution to varietal output is low compared to estimates for other crops in Table 18.11.[3] Historically, the public

Table 18.11. The contribution of IARCs of the CGIAR to varietal output in SSA 1980–2011.

Crop	Number of dated released varieties related to CGIAR activity	Share of CGIAR-related varieties to total dated releases (%)
Chickpea	23	95.8
Lentil	13	86.7
Pigeonpea	14	82.4
Potato	72	75.0
Yam	26	74.3
Maize (WCA)	173	74.2
Cassava	143	68.1
Sweetpotato	59	66.3
Cowpea	88	57.5
Rice	179	51.4
Soybean	69	48.9
Wheat[a]	81	45.0
Groundnut	41	43.6
Pearl millet	45	40.2
Faba bean	10	40.0
Bean	88	39.1
Maize (ESA)	171	22.8
Sorghum	38	24.8
Barley	8	21.1
Banana	1	16.7
Field pea	4	16.7

[a]The share estimate for wheat is understated because data collected in the smaller-producing countries did not contain information on the institutional source of genetic material since 2000.

sector's contribution to varietal research declines when the private sector becomes established in cross-pollinated crops that can be readily hybridized (Fuglie and Walker, 2001). The private sector is well established in Kenya, Zambia and Zimbabwe where hybrids dominate the market.

The 39% estimate for beans approaches the average level of CGIAR contribution in Table 18.11. Multiple smaller institutional providers have added a global perspective to CIAT's primary role as a source of genetic materials for the generation of bean varietal output in ESA. These include the Bean and Cowpea Collaborative Research Support Program (CRSP (recently renamed Innovation Laboratory)) in the USA, Institute of Horticultural Plant Breeding (IVT) in the Netherlands, Escuela Agricola Panamericana Zamorano (EAP) in Honduras, Centro Agronómico Tropical de Investigación y Enseñanza (CATIE) in Costa Rica, National Vegetable Research Station (NVRS)–Wellsbourne Project in the UK and the Tokachi Agricultural Experiment Station in Japan. Genetic materials from the genebank in Beltsville, Maryland, USA, have also figured prominently in several varietal releases.

The Institut de Recherches Agronomiques Tropicales (IRAT) now Agricultural Research for Development (CIRAD) has played a large role in generating materials that have resulted in varietal change in several food crops in West Africa. CIRAD also works on non-staple crops and has historically placed less emphasis on genetic enhancement than has CGIAR. But the relatively low level of CGIAR contribution to sorghum releases in West Africa is not related to strong NARs in centres of diversity or to alternative suppliers of material. The overly aggressive pursuit of a breeding strategy focusing on shorter statured, photoperiod-insensitive material is a plausible explanation of why ICRISAT's contribution is not higher, especially in West Africa (Ndjeunga *et al.*, 2012). Farmers strongly prefer tall, photoperiod-sensitive Guinean types of sorghum.

The commodity centres in the CGIAR mostly date from the late 1960s and the early 1970s. We would expect to see a rising contribution from CG-related materials over time from 1980. That expectation is confirmed here, between the 1980s and the 1990s, as the CGIAR share in varietal output rose from 42–46% (Table 18.12). But, contrary to our expectation, the role of the CGIAR declined in the 2000s compared to the 1990s.[4] This decline could be attributed to the funding crisis in the mid-to-late 1990s and early 2000s when the exchange of germplasm and genetic materials became more constricted. The increasing rate of private sector releases in maize in ESA, especially in Kenya and Zambia with more than 100 releases since 2000, has directly had a dampening effect on the CGIAR share. When maize in ESA is omitted, the revised estimate in the second row of Table 18.12 shows a plateauing of the CGIAR contribution at about 56% in the 1990s and 2000s.

Revisiting the hypotheses about varietal output

Our findings broadly support the hypothesis that the stock of released and non-released improved

Table 18.12. IARC-related percentage share estimates over time with and without maize in ESA.

Basis for the estimation	1980s	1990s	2000s	Average share
All crops and regions in the database	41.5	45.8	41.0	42.8
Without maize in ESA in the estimation	43.6	55.9	56.2	51.9

varieties that are potentially available to farmers for use is increasing over time. Relative to the levels of their production, however, pearl millet, sorghum, groundnut, cowpea and yam lag behind other crops in output of improved varieties.

Looking into the future, we expect that the upward trend in varietal output will continue. The food price crisis in 2008 has led to greater funding for agricultural development in general and agricultural research in particular. Although slow in coming, greater regional harmonization of plant regulations should also stimulate varietal output at the national level.

The evidence is also positive but not as robust for the hypothesis that *output stability is increasing over time because peaks and troughs in varietal generation are less evident than in the recent past.* Although seemingly improving, stability in varietal releases documented in Chapters 6–12 for SSA pales in comparison to what is described in Chapters 13 and 14 for South Asia. Indeed, the hallmark of the release profiles in South Asia has been the stability of varietal release over time. For example, the All-India Coordinated Sorghum Improvement Program of ICAR released an improved variety or hybrid at either the central or state level annually from 1961 to 2010 in all but three years (Chapter 14, this volume). For the five major dryland crops produced in peninsular India, a minimum of 20 releases were registered in each of the five decades from 1960. By way of contrast, rice improvement in Nigeria was the most stable food-crop research programme in the DIIVA varietal release database. Between 1954 and 2005, 57 improved varieties in the public sector FARO (Federal Agricultural Research Oryzae) series were released for cultivation. But, during that period that spanned more than five decades, varieties were released in only 23 years. Multi-year gaps in release were common. The stability of varietal release in crop improvement programmes in South Asia speaks for their durable funding, scientific staffing and research organization.

The last hypothesis in Chapter 3 on varietal output was not supported to the extent that was expected. In general, we did not find persuasive evidence that *varietal output reflected the evolution of plant breeding over time or a lower IARC presence.* The incidence of direct crossing was less than expected in most crops in the DIIVA Project release database. Even large NARS programmes, such as rice in Nigeria, still rely heavily on introduced finished varieties, although they generated and released varieties from direct crosses in-country as early as the mid-1980s. Releases from landraces continue to figure prominently in a sizeable minority of programmes in the 2000s.

IARC presence and contributions seem to be as relevant now as they were in the past. In contrast, the role of IRRI-related germplasm is diminishing in the varietal output of rainfed rice research programmes in South Asia described in Chapter 13. A reduction in IARC influence testifies to the increasing maturity of those programmes that were documented globally by Evenson and Gollin (2003). That this global finding about the maturity of national plant breeding programmes still does not apply to countries in SSA is troubling.

Notes

[1] Several chapters in this volume also report on the scientific strength of relevant CG Centers. Trends in staff strength in crop improvement are described in Walker *et al.* (2014) for specific IARCs. In general, the number of scientists in crop improvement programmes declined sharply in the CGIAR from the mid-1990s through to the early 2000s.

[2] Declining production in a crop can lead to rising research intensity. But among these observations, only wheat in Zimbabwe seems to demonstrate increasing research intensity attributed to secularly decreasing output.
[3] Recently, CIMMYT and IITA have partnered in the release of more than 100 varieties in SSA as part of the Drought Tolerance for Maize in Africa Initiative.
[4] This could be viewed as progress: NARS and the private sector are taking on additional responsibilities, freeing the CGIAR to focus on basic research.

References

Alene, A.D. and Mwalughali, J. (2012) *The Effectiveness of Crop Improvement Programs in Sub-Saharan Africa from the Perspectives of Varietal Output and Adoption: The Case of Cassava, Cowpea, Maize, Soybean and Yam.* Diffusion and Impact of Improved Varieties in Africa (DIIVA) – Objective 1 Report. International Institute of Tropical Agriculture, Lilonge, Malawi.

Alston, J.M., Norton, G.W. and Pardey, P.G. (1995) *Science under Scarcity: Principles and Practice for Agricultural Research Evaluation and Priority Setting*. Cornell University Press: Ithaca, New York (republished in soft cover by CAB International, Wallingford, UK, 1998).

Beintema, N.M. and Stads, G.J. (2006) *Agricultural R&D in Sub-Saharan Africa: An Era of Stagnation*. International Food Policy Research Institute, Washington, DC.

Beintema, N.M. and Stads, G.J. (2011) *African Agricultural R&D in the New Millennium: Progress for Some, Challenges for many*. IFPRI Food Policy Report. International Food Policy Research Institute, Washington, DC.

Centre for Research on the Epidemiology of Disasters (CRED) (2014) *EM-DAT: The International Disaster Database.* Université Catholique de Louvain, Brussels.

De Groote, H., Gitonga, Z. and Mugo, S. (2011) *Diffusion and Impact of Improved Maize and Wheat Varieties in East and Southern Africa.* Diffusion and impact of improved varieties in Africa (DIIVA) – Objective 1 Report. CIMMYT, Nairobi.

Eicher, C.K. (1995) Zimbabwe's maize-based Green Revolution: preconditions for replication. *World Development* 23, 805–818.

Evenson, R.E. and Gollin, D. (eds) (2003) *Crop Variety Improvement and Its Effect on Productivity: The Impact of International Agricultural Research*. CAB International, Wallingford, UK.

Fuglie, K.O. and Walker, T.S. (2001) Economic incentives and resource allocation in US public and private plant breeding. *Journal of Agricultural and Applied Economics* 33(3), 459–475.

ICRISAT (2013) *ICRISAT Annual Report*. Patancheru, India.

Kislev, Y. (1977) A model of agricultural research. In: Arndt, T.M., Dalrymple, D.G. and Ruttan, V. (eds) *Resource Allocation and Productivity in National and International Agricultural Research*. University of Minnesota, Minneapolis, Minnesota, pp. 265–280.

Manyong, V.M., Kling, J.G., Makinde, K.O., Ajala, S.O. and Menkir, A. (2003) Impact of IITA germplasm improvement on maize production in West and Central Africa. In: Evenson, R.E. and Gollin, D. (eds) *Crop Variety Improvement and Its Effect on Productivity: The Impact of International Agricultural Research*. CAB International, Wallingford, UK, pp. 159–182.

Maredia, M.K. and Eicher, C.K. (1995) The economics of wheat research in developing countries: the hundred million dollar puzzle. *World Development* 23(3), 401–412.

Ndjeunga, J., Hash, C.T., Faye, I., Sanogo, M., Echekwu, C.A. *et al.* (2012) *Assessing the Effectiveness of Agricultural R&D in West Africa: Cases of Pearl Millet, Sorghum and Groundnut Crop Improvement Programs.* Diffusion and Impact of Improved Varieties in Africa (DIIVA) Project – Objective 1 Report. ICRISAT, Niamey, Niger.

Walker, T., Pitoro, R., Tomo, A., Sitoe, I., Salencio, C. *et al.* (2006) *Priority-Setting for Public-sector Agricultural Research in Mozambique with the National Agricultural Survey Data*. IIAM–DFDTT Research Report No. 3E. Instituto de Investigacao de Agricultura Mozambicana: Maputo, Mozambique.

Walker, T.S. (2010) *Guidelines for Data Collection for Objective 1 of the DIIVA Project.* Fletcher, North Carolina.

Walker, T., Alene, A., Ndjuenga, J., Labarta, R., Yegezu, Y. *et al.* (2014) *Measuring the Effectiveness of Crop Improvement Research in Sub-Saharan Africa from the Perspectives of Varietal Output, Adoption and Change: 20 Crops, 30 Countries, and 1150 Cultivars in farmers' fields*. Report of the Standing Panel on Impact Assessment (SPIA), CGIAR Independent Science and Partnership Council (ISPC) Secretariat, Rome.

19 Varietal Adoption, Outcomes and Impact

T.S. Walker,* J. Alwang, A. Alene, J. Ndjuenga, R. Labarta, Y. Yigezu, A. Diagne, R. Andrade, R. Muthoni Andriatsitohaina, H. De Groote, K. Mausch, C. Yirga, F. Simtowe, E. Katungi, W. Jogo, M. Jaleta, S. Pandey and D. Kumara Charyulu

Parallel to the preceding chapter, we synthesize the results of Chapters 6–17 here. The focus is on outcomes and impacts. Outcomes centre on varietal adoption and turnover; impacts refer to changes in on-farm productivity, poverty and food security. Hypotheses from Chapter 3 are revisited at the end of each thematic section.

Varietal Adoption

By crop

The area-weighted grand mean adoption level of improved varieties in Sub-Sharan Africa (SSA) across the 20 crops in the project is 35% (Table 19.1). Two-thirds of the crop entries in Table 19.1 fall below the mean estimate. Starting at the bottom of the table, the limited uptake for improved field pea, which is produced primarily in Ethiopia, is not surprising. Internationally and nationally, field pea is arguably the crop species in Table 19.1 that has had the smallest amount of resources allocated to its improvement.

In contrast, both chickpea and lentil have benefited from international agricultural research in the CGIAR (Consultative Group on International Agricultural Research) since the early-to-mid-1970s. Although progress has been made, adoption of improved cultivars of both crops is concentrated in small pockets of production regions in Ethiopia where extension programmes have been active (Yigezu *et al.*, 2012a). This apparent location specificity is typical of pulse crops, but it is surprising in light of improved lentil varieties that have reportedly significantly heavier yields than their local counterparts.

Adoption levels of faba bean and chickpea are buoyed by a reportedly higher penetration of improved varieties in the Sudan. Indeed, chickpea in the Sudan is the only crop-by-country observation to have been at full adoption level in 2010, albeit on a very small area of 21,000 ha

* E-mail: walkerts@msu.edu

Table 19.1. Adoption of modern varieties (MVs) of food crops in sub-Saharan Africa in 2010.

Crop	Country observations	Total area (ha)	Adopted area (ha)	MVs (%)
Soybean	14	1,185,306	1,041,923	89.7
Maize (WCA)	11	9,972,479	6,556,762	65.7
Wheat	1	1,453,820	850,121	62.5
Pigeonpea	3	365,901	182,452	49.9
Maize (ESA)	9	14,695,862	6,470,405	44.0
Cassava	17	11,035,995	4,376,237	39.7
Rice	19	6,787,043	2,582,317	38.0
Potato	5	615,737	211,772	34.4
Barley	2	970,720	317,597	32.7
Yam	8	4,673,300	1,409,309	30.2
Groundnut	10	6,356,963	1,854,543	29.2
Bean	9	2,497,209	723,544	29.0
Sorghum	8	17,965,926	4,927,345	27.4
Cowpea	18	11,471,533	3,117,621	27.2
Pearl millet	5	14,089,940	2,552,121	18.1
Chickpea	3	249,632	37,438	15.0
Faba bean	2	614,606	85,806	14.0
Lentils	1	94,946	9,874	10.4
Sweetpotato	5	1,478,086	102,143	6.9
Banana	1	915,877	556,784	6.2
Field pea	1	230,749	3,461	1.5
Total/weighted average	152	107,721,630	37,969,577	35.25

(Yigezu *et al.*, 2012a). Meanwhile, Ethiopia harvests more than 0.5 million ha of faba bean, yet the perceived adoption of improved cultivars is very low at 3.5%.

Cooking, dessert and beer bananas in Uganda are also characterized by low adoption. This finding is not that surprising. Stimulating varietal change in a clonally propagated crop – and one that is not an annual – is a challenging proposition anywhere in the world. A focus on disease resistance is necessary, but entrenched consumption preferences are potentially major constraints to adoption, which may be only partial in the best of circumstances (Kagezi *et al.*, 2012).

The National Banana Research Program of the National Agricultural Research Organization (NARO) in Uganda also faces the challenge that elite clones for evaluation were only introduced on farms from 1991. NARO has made a considerable commitment to biotechnology in order to exploit to the fullest the opportunity for varietal change and has mobilized several international partners in the supply of elite clonal materials. The potential for harnessing biotechnology in Uganda for regional varietal change is a recurring theme that has been reported in the Diffusion and Impact of Improved Varieties in Africa (DIIVA) Project for other clonally propagated crops such as cassava (Alene and Mwalughali, 2012).

Groundnut, sorghum and pearl millet also fall below the adoption average of 35% in Table 19.1. They are produced extensively in the Sahelian, Sudian and Guinean zones of West Africa. All three crops share the same poor country-specific outcomes in terms of adoption: negligible diffusion of improved varieties in Burkina Faso and no recorded adoption in Senegal where varietal output has paled in comparison to the robust performance in Mali (Ndjeunga *et al.*, 2012). The uptake of improved groundnut varieties is moderately high in several smaller East African countries but that diffusion does not compensate for the lack of adoption in West Africa.

Scientists in West Africa have also gone down some blind alleys. For example, sorghum breeding overemphasized *Caudatum* types that could not compete with the dominant Guinean materials prevalent in the region (Ndjeunga *et al.*, 2012). Photoperiod-insensitive, short-duration *Caudatum* materials were high yielding but they were susceptible to pests, disease and bird damage and did not measure up to the consumption expectations of semi-subsistence producers who also consume a sizeable share of their output.

Additionally, groundnut crop improvement scientists in the Francophone countries have to compete with old improved cultivars grown prior to independence. Groundnut variety 55-437, released some 40 years ago, is still the dominant variety in Senegal and even in Anglophone Nigeria (Ndjeunga *et al.*, 2012). In Mali, groundnut varieties 47-10 and 28-206 released in the 1950s are the most popular cultivars.

In spite of the dearth of investment in the improvement of these crops in West Africa as well as scientists' ageing profiles, some progress has occurred that has been below the radar for some time. SOSAT C88 – an improved, ICRISAT (International Crops Research Institute for the Semi-Arid Tropics) related short-duration pearl millet variety released in 1988 in Mali and Niger, and in 2000 in Nigeria – lays claim to an area slightly exceeding 1 million hectares. This variety is grown in a larger area than any of the over 1000 improved adopted cultivars listed in the DIIVA database. Varietal change in groundnut in East Africa, especially in Uganda, is another success story that was stimulated by an impressive partnership between NARO, ICRISAT and the Peanut CRSP (Collaborative Research Support Program) of the United States Agency for International Development (USAID).

Barley, cowpea and yams also appear in the lower half of Table 19.1. Starting from a very low base of 11% in 1998, the uptake of improved barley varieties in Ethiopia has slowly but steadily increased over time. Both improved food and malting barleys have contributed substantially to modern variety (MV) adoption (Yigezu *et al.*, 2012b).

Cowpea adoption outcomes are dominated by the performance of crop improvement research in Niger and Nigeria, which, when combined, have a harvested area of over 8 million hectares. Niger is characterized by a harsh production environment and variable scientific capacity, featuring donor instability. These conditions have resulted in an adoption estimate of 9% in Niger that has kept cowpea from entering the top half of Table 19.1.

According to FAO production data, yams have the highest calculated value of production of any crop, including cassava and maize, in SSA. This fact seems incredible because maize and cassava are usually considered the staple food crops in SSA but an absence of crop improvement research targeted on a species as spatially concentrated as yams does not seem that surprising. The 30% adoption estimate for yams in Table 19.1 is attributed to a 75% outcome for improved varieties in Côte d'Ivoire, the second largest producer in West Africa (Alene *et al.*, Chapter 6, this volume). C18 is the prevalent variety. Following its introduction in Côte d'Ivoire in 1992, C18 expanded rapidly, covering large areas of yam cultivation where it sometimes represents 100% of the area cultivated in *Dioscorea alata* – otherwise known as 'yellow' or 'water' yam — one of six economically important yam species. C18 is known for making tasty yellow porridge.

Both beans and sweetpotato partially owe their position in the lower half of Table 19.1 to this study's stance on excluding released local landraces from the definition of MVs. The adoption level for beans would rise to 50% with a broadening of this definition, whereas the adoption level of sweetpotato would triple to 24%.

Among grain legumes in Table 19.1, improved varieties of beans rank third in adoption outcomes. Bean MVs are characterized by a substantially higher uptake in Ethiopia than MVs for any other grain legume in the DIIVA Project, presumably because Ethiopia has developed a vibrant export industry for haricot beans.

In 1984, a regional breeding programme was established in the Great Lakes region of SSA. It focused on breeding for resistance to bean pests and diseases in conditions of low and declining soil fertility typical of small rural household production. To meet this challenge, the Pan-African Bean Research Alliance (PABRA) was launched as a CIAT project in 1996. It now consists of three regional genetic improvement networks – the Eastern and Central Africa Bean Research Network (ECABREN), the Southern Africa Bean Research Network (SABRN) and the West and Central Africa Bean Research Network

(WECABREN) – and encompasses 29 countries in SSA. PABRA has a record of sustainability and growth that is matched only by a few other regional International Agricultural Research Center (IARC)-related crop improvement networks (Lynam, 2010).

The sustainability of the PABRA umbrella network has strongly influenced these positive outcomes for adoption in a crop that is often characterized by niche specificity in terms of production conditions and market preferences. Identification of improved bean varieties in farmers' fields is an onerous undertaking. With a few notable exceptions, improved bean varieties are believed to account for only small chunks of area in most countries, thereby making the validation of such spatially fragmented expert opinion a difficult task.

In the 1970s and 1980s, not much research was conducted on sweetpotato in SSA. Sweetpotato owes its rather modest position in Table 19.1 to a stable and sustained breeding effort in Uganda and Mozambique (Labarta, 2012). Interest in orange-fleshed sweetpotato for its high beta-carotene content has also helped to stimulate and marshal investment in what was once a relatively neglected secondary food crop in SSA. The adoption of improved varieties in Table 19.1 is about equally split between white- and orange-fleshed varieties.

Adoption of potato MVs are at the mean level in Table 19.1. Given the crop's market orientation and rapidly increasing growth rate in SSA over the past two decades, an adoption level that approaches the mean value across all crops could not be termed superior performance. Following a longer-term CIP (International Potato Center) presence, Malawi has only recently released improved varieties that are now in the very early phase of adoption. The greater uptake of improved clones in Ethiopia and Kenya has not compensated for the sharp downturn in the use of improved materials in Rwanda since the 1994 Genocide which destroyed not only the potato improvement programme in Ruhengeri – the hub of CIP activities in the Great Lakes Region (Rueda *et al.*, 1996) – but also devastated an effective seed programme. Although potato is a priority food crop, recovery in Rwanda has been slow for improved clones, which were believed to be close to full adoption in the early 1990s, prior to civil war.

Cassava is perhaps the most surprising member of the set of seven crops with above-average adoption in Table 19.1. In spite of low levels of research intensity documented earlier, the performance of cassava crop improvement has been solid and steady with regard to adoption outcomes. The majority of the countries included in this study have substantially higher levels of uptake of improved varieties now than in the late 1990s (Alene and Mwalughali, 2012). A strategy that has emphasized high yield combined with disease resistance in a mostly sweet, rather than bitter, background seems to have yielded good dividends in many countries. Additionally, donors have actively supported public-sector and non-governmental organization (NGO) programmes to propagate and widely distribute improved planting materials.

The location of pigeonpea in the top half of Table 19.1 was also expected. All three study countries in East Africa have a commercial demand for higher yielding medium-duration types that are well adapted to bi-modal seasonal rainfall in Kenya, Malawi and Tanzania (Simtowe and Mausch, 2012).

Maize in ESA benefited from the large number of released varieties stimulated by liberalization policies and private-sector investment in maize breeding. As discussed in the previous section, varietal output borders on prodigious in some countries, such as Zambia, which has enacted policies strongly favouring maize production. Excellent performance in Zambia and Malawi has not, however, compensated for the lack of tangible progress in Angola and Mozambique. In Angola, the dominant released cultivars only account for about 5% of the area planted and date from the mid-to-late 1960s prior to independence.

Adoption outcomes seem to be at a moderately high level for rice, which is grown in well-defined agroecological settings throughout SSA. Aggregate adoption levels still depend heavily on what happens in Nigeria and Madagascar, two countries that together account for more than half of the rice-growing area in the 14 countries studied that had data available on this aspect. Aggregate adoption levels also hinge on adoption outcomes in the rainfed lowlands and the uplands. The aggregate level is also sensitive to adoption outcomes in Guinea, which arguably has released more varieties with less ensuing adoption than any other of the 152 crop-by-country national adoption observations. Recent gains in adoption in several countries appear to have been driven by a positive response from farmers to the New Rice for Africa (NERICA) varieties (Diagne *et al.*, 2012). More than any other crop, rice was negatively affected by the decision to define MVs from 1970 – an earlier starting date in 1960 would have led to higher adoption levels but this points to the continued use of very old varieties.

Maize in West and Central Africa (WCA) secures the second spot in adoption performance in Table 19.1. Improved maize varieties in WCA gained more ground in adoption than any other crop in SSA between 1998 and 2010. And these gains were accomplished without significant private sector input (Alene and Mwalughali, 2012). Most of these gains were recorded via the adoption of open-pollinated varieties (OPVs). Some of these are getting older and undoubtedly not all farmers renew seed in a timely fashion, raising questions about the sensitivity of our definition of improved varieties. Factoring in seed renewal rates would lead to a lower adoption estimate but the uptake of improved maize varieties would still be very impressive (Alene *et al.*, 2009).

Wheat topped the crop adoption table in 1998. The increasing transition in area from durum to spring bread wheat was one of the factors leading to the higher adoption of improved varieties in Ethiopia – by far the largest producer in SSA. Wheat would probably occupy a higher position in Table 19.1 if reliable data on adoption had been collected for Kenya, Tanzania, Zambia and Zimbabwe. These countries were at the level of full adoption of wheat MVs in 1998. Assuming full adoption in 2010 is eminently plausible because wheat in these four countries is mainly produced in large farms with irrigation. The inclusion of these four countries results in a rise in the adoption estimate to 70%, which is still substantially less than soybean in Table 19.1. The limited penetration of improved durum varieties into farmers' fields in Ethiopia is a major constraint to full adoption of wheat high-yielding varieties (HYVs) in SSA.

Soybean ranks first in our crop adoption table. Soybean is a new crop characterized by strong market demand. Genetic materials are mostly imported from abroad; sufficient time has not elapsed to allow many local landrace materials to develop. Although improved soybean adoption levels are not surprising, their varietal age is – as discussed in the next section. Given soybeans' scope for global expansion, the crop seems to be taking its time in finding a home in farmers' fields in SSA. Nigeria still harvests more soybean area than the other 12 countries in Table 19.1 combined.

By country

Aside from the Central African Republic's second place ranking – attributed to the adoption of rice MVs – there are relatively few counterintuitive findings in the adoption estimate by country rankings (Table 19.2). One is the relatively

Table 19.2. Weighted area adoption levels by country in SSA in 2010.

Country	MV adoption (%)	Number of crop observations
Zimbabwe	92	4
Central African Republic	72	1
Cameroon	68	6
Zambia	67	6
Kenya	63	8
Gambia	56	1
Côte d'Ivoire	55	6
Ghana	53	6
Benin	52	6
Malawi	47	8
Senegal	45	6
Sudan	41	4
Nigeria	41	9
DR Congo	36	6
Madagascar	35	1
Mali	35	6
Ethiopia	33	9
Uganda	33	11
Tanzania	32	10
Guinea	29	5
Togo	22	6
Rwanda	21	4
Angola	17	2
Sierra Leone	16	1
Burundi	14	4
Niger	14	4
Eritrea	13	2
Burkina Faso	13	6
Mozambique	13	5

high placing of the DR Congo in achieving an above-average adoption outcome across all crops in spite of stagnating institutional and economic development.

The five countries at the bottom of Table 19.2 all share a weighted adoption estimate below 15%. Burkina Faso is the outlier with a high adoption performance in maize and rice. Burkina Faso is also the first adopter of Bt (*Bacillus thuringiensis*) cotton varieties aside from South Africa. Burkina Faso's position is attributed to negligible adoption of groundnut, sorghum and pearl millet MVs. Other countries, like Angola, Mozambique and Niger in the lower five, have uniformly low rates of adoption of improved cultivars across all crops.

Optimism is warranted about the prospects for enhancing adoption in countries such as Ethiopia, Mali and Uganda that are now characterized by average levels for SSA as a whole. However, attaining a moderately high adoption rate of 50% as a hypothetical development goal by 2020 will be a daunting challenge, unless adoption prospects improve markedly for countries in the bottom half of the table.

By cultivar

About 87% of the MV adopted area is associated with detailed data containing regional and cultivar-specific information. The other 13% refers to aggregate adoption only at the national level.

The regional and cultivar-specific database accounts for slightly over 33 million hectares. Adopted area is attributed to named (where they are available) and unnamed varieties. Unnamed varieties are aggregated into a category called 'other'.[1]

There are 1173 named releases in the cultivar-specific adoption database. They account for 98% of the 33 million hectares described above. The size distribution of area planted with these varieties is heavily skewed, consistent with previous findings in the 1998 Initiative for maize in ESA, potato, rice and wheat. Most of the varieties are grown on small areas; the median-sized variety is cultivated on only about 7000 hectares, whereas 250 entries were adopted on less than 1000 hectares. The 75th percentile of the cumulative distribution occurs at about 22,000 hectares. Only 76 varieties exceed 100,000 hectares of adopted area.

Few, if any, of these varieties could be called mega varietie that cover tens of millions of hectares, such as the rice variety Swarna that is extensively grown in South Asia (Chapter 13, this volume). The most extensively grown variety is SOSAT-C88 – the leading pearl millet cultivar in Nigeria and the second-ranking improved variety in Mali. SOSAT-C88 was one of the subjects of the impact assessment in the DIIVA Project (Ndjeunga *et al.*, 2011).

Most of the more extensively grown or more economically valuable improved varieties are concentrated in a small subset of crops and countries. Value of production estimates complement harvested area in describing the economic importance of adopted varieties.[2] By either criterion, the top 100 varieties account for about 60–65% of the total adopted area and value of production of all adopted varieties.

On the basis of a value criterion, the share of cereals in the top 100 falls and the share of vegetatively propagated crops rises dramatically. According to FAO production data, 1 hectare of cooking banana, yams or potato can be worth the equivalent of 25–30 hectares of sorghum and pearl millet in value. Therefore, it is not surprising to see relatively small areas of improved clones of these crops claim a larger share in the top 100, when value of production is the criterion. Indeed, a small majority of the varieties in the top-value 100 are vegetatively propagated.

The top ten-ranking varieties are listed in Table 19.3. Cereals dominate the area classification, but only pearl millet cultivar SOSAT C88 makes it into the top ten when the categorization is based on value. Under either criterion, Nigeria contributes more varieties than all other countries combined. Aspects of several of these economically important varieties are described in the next section on spill-overs.

Spill-overs in adoption

Although the history of crop improvement research is marked by spill-overs in adoption in SSA, spill-overs are not the first thing that comes to mind when thinking of adoption outcomes in the harsh rainfed production environments of Africa. Adaptability is restricted by low fertility

Table 19.3. Top-ranked varieties by commodity and country by area and value of production.

	Area			Value		
Rank	Name	Crop	Country	Name	Crop	Country
1	SOSAT C88	Pearl millet	Nigeria	TMS 30572 (Nicass 1)	Cassava	Nigeria
2	Wad Ahmed	Sorghum	Sudan	C18	Yams	Cote d'Ivoire
3	Oba 98	Maize	Nigeria	TDr 89/02660	Yams	Nigeria
4	TMS 30572 (Nicass 1)	Cassava	Nigeria	TMS 4(2)1425 (Nicass 2)	Cassava	Nigeria
5	ICSV 111	Sorghum	Nigeria	NR 8082 (Nicass 14)	Cassava	Nigeria
6	Kubsa	Bread wheat	Ethiopia	TDr 89/02602	Yams	Nigeria
7	ICSV 400	Sorghum	Nigeria	TDr 89/02665	Yams	Nigeria
8	Suwan 1-SR	Maize	Nigeria	SOSAT C88	Pearl millet	Nigeria
9	Tabat	Sorghum	Sudan	Sadisa (91/203)	Cassava	DR Congo
10	C18	Yams	Côte d'Ivoire	Afisiafi (TMS 30572)	Cassava	Ghana

in environments characterized by seemingly high levels of location specificity.

Positive evidence for spill-over outcomes was well documented in the colonial era in SSA. For example, in collaboration with the British, scientists in Sierra Leone had been working to increase regional rice production in the difficult mangrove agroecology since 1934. The locus of their activities – curtailed in the 1990s because of the civil war – was the Rokupr Rice Research Station. Before independence this was known as the West African Rice Research Institute and its mandate was to promote spill-overs. Several released ROK rice varieties became popular, not only in Sierra Leone but also in Guinea and Guinea Bissau. They have also been the subject of adoption studies and impact assessments (Adesina and Zinnah, 1993; Edwin and Masters, 1998).

The case of the high-yielding, late-maturing maize hybrid SR 52 – the world's first triple-cross hybrid grown commercially – released in the early 1960s in present-day Zimbabwe is a well-known example of varietal output that generated benefits to neighbouring countries in Southern Africa (Eicher, 1995). A lesser-known example after independence focused on late-blight-resistant potato cultivars in the Great Lakes Region of East Africa. In the early 1970s, three late-blight-resistant varieties – at the time, recently released from Mexico – were imported into Uganda and Kenya via the Rockefeller Foundation. Although these varieties never laid claim to much area in Mexico, they quickly became popular in several smaller countries in East Africa. Before the 1994 Genocide in Rwanda, Sangema was the dominant variety in Rwanda and was arguably the most economically important in the ESA region in the 1970s and early 1980s. Even today Rosita, a synonym for Sangema, is the prevailing potato variety in Malawi and Mozambique.

Confirming the potential for spill-overs, the products of older regional crop improvement programmes are still visible in their respective geographical sphere of influence. The Armani Regional Station now in Tanzania but at one time covering all of East Africa has been the location for research that has led to long-term spillovers since the 1950s and 1960s in cassava and sweetpotato materials as progenitors and in a few cases as finished elite clones. Researchers at Armani developed the sweetpotato variety known as Tanzania in Uganda and Rwanda, as Sinama in Tanzania, as Enaironi in Kenya, as Kenya in Malawi, and as ADMARC in central Mozambique, and Chingovwa in Zambia (Labarta, 2012). In the five countries included in the CIP study (Labarta, Chapter 9, this volume), this variety is estimated to be cultivated on an area approaching 200,000 hectares, equivalent to 13% of the total sweetpotato area. (Because of its age, Tanzania is not considered in the set of improved varieties.) It combines high dry matter, a marked preference in East Africa, with a strong background of virus resistance in the Great Lakes region.

In many of the study crops within the DIIVA Project, researchers were able to identify more recent examples of spill-overs, where investing in varietal improvement in one country has benefited neighbouring countries or other countries in SSA. Spill-overs in adoption are not as common as spill-overs in releases, but they are very visible when they occur.

IITA researchers described in detail the occurrences of spill-overs in adoption for all five of their mandated crops in the DIIVA Project (Alene and Mwalughali, 2012). In cassava, TMS 30573 occupies 17.8% of the area in Nigeria, 17.5% in Uganda, 7% in Benin and 3.2% in Guinea. Though not officially released, the same clone is also being grown extensively in Kenya where it covers 24% of the cassava area and, to a much lesser extent, is produced in Côte d'Ivoire.

In cowpea, popular multi-country varieties are: IT82E-32 covering 23% of the total cowpea area in Ghana, 11% in Benin and 2% in Cameroon; followed by VITA-7, accounting for 22% of total cowpea area in Guinea and 13% in Democratic Republic of Congo (DR Congo) (Alene and Mwalughali, 2012). Adoption of variety IT81D-1137 is estimated at 17% in DR Congo and 14% in Benin. These varieties are attractive to farmers because they feature high yield potential, disease tolerance and short duration.

In maize, Obatanpa – derived from quality protein maize (QPM) materials and TZEE-Y – fit the description of spill-over varieties that have crossed over the borders of several countries in WCA (Alene and Mwalughali, 2012). Two improved soybean varieties are also widely cultivated in the region. Firstly, TG× 1448-2E – a shattering and frog-eye, leaf-spot resistant IITA-bred variety – is sown on more than 60% of soybean area in Nigeria and on more than 20% of harvested area in Cameroon and Ghana. TG× 1835-10E – another IITA-developed variety desired for its early maturity and resistance to soybean rust, pod shattering and lodging – dominates soybean areas in Uganda (50%) and covers 26% of soybean area in Kenya as well as 6% in Nigeria.

In yams, examples of large spill-over effects are harder to find but a few improved cultivars are found in two countries. Florido is planted in Benin and Togo; TDr 89/02665 is propagated in Ghana and Nigeria in 5–10% of the total planted area.

Groundnut seems to be the exception to the finding that the prevalence of wide adaptability and spill-over varieties is less common in ESA than in WCA. In four of the five groundnut study countries in the ESA region, rosette-resistant ICGV-SM 90704 and drought-tolerant ICGV 83708 ranked first or second in the adoption of improved varieties.

Finally, in rice, NERICA 1 is presently grown in five of the 12 producing countries with cultivar-specific information in the DIIVA adoption database. Earlier, BG 90-2 from Sri Lanka was a commonly introduced cultivar that was released by the majority of rice-producing countries in West Africa and later became popular in several countries.

The incidence of spill-over varieties appears to be higher in West Africa than in East Africa. The Sahelian, Sudanian and Guinean zones of West Africa cut across broad swathes of several countries. This makes for more homogeneous agroecological conditions going from west to east across countries than from north to south within the same country. The incidence and size of spill-overs also varies by crop: lower in beans and higher in potatoes in East Africa. In ESA, spill-over events in maize were not as large, although they were probably underestimated because of incomplete and low quality data. SC 627 is a variety that scores well on wider adaptation and is grown extensively in Tanzania and Malawi (De Groote *et al.*, 2011).

In West Africa, spill-overs vary from crop to crop. Spill-over varieties are readily visible in pearl millet and groundnut but less so in sorghum. The pearl millet variety SOSAT C88 mentioned previously has been adopted in four West African countries. Similarly, the groundnut variety Fleur 11 is also spreading in West Africa from Senegal to Mali and Niger (Ndjeunga *et al.*, 2011).

The emphasis on spill-over varieties in this subsection does not detract from the empirical fact that the varieties selected and used solely within a country are still likely to contribute far more to total adopted area in SSA than multi-country varieties. Moreover, as pointed out earlier in this section, none of the identified spill-over varieties can yet be called mega-varieties. The moderate incidence of well-identified spill-over varieties serves as a reminder that small NARS can still reap some benefits from national and international research. A stable crop improvement presence in the region can generate returns that far exceed national benefits for the investing country.

IARC-related adoption

Most IARCs have been heavy contributors to the varietal change that has taken place in their mandated crops in SSA (Table 19.4); about 22% of the crop area harvested is in IARC-related genetic materials. The relative importance of those materials approaches two-thirds of total area in improved varieties.

The crops in Table 19.4 are ordered by the difference between their estimated share in varietal output and adoption. It is interesting to see sorghum, pearl millet and groundnut at the head of this table because they lag behind in overall adoption. Released varieties of these crops may have had somewhat limited acceptance by farmers (Table 19.1) but IARC-related cultivars have had better adoption outcomes than most in a difficult rainfed production environment.

The mean weighted difference between the CGIAR's adoption and release shares is 20%. The crops towards the bottom of Table 19.4 are relatively new to crop improvement research in the CGIAR so we did not anticipate that they would have had high shares of IARC-partnered adoption.

Perhaps more than any other international non-CG institution and in any crop in the DIIVA Project, CIRAD (Institut de Recherches Agronomiques Tropicales – IRAT) has had a marked impact on the adoption of rice MVs in several countries of Francophone Africa, including Madagascar. This important institutional connection is a plausible explanation of why rice does not rank higher in Table 19.4. Likewise, the smallish negative value of maize in ESA could be attributed to the late start by CIMMYT (the International Center for the Improvement of Maize and Wheat) in the region and to alternative suppliers in the burgeoning private sector.

Comparing adoption levels between 2010 and 1998

The 1998 benchmark provides a basis for carrying out a before and after comparison of the level of varietal adoption for the ten continuing crops in the DIIVA Project (Table 19.5).

Table 19.4. The contribution of the CG Centers to MV adoption in SSA in 2010.

Crop	Adoption			Release	Difference between adoption and release shares (%)
	Estimated adoption (%)	IARC-Related (%)	Share IARC (%)	Share IARC (%)	
Sorghum	27.4	20.6	75.0	24.8	50.2
Pearl millet	18.1	15.7	86.6	40.2	46.4
Groundnut	29.2	25.0	85.8	43.6	42.2
Bean	29.0	23.5	81.0	39.1	41.9
Wheat	58.5	37.7	64.5	45.0	19.5
Banana	6.2	2.2	34.9	16.7	18.2
Potato	34.4	31.2	90.8	75.0	15.8
Sweetpotato	6.9	5.6	81.3	66.3	15.0
Cassava	39.7	32.7	82.5	68.1	14.4
Soybean	87.9	55.6	63.2	48.9	14.3
Lentil	10.4	10.4	100.0	86.7	13.3
Cowpea	27.2	18.1	66.7	57.5	9.2
Maize (WCA)	65.7	53.0	80.6	74.2	6.4
Chickpea	15.0	15.0	100.0	95.8	4.2
Barley	32.7	7.5	23.0	21.1	1.9
Pigeonpea	49.9	41.8	83.9	82.4	1.5
Rice	38.0	19.2	50.6	51.4	–0.8
Maize (ESA)	44.0	12.9	29.4	30.3	–0.9
Field pea	1.5	0.0	0	16.7	–16.7
Yam	30.2	15.1	50.0	74.3	–24.3
Faba bean	14.0	0.5	3.7	40.0	–36.3
Weighted average[a]	35.25	23	65.6	45.5	20.0

[a]Weighted by total area, except the share in adoption estimates that are weighted by total adopted area in each crop.

Table 19.5. Change in MV adoption between 1998 and 2010 in ten food crops of SSA.

Crop	Number of paired observations	1998		2010		Relative importance in 2010 (% area coverage of paired observations)
		Area (ha)	MV adoption (%)	Area (ha)	MV adoption (%)	
Barley	1	897,360	11.0	913,863	33.8	86
Bean	6	1,738,000	14.6	1,903,964	35.1	45
Cassava	15	8,777,800	21.0	10,033,995	42.0	81
Groundnut	3	496,517	12.6	724,019	56.7	7
Maize	19	18,566,300	25.6	24,366,088	52.8	91
Pearl millet	1	1,285,540	22.0	1,520,440	31.1	9
Potatoes	4	353,852	49.2	569,921	37.1	60
Rice	7	3,639,110	48.4	3,787,146	36.5	44
Sorghum	4	12,711,129	19.3	13,354,489	32.4	58
Wheat	1	1,330,000	56.0	1,453,820	63.5	84
Total/weighted average	61	49,795,608	25.0	58,627,745	43.9	55

On average, the 61 observations represent about 55% of the area of the crops grown in SSA. Coverage is adequate in eight of the ten crops to draw inferences about varietal change between 1998 and 2010. Coverage is too scanty to say anything definitive about progress in varietal uptake in groundnut and pearl millet.

Two important empirical facts emerge from Table 19.5. First, the average level of varietal adoption was 25% in 1998. Second, and more importantly, average MV adoption increased at a rate equivalent to a linear annual gain of 1.45 percentage points over the 13-year period to almost 44%.

With the exception of rice and potatoes, all crops experienced an expansion in the use of MVs. Uptake was especially robust in barley, beans, cassava and maize, with adoption levels doubling during the period.

The before and after data points for the primary staples, maize and cassava, are arrayed in Fig. 19.1. Maize in the DR Congo was the only crop-by-country observation to experience a steep decline in the estimated adoption rate between 1998 and 2010. Gains in the uptake of maize hybrids were significant in Zambia and Malawi. Hybrids also played an important role in Ethiopia. Increases in the West African countries and in Tanzania and Uganda were almost entirely fuelled by the spread of improved OPVs. In general, the cassava-growing countries were characterized by lower adoption levels in 1998 than the maize-producing countries; but, aside from Tanzania, every cassava-producing country displayed a propensity for the greater uptake of improved clones in 2010 than in 1998.

The difference in adoption between the two periods is negatively associated with the magnitude of adoption in 1998. Countries that commenced with levels of adoption equal to, or below, 40% tended to realize more gains in adoption. Those that started with moderately high rates of adoption of improved varieties were hard pressed to achieve even more positive outcomes in adoption. We expect this type of behaviour when a country approaches full adoption but not when it is at a moderate to high level of MV acceptance such as improved maize cultivars in Burkina Faso, Ghana and Kenya in 1998.

Lack of progress in countries with already moderately high rates of adoption indicates the existence of marginal production regions where MVs do not compete favourably with traditional varieties on a few important characteristics. It will be interesting to see if the new entrants in the moderate-to-high adoption group in Fig. 19.1 will be able to consolidate and expand on their gains.

Comparable before and after data on the remaining crops in Table 19.5 are presented in Fig. 19.2. Many relatively small-producing countries made relatively large gains in the adoption of beans and groundnut. Sorghum in the Sudan was the largest crop-by-country combination to register appreciable gains in adoption.

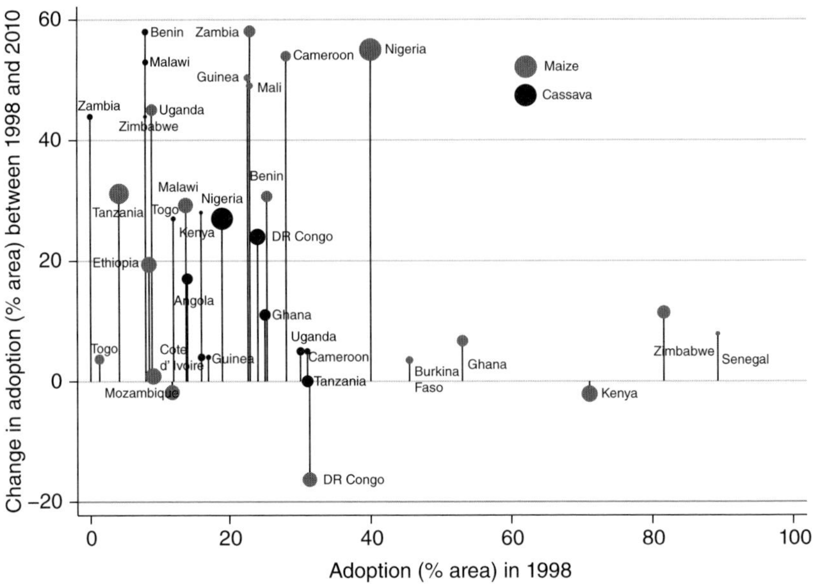

Fig. 19.1. Change in the estimated level of adoption of improved maize and cassava varieties between 1998 and 2010 (balloons in the droplines are weighted by area in 2010).

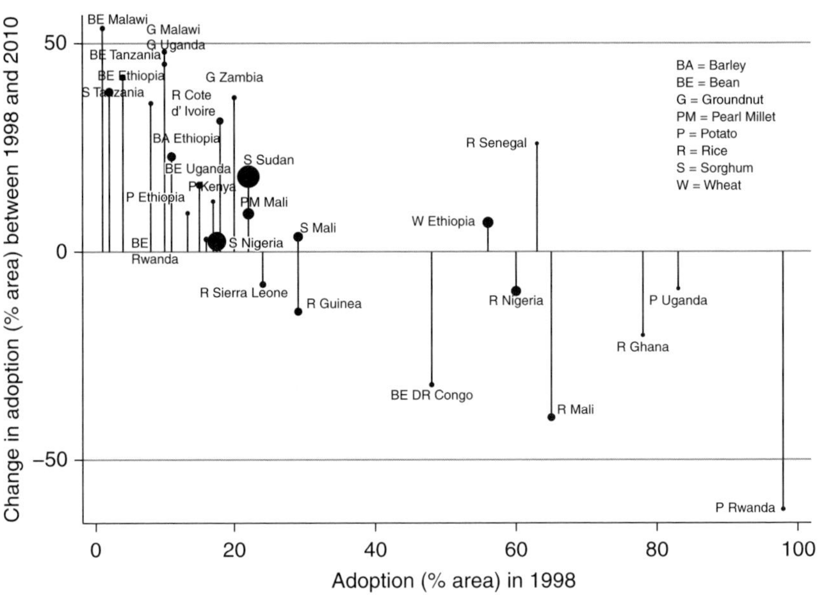

Fig. 19.2. Change in the estimated level of adoption of improved bean, groundnut, pearl millet, potato, rice, sorghum and wheat varieties between 1998 and 2010.

Unlike cassava and maize in Fig. 19.1, the relative importance of MVs declined in several countries between the 1990s and 2010. In particular, the adoption estimate for improved clones of potato decreased sharply from 97% of the harvested area in 1993 to 35% in 2010. As discussed, potato MVs became less important because of the devastation in Rwanda caused by the 1994 Genocide, which did not predate the 1998 Initiative because the adoption estimates for Rwanda referred to 1993.

In contrast, the estimated deteriorating position of MVs in rice could be attributed to a change in methods. Expert opinion panels were used to generate all the estimates for rice MVs in 1998. Surveys funded by the Japan International Cooperation Agency (JICA) were deployed by researchers in AfricaRice to arrive at nationally representative estimates of MV adoption in 20 African countries from 2008–2011.

If progress in MV adoption was slow, switching methods could be sufficient to change a small positive outcome to a meagre negative consequence.

Similar to the evidence presented in Fig. 19.1, countries characterized by moderately high levels of adoption in 1998 had a hard time maintaining these levels, let alone achieving gains in adoption. Rice in Senegal and, to a lesser extent, wheat in Ethiopia are the only two crop-by-country observations that exhibited substantial gains in adoption from 'moderate' to 'high'. Gains in adoption were concentrated at the lower end of the x axis in Fig. 19.2 in much the same manner as very positive outcomes were clustered in the same region of Fig. 19.1.

About 90% of the paired observations in Table 19.5 showed an increase in the uptake of improved varieties (Figs 19.1 and 19.2). Again, disadoption and/or overestimation of MV adoption levels in 1998 occurred mainly in potatoes and rice. The finding of a few cases of disadoption is unexpected because the ending of fertilizer subsidies is frequently mentioned as a motivation for reversion to local varieties. The evidence for disadoption is sparse in maize, which is the most intensive user of fertilizer among the food crops in the DIIVA Project.

Revisiting the hypotheses about varietal adoption

We found widespread support for several of the adoption-related hypotheses in Chapter 3. In particular:

- The level of adoption of improved varieties and hybrids was steadily increasing over time and was substantially higher in 2010 than in 1998.
- Spill-over varieties were found in all food crops and they laid claim to a sizeable share of adopted area.
- The share of materials related to CG Centers was higher in varietal adoption than in varietal output.

Findings on the above hypotheses varied by crop, but, in general, they were largely affirmative for the 20 crops as a whole. The evidence was not as generic in its support for the other two adoption-related hypotheses expressed in Chapter 3. First, disadoption of improved varieties on aggregate was rare and was not caused by economic restructuring and liberalization. We did not encounter support for the contention that increasing fertilizer prices led to the widespread abandonment of maize MVs and a reversion to traditional varieties.

Replacement of improved sorghum cultivars in Nigeria and reversion to local varieties were the most notable example of disadoption in the DIIVA database (Ndjeunga *et al.*, 2011). These varieties were extended to farmers in the late 1990s and were partially accepted by the early 2000s. Gains perceived by farmers in earliness and insect tolerance did not compensate for perceived losses in drought tolerance, stalk strength and head size to result in sustained adoption. Differences in yield and income between adopting and non-adopting households were not statistically significant. The absence of wider impacts was attributed to disadoption (Ndjeunga *et al.*, 2011). In contrast, pearl millet MVs in northern Nigeria were associated with significant differences in yield and income.[3]

Second, we documented sufficient cases to support the proposition that adoption of improved varieties was positively influenced by market demand, the potential of the production environment and the crop's multiplication ratio. The case for market demand was epitomized by haricot bean exports that stimulated greater uptake of improved varieties relative to other pulse crops in Ethiopia and relative to other bean-producing countries. Small, incremental gains in adoption between 1998 and 2010 for countries and crops with good adoption outcomes in 1998 were indicative of ceiling rates of adoption emerging in some subnational regions where production prospects were more marginal than those where diffusion had initially occurred. With a multiplication ratio of only about 15, groundnut was an apt example of crop for which high seed costs dampened diffusion of MVs, even though there is good market demand in many cases.

In spite of the general and specific favourable findings for these adoption-related hypotheses, improved varieties have diffused more slowly in SSA than in other developing countries. The average speed of diffusion was estimated at 0.11, which is considerably below the low threshold benchmark of 0.20 that comes from a survey of relevant studies (see Fuglie and Marder, Chapter 17, this volume).

Moreover, we should not lose sight of the fact that adoption of improved varieties and hybrids in dryland food crops in South Asia is markedly higher than the levels estimated in SSA (Chapters 13 and 14, this volume). Across the five rice-growing countries in Chapter 13, adoption of MVs averaged about 80% in 2010 and is still trending upwards since 1999. For pearl millet, sorghum and groundnut, levels of MV adoption ranged from about 55–70% in peninsular India. A comparable interval for the uptake of improved cultivars in these three important crops in SSA is 20–30% in 2010 (Table 19.1).

The regional estimates by state in India in Chapter 14 suggest that some important growing regions have continued to be bypassed by the Green Revolution. For all intents and purposes, post-rainy-season (rabi) sorghum production on residual moisture in peninsular India is still dominated by the old selected landrace M35-1, although the post-rainy-season crop now contributes to the bulk of sorghum output in India. Likewise, relatively few groundnut cultivars released in South India since independence have been able to compete with the old improved variety TMV-2. These examples of negligible varietal change highlight the observation that the production environment can prove to be a formidable challenge to progress even in a reasonably efficient and stable system of national and international agricultural research and in an institutionally enabling environment. Fortunately, few of these 'dry holes' are visible in the landscape of modern varietal change in South Asia in dryland food crops. And the situation is dynamic. Until recently, the prospects were believed to be bleak that pearl millet hybrids could penetrate into the arid drylands of Rajasthan. Now, more than half of the area is sown to hybrids in India's largest millet-growing state.

Varietal Turnover

The velocity of varietal turnover in 2010 by crop

The average results by crop are tightly clustered in the range of 10–20 years (Table 19.6). This means that there may be few, if any, crops where older-adopted improved materials have substantially eroded the profitability of plant breeding. But, by the same token, there was also little evidence that rapid varietal change is taking place. The area-weighted grand mean is 14 years, indicating that the average MV in farmers' fields in 2010 dated from 1996.

Only 16 of the 117 crop-by-country programmes were characterized by above average adoption combined with a varietal age of less than 10 years. These better-performing crop-by-country

Table 19.6. The velocity of varietal turnover of improved varieties in farmers' fields in SSA by crop.

Crop	Varietal age (years)	Number of country programmes
Banana	10.2	1
Sweetpotato	10.3	5
Groundnut	11.7	5
Chickpea	11.9	2
Cowpea	11.9	16
Lentil	12.5	1
Maize (WCA)	12.8	11
Wheat	12.8	1
Maize (ESA)	13.0	8
Beans	13.8	9
Cassava	14.1	17
Soybean	14.2	11
Pearl millet	14.8	3
Rice	15.8	4
Sorghum	17.4	6
Pigeonpea	17.9	2
Yam	18.4	5
Barley	18.5	2
Field pea	18.9	1
Potato	19.4	5
Faba bean	20.7	2
Weighted mean/Total	14.0	117

entries are a blend of larger-area programmes in maize, cassava and cowpea with several very small programmes in soybean and rice.

The cropwise results on varietal turnover in Table 19.6 are somewhat counterintuitive because crops such as sweetpotato and banana, with low multiplication ratios, are characterized by a younger portfolio of varieties compared with several propagated crops with stronger market demand. However, this is not surprising because of the dearth of earlier research on these clonal crops that translated into few, if any, releases in the 1980s and 1990s.

Table 19.6 contains several other surprises. For example, soybeans should have performed better on area-weighted average age given its emerging and expanding cultivation in SSA. However, the youngest soybean varieties in farmers' fields in Nigeria are 'old' because they were released in the early 1990s.

The lack of difference in varietal age between maize in WCA and ESA is also unexpected. Improved cultivars in WCA are OPVs; hybrids dominate maize production in ESA. Historically, and especially in the last decade, many more hybrids have been released in ESA than OPVs in WCA. Yet, the genetic and seed market-related differences between these two contrasting types of material have not translated into substantial differences in varietal turnover. H-614 is the dominant hybrid in Kenya. It was released in 1986. HB-660 is less dominant but it is the leading improved cultivar in Ethiopia. Both hybrids are closely related with the same parental materials. They trace their roots to the Kitale Station in Kenya from crosses between Kitale Synthetic and Ecuador 573, a landrace from the Andean Highlands collected by the Rockefeller Foundation in 1953 (De Groote, 2013, personal communication). In Kenya, the mean varietal age of hybrids and improved OPVs across the six maize-producing agroecologies was 24 years in a nationally representative adoption survey in 2010 (Swanckaert *et al.*, 2012).

The vintage of adopted varieties

A small majority of the 1145 cultivars in the adopted variety database carry information on the date of release. These varieties account for about 80% of the adopted area and value of production. Their age distribution is presented in Table 19.7. The largest area and value share come from the cohort of varieties that were released in the late 1990s. This finding suggests that CG Centers were able to supply materials for release by their NARS partners during a time of financial crisis in the late 1990s and early 2000s. From this, it is possible to infer that financial constraints did not entirely stop the flow of materials in the pipeline. A 15% value share for varieties released between 2006 and 2011 is encouraging and indicates that materials in the pipeline are finding a home in farmers' fields. A sizeable chunk of the recent difference between the area and value share has been attributed to the release of two promising improved yam clones in Nigeria.

The share estimates in Table 19.7 also hint at the longer-term impact of varietal change. Improved varieties in the early 1980s are still making a substantive contribution that cannot be ignored. A case in point is IITA's release of its important cassava variety TMS 30572 in 1984. In contrast, materials released prior to 1980 in the early years of the CGIAR were relatively limited in number and their impact has eroded over time.

Comparing levels of varietal change in 1998 and 2010

Improved varieties are not getting any younger in farmers' fields. For maize and wheat, age is

Table 19.7. The vintage of varieties contributing to adoption in 2010 by criterion and by release period.

	Criterion	
Release period	Area share (%)	Value share (%)
1970–1975	1.7	1.1
1976–1980	2.7	2.9
1981–1985	8.3	10.6
1986–1990	12.7	12.8
1991–1995	19.4	15.6
1996–2000	27.1	23.9
2001–2005	17.7	17.4
2006–2011	10.3	15.2
Total area ('000 ha)	27,477.4	
Total value in US$ (million)		12,095.20

roughly the same as it was 14 years ago. For three of the four countries producing potatoes, varieties are becoming older. For rice, the average age of MVs was the highest of the cereals in both 1998 and 2010 for the same observations in both benchmark periods. Varietal age of maize in Kenya has increased slowly but surely from 17 years in 1992 to 22 years in 2001 to 24 years in 2010 (Swanckaert *et al.*, 2012). Although age has fallen markedly in the dry transitional zone in response to rapid varietal adoption and change, new private sector seed suppliers have not been able to penetrate into other zones where adoption levels are stagnating.

Revisiting the hypotheses about varietal turnover

The expectation that *varietal turnover is relatively high and is increasing over time* was not supported by the estimates of age of improved varieties in the fields of African farmers. However, in contrast to outcomes on adoption, varietal turnover is not significantly faster in dryland crops or in rice in South Asia. Indeed, improved varieties in rice paddies in South Asia are older than most food-crop varieties adopted by farmers in SSA: their average age varied from 14 to 25 years across the five study countries and the three study states in India in Chapter 13 (Pandey *et al.*, this volume). Very slow varietal turnover in rice has eroded the returns to recent investments in national and international rice improvement and is mainly attributed to the enduring popularity during the past three decades of Swarna, a variety characterized by widespread adaptability and stability.

Four of the five dryland crops in Chapter 14 (Kumara Charyulu *et al.*, this volume) would also fall in the range of 10–20 years shown to be typical for crops in SSA in Table 19.6. Pearl millet is the exception. Indian farmers who first adopted pearl millet hybrids in the late 1960s and early 1970s are now sowing their 4th or 5th hybrid. Because of downy mildew epidemics caused by the breakdown of genetic resistance, pearl millet hybrids need to be replaced every 5–10 years. Failure to replace susceptible hybrids leads to sharp declines in yield and so-called 'boom and bust' cycles in productivity (ICRISAT, 2004). Maintenance breeding is a must and is characterized by high returns. Molecular biology has accelerated the search for sources of genetic resistance to downy mildew that, in turn, should result in a speedier turnover of popular pearl millet hybrids.

Impacts

The substantive results on estimated impacts from the DIIVA impact studies are described in detail in Chapters 15–17. The direction and order of magnitude of these results were in line with expectations at the start of the project in 2010.

Yield

Quantifying differences in productivity in replacing traditional with improved varieties received the lion's share of attention in the DIIVA impact assessment studies. Without reliable estimation of productivity differences, further measurement of impacts of varietal change on poverty, food security and other consequences would have been flawed (Chapters 15 and 16, this volume).

The estimated yield differential from adopted improved varieties over local replaced varieties varied from 0% to 100% in dryland agriculture in the case studies based on nationally or regionally representative surveys that are described in Chapter 4. At one extreme, no significant productivity differences were documented between improved and local sorghum varieties in northern Nigeria (Ndjeunga *et al.*, 2011). The absence of detectable yield differences was believed to be an important determinant in the recent disadoption of these improved varieties.

Pearl millet and groundnut in northern Nigeria reflect the conventional wisdom that productivity gains from 'naked' varietal diffusion – adoption without changing input use or management practices – are likely to be statistically significant but small in rainfed agriculture in SSA. The estimated increase in pearl millet productivity was about 90 kg per adopted hectare, equivalent to a 15–20% yield gain (Ndjeunga *et al.*, 2011). Likewise, improved groundnut varieties yielded about 15–20% over local varieties (Ndjeunga *et al.*, 2013). This relative advantage translated into a higher productivity increase of 150–200 kg per hectare because groundnut is produced in more rainfall-assured production subregions in northern Nigeria than pearl millet.

Higher relative yield gains favouring improved varieties were recorded for beans in Rwanda and Uganda and for maize in Ethiopia. Production of these crops takes place at higher elevations and in regions of higher production potential than pearl millet and sorghum production in the hotter arid and semi-arid zones of West Africa. Improved varieties conferred a yield advantage of 53% in Rwanda and 60% in Uganda in bean production (Larochelle *et al.*, Chapter 16, this volume). In Ethiopia, maize hybrids and improved OPVs out-yielded local landraces by 48–64% in farmers' fields (Zeng *et al.*, Chapter 15, this volume). Farmers in Ethiopia spent, however, about 23–30% more in production costs in inputs such as hybrid seed, fertilizer and herbicide. Maize in Ethiopia was the only case study where adoption of improved varieties was accompanied by substantial investment in complementary inputs.

The aggregate estimate of the contribution of improved varieties to increased productivity in SSA in all food crops from 1980 to 2010 was at the higher end of the spectrum defined in the case studies. The impact of improved varieties on farm productivity in SSA has been significant, raising average net crop yield on adopting areas by around 0.55 tonnes per hectare, or by 47%, from 1976–1980 average levels (Fuglie and Marder, Chapter 17, this volume).

Poverty

Persuasive evidence on the poverty consequences of improved varietal change was presented in the case studies on maize in Ethiopia and on beans in Rwanda and Uganda (Chapters 15 and 16, this volume). The impact on poverty was small in bean production. Annual profits from bean growing (accounting for two growing seasons in each country) increased by about US$75 and US$65 per bean-growing household in Rwanda and Uganda, respectively, compared to what they would have been in the absence of the improved varieties. Without improved varieties, poverty would have been about 0.4% higher in Rwanda and 0.1% higher in Uganda in 2011. A modest poverty impact was attributed to the small area planted to beans – in both countries and cropping seasons the median-sized sown area was only equivalent to about one-sixth of a hectare – and the relatively small contribution of bean consumption and sales in total household income. In Uganda, the poor have not adopted improved bean varieties as widely as households above the poverty line.

The adoption of maize hybrids and improved OPVs in Ethiopia generated large poverty impacts. At 0.85 hectares, the average maize-growing area in Ethiopia was more than five times larger than the mean bean area in Rwanda and Uganda; maize figured more prominently in household income. Lower food prices on poor net consuming households were as or more important than direct income gains to producers in reducing poverty in Ethiopia. Diffusion of improved maize cultivars led to a 0.8–1.3% reduction in the overall rural poverty headcount ratio, and to proportional declines in the depth and severity of poverty. Between 45,000 and 95,000 rural households were no longer classified as poor in 2020 because of the adoption of improved maize genotypes.

As the total cropping area under maize is still expanding in Ethiopia, the poverty impacts of improved maize varieties should continue to increase in the future. Unlike in the case of bean producers in Uganda, the poor were found to be as likely to adopt improved varieties of maize as the non-poor, holding all other factors constant, and they experienced similar yield increments and reductions in the cost per unit of production from adoption. The small size of their land holdings, rather than their inability to adopt, explains why they derived fewer absolute benefits from adoption.

The magnitude of the monetary measure of US$6 billion/year also bears witness to the potential for poverty reduction from improved varietal change. If present adoption rates and per hectare impacts continue, the added value from improved varieties could approach US$12 billion/year by 2020 (Fuglie and Marder, Chapter 17, this volume).

Food security

Bean in Rwanda and Uganda was the only case study to address the impact of improved varietal change on food security (Larochelle *et al.*, Chapter 16, this volume). In Rwanda, 16% more households would have been food insecure without improved bean varieties; in Uganda, 2% more households would have been food insecure. Initially, households in Uganda were characterized by greater dietary diversity and this partially explains why the effect of improved bean varieties in Rwanda on food security was substantially larger than in Uganda.

Revisiting the hypotheses on impacts

As pointed out in Chapter 3, the hypotheses on impacts were not as well formulated as those for other aspects of the DIIVA study. Nonetheless, much of the thinking by authors of the DIIVA proposal and of the impact assessments about the effects of improved varietal change was confirmed by the case studies and by the aggregate analysis in Chapter 17. The net yield gains in the case studies spanned a wide range from 0% to 100%. The quality of the production environment loomed large in conditioning favourable yield gains and in the use of additional complementary inputs that reinforced productivity differences. Large poverty effects for improved maize varieties in Ethiopia and notable food security consequences were documented for improved bean varieties in Rwanda. As expected, the aggregate time-series analysis in Chapter 17 showed that varietal change was an important contributing factor to technological change in food-crop agriculture in SSA.

Although we did not scour the landscape, we did not encounter any evidence for negative unintended consequences. The transfer of improved sorghum cultivars in northern Nigeria could be called the worst-case scenario we encountered. That expenditure on extension now seems to have been wasted because widespread disadoption is reported (Ndjeunga *et al.*, 2011). The strengths of these newer varieties do not appear to compensate for their perceived weaknesses. In contrast, sustained adoption of improved pearl millet and groundnut varieties has taken place in northern Nigeria (Ndjeunga *et al.*, 2011; Ndjeunga *et al.*, 2013).

Notes

[1] Every effort was made to minimize the number of varieties in the 'other' category. Most of the specific entries come from survey data and refer to names that are believed to be MVs but that could not be linked to a specific released variety. A few of the observations based on expert opinion also have a small residual 'other' category.

[2] Value of production is an important criterion because varietal change in crops with more attractive prices and/or higher base yields has the potential to generate greater net benefits per hectare of adopted area.

[3] When SOSAT-C88 was first introduced, its seed sold for six times the market price of pearl millet in northern Nigeria (ICRISAT, 2000). SOSAT C-88 is prized for its early maturity, insect tolerance, grain colour and its quick cooking time (Ndjeunga *et al.*, 2011). Low fodder production and susceptibility to *Striga* are its main weaknesses.

References

Adesina, A. and Zinnah, M.M. (1993) Technology characteristics, farmer's perceptions and adoption decisions: a Tobit model application in Sierra Leone. *Agricultural Economics* 9, 297–311.

Alene, A.D., Menkir, A., Ajala, S.O., Badu-Aprak, B., Olanrewaju, A.S. *et al.* (2009) The economic and poverty impacts of maize research in West and Central Africa. *Agricultural Economics* 40(5), 535–550.

Alene, A.D. and Mwalughali, J. (2012) *The Effectiveness of Crop Improvement Programs in Sub-Saharan Africa from the Perspectives of Varietal Output and Adoption: The Case of Cassava, Cowpea, Maize, Soybean and Yam.* Diffusion and Impact of Improved Varieties in Africa (DIIVA) – Objective 1 Report. International Institute of Tropical Agriculture: Lilongwe, Malawi.

De Groote, H., Gitonga, Z. and Mugo, S. (2011) *Diffusion and Impact of Improved Maize and Wheat Varieties in East and Southern Africa*. Diffusion and Impact of Improved Varieties in Africa (DIIVA) – Objective 1 Report. CIMMYT, Nairobi.

Diagne, A., Kinkingninhoun-Medagbe, F.M., Amovin-Assagba, E., Sanni, K. and Toure, A. (2012) *To Attain a Wider Understanding of Key Aspects of the Performance of Food-Crop Genetic Improvement in Priority Country-by-Commodity Combinations in Sub-Saharan Africa: The Case of Rice Crop*. Diffusion and Impact of Improved Varieties in Africa (DIIVA) – Objective 1 Report. AfricaRice Center, Cotonou, Benin.

Edwin, J. and Masters, W.A. (1998) Returns to rice technology development in Sierra Leone. *Sahelian Studies and Research* 1, 67–74.

Eicher, C.K. (1995) Zimbabwe's maize-based Green Revolution: preconditions for replication. *World Development* 23(5), 805–818.

ICRISAT (2000) ICRISAT Annual Report. Patancheru, India.

ICRISAT (2004) ICRISAT Annual Report. Patancheru, India.

Kagezi, G., Mpiira S., Jogo, W., Karamura, E., Gotor, E. *et al.* (2012) *Adoption of New Banana Varieties in Uganda Report for the DIIVA Project*. Diffusion and Impact of Improved Varieties in Africa (DIIVA) – Objective 1 Report. Bioversity International, Rome.

Labarta, R. (2012) *The Genetic Improvement of Potato and Sweetpotato in Sub-Saharan Africa. Diffusion of Improved Varieties in Africa (DIIVA) Project Report on Objective 1*. International Potato Center (CIP), Nairobi.

Lynam, J. (2010) *Evolving a Plant Breeding and Seed System in Sub-Saharan Africa in an Era of Donor Dependence*. Prepared for the Global Initiative on Plant Breeding. FAO, Rome.

Ndjeunga, J., Umar, J., Ahmed, B., Aba Aba, Ibro, A. *et al.* (2011) *Adoption and Impacts of Modern Sorghum and Pearl Millet Varieties in Northern Nigeria*. ICRISAT, Niamey, Niger.

Ndjeunga, J., Hash, C.T., Faye, I., Sanogo, M., Echekwu, C.A. *et al.* (2012) *Assessing the Effectiveness of Agricultural R&D in West Africa: Cases of Pearl Millet, Sorghum and Groundnut Crop Improvement Programs*. Diffusion and Impact of Improved Varieties in Africa (DIIVA) Project, Objective 1 Report. ICRISAT, Niamey, Niger.

Ndjeunga, J., Ntare, B.R., Ajeigbe, H., Echekwu, C.A., Ibro, A., and Amadou, A. (2013) *Adoption and Impacts of Modern Groundnut Varieties in Nigeria. Diffusion and Impact of Improved Varieties in Africa (DIIVA) Project – Objective 2 Report*. ICRISAT, Niamey, Niger.

Rueda, J.L., Ewell, P.T., Walker, T.S., Soto, M., Bicamumpaka, M. and Berrios, D. (1996) Economic impact of high-yielding, late-blight-resistant varieties in the Eastern and Central African Highlands. In: Walker, T. and Crissman, C. (eds) *Case Studies of the Economic Impact of CIP-Related Technologies*. CIP, Lima, pp. 15–30.

Simtowe, F. and Mausch, K. (2012) *Assessing the Effectiveness of Agricultural R&D in Eastern and Southern Africa: Cases of Groundnuts, Pigeonpea and Sorghum Crop Improvement Programs*. Diffusion and Impact of Improved Varieties in Africa (DIIVA) – Objective 1 Report. ICRISAT, Nairobi.

Swanckaert, J., De Groote, H. and D'Hawese, M. (2012) *Mapping the Effect of Market Liberalisation Policies on the Maize Seed Systems in Kenya Based on Micro-Evidence from 1992 to 2010*. CIMMYT, Nairobi.

Yigezu, Y.A., Yirga, C. and Aw-Hassan, A. (2012a) *Varietal Output and Adoption in Barley, Lentils, Chickpeas, Faba beans, and Field Peas in Ethiopia, Eretria, and Sudan*. Diffusion and Impact of Improved Varieties in Africa (DIIVA) – Objective 1 Report. ICARDA, Aleppo, Syria.

Yigezu, Y.A., Yirga, C., Aw-Hassan, A. and Labarta, R. (2012b) *Barley in Ethiopia: Verifying the Adoption Estimates from a Panel of Experts with Secondary Data and Household and Community Focus Group Ss*. Diffusion and Impact of Improved Varieties in Africa (DIIVA) – Objective 2 Report. ICARDA, Aleppo, Syria.

20 Validating Adoption Estimates Generated by Expert Opinion and Assessing the Reliability of Adoption Estimates with Different Methods

T.S. Walker*
Independent Researcher, Fletcher, North Carolina, USA

Arriving at a comprehensive set of estimates of improved varietal adoption for many important food crops and producing countries was the core activity of the Diffusion and Impact of Improved Varieties in Africa (DIIVA) Project. In Chapter 4, this volume, we spelled out our reasoning for the choice of expert opinion over other methods. Briefly, neither did participants have the time nor did the project have the resources to carry out household- or community-level surveys on varietal adoption for all crops in all countries. Moreover, we needed compatibility with estimates generated in the 1998 Initiative for comparative purposes. That earlier benchmark on varietal adoption was estimated via expert panels.

Estimates from expert panels are only as good as the expert's knowledge and the elicitation protocol. In theory, competing methods, such as seed-sales inquiries and household surveys, seem like better options for generating reliable varietal estimates but, in practice, they are also flawed. Information on seed sales can be extremely useful in documenting the adoption of cereal hybrids; however, companies are often unwilling to share information on seed sales of specific hybrids with the public. As the analysis in Chapter 16 on beans in Rwanda and Uganda makes abundantly clear, estimates from household varietal surveys are only as reliable as the farmer's knowledge of the true identity of the genotypes she/he is sowing. Depending on context, a long list of location-specific varietal names is likely to be generated that contains fuzzy varietal identities. Separating improved or modern varieties (MVs) from local landraces requires considerable effort and may not result in successful resolution of varietal identity for many names on the ill-defined list.

Several of the weaknesses of varietal surveys could be overcome with the massive or even the selective deployment of DNA fingerprinting. In 2010, the commercial application of DNA fingerprinting for genotype identification was routine for well-defined uses related to seed regulations in developed countries such as the USA and Australia. Seed and plant tissue samples have not yet been collected and analysed, however, to draw inferences about varietal adoption within the framework of large-scale nationally representative surveys.[1]

Therefore, our comparative evidence on methods focuses on the validation of expert opinion with data from the surveys that were described in Chapter 4. What did we learn from the

* E-mail: walkerts@msu.edu

DIIVA Project that will improve the accuracy, precision and cost-effectiveness of estimation in assessing the adoption of improved varieties as a group and individually?

Shedding light on the challenges in estimating varietal adoption and arriving at credible orders of magnitude is the intent of this chapter that can be summarized by these queries:

1. Were there systematic differences between adoption estimates from the expert panel and survey sources?
2. Do we need to revise the estimate of a weighted average of 35% of MV adoption for 2010 for sub-Saharan Africa (SSA) as a whole in light of the validation results?
3. What were the main weaknesses in using expert panels and surveys in estimating MV adoption?

Validating Expert Opinion with Survey Data

This section is about the validation of expert opinion (EE) with household (HH) survey data. Later, we validate both of these with community interviews using focus groups. If these three sources give congruent results, then we are reasonably certain that the adoption estimates are credible. However, finding small differences among the three sources is unlikely because the crop context is quite variable; that is, some methods may work better with some crops than with others.

Moreover, it is important to point out that when EE and HH estimates diverge, we cannot necessarily draw the implication that expert opinion overestimates or underestimates the true value. The HH estimates could be as biased as the EE estimates. The implicit assumption throughout this chapter is that convergence among estimates from different sources instills confidence of credibility. But we really do not know the accuracy or precision of estimates without the results of DNA fingerprinting, which is the gold standard.

The evidence presented in Chapters 13 and 14 for South Asia complements the comparative analysis of expert- and survey-derived estimates for SSA. We begin with a brief review of the literature and a description of the protocol for eliciting expert opinion.

Maize hybrids and OPVs in East Africa in 1998

The DIIVA Project was not the first to compare subjective estimates on adoption from expert panels with more objective data. In the spirit of the 1998 Initiative, CIMMYT (Centro Internacional de Mejoramiento de Maiz y Trigo) economists assessed the congruence between expert opinion from national agricultural research system (NARS) scientists, mainly plant breeders, and aggregate adoption estimates from data on seed sales of hybrids and open-pollinated varieties (OPVs) for maize-growing countries in East and Southern Africa (Hassan *et al.*, 2001). Their assessment showed that expert opinion on adoption in countries where hybrids were popular and approaching full adoption was very consistent with estimates derived from seed production data. In contrast, estimates from the two sources diverged as the importance of OPVs increased. Expert opinion in Uganda and Tanzania resulted in markedly higher estimates than those inferred from annual seed-related data.

The protocol for eliciting expert opinion

In the DIIVA Project Implementation Workshop held in Addis Ababa, Ethiopia, in February 2010, a 13-step protocol was described for eliciting expert opinion on the adoption of improved cultivars (Walker, 2010):

1. Ensure that the historical information on varietal release has been updated and is available. In other words, the varietal release database precedes and lays the foundation for the assessment of adoption perceptions.
2. Canvass background evidence on recent adoption studies and variety-specific seed distribution and sales.
3. Convene an expert panel (usually NARS crop improvement scientists of the commodity of interest and other experts with extensive field-level knowledge of varietal adoption).
4. Divide the country into subregions or recommendation domains that the experts are most comfortable with (as few as 2–3 or as many as 10 or more). These subregions should be as fully described as possible in the form of a map or a listing of subnational administrative units.

5. Assign relative areas to each subregion from the subnational HarvestChoice database from the most recent year or a 3-year average of recent years.
6. Assess the correspondence between the HarvestChoice agroecological, socioeconomic classification and the experts' description of subregions.
7. Elicit perceptions on the rank of specific-improved varieties and local varieties as a group in descending order of popularity in each subregion. The reference point for the ranking is 100% of the crop's area in the subregion.
8. Check the congruence between varieties in the expert adoption schedule and the release list.
9. Elicit descriptive information on non-local varieties that are subregionally important (they are on the expert perception adoption schedule) but are not on the release list. Such information relates to the date of first use, institutionally specific classification in the release database, distinguishing characteristics, etc.
10. Translate the cultivar-specific perceptions on ranks into perceptions of a percentage (%) of area for each ranked category. Do this for each subregion and for the most recent cropping year, say 2009–2010. The easiest way to do this may be to start with the aggregate category group of local varieties for a percentage area estimate and then estimate percentage area for the dominant MV, the second most dominate MV, the third ranked MV, etc.
11. Highlight issues of greatest uncertainty in the perceptions of percentage (%) area; note ranges where uncertainty is greatest.
12. Discuss areas of discrepancies between the background information and the elicited perception and revise the perceptions if the discrepancies are large and if revisions are warranted.
13. Draft a brief 1–2 page report documenting the substance and the process (composition of the expert panel; a description of the subregions; background information on adoption; details on how perceptions were assessed; a description of the varieties included in the adoption perception schedule that were not on the release list; and magnitude and reasons for any revisions to expert opinion) for each priority country-by-crop combination.

Some of these steps were not viewed to be as essential as others. For example, the use of the HarvestChoice database in Steps 5 and 6 was optional, depending on circumstances.

What worked and what did not: anecdotal evidence

Tailoring this protocol to crop and country experiences was encouraged. For rice in South Asia, this detailed method for estimating varietal adoption was condensed and generalized into five steps (Pandey *et al.*, Chapter 13, this volume). Elicitation procedures were also honed in a formalized workshop setting where results of a large-scale varietal rice survey in Odisha in East India were compared to expert opinion.

Through trial and error, the DIIVA Project's coordinator and its Steering Committee members expected that some CG Center participants would arrive at a varietal adoption assessment process that was superior to the one described above in terms of cost-effectiveness and precision.

A review of methods deployed by the CG Centers shows several concrete examples of adaptation (J. Stevenson and J. Burgess, 2013, personal communication):

- CIAT (Centro Internacional de Agricultura Tropical) used a very inclusive approach in the formation of its 13 expert panels that were widely attended by knowledgeable people from over 150 institutions, including the private sector and non-governmental organizations (NGOs) (Muthoni and Andrade, 2012).
- Several CG Centers, including the International Potato Center (CIP), adapted the process when they saw that progress depended on increasing the level of hands-on management by conducting more in-country visits with a high level of supervision.
- After an initial failure at generating credible national estimates, the International Crops Research Institute for the Semi-Arid Tropics (ICRISAT) used a geographic information system (GIS) orientation to build estimates from the ground up with subnational experts. Subsequently, ICRISAT convened a larger regional workshop of NARS scientists for validation of the spatially disaggregate estimates (Ndjeunga *et al.*, 2013).

- The International Institute of Tropical Agriculture (IITA) created teams in Francophone and Anglophone SSA to hold workshops, assemble data and canvass information. Perhaps wisely, IITA staff discouraged the review of adoption studies because such research could unduly influence the thinking of participants and keep them from reflecting on their personal experience.

It was also apparent what did not work: mailing out questionnaires to key collaborators and hoping for responses and delegating the lion's share of responsibility to in-country consultants. These approaches resulted in considerable e-mail fatigue but little in the way of reliable information. Because US$6000–7000 were allocated for each country-by-crop observation, a more aggressive supervisory approach could be pursued. That approach usually bore fruit.

Participants now agree that a more geographically decentralized process, featuring wider participation by different institutional actors in society, is needed to arrive at more precise and accurate expert-opinion estimates of MV adoption. Balancing knowledgeable experts with representation from a wider sectoral audience is a challenge when seeking expert opinion on progress in varietal research.

In some contexts, other methods would be eminently more suitable than expert opinion. But what is frequently overlooked is the fact that, to arrive at a significantly better outcome, alternative methods require relatively sophisticated skills in application and energetic, persistent interviewers.

The basis for the validation

The nine large-scale household surveys described in Chapter 4 in Table 4.4 are the basis for evaluation. They provide the raw material for 15 crop-by-country comparisons. They were complemented by a more limited survey that canvassed four regions in Uganda to assess adoption of recently released clonal material in banana (Kagezi *et al.*, 2012). We also used output from a recent IFPRI–CSIR (Council for Scientific and Industrial Research, Ghana) survey on adoption of maize and rice MVs in Ghana (Ragasa *et al.*, 2013a,b). The 12 surveys furnish us with 18 country-by-crop, 34 subregion-by-crop, and 279 variety-specific observations for comparative analysis.

Oral responses on seed usage and on area planted to specific varieties provided the raw material for the subsequent calculation of adoption estimates in the household surveys. The cassava survey team complemented their household interviews with field measurements that featured varietal photographs using mobile phones (Alene and Mwalughali, 2012). These were analysed by research scientists who were able to assess varietal identity from the pictures displaying morphological plant characteristics.

Because of the lack of close supervision and the existence of recent survey estimates on adoption, cultivar-specific estimates could not be elicited from expert panels for rice in Nigeria; therefore, the validation exercise for this survey focuses on the aggregate adoption of MVs as a group in relation to local varieties. AfricaRice also undertook a similar national survey in 2009. If the expert panel had had access to those results, its responses could have been contaminated by that information.

In the comparative analysis that follows, the definition of improved varieties needs to be constant across all sources in comparing estimates. In most contexts, reviewing the national release list was the basis for defining improved varieties. In practice, the estimated adoption levels were based on what experts stated were improved varieties during the elicitation exercise. For beans and sweetpotatoes in Rwanda and Uganda, released local landraces were included in the set of improved varieties. For groundnuts in Nigeria, the old improved variety 55-437 was included. The definition of improved varieties needs to be constant across all sources in comparing estimates. In most contexts, reviewing the national release list was the basis for defining MVs.

Comparing expert opinion with the survey estimates

Congruence between national estimates is described in Table 20.1 where the 18 matching observations in the database are ordered according to mean absolute percentage error (MAPE). In appraising congruence, both the percentage differences in column 5 and MAPEs in column 6 provide complementary information.

Table 20.1. Validating adoption estimates from expert opinion with survey results by crop.

		Estimate of MV adoption (%)			
Crop	Country/ region	Expert opinion	National survey	Difference (%)	Mean absolute percent error (MAPE)
Maize	Ghana	57.0	59.6	–2.6	4
Maize	Ethiopia	26.5	27.9	–1.4	5
Sorghum	Tanzania	42.3	38.7	3.6	9
Rice	Nigeria	50.4	56.2	–5.8	10
Bread wheat	Ethiopia	87.7	77.8	9.9	13
Groundnut	Tanzania	32.2	28.4	3.8	13
Beans	Rwanda	68.2	60.1	8.1	13
Potato	Ethiopia	25.2	22.2	3.0	14
Barley	Ethiopia	29.2	33.8	–4.7	14
Pigeonpea	Tanzania	39.5	49.7	–10.2	21
Banana	Uganda	8.0	6.2	1.8	29
Cassava	SW Nigeria	68.0	52.0	16.0	31
Sweetpotato	Rwanda	41.6	27.9	13.7	49
Beans	Uganda	60.0	40.0	20.0	50
Groundnut	Nigeria	51.2	31.0	20.2	65
Potato	Rwanda	84.9	35.6	49.3	138
Sweetpotato	Uganda	78.8	17.9	60.9	340
Durum wheat	Ethiopia	13.0	0.5	12.5	2500

Adoption estimates elicited by IITA from NARS participants in Ghana closely matched the results of the IFPRI national survey on maize adoption. The fit is also reasonably good for the next ten observations in Table 20.1. MAPEs are less than 30%, and differences are under 10% with the borderline exception of pigeonpea in Tanzania. In contrast, a lack of agreement is apparent in the last seven observations.

Arguably, the most egregious mismatch between expert opinion and survey estimates centers on sweetpotato in Uganda. Estimates were elicited for the four main geographic regions of the country and were aggregated to generate a national estimate. The discrepancy between sources of estimates was most marked in the eastern and northern regions with differences exceeding 75%. Labarta *et al.* (2012) give two plausible reasons for the wide divergence between the expert opinion and the survey estimates. Large quantities of improved sweetpotato vines were transferred to the northern region in response to relief programmes. Interest in orange-fleshed sweetpotatoes has also sparked a massive transfer of planting material in selected districts. Historically, drought tolerance of planting material is a known weakness of improved varieties and improving tolerance is a primary breeding objective. It appears that transfer of large quantities of planting material fueled exuberance and optimism about the prospects for adoption that departed sharply from the reality of propagating sweetpotato in a drought-prone environment where a few well-defined local varieties reign.

The second explanation focuses on varietal invisibility in the sweetpotato crop, which seldom exceeds 0.5 hectares per field, is often planted in association with other crops, and usually is harvested piecemeal. It is a crop characterized by poor road visibility that leads to blurred perceptions in identifying varieties that farmers are growing. As a result, varietal identity and diversity is not apparent without taking the time and effort to make field visits, especially at flowering (Labarta *et al.*, 2012).

Inspection of the regional and cultivar-specific databases also provides clues about the likely reasons for the poor congruence between the estimated sources for the other six observations. Although problematic regions and cultivars can be identified, explanations for what led to these large order of magnitude disparities are mostly speculative. Nevertheless, based on these seven cases, we can say that over-optimism about technology transfer programmes can result in substantial overestimates of technology adoption. The case of potato in Rwanda shows that civil war may lead to the collapse of MV varietal adoption. These are two contextual situations that analysts need to be aware of in measuring the long-term uptake of improved varieties (Labarta *et al.*, 2012).

Knowing what is going on in farmers' fields is desirable when scientists give expert opinion on adoption. But such knowledge is not always recorded at the main research station. Variation in the elicitation process of expert opinion also does not seem to play a significant role in understanding variations in congruence. Nor does more prestigious science make for more congruent estimates.

As mentioned earlier, NARO's sweetpotato breeding programme in Uganda is highly respected and is the hub of sweetpotato improvement in the Great Lakes region. IITA's Center is located in southwest Nigeria where the cassava survey was conducted. The same people and the same process generated the congruent estimates for maize in Ghana and the rather 'disagreeable' estimates for cassava in southwestern Nigeria. With the same elicitation process, CIP was responsible for congruent estimates for potato in Ethiopia and divergent estimates for potato in Rwanda.

Likewise, differences in crops and countries do not seem to feature as explanations of the variation in types of estimates. Ethiopia was associated with convergent estimates for barley, maize, and potato and divergent estimates for durum wheat (Yigezu *et al.*, 2012; Jaleta *et al.*, 2013; Labarta *et al.*, 2012). Adoption estimates for bread wheat in Ethiopia did not vary that much by source but, in relative terms, the estimates for durum wheat were substantially different (Yirba *et al.*, 2012). These estimates were elicited from the same group of wheat improvement scientists.

The simple mean MV adoption level was 48% for expert perceptions and 36.5% for survey estimations. The 11% mean difference in a paired *t*-test is statistically significant at the 5% level. If we designate sweetpotato in Uganda and potato in Rwanda as outliers because of their changing contextual situations and re-estimate the means for the remaining 16 observations, the mean difference between the two estimation methods narrows to 5.5% and is, again, significant statistically at 0.05. Proportionally, the survey estimate is seven-eighths the size of the expert opinion estimate. Reducing our 35% estimate for aggregate adoption of MVs in SSA by the same proportion yields a revised estimate that approaches 31%. This revised estimate incorporates a correction from the finding that expert opinion tends to generate somewhat higher levels of adoption than properly conducted household survey estimation.

Redoing the above calculation on the disaggregated regional data set of 34 observations (excluding potato in Rwanda and sweetpotato in Uganda) gives identical results. The simple-mean, expert-opinion estimate of 36.4 is 4% higher than the survey estimate of 32.4.

The mean adoption estimate of 36.5 in the national surveys was made up of 26.5% from MVs named by the panel and by 10% from unnamed or other named materials believed to be MVs. The size of the second component varies from survey to survey, but it is usually sizeable as there is always a leftover quantity of MV area that cannot be assigned to a specific cultivar. For this reason, the area of specific MVs will typically be proportionally less than total adoption levels. Because the ability to designate specific areas to MVs is imperfect, survey-specific cultivar estimates will often be substantially less than comparable estimates from expert panels.

Thus far, we have presented comparative results from the 18-observation national database. Similar differences in MV adoption between expert opinion and household surveys also were found in the 34-observation regional database. Findings for the 279-cultivar specific database are presented in Table 20.2, which divides the varieties into four categories depending on the level of perceived adoption by experts. For example, of the 279 varieties from the national comparisons in Table 20.2, experts perceived that 44 had a level of adoption that exceeded 10% of cultivated area of the crop. Experts believed that, on average, these were sown to 24% of the area available; the mean survey estimate for the same 44 varieties was about 13% resulting in a difference between the two sources of about 11%. From the previous discussion of the national and regional data, it was likely that expert estimates were higher

Table 20.2. Agreement between expert and survey mean estimates of specific varieties by expert interval for different levels of adoption of MVs.

	Adoption (%) from the expert opinion estimate			
Source	0–1	1.01–5	5.01–10	>10
Expert	0.39	2.97	8.02	24.12
Survey	0.70	1.15	2.84	12.86
Difference[a]	–0.30	1.82	5.18	11.27
Number of observations	105	100	30	44

[a]Between the mean expert opinion and survey estimates.

than the household survey estimates at all levels, except the lowest estimates.

This tendency for systematic differences to emerge between the two sources of estimates applies to all levels of estimates but figures most prominently for expert estimates in the range of 5–10% (Table 20.2). For the lowest level of adoption in the MV cultivar database, the survey estimates are higher than expert opinion, which to some extent neglected these varieties. Restricting the analysis to only positive observations for expert opinion in this lowest interval does not reverse the finding that the survey estimates are higher than those for expert panels.

The message conveyed in Table 20.2 is that probably neither surveys nor expert panels can do a good job in delivering accurate estimates of cultivar-specific adoption. Expert panels will most likely overestimate the importance of specific varieties; surveys will understate their relevance. Although skillful use of both methods may suffice for our purposes, we should be aware of the sources of bias when the focus is on MV-specific adoption. Accuracy in survey estimates depends heavily on whether or not a plethora of names can reliably be assigned to specific varieties.

Of the DIIVA-funded household surveys, bean researchers in Rwanda worked hardest in tracing the identities of farmers and their crop varieties in many locations. They believed they were able to assign successfully 88% of the area available to local, selected and improved varieties (Katungi and Larochelle, 2012). CIAT researchers and their partners had considerable experience in the counting of bean varieties. Their work in the DIIVA Project built on interviews with village focus groups carried out in 2000 and 2005 when respondents ranked the importance of different varieties. With the addition of data from 2010, patterns emerging over time could be seen.

On the other hand, expert opinion tends to focus on a subset of varieties while ignoring the relative importance of other MVs. The otherwise excellent household survey work in southwest Nigeria for cassava was an apt example of not being inclusive enough in eliciting estimates from experts – the elicitation did not mention the leading MV found in the household survey, apparently because it did not appear on the release list.

Being too inclusive can also prove to be a risky strategy. Returning to beans in Rwanda, experts allocated very small areas to 22 improved varieties. Sixteen of these had negligible adoption outcomes in the household survey results. An additional 25 MVs accounted for about 10% of the area. These did not receive an area allocation by the expert panel.

Validation results in South Asia

By far, the closest correspondence between estimates of expert opinion and those from subsequent varietal adoption surveys was obtained for rice in South Asia (Pandey, *et al.*, Chapter 13, this volume). For five of the seven state/country observations, elicitation was conducted at considerable spatial resolution for agroecologies and/or administrative districts.

Six of the seven observations were characterized by MV adoption greater than 75%. With the exception of Bangladesh, the difference between sources in the aggregate estimate of MV adoption was negligible for these high adoption countries and Indian states. Indeed, the expert elicitation suggested slightly lower uptake of improved varieties than the survey estimates.

At an extreme, HH estimates in Bangladesh were 10 percentage points higher than EE estimates. These high-adoption observations were also characterized by mostly small differences ranging from 1 to 5.5% between expert and survey estimates for individual improved varieties. Only the Indian state of West Bengal and Nepal had moderately high inter-source specific varietal differences that exceeded 3 percentage points.

Bhutan, where MV rice adoption is less than 50%, was the outlier. Expert estimates in the high and low altitude rice-growing areas agreed reasonably well with survey estimates, but the survey estimate in the mid-altitude zone fell short of expert opinion by 50%. Pandey *et al.* in Chapter 13 ascribe this lack of congruence to the fact that few people on the expert panel were familiar with varietal adoption in the mid-altitude zone. In other words, the panel was spatially ignorant of varietal change in this zone that, based on the survey estimates, was lagging behind the high and low altitude regions in MV adoption.

Overall, these results point to the conclusion that the investment in household surveys did not generate that much value in terms of additional information on varietal adoption for rainfed rice in South Asia. Expert elicitation was the more cost effective option especially in the states and regions where full adoption of MVs was becoming a reality. Elicitation sub-nationally was also found to be more informative than nationally, mainly because more improved varieties were identified with positive production areas in the district data sets.

The validated findings for the ICRISAT-mandate crops in peninsular India echoed the results in SSA more than for rice in South Asia (Kumara Charyulu *et al.*, Chapter 14, this volume). They highlighted several contexts where expert elicitation gave very imprecise or incomplete results which were inconsistent with the HH survey estimates. Like the situation for maize hybrids in East and Southern Africa, public-sector scientists in national programmes were relatively clueless about the recent adoption dynamics of individual sorghum and pearl millet hybrids in peninsular India. The recent HarvestPlus survey of pearl millet hybrids and improved OPVs in Rajasthan demonstrated that a well-supervised large sample survey combined with farmer information on hybrids from the seed bags they purchased could give credible, up-to-date information on hybrid diffusion (Asare-Marfo *et al.*, 2013). Part of the problem was the absence of private-sector participation on the expert panels but even with such participation it is unlikely that information on varietal adoption could begin to approach the quality of estimates from the aforementioned survey.

The dominance of government and university scientists in the elicitation process resulted in biased estimates on area allocation between hybrids and improved OPVs. For example, expert opinion substantially overstated the importance of improved OPVs (at 20% of sown area) which, for all intents and purposes, were not mentioned in the 60 village focus-group and the 360 household interviews on rainy season sorghum in Maharashtra. And we can be confident farmers there have a good understanding of the difference between hybrids and OPVs.

Input from the public-sector extension service in survey sample design may also predispose the results towards finding more estimated area in improved OPVs than would actually be the case when the private sector is very active in generating and distributing hybrids. (The public sector is charged with the transfer of improved OPVs). Evidence for this type of bias came from the comparison of two surveys on varietal adoption of pearl millet MVs in Maharashtra. The smaller ICRISAT survey of 360 households estimated the area share of the leading improved OPV at about 20%; the larger HarvestPlus survey of about 2000 households estimated the area share of this variety at only 4%; the bulk (96%) of area was planted to hybrids. This speaks to another problem – ensuring high quality sampling procedures when designing surveys.

In addition to the inadequacy of expert elicitation for varietal change conditioned by private-sector hybrids, the results in Chapter 14 (this volume) also point to the potential for marked variation in the quality of estimates from expert opinion. The expert elicitation on the uptake of improved cultivars for pearl millet, sorghum, groundnut, pigeonpea, and chickpea took place during the annual meetings of their respective All-India Crop Improvement

Programs. For each crop, expert panels were convened, and estimates were produced for five to six of the major-producing states. This was an apt time to conduct expert elicitation because the crop is foremost in the minds of scientists during these weekly meetings. However, casual inspection of the estimates suggests that knowledge about the popularity of improved varieties for some crops in a few states was scanty, probably because of the very large size of and regional-variation in production in some states which exceeds output in most producing countries in sub-Saharan Africa. For example, pigeonpea improvement scientists were able to estimate the use of improved varieties as a whole but were unable to allocate that area to specific cultivars in two of the most important producing states.

Challenges in nationally representative adoption surveys

Two challenges have been highlighted in arriving at cost-effective MV adoption levels from survey data. The first is responding to the need to save resources by covering multiple crops in shared agroecologies. There are usually tradeoffs between the potential for cost saving and the reliability of estimates (C. Ragosa, 2013, personal communication). The other is the aforementioned problem of identifying a specific MV from a multiplicity of names that exhibit widespread spatial variation. Judging whether a cultivar is or is not an MV is a corollary to the identification problem. The recycling of seed in cross-pollinated crops is another difficult issue that calls for standardization after three to five years depending on the rate of outcrossing.

Levels of MV adoption can vary widely even in well-conducted surveys. On the basis of a national-level survey of rice in Ghana in 2012, researchers estimated aggregate MV adoption approaching 60% (Ragasa *et al.*, 2013b). However, researchers from an earlier national survey carried out in 2009 arrived at an estimate exceeding 80% (Diagne *et al.*, 2012). The difference is not attributable to the differing survey years – the disparity in estimates emanates from decisions researchers had taken in classifying varieties as 'improved' or 'traditional'.

In both surveys, the leading variety was Jasmine 85, an IRRI variety bred in Thailand in the 1960s. (Jasmine 85 was officially released in Ghana in 2009 after it had already been adopted widely by farmers – now grown on 27% of the area.) But the key question is: what to do about Mandii, the second leading variety laying claim to 19% of area? Researchers in the 2012 survey classified it as a 'local variety', whereas researchers in the 2009 survey designated it as 'improved'. Their list of improved varieties contains 104 names with only seven dated released varieties with adopted area. Mandii seems to be expanding; its area in 2009 was estimated at 7%. Given the uncertainty about its origins, the 'What to do about Mandii?' question can most likely only be addressed by DNA fingerprinting. This also points to the need to clearly define the parameters determining 'improved varieties'.

The DIIVA Project has also reconfirmed the need for field measurement in cases where varieties are difficult to distinguish morphologically. The survey of cassava in south-western Nigeria epitomizes this case (Alene and Mwalughali, 2012). Farmers knew improved varieties by a group name but could not distinguish relatively small morphological and phenotypic differences that allowed for the elicitation of reliable data on specific MV cultivars. In this case, there was no substitute for field measurement, which is more doable in cassava because it is in the field for a longer time in a mature state than other food crops.

Survey performance could be improved if focus groups generated reliable information on varietal adoption. The use of focus group interviews in a community questionnaire was one of the features of the surveys supported by the DIIVA Project (see Chapter 4, this volume). In their validation reports, project participants formally compared responses from focus groups and household questionnaires. Although these results have not been rigorously analysed, reading the reports suggests that focus groups can provide useful information about the relative importance of the variety in the village and the adoption levels of individual farmers; but household data are strongly preferred if cultivar-specific area estimation is the goal (Mausch and Simtowe, 2012).

Researchers from AfricaRice were more optimistic about the use of community-based instruments than most researchers from other CG Centers for cost-efficiency reasons (Diagne *et al.*, 2013). They still opted for household schedules over focus group interviews, however, where it was necessary to collect cultivar-specific information. They also found ways to collect area data at the community level, which compared favourably, i.e. was consistent, with information gathered from household interviews (Table 20.3).

Although there is an 11% gap between the two estimates on the level of adoption of MVs as a whole, the matched ranking between community and household surveys for rice in Nigeria is characterized by widespread agreement on the relative importance (i.e. ordinal ranking) of specific cultivars. The simple correlation between the community and the household rankings approaches 0.80 in Table 20.3.

Common bean cultivars show even stronger agreement in Rwanda and Uganda between a community estimate in the percentage of villages where a cultivar was mentioned in a focus group and the percentage area of the same cultivar from the household survey (see Chapter 16, this volume). For 67 common cultivars in Rwanda, the estimated correlation coefficient exceeds 0.80; for the leading 19 common varieties in Uganda, the estimated correlation coefficient exceeds 0.95. In both of these meticulously carried out surveys, the community survey focused on the top three cultivars at three points in time. All the cited focus-group varieties could be paired up with household responses but not all the household responses could be matched to the top varieties perceived by focus group respondents. More than 100 varietal names generated in the household survey could not find an identical partner in the community focus-group enquiries that also embraced a relatively large set of varieties. Most of these unmatched household varieties were planted in very small areas.

Table 20.3. Comparison of village-level and household-level interview data on varietal adoption using area grown under these varieties for rice in Nigeria.

	Village interview		Household interview (2009)		
Variety	Percentage	Rank	Percentage	Rank	Gap
Traditional	54.75	1	43.73	1	11.02
Modern	45.25	2	56.27	2	−11.02
FARO 44 / SIPI 4	8.04	1	12.35	1	−4.31
CHINA	7.03	2	8.76	2	−1.73
IMPROVED	3.48	3	4.83	3	−1.35
NERICA (others)	2.94	4	4.36	4	−1.42
FARO 15	2.80	5	4.03	5	−1.23
FARO 46 / WITA 1	2.39	7	2.77	6	−0.38
FARO 52 / WITA 4	2.68	6	2.45	7	0.23
FARO 55 / NERICA	1.77	8	2.15	8	−0.38
FARO 37 / WITA 3	1.59	9	2.01	9	−0.42
FARO 29 / BG 90-	1.48	10	1.58	10	−0.10
FARO 61 / WAB 18	1.38	11	1.32	11	0.06
BUTUKA	0.17	31	1.20	12	−1.03
FARO 51 / CISADA	0.95	12	1.09	13	−0.14
TURN 2	0.53	14	1.07	14	−0.54
ECWA	0.28	21	1.03	15	−0.75
IR 8	0.57	13	0.71	16	0.14
CAROLINA	0.50	16	0.66	17	−0.16
WILLY RICE	0.33	20	0.59	18	−0.20
FARO 21	0.34	17	0.55	19	−0.21
FARO 35 / WITA 2	0.52	15	0.38	20	0.14
YARJOHN	0.27	22	0.28	21	−0.01
FADAMA2	0.06	52	0.27	22	−0.21
Other improved	5.15		1.83		

Source: Diagne *et al.*, 2013.

The high estimated correlation coefficients for bean cultivars in Rwanda and Uganda suggest that well-constructed community focus groups can provide valuable information on the relative importance of leading varieties in a well-defined regional setting. The three ICRISAT validation surveys on dryland crops in peninsular India also indicated that focus-group and household estimates on varietal adoption gave very similar results (Chapter 14, this volume). They further illustrated the importance of forming the focus group with farmers who planted in the most recent cropping year.

Summary

The DIIVA Project focused on quantifying the adoption of improved genotypes in food crops and in dryland regions stereotyped as having been bypassed by the Green Revolution. For reasons related to consistency with previous benchmark research (Evenson and Gollin, 2003), the higher costs of competing methods, the scarce time of participating investigators, and the comprehensive nature of crop and country coverage, elicitation of expert opinion was the method of choice for estimating varietal adoption. Expert opinion accounted for nearly 75% of the 152 crop-by-country estimates of varietal adoption in Chapters 6–12 in SSA, all seven estimates in rice in South Asia in Chapter 13 and for all 25 estimates in five major dryland crops in India in Chapter 14.

In carrying out the process of expert elicitation from a standardized protocol, participating researchers generated considerable anecdotal evidence on what worked. The protocol was adapted to regional- and crop-specific circumstances that featured a considerable amount of 'learning by doing' by CG Center staff conducting the expert panels. In general, more effective elicitation was characterized by:

- Close and intensive supervision by project-related staff;
- Organization of and attendance by project staff at time-bound workshops featuring direct interaction with expert panel members;
- Greater spatial resolution in the elicitation of estimates that were subsequently aggregated to regional and national levels;
- Including more members from the informal sector and from NGOs with geographic-specific expertise in technology transfer on the panels; and
- Feedback from CG Center breeders in the final stages of the process.

Lessons on what did not work were transparent. The CG Center that relied solely on NARS scientists as consultants to carry out expert elicitation on their behalf was only able to provide quality cultivar-specific adoption estimates for two of their 14 assigned crop-by-country observations. Much more intensive supervision was needed.

Validation exercises that compared the expert estimates to those derived from representative national and regional surveys enhanced the potential for learning. Eighteen validation surveys, mostly supported by the DIIVA Project, were carried out in SSA, six for rainfed rice in South Asia and three for dryland crops in peninsular India.

In SSA, household survey estimates and those from expert opinion panels were reasonably close for 10 of the 18 observations. For these 10 'consensus' observations, the simple mean of both the expert elicitation and household survey results averaged 45%.

The other eight estimates did not agree nearly as well and could be labelled imprecise or 'controversial.' Survey estimates were lower for these eight observations, and for two they were markedly lower. Ignoring these two outliers, survey estimates were about seven-eighths the size of expert elicitations. Applying the seven-eighths finding from the validation exercise gives a more conservative estimate of about 31% for MV adoption if surveys had replaced expert opinion panels.

What explains the differences between the relatively convergent ten and the controversial eight observations? The controversial estimates did not seem to be associated with variations in the elicitation approach or specific to a crop or country. The same scientists using the same approach in the same countries were identified with both consensus and controversial estimates. For the two extreme outliers, controversial estimates had more to do with the extenuating circumstances of rapid change, that is, disruption associated with rampant over-optimism about the prospects for large technology transfer efforts or with civil war that also can be devastating for the applicability of prior knowledge in conditions where confirmation is difficult.

Other lessons were more subtle. For all but the smallest positive category of adoption, expert estimates were systematically higher than survey estimates for individual improved varieties. Slightly over 70% of the mean adoption estimates in the national surveys were composed of MVs for which the panel held positive adoption beliefs; the other 30% came from unnamed or other named materials believed to be MVs. The size of the second component varies from survey to survey but it is usually sizeable because there is always a leftover quantity of MV area that cannot be assigned to a specific cultivar. For this reason, the summed area of well-identified MVs will typically be less than an aggregate adoption level. Household surveys are likely to understate the importance of specific improved cultivars; detailed estimates from expert opinion that feature few if any varieties in a residual 'other' category are likely to overemphasize the uptake of specific MVs, especially those in the 5–10% range of adoption. Accuracy in household survey estimates depends on whether or not numerous regional- and location-specific names can reliably be assigned to specific varieties.

The validation results in South Asia in Chapters 13 and 14 illustrated the strengths and weaknesses of expert elicitation. For rice, the differences between expert and household survey estimates were minor; so much so that if the sole motivation was to estimate varietal adoption, expert opinion provided the most cost-effective option.

Why is expert elicitation a superior alternative to a survey in rice in South Asia in measuring varietal adoption? Plausible explanations include the dominance of rice as a food staple, the large number of government and university scientists working on the crop, the existence of well-defined production ecologies that provide a framework for thinking about the crop, and the abundance of secondary data and adoption studies that can be incorporated into and update experts' beliefs.

In contrast, the findings for sorghum and pearl millet hybrids in peninsular India show that expert opinion will not generate credible results in cereals where the private sector is very active in varietal development and distribution. Without private-sector participation in the form of information on seed sales, expert estimates on the leading hybrids will be outdated and incomplete.

An encouraging development for future estimation was the good agreement between the focus-group community and the household estimates in several of the surveys in SSA and in South Asia. Community focus groups could provide a valuable means to ground-truth expert opinion in a rapid rural appraisal format. Or they could stand alone as an independent source of adoption estimates if funding were sufficient to cover representative communities on a timely and routine basis. Given that travel to the community is usually the largest cost component of any rural survey in SSA, the issue of relative costs is relevant. Nonetheless, in future large-scale adoption studies, we need to find a cost-effective alternative to the representative household surveys that require a sample size of 500–700 households in order to validate expert opinion from the more qualitative perspective of 'Do the elicited estimates roughly reflect reality or not?' Well-structured, community focus-group discussions combined with field visits and the selective use of DNA fingerprinting could be an attractive alternative.

Costs of DNA fingerprinting are declining rapidly. The size and shape of its role in the analysis of varietal adoption in developing countries are the subjects of several pilot studies financed by the Bill & Melinda Gates Foundation. Preliminary results presented at a workshop in 2014 suggested that farmers' oral responses in surveys on the identity of sown varieties could be very imprecise. Specialists in DNA fingerprinting and some economists attending the workshop believed that employing this technique in more costly specialized adoption surveys focusing on the collection of relatively large numbers of plant-tissue samples in farmers' fields was the only way forward to reliably identify improved varieties in developing country agriculture. Similar to randomized control trials (RCTs) that are discussed in Chapter 21 for the purpose of impact assessment, DNA fingerprinting was viewed as a perfect and increasingly cost-effective technique to evaluate improved varietal adoption. Others saw the role of DNA fingerprinting as highly complementary to existing methods in playing a more selective role to shed light on well-identified issues and questions such as the one posed earlier in understanding the genetic background of the second leading rice variety in Ghana: 'Is Mandii a local landrace or an improved variety and what in its genetic composition gives rise to its increasing popularity?' Irrespective of whether DNA fingerprinting becomes an essential or a complementary component in future inquiries on varietal adoption, its emergence as a viable tool for genotypic identification in farmer fields is a most welcome development.

Note

[1] Very selective use of DNA fingerprinting was contemplated in the DIIVA Project, and a proposal for the funding of a pilot DNA fingerprinting application was drafted early in 2010; however, it was not carried out. After the completion of the DIIVA Project in 2012, the Bill & Melinda Gates Foundation (BMGF) funded several pilot applications of DNA fingerprinting to measure varietal adoption. Preliminary results were presented at a workshop at the BMGF headquarters in Seattle, Washington, in the summer of 2014.

References

Alene, A.D. and Mwalughali, J. (2012) *Adoption of Improved Cassava Varieties in Southwestern Nigeria, Objective 2 Technical Report*. International Institute of Tropical Agriculture (IITA), Lolongwe, Malawi.

Asare-Marfo, D., Birol, E., Karandikar, B., Roy, D. and Singh, S. (2013) *Varietal Adoption of Pearl Millet (Bajra) in Maharasthra and Rajasthan, India: a Summary Paper*. HarvestPlus, International Food Policy Research Institute, Washington, DC.

Diagne, A., Kinkingninhoun-Medagbe, F.M., Amovin-Assagba, E., Sanni, K. and Toure A. (2012) *To Attain a Wider Understanding of Key Aspects of the Performance of Food-crop Genetic Improvement in Priority Country-by-Commodity Combinations in Sub-Saharan Africa: the Case of Rice Crop*. DIIVA, Objective 1 Report. AfricaRice Center, Cotonou, Benin.

Diagne. A., Kinkingninhoun-Medagbe, F.M., Ojehomon, V.T., Abedayo, O.O., Amovin-Assagba, E. and Nakelse, T. (2013) *Assessing the Diffusion and Adoption of Improved Rice Varieties in Nigeria*. DIIVA, Objective 2 Report. Africa Rice Center, Cotonou, Benin.

Evenson, R.E. and Gollin, D. (eds) (2003) *Crop Variety Improvement and its Effect on Productivity: the Impact of International Agricultural Research*. CAB International, Wallingford, UK.

Hassan, R.M., Mekuria, M. and Mwangi, W. (2001) *Maize Breeding Research in Eastern and Southern Africa: Current Status and Impacts of Past Investments made by the Public and Private Sectors 1966–1997*. International Maize and Wheat Improvement Center (CIMMYT), Mexico City.

Jaleta, M., Yirga, C., Kassie, M., De Groot, H. and Shiferaw, B. (2013) *Knowledge, Adoption and Use Intensity of Improved Maize Technologies in Ethiopia*. CIMMYT's Report on Objective 2, Part 1. International Maize and Wheat Improvement Center (CIMMYT), Mexico City.

Kagezi, G., Mpiira, S., Jogo, W., Karamura, E., Gotor, E. *et al.* (2012) *Adoption of New Banana Varieties in Uganda: report for the DIIVA Project*. DIIVA Objective 1 Report. Bioversity International, Rome.

Katungi, E. and Larochelle, C. (2012) *Validating Adoption of Modern Bean Varieties in Rwanda and Uganda Using Household and Community Surveys*. DIIVA, Objective 2 Report. International Center for Tropical Agriculture (CIAT), Kampala, Uganda.

Labarta, R., Wambugu, S., Yirga, C., Mugabo, J., Dieu Nsabimana, J. *et al.* (2012) *The Adoption of Improved Potato and Sweetpotato Varieties in Ethiopia, Rwanda and Uganda: Comparing Expert Opinion Estimates and the Results of Nationally Representative Surveys*. DIIVA, Objective 2 Report. International Potato Center (CIP), Nairobi.

Mausch, K. and Simtowe, F. (2012) *Adoption Rates and their Reliability Evidence Based on Expert Panel Discussions, Community Focus Groups and Household Survey in Tanzania – Groundnuts, Pigeonpea and Sorghum*. DIIVA, Objective 2 Report. International Crops Research Institute for the Semi-Arid Tropics (ICRISAT), Nairobi.

Muthoni, R. and Andrade, R. (2012) *The Impact of Bean Genetic Improvement Research*. DIIVA, Objective 2 Report. International Center for Tropical Agriculture (CIAT), Kampala, Uganda.

Ndjeunga, J., Ntare, B.R., Ajeigbe, HI, Echekwu, C.A., Ibro, A. and Amadou, A. (2013) *Adoption and Impacts of Modern Groundnut Varieties in Nigeria*. DIIVA, Objective 2 Report, International Crops Research Institute for the Semi-Arid Tropics (ICRISAT), Niamey, Niger.

Ragasa, C., Dankyi, A., Acheampong, P., Nimo Wiredu, A., Tripp, R. *et al.* (2013a) *Patterns of Adoption of Improved Maize Technologies in Ghana*. Draft Report. International Food Policy Research Institute (IFPRI), Washington, DC. The DIIVA Project. Fletcher, North Carolina.

Ragasa, C., Dankyi, A., Acheampong, P., Nimo Wiredu, A., Tripp, R. *et al.* (2013b) *Patterns of Adoption of Improved Rice Technologies in Ghana*. Draft Report. International Food Policy Research Institute (IFPRI), Washington, DC.

Walker, T.S. (2010) *Guidelines for Data Collection for Objective 1 of the DIIVA Project*. Fletcher, North Carolina.

Yigezu, Y.A., Yirga, C., Aw-Hassan, A. and Labarta, R. (2012) *Barley in Ethiopia: Verifying the Adoption Estimates from a Panel of Experts with Secondary Data and Household and Community Focus Group surveys*. DIIVA Objective 2 Report. International Centre for Agricultural Research in the Dry Areas (ICARDA), Aleppo, Syria.

Yirba, C., Jaleta, M., De Groote, H. and Gitonga, Z. (2012) *Diffusion and Impact of Improved Maize and Wheat Varieties in East and Southern Africa*. CIMMYT's Report on Objective 2, Part II – Wheat in Ethiopia. International Maize and Wheat Improvement Center (CIMMYT), Nairobi.

21 Implications for Monitoring Progress and Assessing Impacts

J. Alwang*
Department of Agricultural and Applied Economics, Virginia Tech, Blacksburg, USA

The Diffusion and Impact of Improved Varieties in Africa (DIIVA) studies provide a detailed portrait of the attributes and spread of new crop varieties in Africa. They give estimates of economic impacts and the distribution of these impacts among different population groups in case-study countries[1] and for the sub-Saharan Africa (SSA) region as a whole. The previous chapter summarizes lessons learned from the DIIVA experience for using expert opinion to infer the pattern and extent of diffusion of new crop varieties. This chapter summarizes lessons learned from the DIIVA experience about ongoing monitoring and measurement of varietal diffusion and its impacts. Messages from DIIVA include: (i) better use of data for ongoing and future assessment; (ii) information on use of observational data and techniques for cleanly identifying impacts of technical change; and (iii) improved processes for measuring non-efficiency economic impacts.

Overview of Impact Assessment and Its Various Audiences

Impact assessment itself requires substantial effort and resources. Research managers and donors exhibit increased interest in knowing the impacts of their investments, but it is important to recognize that optimal application of impact assessment tools must consider their costs and seek a balance between benefits from the assessment and resources allocated to alternative uses. Impact assessment encompasses a wide continuum of practices from 'quick and dirty' low-cost assessments to multi-year highly complex evaluations and the techniques employed and required data are correspondingly diverse. Data needs for a relatively rough assessment can usually be obtained from variety trials conducted as a normal part of the technology validation process. Alternative estimates of adoption can be obtained from expert opinions, from seed distribution systems or from ongoing monitoring efforts – where they exist.[2] Such data can be incorporated into economic surplus models to obtain estimates of market conditions with and without the technology. For impact assessments where convincing causal evidence is desired, however, time and resource requirements can be substantial. The DIIVA study of the impacts of improved maize on poverty in Ethiopia took approximately 2 years to complete (including the collection of specialized survey data) and cost more than US$200,000. The cost of a study is increased by having a distributional focus because more effort is needed to collect appropriate variables and the time needed to analyse data may be increased. The overall

* E-mail: alwangj@vt.edu

cost, however, of a credible impact study of the DIIVA type exceeds US$150,000. Each of the DIIVA impact studies uses cross-sectional data collected for the express purpose of estimating impacts; panel data, which would better enable identification of the technology effect, would require more time and resources to collect. An obvious avenue for reducing costs is to expand existing multi-purpose surveys to include variety-specific information.

Two broad categories of impact assessment of varietal technology have been identified:[3] (i) assessments of impacts of long-term research investments producing technologies/varieties that have been released and spread over time; and (ii) assessments of specific (micro-level) impacts on yields and other outcome parameters when the technology is still being considered for release. The first category often uses observational (household- and field-level) data and employs econometric or other statistical techniques to 'identify' the effect of technology adoption at the household level. Household-level outcomes from adoption include changes in productivity, unit costs of production, input use, etc. These outcomes contribute to changes in household income and have indirect impacts on food consumption, nutrition and health. At the market level, changes in supply owing to lower per-unit cost of production induce changes in market prices and quantities transacted in commodity markets, which are indirect outcomes from aggregate adoption. These changes create further indirect impacts on producers and consumers and these impacts can be measured as changes in economic surplus or more theoretically consistent measures, such as compensating or equivalent variation. The detailed DIIVA impact studies clearly indicate that these market-level factors contribute significantly to poverty reduction but the impacts depend on the structure of the corresponding markets. Because field sizes in Africa are relatively small and because market prices respond rapidly to supply responses, measured farm-level effects (particularly on crop income) are relatively modest. Induced changes are also felt in market-mediated outcomes such as changes in labour demand.

Many audiences are interested in the distributional effects of new crop varieties and the analysis should consider potential heterogeneous direct effects across producers and the distribution of indirect effects among heterogeneous market participants. A distribution-sensitive analysis would have to account for potential differences in adoption and variability in the per-unit cost of production or productivity changes associated with technology adoption. However, the agricultural technology literature shows clearly, for example, that poverty reductions from new agricultural technologies are felt over time as diffusion proceeds and market effects are played out.

The second category of impact assessment may use information from experimental trials to measure how the technology affects input use and productivity at the field or farm level. Randomized control trials (RCTs) may be employed to directly identify impacts of technology adoption using random assignment of the technology (possibly at the village or similar level) to address problems with endogenous adoption of technologies (de Janvry *et al.*, 2012). Of course, once per-unit impacts (at farm or field level) are measured, aggregation is needed to understand potential household level or more aggregate or economy-wide impacts. These impacts are played out over time as farmers adjust input levels (and learn about management of the variety), adjust land shares under the new technologies, etc. Assumptions about future adoption rates are required along with additional information on market-related behaviour.[4] Because of their micro-economic focus, such techniques are not readily applicable for the measurement of impacts of widely diffused technologies, most of which have probably undergone a rigorous (randomized)[5] assessment of productivity gains and other benefits, such as resistance to specific biotic constraints prior to their release (Norton and Alwang, in press).

Data for Crop Variety Impact Assessment

The DIIVA studies demonstrate the utility of different types of data. Ultimately, the economic impact of a crop variety release depends on its per-land unit increment to productivity net of cost (compared to what had been planted prior or would be planted in its stead) and its spread – the number of land units under production. Methods to measure spread were covered in prior chapters and we focus on measurement of productivity increments here. Some analyses examine higher-level outcomes, such as crop income, total household income and indirect outcomes such as nutritional well-being. These outcomes clearly depend on the plot-level productivity effects. We begin by focusing on the measurement of productivity changes and then turn to the measurement of higher-level outcomes.

Experiment station and technology validation trial data

A potentially important source of information on productivity effects is data from experimental trials conducted both on agricultural experiment stations and in farmer fields prior to variety release. Pre-release experimental trials include, depending on the crop variety technology and country, on-station, farmer-field and regional technology trials. These trials are a normal part of technology validation and are used by national agricultural research systems (NARS) to assess and validate the productivity, input use, drought resistance and other attributes of a variety prior to its release. They are designed to measure the statistical significance of differences with respect to a control (usually the farmer's preferred variety), and within-field randomization is used to control for potentially confounding effects (such as variability in soil quality). Trial results could provide important inputs into efforts to monitor impacts of diffused improved varieties over time. The main challenge in using these experimental results for inferring impacts beyond the immediate context of the experiment involves overcoming questions of external validity.

Two broad factors limit the external validity of such experiments: heterogeneity of performance of the variety under diverse agroecological conditions, and variations in yield and other performance attributes under typical adopting farmer management practices. Both factors contribute to the well-known 'yield gap' between variety performance in controlled trials (or under optimal conditions) and those observed in farmer fields (van Ittersum and Cassman, 2013). In SSA, marked heterogeneity in agroecological conditions and management ability, uncertainty about variety names and pedigrees, and wide variation in access to inputs complicates quantification of both factors. These factors are highlighted in the Ethiopia and Rwanda/Uganda DIIVA case studies. On-station trials are managed by professionals and soil fertility and input applications are optimal. When variety trials are moved to farmer fields the common practice is to select those farmers who are known to be good managers, who possess reasonably fertile fields and who follow researcher instructions. These and other factors contribute to the questionable external validity associated with these data. An assessment of impacts of new varieties using trial data combined with information on variety spread is likely to overstate impacts as yield gaps are widespread and persistent.

The central issue revolves around the degree of non-representativeness associated with variety validation and testing. External validity can be enhanced by making minor adjustments to normal testing processes. Experiments could be spread over wider geographical areas including those areas where the technology is expected to be only marginally more productive than existing technologies. Such results could be used to build agroecological heterogeneity into the assessments. Participating farmers might be selected at random or villages might be selected at random, as suggested by de Janvry *et al.* (2011) to enhance the realism of the trial. Relaxed evaluation protocols (instructions) can be provided to subjects, allowing the experiment to mimic normal processes for learning about technologies (careful measurement of inputs, including labour applications, is needed to conduct such an analysis). All these alterations could contribute to improved external validity of experimental trial results and recent lessons from RCTs could be used to help cleanly identify the impacts of new varieties.

With access to high-quality trial data, low-cost impact assessment could focus on measuring the spread of adoption, possibly using methods described in this volume, and quantification of the yield gap and its variation over space. The data could be used in cost-benefit calculations or to compute a k-shift in an economic surplus analysis. Alternatively, existing trial data could be used as a jumping-off point and the resulting estimates of impact could be treated as upper bound assuming positive bias in the selection of participating farmers. One challenge in using trial data is that they are plot/field level and provide no information about how changes in productivity associated with the adoption of a variety affect household decisions such as land allocation or amounts offered for sale, or broader outcomes such as changes in factor and product markets. An additional challenge for use of such data for ongoing monitoring and impact assessment is ensuring that they are comparable across varieties and countries and are collected and systematically archived. These data, possibly supplemented with results from RCTs, where available, would provide a useful approximation of aggregate direct impacts of varietal technology adoption but their use will clearly require good estimates of diffusion.

An important lesson from DIIVA is that locating good-quality data from experimental trials, even at the CG Centers, is a challenge because experimenters collect different types of data, quality is variable and the results are often not stored electronically.[6] Scientists lose track of them and prefer not to expend resources finding and compiling them. Few examples exist in the CG system of systematically archiving variety trial data and, where the data do exist, they are inaccessible, difficult to work with or both. Standardized practices have not been developed.[7] As a result, this rich potential source of data for impact assessment is under-developed and its widespread use is compromised.

A second lesson from DIIVA, highlighted elsewhere in this volume, is that measuring adoption is a complicated process and that even tested varieties with initial acceptance may not take off. The DIIVA study on impacts of improved sorghum in Nigeria (Ndjeunga *et al.*, 2014) showed that disadoption occurred following a multi-year extension effort and some well-documented initial acceptance. Evidence from this case shows that the improved sorghum cultivars do not have enough of the traits (demanded by farmers) that are markedly superior to traditional varieties. In these and many other cases, technology diffusion needs frequent measurement.

Randomized control trials

Controlled trials are now preferred by many economists for measuring micro-economic impacts of development interventions. By randomizing the treatment, selection bias can be eliminated and a causal treatment effect can be directly inferred. In the case of agricultural technologies such as improved varieties, randomization faces a number of challenges including ethical considerations, the fact that released varieties, as noted above, have already undergone a rigorous evaluation, and the difficulties of ensuring compliance and controlling for spill-over effects, and endogenous adjustments in input use farmers make when they adopt technologies (Barrett and Carter, 2010; de Janvry *et al.*, 2011; Bulte *et al.*, 2014; Norton and Alwang, in press). RCTs tend to take several years to produce meaningful results and much of the agricultural research complex is designed to reduce time to variety release because many benefits from improved varieties are time sensitive. RCTs also focus on micro-economic level outcomes such as short-term impacts on yield or household income; occasionally they produce village-level estimates of impact but the focus is on the short-term and immediately measureable impacts. As noted, many of the more important impacts of variety technology adoption play out over a longer time. As a result, RCTs for technology assessment are not likely to be especially fruitful in terms of documenting the large-scale adoption of improved technologies ex-post.

Given these limitations and the fact that many assessments wish to examine impacts of broad research programmes and technologies that have been diffused over time (the DIIVA studies focus on impacts of broad lines of research such as improved maize, beans, potatoes, rice, sorghum and pearl millet), a practical feasible way to measure micro-economic impacts of technology adoption is through household survey (observational) data. Use of observational data is complicated by the difficulty in convincingly identifying the treatment effect (discussed below) and by the fact that survey data set the focus on a single crop cycle or year and the dynamics of impact may play out over many years. When micro-level impacts depend on the time since adoption it is extremely important to collect information on and analyse the effects of adoption history.

Measuring Impacts Using Observational Data

Because RCTs and experimental trial data are not optimal for measuring economic impacts of technologies after they have been released and diffused, most assessments of impacts of agricultural technologies rely on observational data. An important advance in measuring micro-economic impacts of adoption of agricultural technologies comes from the notion that technology adoption is conceptually similar to a 'treatment effect'. All the DIIVA impact assessments used a treatment effect (TE) approach and applied the approach using different estimators. Micro-economic effects of adoption,[8] in the TE framework, are measured as differences between outcomes for adopters compared to what they would have experienced in the absence of the technology. These outcomes might include yields, total factor productivity, or input uses and costs of production.[9] Higher-level outcomes include household income, investment and food security. When economy-wide impact measurement is desired, the TE approach would require simulating differences between the current state of the economy and that which would have existed had the varieties not been diffused. A market model must be combined with the measured micro-economic effects (Zeng *et al.*, Chapter 15, this volume).

In the TE literature, a credible assessment of impacts requires clear identification of the counterfactual – what would have happened in the absence of the technology? As is well known, no one (or nothing) is observed in both the treated and untreated state at the same time so construction of the counterfactual is not a trivial exercise. In particular, when treatment is not assigned experimentally, unobserved factors affecting the individual decision to adopt may be correlated with outcomes and selectivity bias may result. Selectivity makes it difficult to disentangle the separate effects of the technology and other factors associated with the outcome. Recent advances in assessment techniques contribute to a broad menu of techniques to address this bias but the general message in the literature is that care is needed in development of the counterfactual.

Use of the TE approach has an important consequence for estimation: the treatment unit and the outcome of interest might be specified at different levels (e.g. field, household or village level). Longer impact pathways (that is, the number of links between the treatment and the outcome) have more opportunities for confounding factors to be present and this confounding can weaken the interpretation of estimated TEs as causal parameters. For example, the DIIVA impact studies used the field as the treatment unit when estimating technology productivity effects.[10] Within an individual field there are few opportunities for confounding, but it is important to control for differences in soil productivity, rainfall and other agroecological factors even at the field level.

If outcomes at a higher level, say at the household level, are of interest, additional opportunities for confounding emerge[11] and the estimated TE may be attenuated at the higher level because the unit-level outcome (e.g. increased yield) may be a small component of the higher level. For example, Zeng *et al.* (Chapter 15, this volume) find that whereas adoption of improved maize substantially raises field-level productivity, the impacts of adoption on household income are relatively small. The small impact on income results from the small size of maize and the relatively small shares of household income coming from maize. Similar findings emerged from the study of improved pearl millet in Nigeria where substantially improved productivity was found to have relatively small impacts on household income (Ndjeunga *et al.*, Chapter 7, this volume).

Estimates, for example, of higher-level outcomes such as changes in household income due to adoption, obtained by comparing a measure of income to the counterfactual,[12] generally show larger effects on these outcomes than would be plausible given the field-level outcomes (yield gains). For example, the DIIVA study of impacts of new rice varieties in Nigeria and Tanzania (Diagne *et al.*, Chapter 10, this volume) showed effects on household-level outcomes of income and food security that were not plausible considering the rather modest yield gains associated with technology adoption. With such findings it becomes important to have a theory of change within the household and use the data to investigate such change. Without a plausible explanation for findings of large impacts at the household level, the results suggest the presence of confounding effects.

Because yield increases are usually accompanied by more intensive use of inputs, measurement of the productivity effect must account for adjustments in inputs and changes in acreage planted to the improved variety following adoption (Bulte *et al.*, 2014). The DIIVA impact studies in Ethiopia and Rwanda/Uganda estimated both yield and cost effects per hectare and the net value of production per hectare. Land area planted to improved crops was also measured. Thus, adjustments associated with adoption were included. Adoption may displace an existing (say, landrace) variety of a given commodity or a different crop. Alternatively, new land can be brought into production. As observational data reflect all adjustments at the time of the survey, the assumption made in the DIIVA studies was that the land under the improved variety had displaced traditional varieties of the same crop. This assumption should be checked but it is probably associated with mild under-measurement of the true TE.[13] The field-level effects from the DIIVA impact studies from Ethiopia and Uganda/Rwanda range from non-significant to 30–40% productivity gains (accounting for input adjustments) associated with adoption.

Addressing heterogeneous impacts

When distributional impacts of the technology are of interest, alternative techniques need to be used and the econometrics of measurement of heterogeneous treatment effects substantially lag those used to measure homogeneous effects. Heterogeneous effects might emerge because poorer farmers or those in areas where agroecological conditions are limiting may be less likely to benefit from technology adoption than better off farmers.[14] The challenges in the estimation of TE when the effects differ for different groups of farmers revolve around the choice of modelling strategy and the source of the heterogeneity. The DIIVA case studies employ different approaches to allow for TE heterogeneity and all four found heterogeneity to be important in a statistical sense.

Larochelle *et al.* (Chapter 16, this volume) present two estimates associated with heterogeneity in impacts of adoption of improved beans: (i) using interaction effects in the regression model and assuming that heterogeneity comes from inputs, soil and plot characteristics, and selected household characteristics; and (ii) heterogeneity from observed and unobserved factors. The estimates provide evidence that adoption of improved beans affect different farmers in different ways but the differences are not particularly important in a quantitative sense. Zeng *et al.* (Chapter 15, this volume) explore heterogeneity in outcomes of maize adoption in Ethiopia across two dimensions: (i) across the distribution of outcomes (yield and input cost changes) using a quantile regression (Chernozhukov and Hansen, 2005); and (ii) across different probabilities of adoption using a marginal treatment effects approach (Heckman *et al.*, 2006). For effects of adoption on yield, the estimates show important heterogeneity but not in the direction expected. In particular, those fields where adoption was most likely to occur (based on a probit model) enjoyed a smaller yield gain from improved varieties than those where adoption was less likely. The finding of negative selection has been observed elsewhere in the literature (Suri, 2011; Larochelle *et al.*, 2014). Suri, who examined impacts of improved maize in Kenya, speculates that negative selection is associated with limited access to inputs, owing, for example, to poor infrastructure in otherwise favoured areas.

Zeng *et al.* (Chapter 15, this volume) further explore the effects of heterogeneity on input costs and found, like Suri, that cost heterogeneity partly explains the estimates of yield heterogeneity. Cost increases associated with planting improved maize varieties are highest for those fields where adoption is most likely to occur so that the adoption decision and its impact on household income from maize reflects a balance between expected changes in yields and costs of production. This result is clearly consistent with Bulte *et al.* (2014) and illustrates the importance of measuring both the yield change and input adjustments associated with technology adoption. The DIIVA impact studies of rice improvement (Diagne *et al.*, Chapter 10, this volume) also use a local average treatment effect approach and find evidence of heterogeneous impacts of adoption on all their outcome variables – yield, rice revenue, household income and food share. They conclude that failure to account for treatment heterogeneity leads to a problematic interpretation of treatment effect estimates. An important lesson from the DIIVA impact studies is that heterogeneity in impacts of technology adoption is common and estimates of impacts need to account for possible heterogeneity. Especially in SSA where numerous authors have discussed the challenge of estimation of technology impacts given marked heterogeneity, reliance on mean estimates of a TE might mislead.

What to measure?

When economic outcomes are the primary focus, the decision about what to measure is relatively straightforward, but when distributional, consumption or nutritional impacts are of interest, key choices need to be made. All the DIIVA studies involving household surveys employed now-standard survey instruments asking farmers to recall for the last production season:[15] land under production and varieties being used, input use and costs, and yields. Two challenges are worth noting. First, because African farmers frequently plant multiple plots, associating input use with individual plots can be a challenge. The DIIVA survey instruments referred to inputs applied to particular fields but farmer recall of these inputs was subject to error and the research teams were not happy with many of these responses. Special efforts are needed to record properly the differences in input uses across plots within the household.[16] When households plant on multiple fields, error clustering at the household level is appropriate during estimation. Second, specific varieties are difficult to identify through field surveys, especially when variety improvement involves the release of a stream of improved varieties over time. In Rwanda, more than 400 local names for bean varieties were reported and, in Uganda, the list had more than 500 names (Larochelle *et al.*, 2014). Substantial effort was subsequently placed on categorizing named varieties as improved or traditional (see the case study in this volume, Chapter 16). In such cases, the relevant treatment might be adoption of an(y) improved variety and the counterfactual would be outcomes from planting traditional varieties. The impact would be a static estimate of the value of the breeding programme. Alternatively, if the interest was in evaluating the impacts of a specific variety, the treatment would be adoption of it and the counterfactual might be outcomes from planting other improved varieties.[17]

Genetic testing is now relatively low cost and will surely be a central part of subsequent variety identification efforts. In cases where improved technologies are hybrids, identification of whether a variety is improved or not improved is not a particularly vexing problem. However, even for maize, improved open-pollinated varieties are common (e.g. Zeng *et al.*, Chapter 15, this volume) and seeds from these varieties can be recycled. Potential for recycling seeds contributes to uncertainty about the variety because farmers may have different recollections about the original source of the germplasm. In addition, for some crops such as maize, continued recycling of improved seeds leads to degradation, gradually lowering yields and increasing susceptibility to disease. Simply categorizing a variety as improved is difficult. DIIVA used expert opinion to circumvent this problem but it is not a trivial one. As with any varietal change, the estimated productivity effect can be attributed to a pure varietal effect and a seed effect. Crop improvement is about the pure varietal effect. It is important to hold the vintage of the seed constant (between the introduced variety and the variety that is replaced) to make it possible to measure the pure varietal effect. Potentially, confounding the two effects can be a problem in vegetatively propagated and cross-pollinated crops where seed programmes are not institutionally effective (and they almost never are in developing countries).

Impacts of agricultural technologies are obviously unevenly spread over a population; even the most basic economic surplus model embodies the assumption that producers and consumers are affected differentially. Within producer groups, impacts may vary across many dimensions including farm size, assets, farmer skill levels, agroclimatic conditions and over space. All these sources of heterogeneity need to be considered during sample design (over which of these dimensions does the study aspire to measure impacts on a representative basis?) and during estimation, where heterogeneous impacts may need to be accounted for. When measurement is desired over a continuous variable such as household well-being, clear efforts are needed to measure and account for this continuity. In fact, most donors are interested in the impacts of agricultural technologies on the poor, so measurement of the impact of technology adoption beyond yield and cost measurements and on outcomes such as household well-being is critical. Evidence from studies of technology adoption on household-level outcomes from around the world shows clearly that adoption has a dynamic impact on producers; the timing of adoption relative to the availability of new technologies affects individual outcomes, and individual outcomes evolve over time as investments in human and physical capital are realized. Long-term studies of impact dynamics, such as those using panels of households, are needed. Unfortunately, the high cost of maintaining representative panels has led to a decline in their availability. This is an area where additional investments are clearly important. The World Bank initiative expanding Living Standard Measurement Survey (LSMS) data collection to include Integrated Surveys on Agriculture is an important step in the right direction. LSMS panels tend to be of short duration, however, and, if the goal is to use them to measure impacts of varietal change, such surveys still face the challenge of properly identifying the variety.

The measurement of well-being adds an additional complication to already highly detailed survey requirements because, in addition to measuring production attributes, either income or consumption expenditures (or a less satisfactory proxy of well-being) must be captured. Measuring either of these variables introduces increased survey complexity. The DIIVA studies focusing on poverty impacts (Ndjeunga *et al.*, 2011; Larochelle *et al.*, 2014; Zeng *et al.*, 2014) use different approaches. Zeng *et al.* and Ndjeunga *et al.* focus on household income, which, in agricultural contexts, is difficult to measure owing to seasonality, diverse sources of income generation and potential for strategic behaviour on the part of respondents (Deaton, 1997; Deaton and Zaidi, 2002). Larochelle *et al.* use household consumption expenditures, these are preferred to income as a measure of well-being but also require often complex questionnaires (Ravallion, 1992; Deaton, 1997).[18] They reduced the cost associated with collecting detailed expenditures by administering the consumption module to one-half (randomly selected) of the surveyed households. This experience indicated that, while cost savings were considerable and only minimal power was lost in detecting differences across the expenditure distribution, administration of the survey was complicated. An increasing body of evidence shows the utility of applying hybrid surveys that collect non-consumption data from all households and consumption data from only a subset. These data allow the analyst to estimate a consumption model where consumption expenditures are estimated as a function of household assets and other characteristics. The model estimates are then used to predict consumption expenditures for the entire sample, saving survey costs without unduly sacrificing estimate precision (Ahmed *et al.*, 2014).

Additional short-cuts to analysing impacts along the well-being distribution include the use of an asset index (Larochelle *et al.*, 2014) or a multi-dimensional poverty index approach (Stoeffler *et al.*, 2015).

Measurement of outcomes such as food security, individual consumption or nutritional status requires additional consideration of what to measure and how measurement affects sample size and survey costs. The Uganda and Rwanda case studies use a simple measure of household dietary diversity, i.e. specific counts of food consumed over a reference period (Larochelle *et al.*, Chapter 16, this volume). This measure was easily computed using the survey's consumption module, and this choice reflected a tradeoff between cost and focus of survey. Recent literature shows that this simple measure is a relatively reliable substitute as a measure of food security compared to more complex and costly individual dietary recall methods (Hoddinott and Yohannes, 2002; Kennedy *et al.*, 2007, 2011). In the bean impact study, results show that adoption of improved bean varieties is statistically significantly associated with improvements in this measure of food security. The Ethiopian maize survey included an anthropometry module for children under 5 years old. Anthropometry has been shown in the literature to be a good indicator of child nutritional status but collecting anthropometry adds substantial survey costs. The data show that maize adoption has led to statistically significant increases in child nutritional status and that the primary pathway through which this impact is felt is through increased own-consumption of maize (Zeng *et al.*, 2014).

The DIIVA experience clearly demonstrates that convincing assessments of non-economic impacts of agricultural technologies is possible but can involve substantially higher survey costs than those for economic impacts (which are already relatively costly).

Estimation Issues

Constructing the counterfactual

As is well known, no entity can be observed in both the treated (i.e. adopted the technology) and the untreated state at the same time, so artificial means are needed to construct a credible counterfactual. When observational data are used, the main options for the counterfactual are matching techniques and structural econometric models such as instrumental variables.[19] Examples using both techniques are readily available in the impact assessment literature.[20] The techniques rely on fundamentally different assumptions. Matching requires the analyst to identify in the data set observational units that are very 'close' to adopting units and assume that the outcome differences between the adopter and the matched unit(s) are due to adoption.[21] Thus, 'selection' (i.e. adoption of the technology) is assumed to be on observable factors and this assumption means that, once observable factors are controlled for, adoption is assumed to be 'as good as random' (de Janvry *et al.*, 2011). When matching processes are used, the counterfactual for adopters is assumed to be the outcome for the matched (non-adopting) unit(s).[22]

Econometric modelling techniques to correct for endogenous technology adoption include Heckman-like selectivity adjustments and assorted instrumental variables (IV) approaches. The counterfactual under these approaches is arrived at by making a statistical adjustment to observed outcomes. Such models assume that selection is on unobservables, that is, there is an (or are many) unobserved factor(s) (such as management ability or unobserved variability in soil productivity) that affect both the decision to adopt and the outcome. Without controlling in some way for these factors, selection bias will emerge. The solution requires making assumptions about distributions of variables and about the functional form of the relationship in question. It also requires identifying variables (instruments) that affect the adoption decision and only affect the outcome through their effect on adoption. The search for instruments will be discussed in greater detail below but a major weakness of IV approaches is the need to find good instruments.

Recent exchanges in the economics literature (e.g. Deaton, 2010; Imbens, 2010) point to the need for credible instruments. For measuring impacts of technology adoption, credible instruments are those that affect the access to the new technology but only affect the outcome (yields, income, etc.) through their impact on adoption. These instruments must be carefully justified using a plausible theory about why they directly affect only one part of the impact pathway. Use of statistical tests – although important to establish the statistical validity of an instrument – without a plausible theory is not recommended.

There is an ongoing robust discussion in the literature about the advantages and disadvantages of matching versus instrumental variables approaches for technology impact studies. Both techniques rely on assumptions that may or may not be acceptable.[23] Resolution of this debate is far beyond the scope of the DIIVA studies but the DIIVA experience provides a few lessons. First, because we are dealing in a world with duelling assumptions, it is important to check carefully the robustness of estimates. When findings about impacts depend critically on a few model assumptions, it is important to note this. For example, Rosenbaum (2002) developed a method of checking propensity score matching (PSM) matches to assess if the matched results are robust to the possible presence of unobserved factors. The method provides a specific measure of the bias that would need to be present in order to explain the observed associations. Such sensitivity analyses should be undertaken. All four DIIVA impact studies employed alternative estimation methods and robustness was systematically assessed.

A second lesson from DIIVA is that efforts to construct a counterfactual should include an earnest search for alternatives. For example, the presence of multiple maize plots in Ethiopia provided an alternative counterfactual: within a household, fields with improved varieties could be matched with those planted with unimproved varieties. This simple counterfactual controls for differences across households in managerial ability and other household-level characteristics.

Its validity rests on the assumptions that soil productivity is homogeneous within households but validation of primary results with these kinds of alternatives furthers confidence in the robustness of findings.

Third, creation of the counterfactual is complicated when impacts of interest are indirect or induced. Adoption of new varieties stimulates indirect and induced impacts in labour and other input markets and for the economy as a whole. In fact, as will be discussed further below, broader impacts on, say, poverty reduction occur mainly through these indirect effects. The TE approach is best suited for the measurement of direct, micro-economic outcomes and, when these are the primary concern, the methods discussed above can be used to develop the counterfactual, notwithstanding the major challenge already alluded to. If information on indirect or induced impacts is of interest, the TE estimates can be used as a building block to construct a counterfactual market model. Both Zeng *et al.* and Larochelle *et al.* (Chapters 15 and 16, this volume) aggregate micro-level TE estimates to the market level by creating a counterfactual market supply curve. These counterfactuals were used in partial equilibrium surplus models to examine indirect impacts on consumers and producers. Zeng *et al.* went further by allocating this surplus change back to individual households and constructing a counterfactual distribution of well-being. Alternative market models, such as general equilibrium models, can similarly be used to construct a market counterfactual.

Identification issues

A key challenge to the use of observational data is to *identify* the causal impact of adoption on the outcome. Matching methods use the assumption that the matched entities are the same as the adopters except for the fact that they have not adopted. This is the implication of the assumption of selection on observables upon which matching methods rest. When IV estimates are used, the analyst must find instruments that only affect the outcome through their effect on adoption. Essentially, the instruments are assumed to induce exogenous variability in the adoption decision allowing the analyst to disentangle the impact of the decision from other factors affecting the outcome. The importance of identification is immediately apparent when we think about the power of a randomized experiment: randomization ensures that the only difference between treated and untreated units is the treatment itself.

Because IV methods have been so widely used, statisticians have developed a number of tests for these instruments but none of them is entirely satisfactory because they rely on the assumption that the model itself is identified; that is, there are tests of over-identification, weak instruments (instrument relevance), of exogeneity, etc. but all these tests require that the model contains at least one good instrument. This assumption is itself untestable.

The DIIVA case studies of impacts in Rwanda/Uganda and Ethiopia used logic and expert opinion to find appropriate instruments. Essentially this process involved a search for exogenous variation in factors affecting adoption – variation that is as close to possible to being 'as if' randomly assigned in an experiment. The research team began by identifying promising instruments on the basis of literature and knowledge of the seed system in the respective country. These potential instruments were discussed further during in-country project planning meetings. They were subsequently discussed with breeders, seed suppliers, extension agents and farmers. Most promising instruments were then incorporated into the household or community questionnaires. Discontinuity in improved seed availability due to a natural disaster (such as a drought or flood) or other factors was one such instrument; in all three countries, experts agreed that these random events could plausibly affect seed availability and adoption but would be unlikely to affect other outcomes except through their impacts on adoption. In the Rwanda/Uganda study of improved bean varieties, experts also discussed the use of transaction costs associated with obtaining seeds, and distances to paved roads, input distribution and population density, as a reflection of access to information. Some concern was voiced that these factors may also be directly related to profitability via their impact on access to other inputs. As noted in the study (see Larochelle *et al.*, Chapter 16, this volume), suitability of these variables as instruments depends on the specific relationship being modelled.

For example, in a production function where inputs are being controlled for, the use of seed transaction costs as an identifying instrument is justifiable.

The history of adoption in a village was also discussed as a possible instrument but knowledge that improved seeds were probably first disseminated into areas where profitability is highest mitigates against the use of this variable. In the Ethiopia study, distances to seed-related infrastructure (seed dealer, agricultural extension office, farmer cooperative and main market) and the perceived quality of roads were included as instruments. Use of such instruments can only be justified when examining certain relationships (e.g. effect of adoption on yields in a production function framework). Because these instruments passed all appropriate statistical tests, they were deemed to be acceptable. Since statistical testing of instruments is dependent itself on assumptions (e.g. the model is identified), it is important that IVs be subjected first to a logic test (do they make sense?). A generic recommendation from the DIIVA studies is that researchers employing IV approaches need to pay close and careful attention to their identifying assumptions. Much of the existing literature on variety technology impacts lacks rigorous discussion of identification.

Indirect and Induced Effects of New Agricultural Technologies

The DIIVA studies acknowledge that technology adoption has direct (field-level and, in some cases, household-level), indirect (market-level) and induced (changes in economic structure) effects. The indirect and induced impacts (termed spill-overs by de Janvry *et al.*, 2011) are important outcomes of the technology diffusion process. Overall impacts of new technologies on, for example, the incidence and depth of poverty in a population, cannot be measured convincingly if spill-over effects are not carefully considered. Yet these effects often take time, often many years, to materialize. This time lag is one reason why observational data, as opposed to RCTs or experimental trial data, are preferred for poverty assessments. However, assessment of long-term effects across a broad population makes it difficult to construct an appropriate counterfactual.

Spill-overs are difficult to quantify and are not easily amenable to the TE approach. Yet many impacts of interest to donors result from spill-overs. For example, evidence from the Green Revolution (GR) shows that reductions in poverty in India came mainly through increased demand for labour and lower food prices (Pinstrup-Andersen and Hazell, 1985; Pingali, 2012). Similarly, early evaluations of the environmental impacts of GR technologies found increased environmental degradation owing to spill-overs, such as water management challenges and runoff from fertilizer (Pingali, 2012). Without including indirect environmental effects, mainly land clearing avoided due to intensification on existing farmland (further spill-overs), estimates of environmental impacts would not tell the full story of the environmental impacts of the GR, however (Pingali, 2012).

Direct measurement of spill-over impacts is challenging because the counterfactual will never be observed (what would be the state of the economy or environment in the absence of technology adoption?).[24] As a result, inference about spill-over effects must rely on additional assumptions about the structure of the economy or the spill-over process, and about how to appropriately aggregate observed induced changes in micro-economic behaviour. The DIIVA studies used simple models – partial equilibrium and only focusing on the product market – to infer indirect impacts on prices and market participants. Even these models, however, rely on assumptions about relevant elasticities, dynamics of market shares and integration of spatially separated markets. They show that indirect effects on income and poverty are important, which further confirms the notion that lessons about indirect effects require relatively long gestation between variety release and realization of the full effects.

Lessons Learned about Poverty Impacts of New Varieties

DIIVA produced important lessons about distributional impacts of new agricultural technologies in Africa. Estimates from observational data showed significant increases in field-level yields and productivity; household incomes rose as a result. In Rwanda/Uganda and Ethiopia, poor households and those with small landholdings and few agricultural assets were as likely to adopt as the non-poor and relatively wealthy, but the small size of their holdings restricted them from benefiting substantially from the new technologies. For pearl millet in Nigeria (Ndjeunga *et al.*, 2011), poor households were found to benefit more from adoption of the new technology. Few significant household-level obstacles to adoption of improved varieties by poor farmers were identified in any of these studies. Availability of appropriate seed, rather than its price or the economic ability to purchase it, tends to determine adoption. In none of these cases was a strong distributional bias in adoption observed.

Poor African farmers tend to farm smaller plots than their non-poor neighbours; across agroecologies and recognizing heterogeneity in soil quality, the poor have less access to land than others. The small extensions combined with relatively modest productivity gains associated with the varieties examined in the DIIVA case studies mean that the net impact of adoption on on-farm income is relatively small.

Because of the relatively small holding sizes, adoption alone is not associated with substantial reductions in farm-household poverty through the direct income effect. Over time, it is expected that enhanced productivity will lead to asset accumulation, greater ability to bear risk and investments in human capital. Improved nutritional status and greater food security will further foster these outcomes. Growth in demand for labour will increase wage rates and contribute to structural changes in the rural economy. Growth in demand for purchased inputs and output sales will further contribute to this change. Increased food supplies lower prices to consumers and contribute to poverty reduction. All these factors will contribute to development of the rural economy in areas where diffusion is most generalized and will, over time, contribute to broad reduction in poverty; however, conclusively identifying such impacts in a TE approach is not possible.

Studies in this volume did not examine other gains such as improved nutrition owing to consumption of better-quality home-produced foods. In the case of sweetpotatoes, for example, orange-fleshed varieties are expected to produce substantial nutritional benefits through increased dietary vitamin A. The DIIVA surveys in Uganda and Rwanda, however, showed very little adoption of these varieties (as of 2012) and, in any case, the surveys were not designed to detect differences in dietary intakes or outcomes such as vitamin A consumption.

As noted, indirect impacts of improved varieties are expected to be important engines of poverty reduction over time. The relatively long gestation period between release and broad poverty reduction provides support for use of observational data to measure these effects. It also justifies a more careful look at the spillover mechanisms: how adoption affects household-level decisions such as labour supply, investments in child schooling, ability to bear risk, etc. The difficulty is that multi-purpose surveys are exceedingly expensive, particularly when combined with the degree of detail needed to measure field-level outcomes.

The upshot of the DIIVA experience is that poverty impact assessment might take a two-pronged approach. It is clearly important to monitor over time progress in the diffusion of improved varieties. This information could be used to aggregate sample-based studies such as in DIIVA. It also could be used to conduct rough assessments of impact. The area under an improved variety is a principal determinant of its impact. Experimental trial results could be combined with this information in a market model to generate rough estimates of indirect (market price) effects. These data could be combined with a household survey containing consumption information to estimate the distribution of the aggregate benefits to consumers and use this information to measure changes in poverty (Alwang and Siegel, 2003). Secondary information from agricultural statistics might also be used to infer the distribution of benefits to producers. To conduct an adequate evaluation of impact, however, whether for a single variety or for an entire research programme, an assessor needs information on diffusion and its patterns over time.

The second approach would be the one taken in the DIIVA case studies: to make inferences about household-level treatment effects and combine these with aggregate adoption information. Whereas such studies would not provide conclusive causal evidence of long-term reductions in poverty associated with adoption, they would provide evidence of the magnitude and spread of the direct effect. Information on diffusion over time is an essential input into such inferences.

A persuasive impact assessment needs to begin from the ground up by documenting productivity effects at the field level at the start of the impact pathway. Before any conclusions can be made about aggregate impacts, or distributional impacts, the micro-level changes must be clearly established. Micro-level impacts depend on farmer demand for and subsequent documentation of variety characteristics and an evaluation of these should be the starting point for impact assessment. Knowledge about perceived strengths and weaknesses of varieties sheds light on what to look for and emphasize in impact assessment and is also of primary interest to crop improvement scientists. Simple comparative methods are available to frame characteristic demand in surveys.

Notes

[1] The DIIVA impact studies include Ethiopia/maize (Zeng *et al.*, 2014), Rwanda/Uganda/beans (Larochelle *et al.*, 2014) and the SSA-wide study (Fuglie and Marder, Chapter 17) presented in this volume. In addition, DIIVA impact case studies include an evaluation of the impacts of improved rice in Tanzania and Nigeria (Diagne *et al.*, 2014) and of improved sorghum and millet in northern Nigeria (Ndjeunga *et al.*, 2011). In all of these case studies, the technology in question is suitable for rainfed agriculture. This focus makes sense because roughly 95% of agriculture in SSA is rainfed. Under rainfed conditions, however, the cropping year will have an obviously large impact on productivity and measured impacts. Evidence from the six countries included in the case studies is that the cropping year in question was an average one. In specific studies, rainfall was accounted for as much as possible, for example, by including rainfall as covariates in econometric models.

[2] Assessment of diffusion of improved varieties is complex but assessment of diffusion of *specific* improved varieties, particularly when multiple improved varieties have been released over time, introduces additional complications. Although experts may be able to accurately estimate the spread of generic improved varieties, the ability to identify land planted to specific varieties is questionable, except where a single improved variety dominates the landscape. In the Ethiopia, Rwanda/Uganda, and Tanzania and Nigeria (rice) DIIVA studies, the focus is on a large number of varieties released over a relatively long time period. The Nigeria sorghum and pearl millet study focuses on two improved sorghum varieties and a single pearl millet variety (SOSAT C88, which accounts for an estimated 95% of area planted to improved pearl millet varieties).

[3] An important tradition of ex-ante assessment forecasts the longer-term impacts of technologies prior to or soon after their release. These assessment methods usually use market models, such as partial equilibrium models, to forecast impacts on market participants. They are frequently used in priority setting exercises to estimate expected returns from alternative research investments.

[4] In such circumstances, the RCT could be viewed as an input into the ex-ante types of studies identified in note 3. In an ex-ante framework it is desirable to measure the effect of the technology on future adopters, and, since adoption is a result of household decision making, randomized assignment (under an RCT) may not produce the effect of interest. When randomization is at the village level (villages are selected at random for distribution of the new variety) the adoption decision still remains in the hands of the household and decomposing the aggregate (village-wide) effect into the effect on specific adopters remains problematic (Miguel and Kremer, 2004).

[5] Here 'random' refers to a study design similar to those used by Fisher (the father of agricultural statistics) to conduct exact significance tests of differences in yields and other variables. In such studies, experimental protocols are carefully spelled out and adhered to, but a randomized design is essential to draw inferences from small samples.

[6] Under the original objective 4 of the DIIVA studies, substantial effort was devoted to finding experimental trial results (conducted during initial evaluation of the technology) from IARC and NARS scientists. This exercise revealed that few, if any, of these results are available; in the end, the idea of benchmarking unit cost savings from historical trial data was abandoned. Even if such data were available, their usefulness would have been limited by the factors described here.

[7] A new CG-wide initiative called agtrials (www.agtrials.org) is intended to systematically collect and archive yield trial data globally.

[8] The treatment may be conceived of as adoption of the new variety or exposure to the variety. In the latter case, adoption would be an intermediate outcome.
[9] Adoption and subsequent impacts of adoption depend, among other factors, on farmer demand for characteristics and the perceived characteristic of the variety in question. Any credible impact assessment should begin at the variety and its characteristics. The DIIVA analysis of improved sorghum and pearl millet in Nigeria is distinguished by its careful attention to variety characteristics and farmer perceptions of them. In the study, disadoption of improved sorghum (and its subsequent negligible impact) was found to be driven by unhappiness with the variety's characteristics. Similar stories abound in the literature and it is important that impact assessment start with a clear understanding of the expected benefits of the variety and farmer perceptions about how the variety will contribute to these benefits.
[10] Even this seemingly mundane decision is not trivial. At the field level, adoption may be associated with changes in yields, in costs of production and in total factor productivity. Measurement of each of these, in turn, requires a number of decisions. Further, at the field level, adoption may be uneven – particularly in the case of beans – multiple varieties may be planted on a single field. With multiple varieties in a single field, the analyst must choose between a binary versus a continuous TE approach.
[11] At the household level, decisions must be made about who is or is not a 'technology adopter'. Many households plant multiple fields of the same crop, often using improved varieties on one and local landraces on another. Others plant multiple varieties, some improved and others unimproved, on the same plot. These households are 'partial adopters' and, when estimating a TE at the household level, partial adoption should be accounted for (Larochelle *et al.*, Chapter 16, this volume).
[12] This comparison was made using regression techniques or statistical matching procedures in the different DIIVA impact studies.
[13] 'Mild' because virtually all the farm households in all the DIIVA surveys planted mixtures of improved and local varieties; thus the adjustment in acreage is likely to be infra-marginal where the margin is the total land planted to the crop in question. If the adjustment is infra-marginal, the true counterfactual is the returns under the local variety. If the adjustment is extra-marginal, the counterfactual would be the returns under the variety that was displaced.
[14] Limited resource farmers may be less likely to adopt due to various factors and, conditional on adoption, the effect of the new technology on yields and costs of production may differ. Each of these considerations might be important if the focus is on distributional effects.
[15] A recent RCT has shown that recall methods for collecting these types of agricultural data produce acceptable results (Beegle *et al.*, 2012). This paper conducted an RCT to examine potential biases with different lengths of recall period and generally showed that the biases are not great.
[16] There is no known evidence in the literature about the degree of mis-measurement of input intensities associated with use of recall techniques when multiple fields are cultivated. This is clearly an area where further research, probably in the form of an RCT, is needed.
[17] The DIIVA studies examined the impacts of a continuous line of improved varieties (except for sorghum and millet where only one or two varieties accounted for the vast area of improved germplasm). The samples were sized to be nationally representative of, for example, maize producers. If impacts of individual varieties are the target of the research, the sampling strategy will need to be adjusted to ensure that the data set contains sufficient observations for detecting changes in the population of interest (adopters of the specific variety).
[18] Issues such as use of recall versus diary methods, recall period, etc. are beyond the scope of the DIIVA studies.
[19] The DIIVA impact studies were all based on cross-sectional data sets. If panel data are available, other options for constructing the counterfactual are available, but panel data can introduce additional complications that need to be dealt with during estimation. When multiple varieties including a traditional and an improved variety are planted on a single plot, differences in outcomes can be measured under fairly straightforward assumptions and the counterfactual could be created using outcomes from the traditional variety. Such a comparison would, however, involve specific measurement of inputs (e.g. fertilizer) and careful attention to small-scale measurement is required.
[20] See, for example, Becerril and Abdulai (2010) and Kassie *et al.* (2011) for assessments using propensity score matching (PSM); Mathenge *et al.* (2014) and Sanglestsawai *et al.* (2014) are examples using IV approaches.
[21] In PSM, a first-stage regression (usually a probit or logit model) is used to estimate the propensity score (or the probability of adopting) as a function of explanatory variables. Matching is then done based on the propensity score – adopting fields are matched with non-adopting fields based on 'close' values of the propensity score.

[22] This assumption highlights the main problem with a matching approach – it is impossible to explain why one household adopts, while its matched unit does not adopt – without resorting to explanations involving unobserved factors or claims about differential access to the technology in question.
[23] Jalan and Ravallion (2003) note that, while matching methods rely on the rather implausible assumption that matched units are identical except with respect to adoption status (selection on observables), IV methods also rely on different (many untestable) assumptions. An important advantage of matching techniques is their lack of reliance on functional form assumptions (not the case for IV methods), but the assumption of selection on observables is a liability. At a minimum, use of PSM to construct a counterfactual to adoption of new varieties requires a rich data set that includes measurement of all conceivable factors affecting adoption as well as the history of adoption. Even then, the question raised in note 22 needs to be addressed.
[24] There is a long-standing tradition of using aggregate national or cross-country time series data to estimate impacts of technology (mainly investments in agricultural research) on outcomes such as agricultural growth or productivity (see Evenson and Gollin, 2003). Studies have examined how agricultural growth contributes to poverty reduction (see Christiaensen *et al.*, 2011). However, none of these types of studies search for credible evidence of causal links.

References

Ahmed, F., Dorji, C., Takamatsu, S. and Yoshida, N. (2014) *Hybrid Survey to Improve the Reliability of Poverty Statistics in a Cost-Effective Manner*. Policy Research Working Paper 6909. The World Bank, Washington, DC.

Alwang, J. and Siegel, P.B. (2003) Measuring the impacts of agricultural research on poverty reduction: an application to Malawi. *Agricultural Economics* 29(1), 1–14.

Barrett, C.B. and Carter, M. (2010) The power and pitfalls of experiments in development economics: Some non-random reflections. *Applied Economic Perspectives and Policy* 32(4), 515–548.

Becerril, J. and Abdulai, A. (2010) The impact of improved maize varieties on poverty in Mexico: A propensity score-matching approach. *World Development* 38(7), 1024–1035.

Beegle, K., Carletto, C. and Himelein, K. (2012) Reliability of recall in agricultural data. *Journal of Development Economics* 98, 34–41.

Bulte, E., Beekman, G., DiFalo, S., Halla, J. and Lei, P. (2014) Behavioral responses and the impact of new agricultural technologies: evidence from a double-blind field experiment in Tanzania. *American Journal of Agricultural Economics* 96(3), 813–830.

Chernozhukov, V. and Hansen, C. (2005) An IV model of quantile treatment effects. *Econometrica* 73, 245–261.

Christiaensen, L., Demery L. and Kuhl, J. (2011) The (evolving) role of agriculture in poverty reduction: An empirical perspective. *Journal of Development Economics* 96(2), 239–254.

Deaton, A. (1997) *The Analysis of Household Survey Data*. Johns Hopkins University Press, Baltimore, MD.

Deaton, A. (2010) Instruments, randomization and learning about development. *Journal of Economic Literature* 48, 424–455.

Deaton, A. and Zaidi, S. (2002) *Guidelines for Constructing Consumption Aggregates for Welfare Analysis*. World Bank Publications, Washington, DC.

Diagne, A., Groom, B., Coruche, M.P., Kajissa, K. and Nakano, Y. (2014) *Measuring and Assessing the Impacts of Crop Genetic Improvement in Africa: The Case of Pearl Millet, Rice and Sorghum*. Objective 3 Report for the Diffusion and Impact of Improved Varieties in Africa (DIIVA) Project. AfricaRice in partnership with ICRISAT, IRRI, and SOAS, Cotonou, Benin.

Evenson, R.E. and Gollin, D. (2003) Assessing the impact of the Green Revolution. *Science* 300, 758–762.

Heckman, J.J., Urzua, S. and Vytlacil, E. (2006) Understanding instrumental variables in models with essential heterogeneity. *The Review of Economics and Statistics* 88, 389–432.

Hoddinott, J. and Yohannes, Y. (2002) Dietary diversity as a food security indicator. *Food Consumption and Nutrition Division*. Retrieved from http://www.ifpri.org/sites/default/files/publications/fcndp136.pdf (Accessed 18 May 2015).

Imbens, G.W. (2010) Better LATE than nothing: some comments on Deaton (2009) and Heckman and Urzua (2009). *Journal of Economic Literature* 48, 399–423.

van Ittersum, M.K. and Cassman, K.G. (2013) Yield gap analysis – rationale, methods and applications – Introduction to the Special issue. *Field Crops Research* 143, 1–3.
Jalan, J. and Ravallion, M. (2003) Does piped water reduce diarrhea for children in rural India? *Journal of Econometrics* 112(1), 153–173.
de Janvry, A., Dunstan, A. and Sadoulet, E. (2011) *Recent Advances in Impact Analysis Methods for Ex-post Impact Assessments of Agricultural Technology: Options for the CGIAR*. Report prepared for the workshop: Increasing the rigor of ex-post impact assessment of agricultural research. A discussion on estimating treatment effects, organized by the CGIAR Standing Panel on Impact Assessment (SPIA), Berkeley, California.
Kassie, M., Shiferaw, B. and Muricho, G. (2011) Agricultural technology, crop income, and poverty alleviation in Uganda. *World Development* 39(10), 1784–1795.
Kennedy, G.L., Pedro, M.R., Seghieri, C., Nantel, G. and Brouwer, I. (2007) Dietary diversity score is a useful indicator of micronutrient intake in non-breast-feeding Filipino children. *The Journal of Nutrition* 137(2), 472–477.
Kennedy, G., Ballard, T., Dop, M.C. and Union E. (2011) *Guidelines for Measuring Household and Individual Dietary Diversity*. Food and Agriculture Organization of the United Nations, Rome.
Larochelle, C., Alwang, J. and Taruvinga, N. (2014) Inter-temporal changes in well-being during a period of hyperinflation: evidence from Zimbabwe. *Journal of African Economies* 23(2), 225–256.
Mathenge, M., Smale, M. and Olwande, J. (2014) The impacts of hybrid maize seed on the welfare of farming households in Kenya. *Food Policy* 44, 262–271.
Miguel, E. and Kremer, M. (2004) Worms: identifying impacts on education and health in the presence of treatment externalities. *Econometrica* 72, 159–217.
Ndjeunga, J., Umar, J., Ahmed, B., Aba, A., Ibro, A. *et al*. (2011, updated 2014) *Adoption and Impacts of Modern Sorghum and Pearl Millet Varieties in Northern Nigeria*. Working Paper. International Crops Research Institute for the Semi-Arid Tropics, Patancheru, India. http://oar.icrisat.org/6794/ (Accessed 18 May 2015).
Norton, G.W. and Alwang, J. (in press) The role of impact assessment in evaluating agricultural R&D. In: Beintema, N. Lynam, J. and Badiane, O. (eds) *The Future of Agricultural Research in Africa South of the Sahara: Synergyzing Resources, Innovation, and Partnerships*. IFPRI/ASTI, Washington, DC.
Pingali, P.L. (2012) Green Revolution: impacts, limits and the path ahead. *Proceedings of the National Academy of Sciences USA* 109(31), 12302–12308.
Pinstrup-Andersen, P. and Hazell, P. (1985) The impact of the Green Revolution and prospects for the future. *Food Reviews International* 1 (1), 1–25.
Ravallion, M. (1992) *Poverty Comparisons: A Guide to Concepts and Methods*. World Bank Living Standards Measurement Survey Working Paper No. 88. World Bank, Washington, DC.
Rosenbaum, P.R. (2002) *Observational Studies*. 2nd edn. Springer New York.
Sanglestsawai, S., Rejesus, R. and Yorobe, J.M. (2014) Do lower yielding farmers benefit from Bt corn? Evidence from instrumental variable quantile regressions. *Food Policy* 44, 285–296.
Stoeffler, Q., Alwang, J. and Mills, B.F. (in press) Multidimensional poverty in crisis: lessons from Zimbabwe. *Journal of Development Studies*.
Suri, T. (2011) Selection and comparative advantage in technology adoption. *Econometrica* 79, 159–209.
Zeng, D., Alwang, J., Norton, G.W., Shiferaw, B., Jaleta, M. and Yirga, C. (2014) *Agricultural Technology Adoption and Child Nutrition Enhancement: Improved Maize Varieties in Rural Ethiopia*. Department of Agricultural and Applied Economics. Virginia Tech, Blacksburg, Virginia.

Index

Page numbers in **bold** type refer to figures and tables.